# Materials Horizons: From Nature to Nanomaterials

**Series Editor**

Vijay Kumar Thakur, School of Aerospace, Transport and Manufacturing, Cranfield University, Cranfield, UK

Materials are an indispensable part of human civilization since the inception of life on earth. With the passage of time, innumerable new materials have been explored as well as developed and the search for new innovative materials continues briskly. Keeping in mind the immense perspectives of various classes of materials, this series aims at providing a comprehensive collection of works across the breadth of materials research at cutting-edge interface of materials science with physics, chemistry, biology and engineering.

This series covers a galaxy of materials ranging from natural materials to nanomaterials. Some of the topics include but not limited to: biological materials, biomimetic materials, ceramics, composites, coatings, functional materials, glasses, inorganic materials, inorganic-organic hybrids, metals, membranes, magnetic materials, manufacturing of materials, nanomaterials, organic materials and pigments to name a few. The series provides most timely and comprehensive information on advanced synthesis, processing, characterization, manufacturing and applications in a broad range of interdisciplinary fields in science, engineering and technology.

This series accepts both authored and edited works, including textbooks, monographs, reference works, and professional books. The books in this series will provide a deep insight into the state-of-art of Materials Horizons and serve students, academic, government and industrial scientists involved in all aspects of materials research.

More information about this series at http://www.springer.com/series/16122

Bibhu Prasad Swain
Editor

# Nanostructured Materials and their Applications

Springer

*Editor*
Bibhu Prasad Swain
Department of Physics
National Institute of Technology Manipur
Imphal, Manipur, India

ISSN 2524-5384 ISSN 2524-5392 (electronic)
Materials Horizons: From Nature to Nanomaterials
ISBN 978-981-15-8306-3 ISBN 978-981-15-8307-0 (eBook)
https://doi.org/10.1007/978-981-15-8307-0



This Springer imprint is published by the registered company Springer Nature Singapore Pte Ltd.
The registered company address is: 152 Beach Road, #21-01/04 Gateway East, Singapore 189721, Singapore

# Contents

# About the Editor

**Dr. Bibhu Prasad Swain** received his Ph.D. in the field of semiconductor thin films in 2006 from the Indian Institute of Technology Bombay. He has twenty years of experience in the field of nanotechnology and nanomaterials. Currently, he is working as an Associate Professor at the Department of Physics, National Institute of Technology Manipur. He has served as Dean (Academic Affair) and IIC President at NIT Manipur. He was a postdoctoral fellow of Japan Society for the Promotion of Science, (Govt. Of Japan), Brain Korea 21 (Republic of Korea), National Research Fellow (University of Cape Town). He is an active member of Marquee, Who is Who, IEEE, American Nano Society, International Association of Engineers, International Association open professional, International Association of Advanced Materials and, International Engineering and Technology Institute. He has published papers in more than 100 SCI/SCIE/scopus indexed journals and several conference proceedings. He is also a reviewer of 25 SCI-indexed journals and successfully executed four sponsored projects from Government of India.

# Chapter 1
# 2-D Nanostructures of Advanced Hybridized $WO_3$ Nanocomposites for High Performance of Supercapacitor Application

**P. Periasamy, T. Krishnakumar, M. Sathish, Murthy Chavali, Prem Felix Siril, and V. P. Devarajan**

## Introduction

### *Overview of Supercapacitors*

The twenty-first century is the victim of the numerous developments in the technological sector and in response to that it is also a victim for the demand of clean and renewable energy source to meet the depletion of natural energy resources like

P. Periasamy (✉)
Gnanamani College of Engineering, Pachal, Namakkal, Tamil Nadu 637 018, India
e-mail: periasamy.nano@gmail.com

Department of Physics, Nehru Institute of Engineering and Technology, T.M Palayam, Coimbatore, Tamil Nadu 641 105, India

T. Krishnakumar
Tagore Institute of Engineering and Technology, Attur, Salem, Tamil Nadu 636 112, India

M. Sathish
Functional Materials Division, CSIR-Central Electrochemical Research Institute, Karaikudi, Tamil Nadu 630 003, India

M. Chavali
Shree Velagapudi Ramakrishna Memorial College (Autonomous), Nagaram, Guntur District, Andhra Pradesh 522 268, India

MCETRC, Tenali, Guntur, Andhra Pradesh 522 201, India

P. F. Siril
School of Basic Sciences, Indian Institute of Technology, Mandi, Himachal Pradesh 175 001, India

V. P. Devarajan
KSR College of Arts and Science for Women, Tiruchengode, Namakkal, Tamil Nadu 637 215, India



B. P. Swain (ed.), *Nanostructured Materials and their Applications*, Materials Horizons: From Nature to Nanomaterials, https://doi.org/10.1007/978-981-15-8307-0_1

fossil fuels and required subsequent developments in the energy storage device with the superior storage and retention of high capacity of electrical energy in the view of global economies growth [1]. An effective technique for storing energy would significantly reduce the scale of materials and would offer reasonable cost-effective energy for everybody. It seems that nearly most of the research people's attention and interest are directed toward energy generation rather than in the direction of the energy storage issue.

Various kinds of mechanisms are proposed for the ideal energy storage device with the intention of achieving high energy and power density. These are charged by fossil fuel or nuclear energy which would be the prospective solution to flattering the irregularities inaccessibility and delivering the energy to the required place. The term energy storage device is used here in the sense of some sort of mechanism for the practical energy storage mechanism [2]. The purpose of energy storage is, in general, to deliver the energy where the energy source is non-portable, to store the energy for later usage, and to alter the ratio of power-to-energy, as achieved by flywheels, capacitors, etc. All the energy storage applications can be put at least any one of these categories.

The general concept of an electrical energy storing mechanism was proposed in the late 1800s. After 150 years, Becker [3] from General Electric claimed his patent (U.S. Patent 2,800,616) for his double layer storage by means of porous carbon electrodes in the aqueous electrolyte in 1957. After Becker, in late 1969, SOHIO Corporation was manufactured for commercial purpose, the first electric double layer capacitor [4] which consisting of a porous carbon electrode in a non-aqueous electrolyte. Another research fellow from the same corporation, Donald L. Boos in 1970, structured the foundation for the several subsequent patents and reports covering all features of Electrochemical technology. In 1978, Nippon Electric Company (NEC) tagged the term "supercapacitor" with the electric double layer capacitor and used as a memory back up in computers. Early commencements of electrochemical supercapacitors were measured as lower specific energy values. The same year Panasonic stepped in double-layer capacitor manufacturing in the name of "Goldcap." The foremost differences between the Goldcap (Panasonic product) and a supercapacitor (NEC products) were that formerly used a non-pasted electrode in non-aqueous electrolyte while the later one used a pasted electrode immersed in the aqueous electrolyte with the construction of the bipolar cell. Overall, the electrochemical supercapacitor (ECs) is exclusively attributed to the complement of batteries and fuel cells' weaknesses. Thereafter, several efforts have been devoted to re-design the ECs to accomplish higher efficiency. There has been no significant improvement was observed that, up to 1990, to enhance the material design and technological development of ECs for cost-effective supercapacitor manufacturing. The Conway research group has propounded a new type of capacitance which is known as "Pseudocapacitance" around in 1991 and initiated widespread primary and extensive research work on $RuO_2$ compound.

This new type of electrochemical capacitance was observed due to some electrosorption process and reversible redox reaction (faradic) on the surface of metal oxide materials, e.g., $RuO_2$, $MnO_2$, $IrO_2$ and $CO_3O_4$, etc. Those pseudocapacitive materials recorded 100 times higher energy storage performance than with

EDLCs. Moreover, both types of faradic and non-faradic charge storage mechanisms are involved in pseudocapacitors thereby enhancement in high energy storage was achieved. After 1990, there was a tremendous development in this field and very active research and development were taken to enrich more in fundamental and designing direction. Nowadays a number of new companies like CAP-XX, ESMA Maxwell Technologies, NCC, and NessCap have invested in the improvement of supercapacitors. Apart from $RuO_2$, new types of electrodes like activated carbon, transition metal oxides are utilized in the present day's capacitor which is available in the market. The ideas about the hybrid energy storage system have been derived from high powered supercapacitors and these are integrated with a battery or fuel cell system.

## *Classification of Electrochemical Supercapacitor*

Generally, based on the electrical charge storage process, it is classified as electrochemical double-layer capacitors (EDLC) and pseudocapacitors.

### Electrochemical Double-Layer Capacitors (EDLC)

In 1879, Helmholtz proposed a new type of capacitance that can be obtained from the double-layer mechanism [5] which consisting of charged electrodes and ions in the electrolyte. Here, electrical energy is stored by a non-faradic process where there is no limit for electrochemical kinetics. Owing to this, EDLCs have remarkable characteristics like higher power density and higher energy density than pseudocapacitors and electrostatic capacitors. The charges are accumulated over the interface of the surface of the electrode and electrolytes. Hence, it forms a double-layer with the average separation of a few angstroms in each layer (Fig. 1.1).

During the period of the charging process, the negative electrode liberates an electron and it starts to travel towards a positive electrode by means of external load.

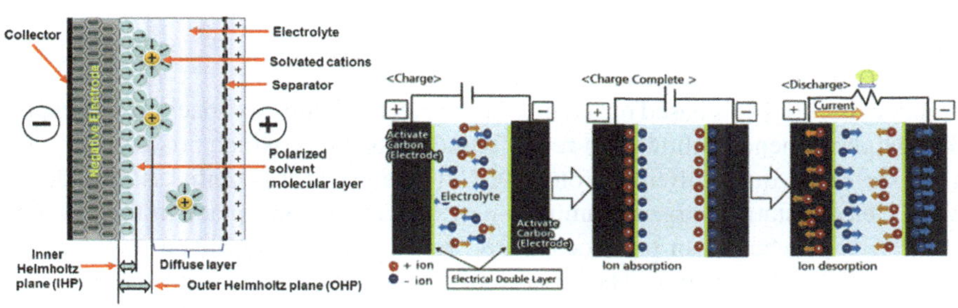


**Fig. 1.1** System of electrochemical double-layer capacitors (EDLCs)

**Table 1.1** List of the commercially available double layer in market

| Make | Device name | Capacitance (F) | Voltage (V) | Type |
|---|---|---|---|---|
| Asahi glass | EDLC | 500–2000 | 3 | Carbon/nonaqueous |
| AVX | Bestcap | 0.022–0.56 | 3.5 | Carbon/polymer/aqueous |
| Cap-XX | Supercapacitor | 0.09–2.8 | 2.3 | Carbon/nonaqueous |
| ELNA | Dynacap | 0.33–100 | 2.5 | Carbon/nonaqueous |
| Evans | Capattery | 0.01–1.5 | 5.5 | Carbon/aqueous |
| Maxwell | Boostcap | 1.8–2600 | 2.5 | Carbon/nonaqueous |
| NEC | Supercapacitor | 0.01–6.5 | 3.5 | Carbon/aqueous/organic |
| Nippon chemicon | DLCAP | 300–3000 | 2.5 | Carbon/nonaqueous |
| Matsushita | Gold capacitor | 0.1–2500 | 2.3 | Carbon/organic |
| Tavrima/ECOND | Supercapacitor | 0.13–160 | 14 | Carbon/aqueous |

At the same time, inside the electrolyte, cations, and anions are traveling towards the opposite direction. The electrochemical process in the positive electrode and a negative electrode, for the charging process, can be expressed as following.

$$E_1 + A^- \underrightarrow{\text{chargeing}} E_1^+ \therefore\because A^- + e^-$$

$$E_1 + C^- + e^- \underrightarrow{\text{chargeing}} E_1^+ \therefore\because C^+$$

$A^-$&$C^+$ is the anions and cations in the electrolyte; $E_1$&$E_2$ is the positive and negative electrode and $\therefore\because$ refers to the interface of electrode and electrolyte. Simply reversing the above equations gives the meaning of discharge mechanisms. The EDLC electrodes materials are prepared with as high surface-area, in order to enhance the surface area of double-layer such as carbon aerogels and porous carbon materials. The pours nature of electrolyte reduces the distance between electrode and electrolyte and so the ions intercalation may increase (Table 1.1).

## Pseudocapacitor

Pseudocapacitor [6] is based on a redox-based electrochemical capacitor. The charge storage phenomenon followed in redox-based capacitance which arises at the electrode surface is quite different from EDLC. Once external potential is provided in EDLC, electrostatic charge accumulation or deficiency takes place in the interface of electrodes, whereas in pseudocapacitor, redox reactions occur on the electrode which is fast and reversible. The passage of electrostatic charge is involved.

The reaction at the positive and negative electrode, during charging, can be expressed as follows.

$$E_1 + A^- \underrightarrow{\text{chargeing}} E_1^{\delta+} \therefore\because A^- + \delta e^- \quad (+\text{ve electrode})$$

$$E_1 + C + \delta e^- \underrightarrow{\text{chargeing}} E_1^{\delta+} \therefore\because C^- \quad (-\text{ ve electrode})$$

$E_1$&$E_2$ represent the positive and negative electrode and $A_-$&$C_+$ represents the anions and cations in electrolyte $\therefore\because$ means interface of electrode and electrolyte. δ e is electrosorption valence parameter due to the oxidation–reduction process at the electrode.

In 1991, Conway [7] represented the consequence of some faradic mechanisms that resulted in Pseudocapacitance features: (1) intercalation pseudocapacitance, (2) redox pseudocapacitance, and (3) underpotential deposition. Intercalation Pseudocapacitance is a consequence of ion interpolation into the porous layers of electrode material, followed by a faradic charge transfer process without crystalline phase change.

Redox pseudocapacitive process arises, when ions are absorbed by electrochemical reaction onto or/near-surface of a redox-active material with associated faradaic charge-transfer. Metal oxides that are under extensive research for a decade, such as $Co_3O_4$, $MnO_2$, NiO and $RuO_2$, in addition to that, conducting polymers (polypyrrole and polyaniline) are typical redox pseudocapacitive materials. Underpotential deposition takes place when metal ions create an adsorbed monolayer at a different metal's surface well above their redox potential. The Langmuir-type electrosorption of H on the substrate of a noble metal such as Pt, Rh, Ru is the classic example for the underpotential deposition. These are depicted in Fig. 1.2

In the above three types of pseudocapacitive behavior, the capacitance is proportional to differentiated $R$ with respect to potential ($V$) such as $\left(C\alpha \frac{\mathrm{d}R}{\mathrm{d}V}\right)$, where C is the properties of capacitive electrode materials. For intercalation pseudocapacitance

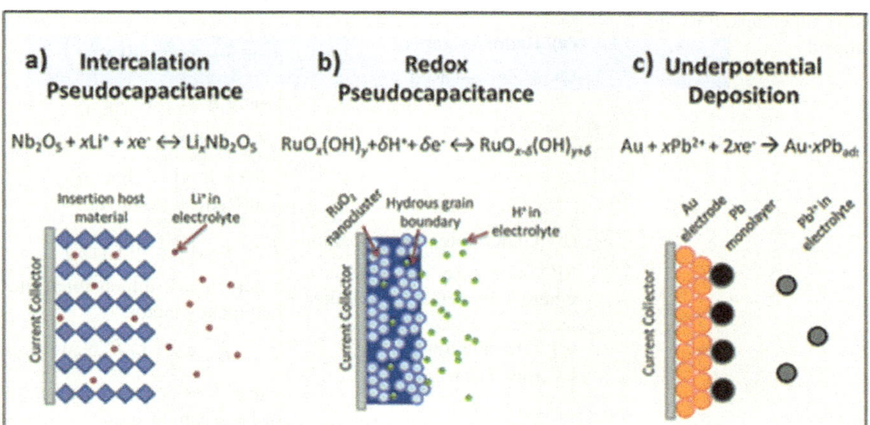


**Fig. 1.2** Various types of pseudocapacitive behavior

and redox pseudocapacitance, $R$ may perhaps be the transformation of an oxidized species to a reduced species and fractional absorption, respectively. In case of the underpotential deposition, $R$ is the measure of fractional coverage of an electrode surface.

From the perspective of thermodynamics, R is correlated to potential by an equation:

$$\frac{R}{1-R} = K \exp\left(\frac{VF}{RT}\right)$$

Thus, the electrode capacitance is dependent on redox potential. The redox potential is expressed as

$$E = E_o + \frac{RT}{nF} \ln\left(\frac{R}{1-R}\right)$$

where $E_0$ refers to the electrostatic potential and $F$ is the electrode material capacitive property (Table 1.2).

In summary, EDLCs techniques paved the foundation to achieve the high specific capacitance ECs and often pseudocapacitance. The peculiar characteristics of EDLC such as the highly active area of electrodes and charge separation at the atomic dimension could offer superior charge storage capability. Pseudocapacitance that emerges from redox or ion adsorption/absorption reactions promoted the achievable capacitance of ECs. Table 1.3 shows the Comparison of EDLCs and pseudocapacitors.

**Table 1.2** The basic thermodynamic relations for redox supercapacitors are enlisted

| Type of system | Basic thermodynamic relations |
|---|---|
| (a) Redox systems<br>$Ox + Ze^- \rightarrow Red$ | $E = E_o + \frac{RT}{zF} \ln\left(\frac{R}{1-R}\right)$<br>where $R = \frac{[OX]}{([Ox]+[Red])}$<br>$\frac{R}{1-R} = \frac{[Ox]}{[Red]}$<br>where, $[Ox]$ = Oxidation and$[Red]$ = Reduction |
| (b) Under potential deposition<br>$M^{z+} + S + Ze^- \rightarrow S^*M$<br>where $S$ = surface lattice sites | $E = E_o + \frac{RT}{zF} \ln\left(\frac{\theta}{1-\theta}\right)$<br>where, $\theta$ = 2-dimensional site occupancy factor |
| (c) Intercalation system<br>$Li^+$ into $MA_2$ | $E = E_o + \frac{RT}{zF} \ln\left(\frac{X}{1-X}\right)$<br>where $X$ = occupancy fraction of layer lattice sites |

**Table 1.3** Comparison of EDLCs and pseudocapacitor

| EDLCs | Pseudocapacitor |
|---|---|
| Non-faradaic process | Faradaic process |
| 20–50 μF $cm^{-2}$ | 2000 μF $cm^{-2}$ for single-electron transferred process; 200–500 mF $cm^{-2}$ for multi-electron transferred process |
| Capacitance is constant with potential, excluding for point of zero charges (PZC) | Some single-state materials exhibit marked maximum capacitance |
| charging/discharging process is Highly reversible | Quite reversible but has intrinsic electrode kinetic rate limitation |
| High power compared to pseudocapacitor | High energy compared to EDLCs |

### Hybrid Supercapacitor

The hybrid supercapacitors [8] are the combination of the existing redox or pseudocapacitors and double layer super-capacitor and retains the benefits of high energy and power density than existing supercapacitor (<3.5 V). Hence, owing to the excellent characteristics, this type of supercapacitor looked for a high pulsed power system and electric vehicle. It consisting of two electrodes such as a negative electrode which acts as the electric double layer capacitor and the positive electrode which acts as alike of battery.

## *Components of Supercapacitor*

The important component of the supercapacitor is electrode material and electrolyte for delivering high power and energy densities [9]. These components must satisfy a few conditions to be used in supercapacitor which is the conductivity of the electrolyte and high porous surface morphology of electrode materials. Since the high surface area of electrode materials plays a significant role which influences the capacitance performance upon the architecture of nanostructured electrode materials. The high surface area of electrode materials helps the ions in the electrolyte to intercalate into the interior molecules of the active electrode material. The performance of electrode materials varies with the type of electrode materials such as conducting polymers or metal oxides in a supercapacitor. Similarly, of electrode materials, electrolytes also affect the characteristic of supercapacitor; the main characteristics of electrolytes are corrosion, ions radius, ions mobility, and potential window.

## Electrode Materials

### Transition Metal Oxides

There are three types of materials used as candidate materials for supercapacitors including transition metal oxides [10], conducting polymers [11], and carbon nanotubes [12]. Generally, Carbon-based materials are treated as electrode materials but recently extensive research is being carried on conducting polymers and metal-oxides. The electrochemical capacitance of metal oxides is influenced by its surface morphology and its excellent conducting properties. The electrode materials with 1-D nanostructures [13] like nanoneedles, nanotubes, and nanowires have proved its excellent charge transfer between electrode and electrolyte owing to its attractive morphology. The reversible redox reaction is highly supported by the presence of more than one oxidation state of metal oxide makes it more suitable for supercapacitor applications.

### Conducting Polymers

Generally conducting polymers (CP) has been used as a common material in electrochemical supercapacitor due to its high conductivity, fast reversibility, large potential window, and considerably low cost which makes the CP are candidate materials for ECs applications. The chemical redox reaction that occurs in CP is not only on the surface of electrode material but also on the interior side of CP. The conducting polymers are distinguished as positive or negatively charged based on the ions insertion into the conducting polymers. The electronic conductivity of CPs is accounted for the oxidation and reduction process. This is ascribed to a couple of double bonds; a couple of double bonds are termed as $\rho$ and $\pi$. The former one ($\rho$) is low mobility and later one ($\pi$) is high mobility which is linked to single or double bonds formed which is more suitable to formulate the redox reaction. Various types of conducting polymers such as Polyaniline (PANI), Polypyrrole (PPy), and Polythiophene (PTh) has proved its suitability for supercapacitor application.

## Electrolytes

To date, different types of electrolytes have been proposed and they can be generally classified as liquid electrolytes, solid-state electrolytes, and redox-active [14] as shown in Fig. 1.3.

The electrolyte is a central part of the electrochemical supercapacitor for providing high ionic conductivity, supporting for electrical double-layer (EDL) formation in electrical double-layer capacitors, reversible redox reactions in pseudocapacitors. The characteristics of electrolytes are such that, the ion concentration and solvent; ion type and size; the interaction between the ion and the solvent; and the intercalation between the electrolyte and the electrode, all have a vital role in the formation of

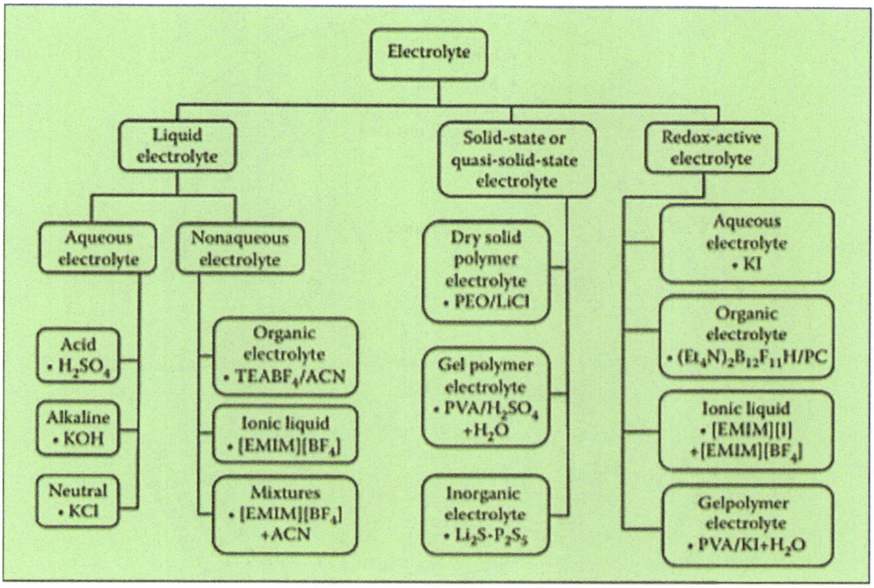


**Fig. 1.3** Classification of electrolytes for supercapacitors

the EDL capacitance and pseudocapacitance. The operational voltage of the supercapacitor depends on the stability of the electrolyte, which is favorable for attaining the high energy and power densities. The general requirements of Electrochemical Supercapacitor electrolytes takes the account of (a) availability at high purity (b) environmental friendliness (c) high ionic conductivity(d) low volatility (e) a low viscosity (f) a wide operational temperature range; (g) a wide electrochemically stable window; and (h) low cost. Besides, the other significant properties of supercapacitors such as cycling stability, equivalent series resistance (ESR), lifetime, operational temperature range, power density, rate capability, and self-discharge rate are greatly affected by electrolytes (Fig. 1.4).

However, there is no electrolyte that meets the entire requirement perfectly so far. Since all the electrolytes have some kind of shortcomings.

### Separator

In a supercapacitor, the separator is used to keep away from the electrical contact and charge transfer between the electrode materials. The requirements of the good separator are, (1) separator must have high electrochemical stability in the electrolyte used, (2) high insulating property to avoid short circuits, (3) good conductivity for electrolyte ions(4) to provide adequate mechanical strength to accomplish

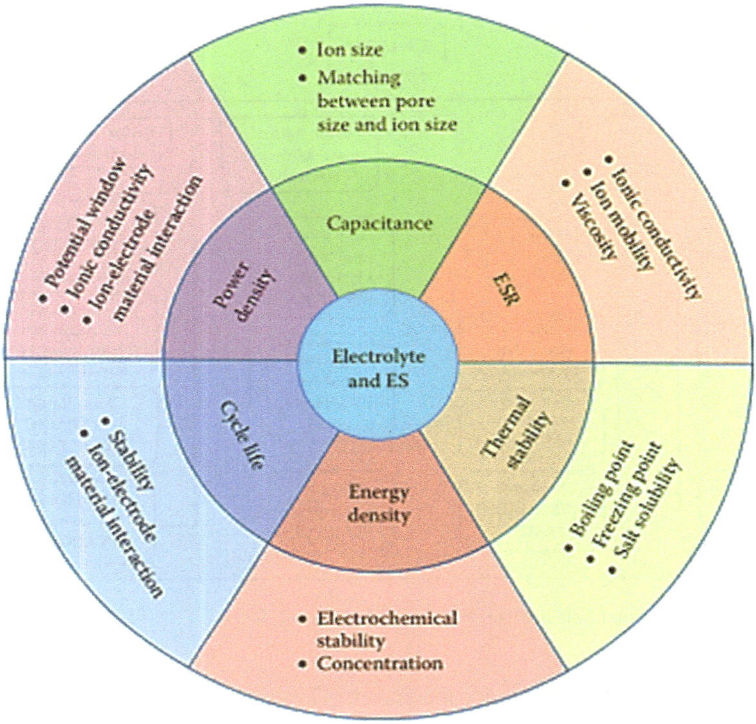


**Fig. 1.4** Effects of the electrolyte on the electrochemical supercapacitor performance

to ES durability. These requirements are fulfilled by thin and highly porous insulating film separators. While selecting the suitable separator for supercapacitor, we must consider the electrode materials, electrolytes, type of ES, operative cell voltage, and working temperature. Generally, Cellulose, Glass fiber, paper, and a Porous Polymer membrane are typically used as a separator in ECs. Recently, the researcher has found the suitability of Graphene oxide film and eggshell membrane as a separator in supercapacitor and achieved higher specific capacitance, lower resistance, quicker charge/discharge ability, good cyclic stability compared with the polyethylene separator.

## Binders

Binders are used to maintain the structural integrity of electrode materials and support the electrode materials to stick on homogeneously onto the current collector. Generally, electrode materials are prepared by mixing the appropriate ratio of active particles/powders and binders. The performance of electrode material and specific capacitance of supercapacitor is affected by the amount of binder added with active particles and also the nature of the binder. Till date, Nafion binder and fluorinated polymeric binders such as polyvinylidene fluoride (PVDF) and polytetrafluoroethylene (PTFE) which contain good adhesive properties have been mostly used for ECs. Considering environmental friendliness, fluorine-free binders have attracted the researcher's attention. For example, natural cellulose, PVP, and polyacrylic acid (PAA) also been proposed as a binder and attained significant capacitance. However, to reduce some negative influence related with the binders (e.g., increases in electrode resistance and decreases in the effective surface area), the investigation has been diverted towards binder-free electrode materials for ECs applications.

## *Challenges in Supercapacitors*

(i) **Low Energy Density**:
The energy density of ECs is lower than 10 Wh/kg, which is very below the energy density of the battery (>50 Wh/kg). For many applications, a high voltage is required for its operation. Since single ECs have a restriction of low voltage; several dozens of ES have to be connected in series to achieve high voltage. Therefore, the major challenge in an electrochemical supercapacitor is low energy density which has to be resolved through research.

(ii) **Cost Efficiency**:
In the electrochemical supercapacitor, the main parts are electrode materials and electrolytes. To obtain high energy density, the development of highly porous surface area carbon materials is required and which is too expensive. A pseudocapacitive material such as $RuO_2$ is high cost and potentially harmful to health. The organic electrolyte cost is far from negligible. The primary reason for ES is not used as energy storage technology is simply too expensive than other available devices.

(iii) **Self-Discharge Rate**:
The self-discharge rate of ES is 10–40% per day. The primary challenges in front of the researcher are to increase energy density and lowering cost. This must be encountered without sacrificing the high cycle life and exceptionally high rate performance that sets ECs apart from batteries.

## *Characteristic, Advantages, and Disadvantages of Supercapacitors*

### Characteristics of a Supercapacitor

- Capacitance range: 1 mF to greater than $10^4$ F.
- Charge/discharge time: Milliseconds to seconds.
- Energy density: 0.05–10 Wh/kg.

- Operating temperature: −40 to + 85 °C.
- Operating voltage: ~ 1 V for aqueous electrolytes; 2–3 V for Organic electrolytes.
- Operating life: 5000–50,000 h (depends on temperature and voltage).
- Power density: 0.01–10 kW/kg.
- Pulse load: 0.1–100 A.
- Pollution potential: No heavy metals.

**Advantages**

- Charges in seconds; No danger of overcharging.
- Excellent power density (>$10^6$ W/Kg) and helps to extend battery runtime and battery life.
- High rates of charge and discharge.
- High cycle efficiency (95% or more).
- Improve load balancing when used in parallel with a battery.
- Lesser RF noise by eliminating DC/DC.
- Meet environmental standards.
- Provide peak power and backup power.
- Reduce battery size, weight, and cost and minimizing the space requirements.
- Virtually unlimited life cycle.

**Disadvantages**

- High self-discharge—higher than that of an electrochemical battery.
- Low voltage cells—to obtain higher voltages, serial connections are needed.
- Linear discharge voltage restricts the use of the full energy spectrum.
- Low energy density, usually, the retention of the energy is $\frac{1}{5}$ to $\frac{1}{10}$ of an electrochemical battery.
- Poor energy density.
- Sophisticated electronic control and switching equipment needed.
- Voltage balancing element is needed if more than three capacitors are connected in series.

## *Applications of Supercapacitors*

i. **Power electronics**—supercapacitors are used as an energy storage device to provide energy pack-up for a portable device in aerospace, commercial, industrial residential, and military applications.
ii. **To regulate voltages** throughout electronic circuit boards, ECs are used.
iii. **Battery Enhancement**—To enhance the performance of battery-based utilities, where the battery provides the energy and the supercapacitor supplied short-term power to increase the run-time of the device.

iv. **Memory backup**—Many electronic devices are now assimilated digital components with memory, a brief interruption in electric power supply would otherwise cause data loss. A supercapacitor would act as the power supply in such situations, thereby retaining data safely.
v. **Electric Vehicles—Since**, Fuel cells are having extremely high energy density but they are limited in their power specifications. Both the power and energy requirements of an Electric Vehicle can, therefore, be fulfilled by combining the fuel cell with supercapacitor technology. This combined power source design allows the high-energy density device. For instance, a fuel cell would provide the average load requirements and peak load requirements that result from accelerating or climbing up hills can be supplied by the high-power device from supercapacitor bank. The usages of a supercapacitor in electric vehicles have gained its momentum; the fast charge–discharge ability of ECs has started to replace the fuel engine in vehicles, thereby reduced the fuel consumption. The Chinese Company 'Sunwin' has produced an electric bus in 2010. In each bus stop, supercapacitors are charged with a pantograph. Within 30 s supercapacitors are charged to 50% and it takes 80 s to charge to 100%.
vi. **Electromechanical Actuators**—Electromechanical actuators can execute reliable vector control for the launch of space vehicles and act as flood-control actuators on submarines.

## Classifications of Nanomaterials

Nanomaterials are those materials whose sizes fall within the range of 1–100 nm at least any one of its dimensions. Several hundreds of novel nanomaterials have been synthesized via various physical and chemical methods and it was demanded to classify the as-prepared nanomaterials. In 1995, Gleiter [15] proposed the idea about classifications and Skorokhod explained further in 2000 [16]. But their classification was not in full form because they have not considered the fullerenes, nanotubes, and nanoflowers into account. The modified classifications of nanostructures were proposed later by Pokropivny and Skorokhod [17].

### *Zero-Dimensional Nanomaterials*

Several research groups across the globe have focused on synthesis zero-dimensional nanostructures like core–shell quantum dots, heterogeneous particles arrays, hollow spheres and nanolenses [18–20]. The simple and fundamental building block is 0D nanostructures and hence it can be effectively utilized for manufacturing 1D, 2D, and 3D nanodevices. Besides, the quantum dots which having zero dimensions are broadly investigated for various applications such as solar cell [21], light-emitting diode [22], laser [23], and single-electron transistors [24].

### *One-Dimensional Nanomaterials*

In 1D nanomaterials, the electron confinement is two dimensional and they possess large surface is which helps to reduce the diffusion time of electrolyte ions. 1D nanostructure with different morphology and structures such as nanorods [25], nanowire [26], nanotubes [27], and nanofibers [28] have been synthesized and extensively studied for the energy storage system. The excellent electronic and ionic conductivity is supported by unique structures of 1D nanostructure and contributing to high performance. The electrode material with 1D offers many advantages to achieve high capacitance, long cycle stability, and provide a direct pathway for electronic movement.

### *Two-Dimensional Nanomaterials*

In 2D nanomaterials, the electron confinement is one dimensional and very thin material because its thickness is in nanoscale. Further, 2D nanomaterials exhibit excellent electrochemical properties and attracted researchers particularly for supercapacitor applications [29]. Currently, researchers have started to focus on 2D nanostructures of graphene [30], transition metal oxides [31, 32], hexagonal boron nitride for catalyst [33, 34], Li-Ion batteries [35], optoelectronics [36, 37], solar cells [38, 39] and sensor [40, 41].

### *Three-Dimensional Nanomaterials*

For the past couple of decades, researchers have devoted to the synthesis of various 3D nanomaterials owing to the peculiar characteristics and large surface area. The performance of nanomaterials highly depends on the surface to volume ratio, dimensionality, shape, and size. Hence, research people paid much more attention to synthesis shape and size-controlled nanomaterials [42, 43].

## Various Synthesizing Methods

### *Inert Gas Condensation Synthesis Method*

Inert gas evaporation is the earliest method used to synthesize nanoparticles. It is the evaporation of material in a cool inert gas (Helium or Argon) at low-pressure conditions in the order of 1 mbar. The convective flow of inert gas passes over the evaporation source and transports the nanoparticles formed above the evaporative source via thermophoresis towards a substrate with liquid nitrogen cooled surface.

Common vaporization methods are resistive evaporation, laser evaporation, and a sputtering method. A modification that consists of a scraper and a collection funnel allows the production of relatively large quantities of nanoparticles, which are agglomerated but do not form hard agglomerates and which can be compacted in the apparatus itself without exposing them to air.

This method was pioneered by the group of Birringer and Gleiter. Increased pressure or increased molecular weight of the inert gas leads to an increase in the mean particle size. This so-called Inert Gas Condensation method is already used on a commercial scale for a wide range of materials. Reactive condensation is possible by adding $O_2$ to the inert gas in order to produce nanosize ceramic particles. Another method replaces the evaporation boat by a hot-wall tubular reactor into which an organometallic precursor in a carrier gas is introduced. This process is known as chemical vapor condensation referring to the chemical reactions taking place as opposed to the inert gas condensation method. Typical Inert Gas Condensation Technique [44] Setup is shown in Fig. 1.5.

### *Electrochemical Method*

In the electrochemical synthesizing method, a certain electric current is allowed to pass between two electrodes which are separated by an electrolyte. Here electricity is used as the driving force and the synthesizing process takes place at the electrode–electrolyte interface. The major advantages of electrochemical techniques include low costs, simple operation, high flexibility, avoidance of vacuum systems as used in physical techniques, easy availability of equipment and instruments, less contamination, and environment-friendly process. More research work has been done on the electrochemical technique in advancing the basic understanding and industrial applications [45, 46]. Typical Electrochemical synthesizing [47] setup is shown in Fig. 1.6.

### *Templating Method*

The template method has become extremely popular during the last decade for preparing various nanostructured materials. In order to construct materials with a similar morphology of known characterized materials (templates), this method utilizes the morphological properties with reactive deposition or dissolution. Hence, it is possible to prepare numerous new materials with a regular and controlled morphology on the nanoscale level by simply adjusting the morphology of the template material [48, 49] (Fig. 1.7).

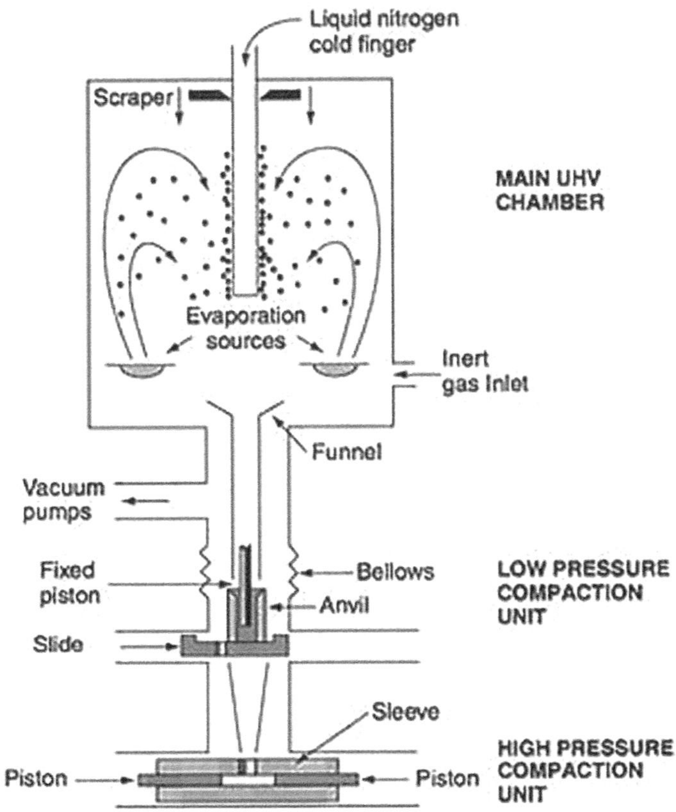


**Fig. 1.5** Inert gas condensation technique setup

## *Chemical Vapor Deposition (CVD)*

To produce high-purity and high-performance of thin films, CVD plays a major role in the semiconductor industry. In a typical CVD process, the substrate is exposed to volatile precursors, which react and/or decompose on the substrate surface to produce the desired film. Frequently, volatile by-products that are produced are removed by gas flow through the reaction chamber. The quality of the deposited materials strongly depends on the reaction temperature, reaction rate, and the concentration of the precursors [50]. The major advantage of this method includes the uniform coating of the nanoparticles or nanofilm. However, the limitations of this process were explained by Sudarshan [51] in (2003). He also pointed out that the process is difficult to scale up and requires a higher temperature. Typical chemical vapor deposition method [52] setup is shown in Fig. 1.8.

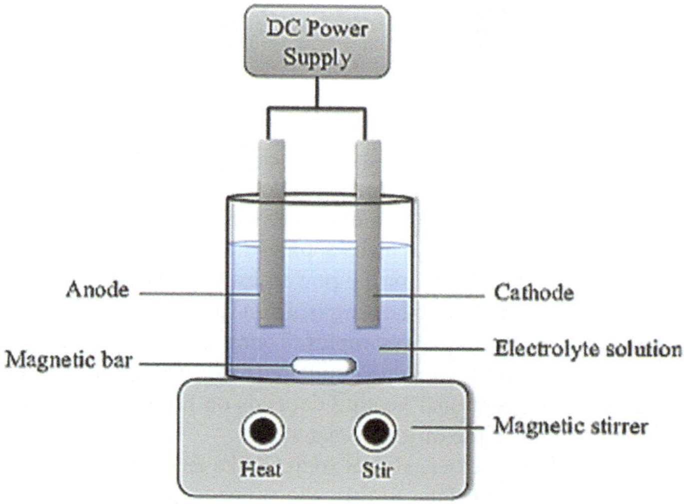


**Fig. 1.6** Electrochemical synthesizing setup

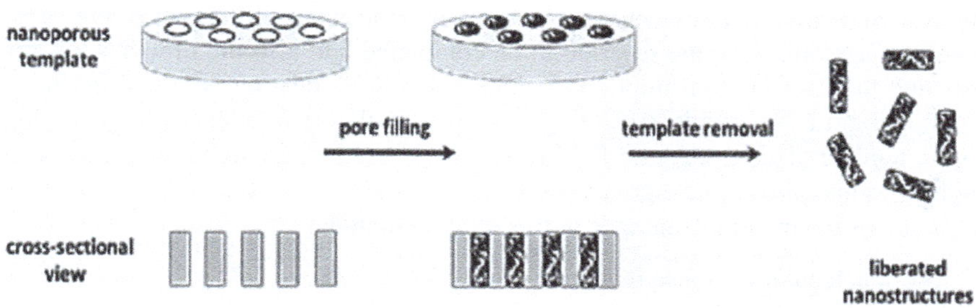


**Fig. 1.7** Schematic representation of template synthesis method

**Fig. 1.8** Chemical vapor deposition method setup

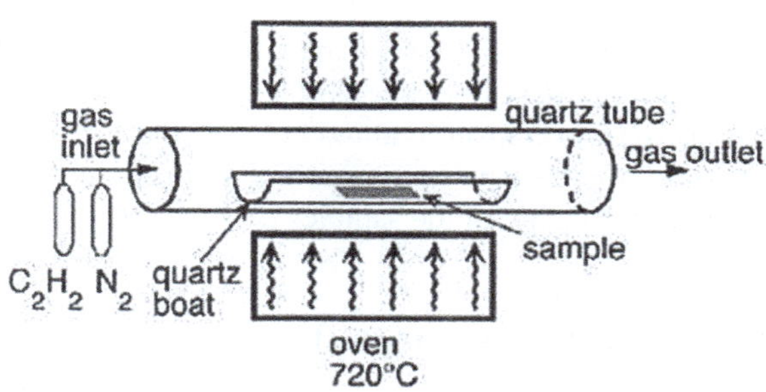

## *Microwave-Assisted Wet Chemical Synthesis*

Microwaves are defined as electromagnetic waves with vacuum wavelength ranging between 0.1 and 100 cm with frequencies between 0.3 and 300 GHz. Microwave-enhanced chemistry is based on the efficient heating of materials by "microwave dielectric heating" effects. This phenomenon is dependent on the ability of a specific material (solvent or reagent) to absorb microwave energy and convert it into heat. With microwave heating, the energy can be applied directly to the sample, and heating can be started or stopped instantly, or the power level can be adjusted to match the required. The microwave dielectric heating is a non-quantum mechanical effect and it leads to volumetric heating of the samples. Therefore, it is necessary to analyze whether it has any significant advantages compared to the thermal heating of chemical reactants. In general, conventional heating depends on both the convection current and thermal conductivity of the different materials.

For such kind of heating, external heat sources like Bunsen burner, electric plate heater, and oil bath are used. Due to this, the process is slow and it is an inefficient method for transferring energy into the desired mixture. The interest in this microwave-assisted organic synthesis has been growing during recent years. With microwave heating energy can be directly applied to the reaction, not to the vessel where it takes time for the reaction to be completed and also the time has taken is less and there is the consumption of time. Microwave heating is based on dielectric heating, i.e., molecule exhibiting a permanent dipole moment will try to align to the applied electromagnetic field resulting in rotation, friction, and collision of molecules and, thus in heat generation [53].

Some of the major advantages of this method include:

- The spectacular decrease in reaction time.
- Improved conversions.
- Clean product formation and.
- Wide scope for the development of new reaction conditions.

Recent reports have shown that microwave heating can be very convenient for use in a large number of organic synthetic methods. Microwave heating is very specific and there is no contact required between the energy source and the reaction vessel. Microwave dielectric heating is a non-quantum mechanical effect and it leads to volumetric heating of the samples [54] (Fig. 1.9).

### Conventional Sol–Gel Method

The sol–gel method [55] is a versatile process used for synthesizing various oxide materials. This synthetic method generally allows controlling the texture, chemical, and morphological properties of the solid. This method also has several advantages over other methods, such as allowing impregnation or co-precipitation, which can be used to introduce dopants. The major advantages of the sol–gel technique include

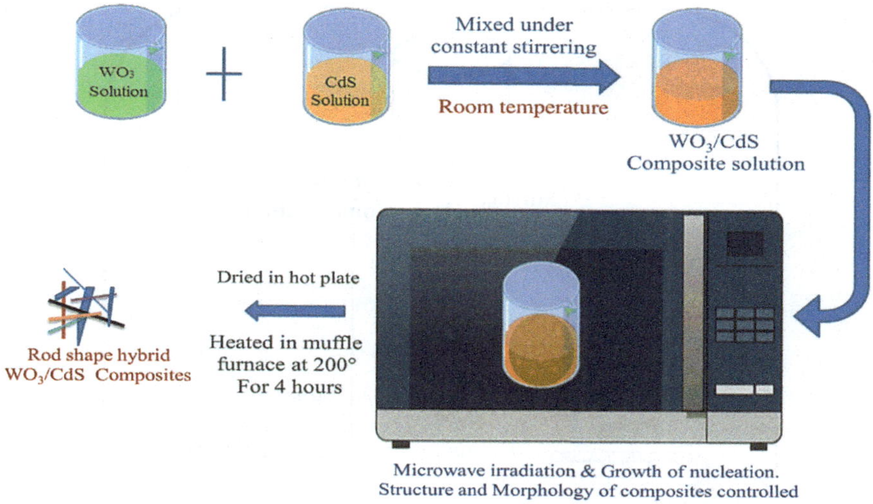


**Fig. 1.9** Typical microwave-assisted wet chemical method

molecular-scale mixing, high purity of the precursors, and homogeneity of the sol–gel products with a high purity of physical, morphological, and chemical properties.

In a typical sol–gel process, a colloidal suspension, or a sol, is formed from the hydrolysis and polymerization reactions of the precursors, which are usually inorganic metal salts or metal–organic compounds such as metal alkoxides. Any factor that affects either or both of these reactions is likely to impact the properties of the gel.

The sol–gel parameters, include the type of precursor, type of solvent, water content, acid or base content, precursor concentration, and temperature. These parameters affect the structure of the initial gel and, in turn, the properties of the material at all subsequent processing steps. After gelation, the wet gel can be optionally aged in its mother liquor, or in another solvent, and washed. The time between the formation of a gel and its drying, known as ageing, is also an important parameter. It is also noted that a gel is not static during ageing but can continue to undergo hydrolysis and condensation (Fig. 1.10).

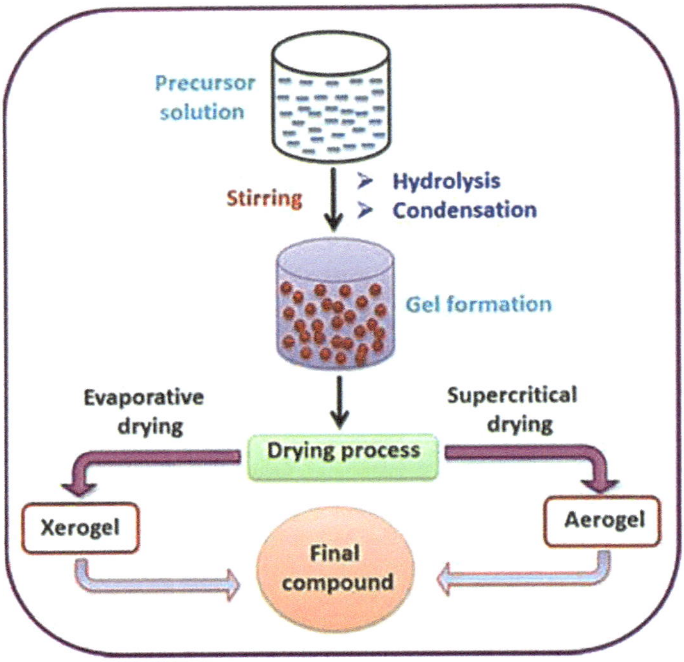


**Fig. 1.10** Typical setup of the sol–gel method

## Hybrid Nanocomposites of $WO_3$ and Its Supercapacitor Performance

### *Graphene/$WO_3$ Hybrid Nanocomposite*

Chaitoglou et al. [56] reported the evaluation of Graphene/$WO_3$ as electrode material for supercapacitor applications. Graphene layers were developed by the chemical vapor deposition (CVD) method. Further, stacks of graphene single and layer-by-layer were prepared by combining metal oxides grown by magnetron sputtering and graphene transfer techniques. The short time sputtering deposition was adopted in order to avoid damage in electrode materials The electrochemical performance of the Graphene/$WO_3$ were investigated by using a Swagelok cell and organic electrolyte which consisting of 1M $LiClO_4$ solved in ethylene carbonate (EC) and diethyl carbonate (DEC) mixed in 1:1 volumetric proportion.

The SEM analysis showed the continuity of the graphene film and all the area was covered with single-layer graphene and TEM images showed the larger particles of $WO_3$ with the d spacing of 0.31 nm. In addition to that, the single layer of graphene was confirmed by the Raman spectrum (Fig. 1.11).

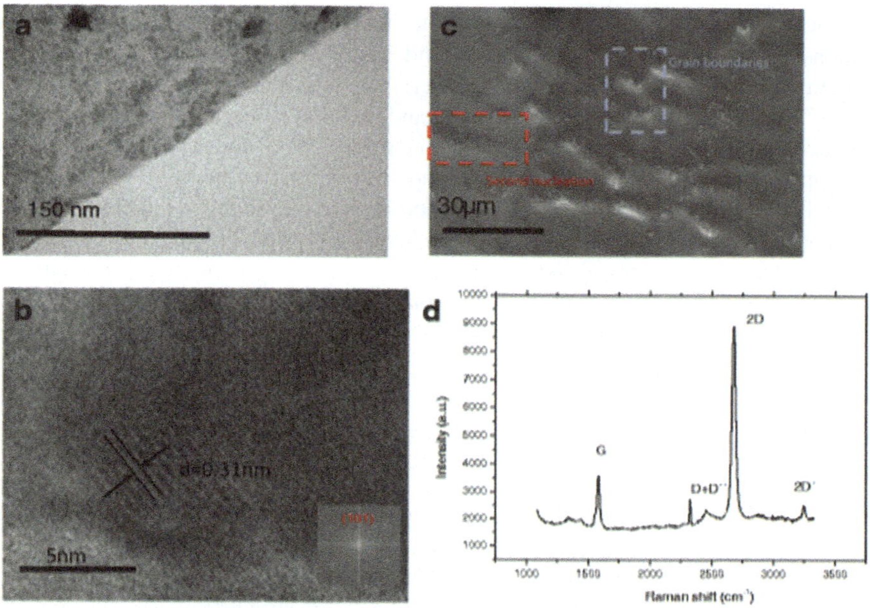


**Fig. 1.11** The TEM images of Gr/$WO_3$ nanofilm structure (**a**, **b**, **c**) and Raman spectra (**d**)

The electrochemical analysis of an as-prepared hybrid nanocomposite was conducted to study their suitability for supercapacitor applications. Graphene/$WO_3$ film exhibited a high capacitance of 2.69 $mF/cm^2$ at the scan rate of 10 mV/s. Further, the capacitance was increased slightly upon increasing the number of graphene layers such that the capacitance of Graphene/$WO_3$ film was increased from 2.69 to 4.15 $mF/cm^2$ during the addition of two more layers (Fig. 1.12).

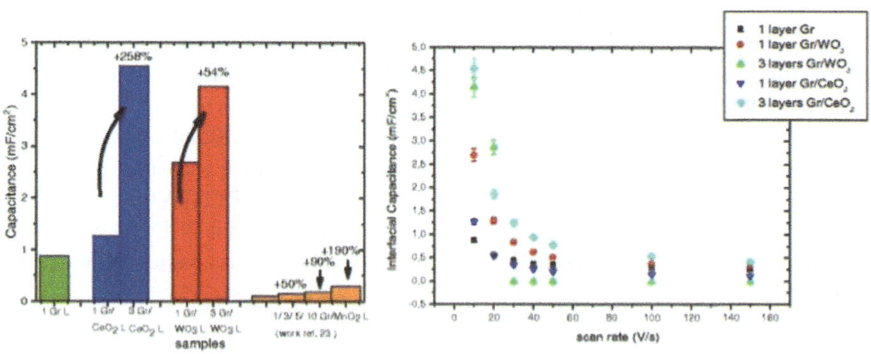


**Fig. 1.12** Histogram with the percentage increase of capacitance with respect to a number of layers and interfacial capacitance of the different hybrid electrodes at different scan rates

The surface area of hybrid composite would increase proportionally with increasing graphene/metal oxide layers and interlayer distance may decrease for benefiting multilayer ion absorption. Further, the power density was increased to 4.5 $\times 10^{-8}$ W-h/cm$^2$ for three-layer graphene and they have observed that capacitance of electrode material was much higher at nine times. The cyclic stability test showed the capacitance retention of between 70 and 90% during the first 850 cycles. From the galvanostatic charge/discharge profile, more time was required for charging and discharge when more metal oxide/graphene layer was added. Finally, they concluded that the deposition of metal oxide nanoparticles over the graphene layer enhances the total specific capacitance of the hybrid material, an additional pseudocapacitance was contributed by metal oxide particles.

## *PEDOT/WO$_3$ Composite Films*

The PEDOT/$WO_3$ composite films were prepared by the electrochemical deposition technique by Zhuzhelskii et al. [57] and investigated for supercapacitor suitability. The surface morphology of the composite film was studied by SEM analysis and film showed the highly porous with irregular spatial network structure of 20–200 nmpores ranges. This irregularity was due to intergrowth of individual globules and about 20–50 nm size individual globules were visualized on the surface of the substrate. During the increase in composite film thickness, cauliflower-like structures and individual globular was observed for PEDOT (1000) films (Fig. 1.13).

To assess the charge storage performance of PEDOT (200)/$WO_3$, Galvanostatic charge–discharge (GCD) studies were conducted. The GCD curves showed linear in the range of potentials afterward redox process have a plateau in the range of −0.3–0.0 V and this plateau corresponded to redox reaction in the $WO_3$ compound. The long cycle test reveals the prolonged cycling capability about 96% retention of initial capacitance following 260 charge–discharge cycles and confirmed the stable capacitive nature *f* composite films. From GCD curves, the specific capacitance of PEDOT (200) / $WO_3$ thin films was calculated as 615 F/g for 1.4 A/g, 375 F/g for 2.8 A/g and of about 308 F/g for 10 A/g. further, their specific capacitance was in concord with cyclic voltammeter data in $H_2SO_4$ electrolyte. The ion diffusion and electron transfer in this hybrid composite facilitated by the presences of conducting porous matrix which lead to diminishing the internal resistance and of the composite and helped to enhance the electrochemical performance of PEDOT/$WO_3$ composites. The maintenance of high current and capacitance specified that PEDOT/$WO_3$ electrode materials could be promising electrode material for supercapacitor applications (Fig. 1.14).

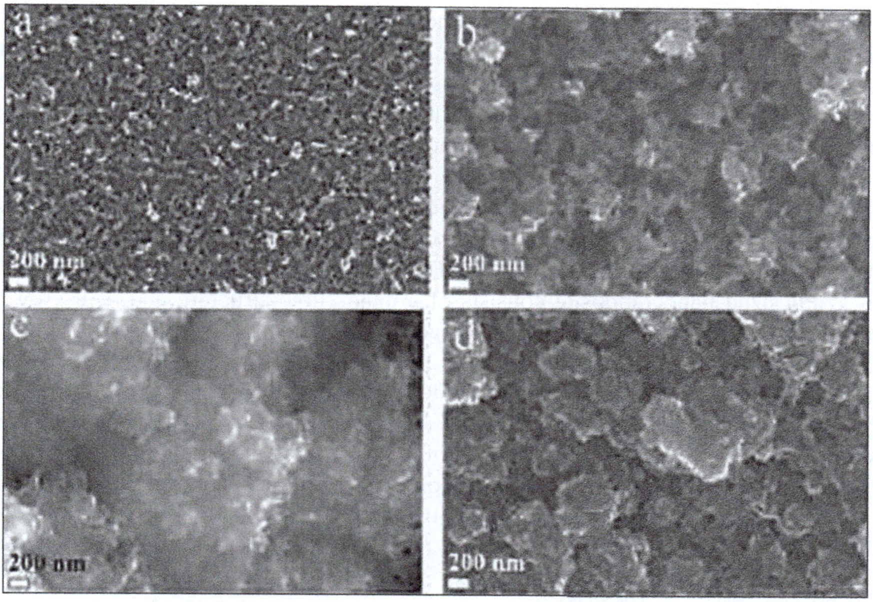


**Fig. 1.13** SEM images of: **a** PEDOT (200) film. **b** PEDOT (200)/$WO_3$ film. **c** PEDOT (1000) film. **d** PEDOT (1000)/$WO_3$ film

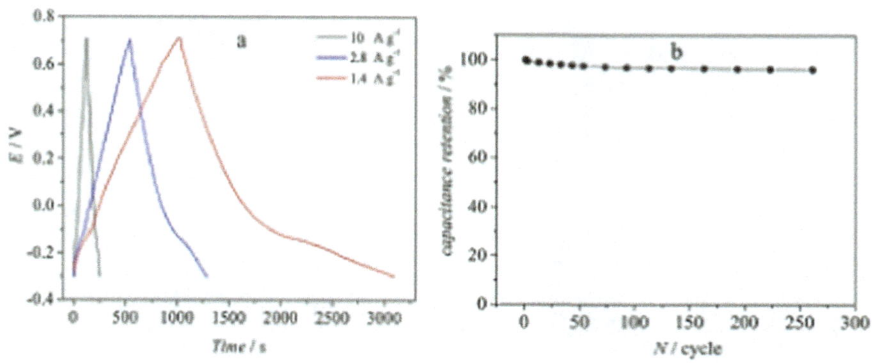


**Fig. 1.14** **a** Galvanostatic charge–discharge curves at different current densities. **b** Capacitance retention from the cycle number

## *$WO_3$-CdS Nanocomposite*

To account for the benefits of the synergetic effect of hybridization, we proposed the novel hybridized $WO_3$-CdS nanocomposites as electrode materials for electrochemical supercapacitor [57]. Based on the literature report, The CdS compound was loaded at different wt% with $WO_3$ and their benefits are reflected in $WO_3$-CdS nanostructures' performance in different applications. Further, they have adopted different synthesizing methods with sophisticated instrumentations. In this work, we have adopted a simple and environmentally friendly microwave-assisted wet chemical method which is a fast and efficient method for nanostructure preparation in our laboratory.

The High-Resolution Transmission Electron Microscopic (HRTEM) analysis was carried out to get in-depth knowledge about the morphology of $WO_3$-CdS hybrid nanostructures and shown in Fig. 1.15. The HRTEM images clearly revealed that as-synthesized nanocomposite consists of a nano cubic structure and well-grown nanorod shape morphology with an average length of 28 nm.

Cyclic Voltammogram (CV) analysis of pure $WO_3$ (a), $WO_3$-CdS (b, c, and d) composites in 20% KOH electrolyte at 10–50 mV/s scan rates were shown in Fig. 1.16. The shape of the CV curves of pure $WO_3$ and $WO_3$-CdS composite samples have demonstrated the behavior of the electrochemical double-layer capacitor with slight pseudo-capacitive behavior. The specific capacitances were calculated as 44 F/g for pure $WO_3$ samples, followed by 48, 170, 172 F/g for $WO_3$-CdS

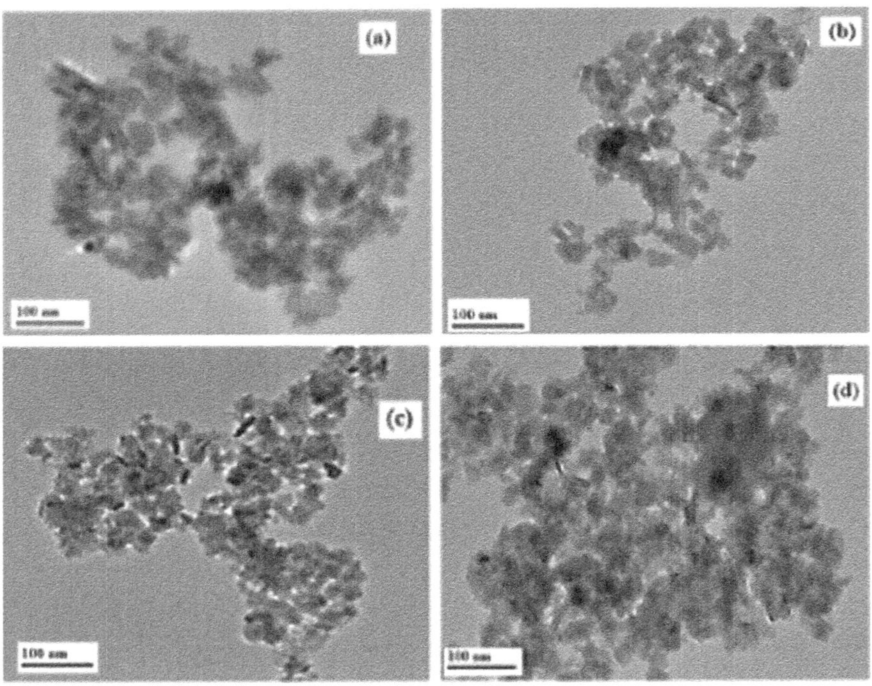


**Fig. 1.15** HRTEM images of $WO_3$ **a** and $WO_3$-CdS **b**, **c**, **d** nanocomposites

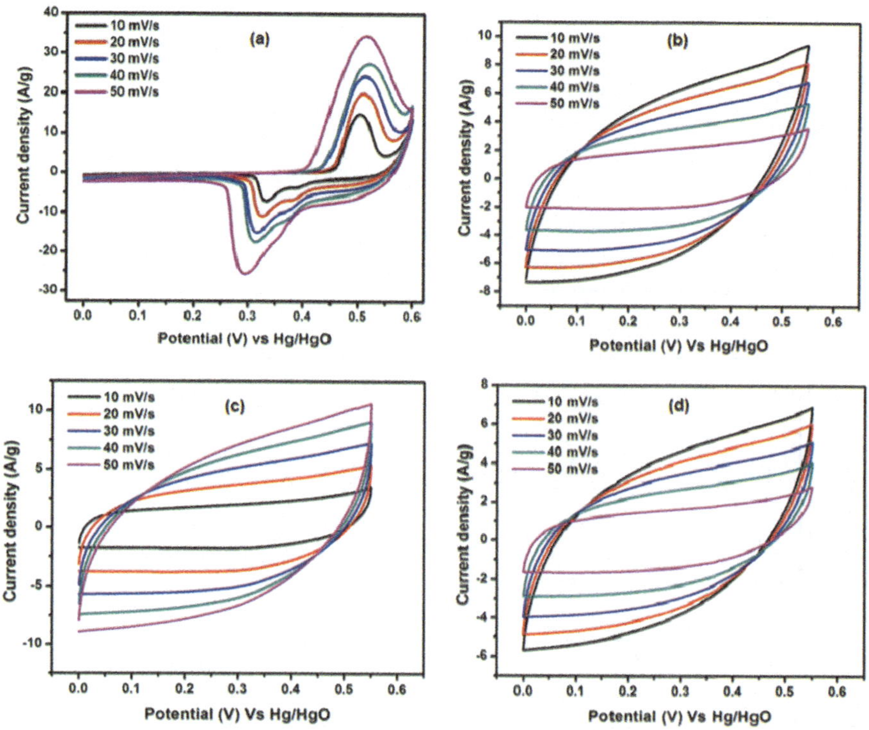


**Fig. 1.16** Cyclic voltammetric curves of $WO_3$ (**a**) and $WO_3$-CdS (**b, c, d**) nanocomposites in 20% KOH electrolyte

composites. High Pseudocapacitive performance of $WO_3$ was achieved by the electrochemical intercalation/de-intercalation process by means of protons exchange and subsequent application of a low voltage [58].

Galvanostatic Charge–Discharge (GCD) analysis of $WO_3$ particles and $WO_3$-CdS composites were carried out in the current density range of 0.5–5 A/g and are depicted in Fig. 1.17. We are interested to check the suitability of as-prepared composites for electrochemical supercapacitor applications. Owing to the internal resistance of as-prepared electrode materials, IR drop appeared at the early discharging time and it is ascribed to the contact resistance between the electrode and the current collector.

The capacitance of active materials highly depends on the intercalation ratio between the active material's surface area and the ability of electrolyte diffusion onto active materials. Actually, part of the electrodes only utilized for intercalation which is nearer to the geometric electrode surface and electrolyte. As a result of the poor diffusion of electrolytes slowed the practicality of thick electrode active nanomaterials being used for supercapacitors. In semiconductor composite systems, the way of different energy levels, the charge injection from one material into another

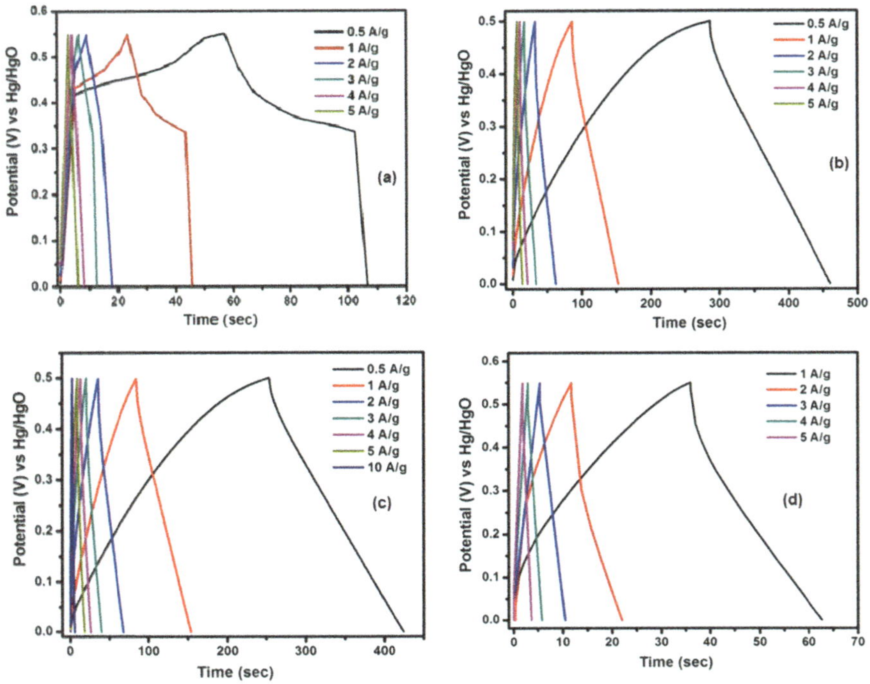


**Fig. 1.17** Galvanostatic charge–discharge curves of $WO_3$ (**a**) and $WO_3$-CdS (**b**, **c**, **d**) nanocomposites in 20% KOH electrolyte

can make possible efficient charge separation by minimizing the electron–hole pair recombination [59].

The same electrode materials were investigated for electrochemical performance in the 1M $H_2So_4$ electrolyte for comparison. Interestingly, its electrochemical performance was found to be enhanced than in 20% KOH electrolyte. The cyclic voltammogram and charge–discharge analysis of $WO_3$-CdS nanocomposites were recorded and presented in Figs. 1.18 and 1.19. The maximum specific capacitances of as-prepared nanocomposites were achieved as 180 F/g (b), 388 F/g (c), and 650 F/g (d), respectively. The loading of the CdS compound has contributed to higher capacitance by providing the excellent host matrix for ions intercalation and exchange.

We have attempted to propose the possible mechanism of $WO_3$-CdS hybrid structures during the oxidation-redox process. In the $WO_3$-CdS hybrid composites, the conduction band and valence band level of CdS are more negative than the $WO_3$ corresponding bands [59], and the electron–hole pairs are generated in a CdS upon increasing the applied current and concentration of CdS ratio. These excited electrons in CdS are optimistically transferred from the conduction band of the CdS to that of $WO_3$ because of the potential difference, consecutively the recombination of

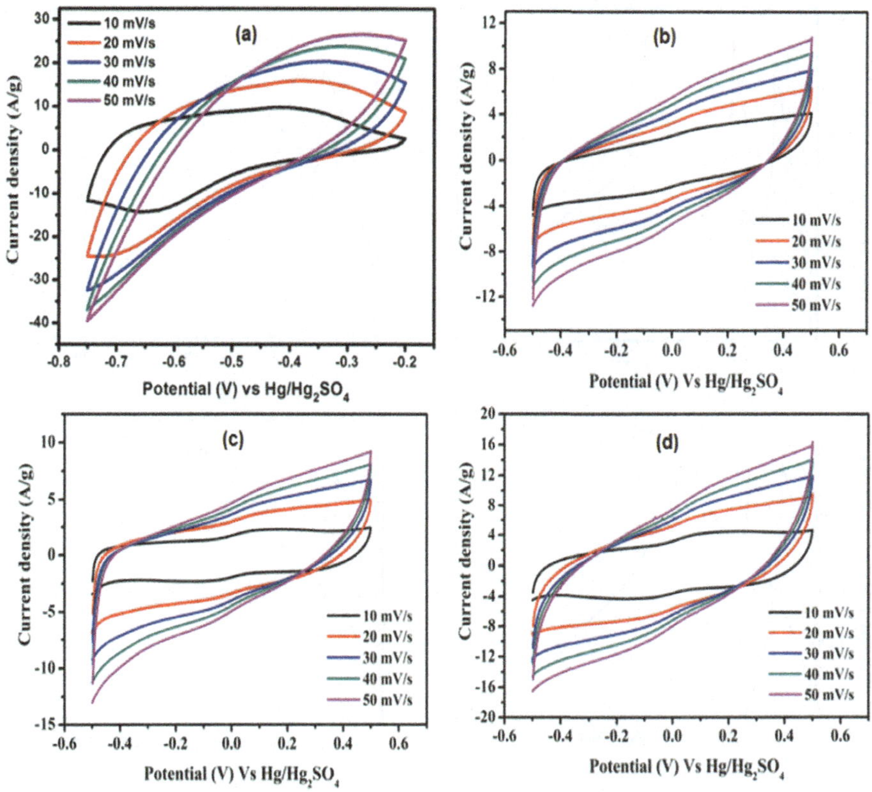


**Fig. 1.18** Cyclic voltammetric curves of $WO_3$ (**a**) and $WO_3$-CdS (**b**, **c**, **d**) nanocomposites in 1M $H_2SO_4$ electrolyte

the electron–hole pairs is promoted, and they perform as a catalyst for the reductive reaction. Alternatively, generated holes move in the opposite direction from the valance band of $WO_3$ to that of CdS and take part in the oxidation reaction, and the electron–hole recombination possibility can be reduced. The efficient charge transfer at the interface of the electrode–electrolyte decreases the charge recombination and resulted in the simplistic charge transport of electrons through the active material [54, 60–63].

Since, both the valence band and conduction band level of CdS is higher than that of $WO_3$ in the hybrid $WO_3$-CdS composites, the energy gradient presenting at the interfaces have a tendency to separate electrons and holes spatially on the different sides of the heterojunction and which will encourage the reduction of electron–hole pairs combination.

Furthermore, the enhanced supercapacitor performance may be attributed to as follows: (i). There was good retention of capacitance in $WO_3$-CdS nanocomposite samples (b, c, d). The CdS additions do not affect the surface pore structure and

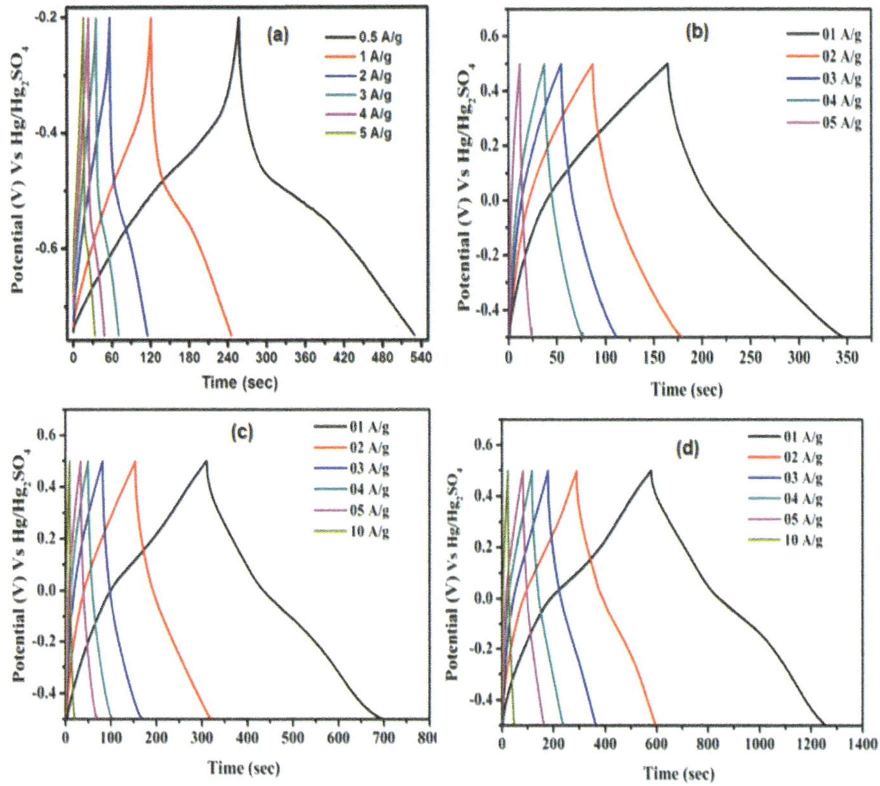


**Fig. 1.19** Galvanostatic charge–discharge curves of $WO_3$ (**a**) and $WO_3$-CdS (**b**, **c**, **d**) nanocomposites in 1M $H_2SO_4$ electrolyte

the electrolyte didn't find it harder to penetrate into active materials. (ii). Further increasing the CdS into $WO_3$ might have shared the generated electron–hole pair mutually with each other and helped the fast diffusion of electrolyte ions. (iii). A higher concentration of CdS with $WO_3$ may be favorable for the long-lasting charge separation and the carrier transport between $WO_3$ and CdS. Consequently, as generated charge carrier lifetime in the hybrid $WO_3$-CdS composites are appreciable.

Further, the long cycle test was conducted and presented in Fig. 1.20 which shows the highest capacity retention of 133% and 100% Coulombic efficiency of the electrode material. These long-lasting charge separations of charge states in the hybrid $WO_3$-CdS (d) composites, where 6 wt % of CdS was incorporate with $WO_3$ turn out to the most suitable for electrochemical double layer supercapacitor application.

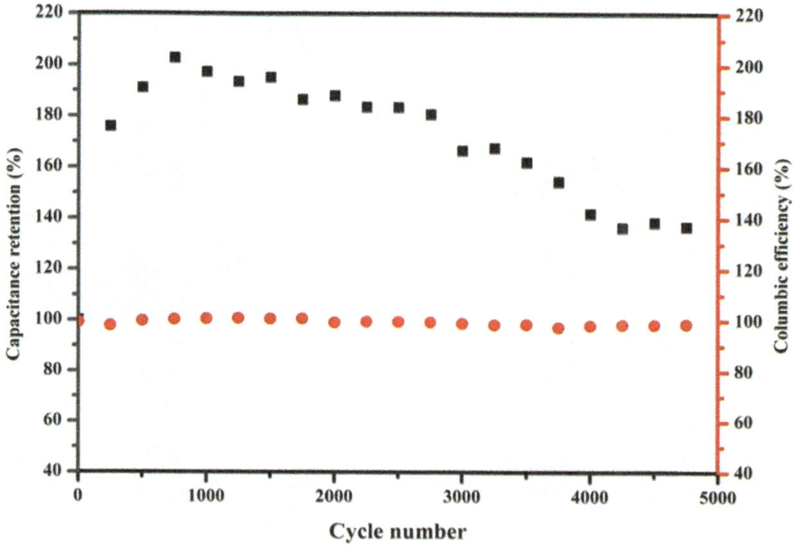


**Fig. 1.20** Cycle performance and Coulombic efficiency of $WO_3$-CdS composites in 1M $H_2SO_4$ electrolyte

## *$WO_3$–ZnS Nanocomposites*

The $WO_3$–ZnS hybrid nanocomposites were prepared by the simple and eco-friendly microwave-assisted wet chemical method [54]. The rational analysis of HRTEM images revealed that the samples contained the small rod shape surface morphology. It can be seen that the $WO_3$ (a) has a glossy surface and pointed to the tip end. The apparent diameter of the $WO_3$ structure was 12 nm, and its length was relatively 17.5 nm to some hundred nanometers. In addition, for comparison, the morphology of the $WO_3$–ZnS composite showed in Fig. 1.21 which has significant inflation by loading different ratios of ZnS into $WO_3$. The considerable length and diameter of $WO_3$–ZnS (b, c, d) composites were disturbed by microwave irradiation and increasing the loading of ZnS.

Cyclic Voltammetry (CV) analysis was performed in the scan rate of 10–100 mV/s in 20% KOH electrolyte for $WO_3$ and $WO_3$–ZnS nanocomposites and depicted in Fig. 1.22. The Galvanostatic Charge–Discharge (GCD) analysis of as-synthesized composites was shown in Fig. 1.23. The distinctive shape of CV curves deviated at a higher scan rate, which represents the decrement of stability of $WO_3$–ZnS nanocomposites (electrode materials) at a high charging state. Initially, the CV curve convincingly demonstrated a typical quasi-rectangular shape without a redox peak. Further, upon increasing the load of ZnS showed obvious redox peak and pseudocapacitive behavior. The redox peaks could be adequately explained by the valence state conversion between $WO_3$ and ZnS as propounded by Pu et al. [64]. In addition, the

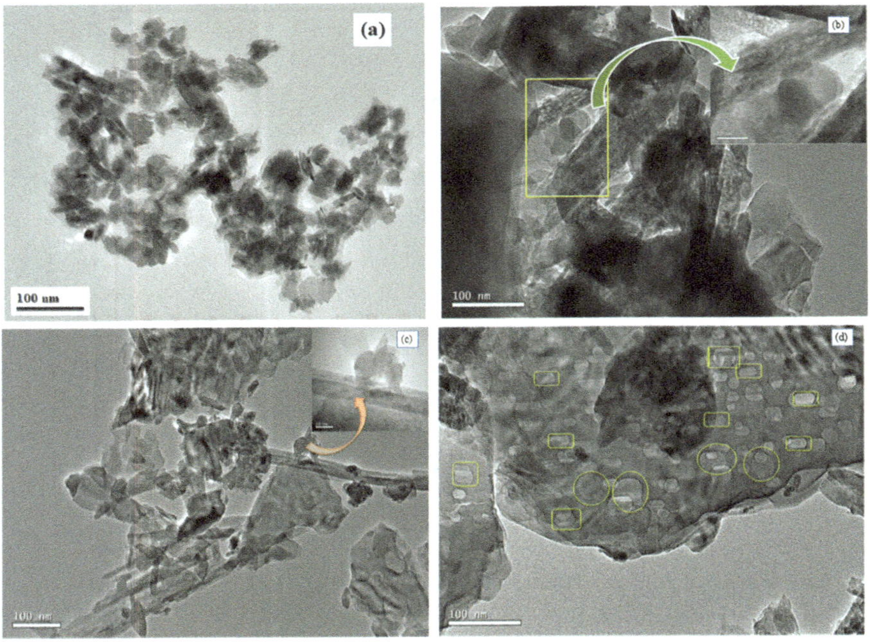


**Fig. 1.21** HRTEM images of $WO_3$ (**a**) and $WO_3$–ZnS (**b**, **c**, **d**) nanocomposites

integrated area under the CV curve seems to be growing when increasing the ZnS content, which reflected in the capacitance decreases. Since the integrated area is directly proportional to the capacitance of electrode materials [65].

The pure $WO_3$ nanostructure instantly shows the highest specific capacitance of 44 F/g. The specific capacitances of $WO_3$–ZnS (b, c, d) nanocomposites were calculated as 105 F/g, 120 F/g, and 215 F/g, respectively. The considerable advancement in the electrochemical capacitance of metal oxides is relay not only on the porous sites and nanoparticles' higher specific surface area, but it seems to represent an important parameter. Several transition metal oxides have abundantly shown excellent capacitance [66, 67].

In our case also the loading of ZnS into $WO_3$ has supported to achieve better porous sites and to precisely control the particle size growth. Each composite sample has shown a decrement in its capacitance at the higher scan rate. This possible deviation of capacitance was reasoned as follows, due to the lacking of accessibility and diffusion limitations of the KOH electrolyte ions into the inner active sites of $WO_3$–ZnS composite at a higher scan rate. It is noteworthy from the GCD curves that the $WO_3$–ZnS (b) composite electrode shows the longest discharge time which again proves the higher electrochemical performance at current density 1 A/g. Remaining $WO_3$–ZnS (c and d) composite showed the long discharge time at the current density of 2 A/g, which is ascribed to the excess content of ZnS. The long cycle test showed in Fig. 1.24 revealed that as-prepared composites capacitance is retained about 71%

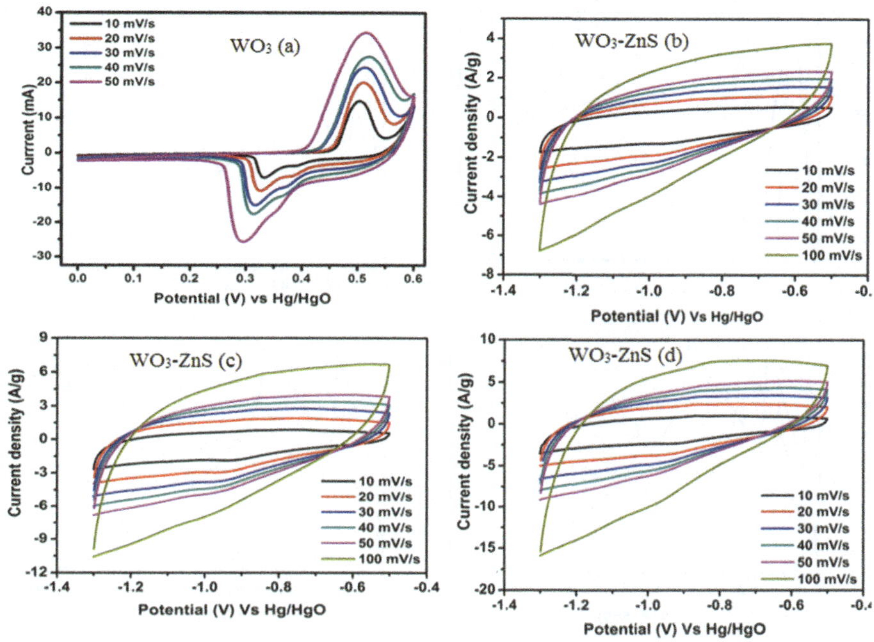


**Fig. 1.22** Cyclic voltammetric curves of $WO_3$ (**a**) and $WO_3$–ZnS (**b, c, d**) nanocomposites in 20% KOH electrolyte

in KOH electrolyte and its coulomb efficiency is more than 100% after 4000 cycles.

The cyclic voltammetry and charge–discharge analysis was performed for the same electrode materials in 1M $H_2SO_4$ electrolyte and presented in Figs. 1.25 and 1.26. The pure $WO_3$nanostructures show the maximum capacitance of 246 F/g. controversially, the hybridizes$WO_3$–ZnS (b, c, d) composites have shown poor performance. The obtained maximum capacitance was 17 F/g for the loading of 6 wt% ZnS and the remaining samples have shown less than 10 F/g. The performance of $WO_3$–ZnS composites might have been affected by several properties of electrolytes such as the concentration, ion's nature, and operating temperature.

In KOH electrolyte, the specific capacitances of $WO_3$–ZnS nanocomposites were increased which represents the packing of ions on the site of fine pores is promoted admirably and facilitated to charge transfer. Whereas in $H_2SO_4$ electrolyte, pure $WO_3$ showed the maximum capacitance of 246 F/g but $WO_3$–ZnS composites showed poor capacitance. This could be attributed to the decomposition of electrolyte during test cycles and evaluations of gases on the electrode or current collector interface. This decomposition of electrolyte and gas evaluation might increase the internal resistance.

It is well-known that the $H_2SO_4$ electrolyte is corrosive for current collectors and electrode materials. The sever deviation of CV and GCD curve shape was attributed to that the as-prepared composites were rigorously affected by corrosion

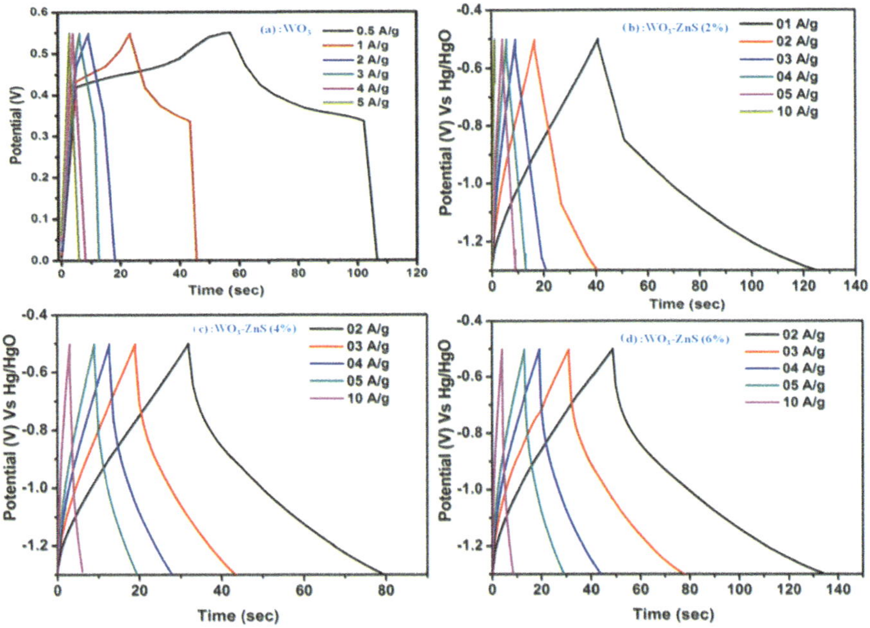


**Fig. 1.23** Galvanostatic charge–discharge curves of $WO_3$ (**a**) and $WO_3$–ZnS (**b**, **c**, **d**) nanocomposites in 20% KOH electrolyte

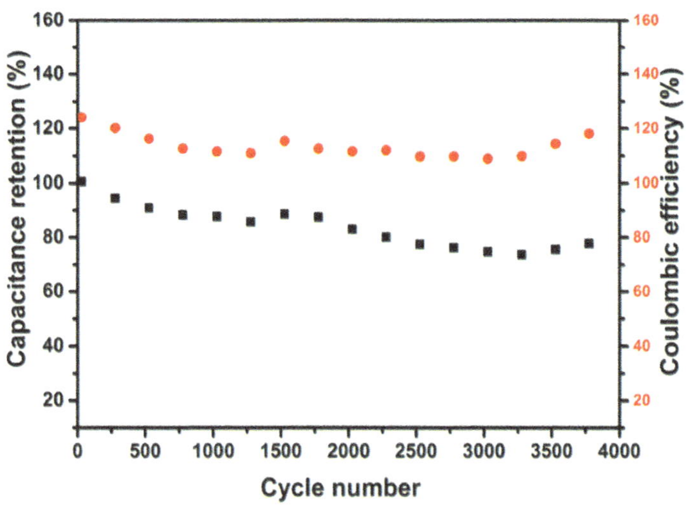


**Fig. 1.24** Cycle performance and Coulombic efficiency of $WO_3$–ZnS (**b**, **c**, **d**) nanocomposites KOH electrolyte

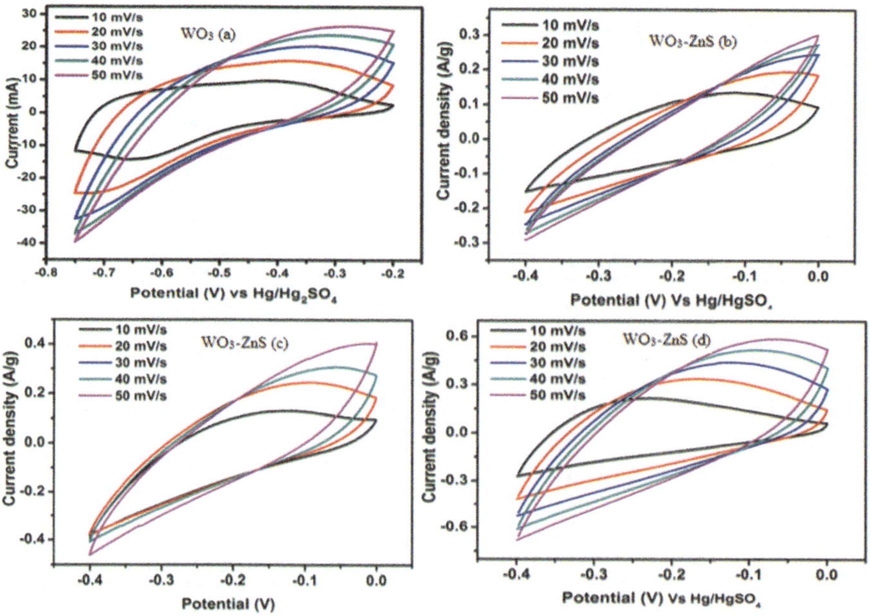


**Fig. 1.25** Cyclic voltammetric curves of pure $WO_3$ (**a**) and $WO_3$–ZnS (**b**, **c**, **d**) nanocomposites in $H_2SO_4$ electrolyte

in $H_2SO_4$ electrolyte which prevented the ion intercalation between electrodes and electrolytes. To overcome this problem, gold (Au) and Indium Tin Oxide (ITO) have been recommended. However, due to the high price, it is less used for practical purposes.

Generally, the electric energy is stored efficiently in a supercapacitor at the electrode/electrolyte interfaces in the electric double layer form, where excess or a deficit of electric charges is accumulated on the electrode surfaces. The concentration of the electrolyte and size of the ions is the deciding factor about the thickness of the double layer. It was acknowledged that the enhancement of the capacitance activity of the hybrid $WO_3$–ZnS nanocomposite was endorsed to the efficient charge carrier separation. Therefore, it was understood that the charge transfer at the heterojunction boundaries of $WO_3$ and ZnS was the most key part of the imperative factor influencing the electrochemical super capacitance performance.

On account of the contradictory positions of the valence band and conduction bands of $WO_3$ and ZnS, the bandgap was narrowed. The excited electrons in the conduction band of ZnS can be able to transfer effortlessly to the conduction band of $WO_3$, whereas the holes in the valence band of $WO_3$ can able to transfer easily to the valence band of ZnS. Therefore, the recombination of the charge carriers was intentionally concealed, which promote the lifetime of the electron–hole pairs. During the process of charging, the electrons travel from the negative electrode to

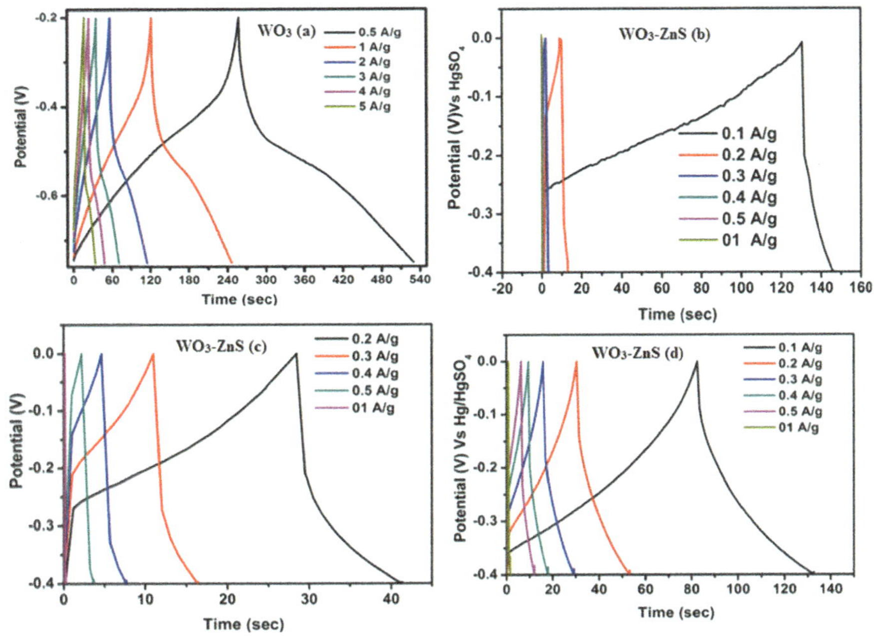


**Fig. 1.26** Galvanostatic charge–discharge curves of $WO_3$ (**a**) and $WO_3$–ZnS (**b**, **c**, **d**) nanocomposites in $H_2SO_4$ electrolyte

the positive electrode through an external load. Within the electrolyte, cations move towards the negative electrode while anions move towards the positive electrode. During discharge, the reverse processes invariably take place.

## *$WO_3$–$V_2O_5$ Hybrid Composite*

For the first time, $WO_3$–$V_2O_5$ composite was prepared for electrochemical supercapacitor application [68]. Electrochemical analysis of $V_2O_5$ strongly suggested that it could be legitimately the candidate material for lithium-ion batteries and a supercapacitor. The careful observation of HRTEM micrographs as shown in Fig. 1.27 has revealed that as-synthesized nanostructures consisting of the irregular shape of $WO_3$ which were altered into a deeply interconnected cluster of small grains. The addition of 4 and 6 wt% of the $V_2O_5$ compound has greatly affected the morphology of $WO_3$ into the densely aggregated grain and rapid grain size growth of $WO_3$–$V_2O_5$ nanocomposites. Along with the grain size growth, the larger number of intragranular pores was also expected to increase, which was anticipated to provide more sites for ions absorption during the electrochemical charge/discharge process.

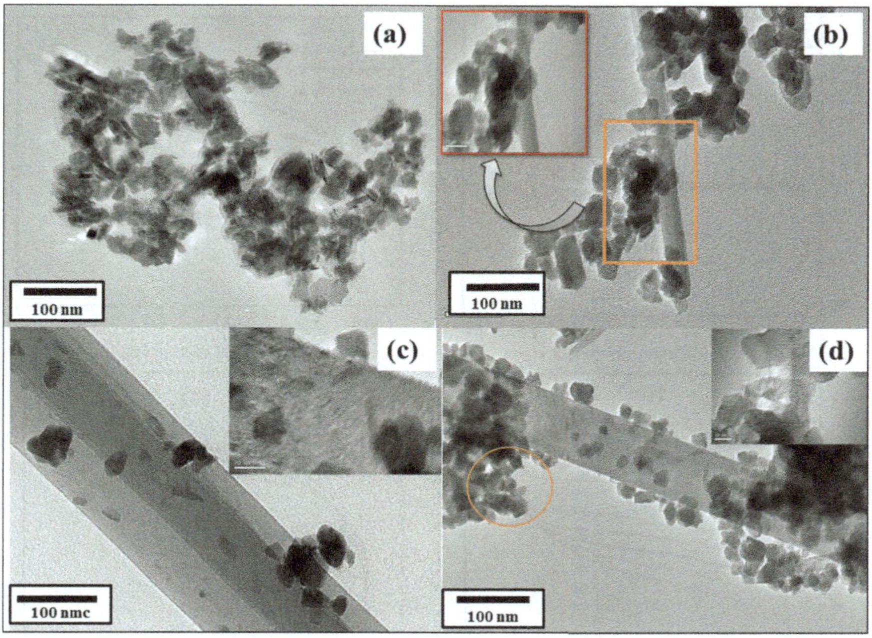


**Fig. 1.27** HRTEM images of $WO_3$ (**a**) and $WO_3–V_2O5$ (**b, c, d**) nanocomposites

Cyclic Voltammetry (CV) analysis of $WO_3$ and $WO_3–V_2O_5$nanocomposites were performed in the potential sweep rate of 10–50 mV/s in 20% KOH electrolyte and presented in Fig. 1.28. During the charge/discharge process (GCD), the resulted current by ions intercalation and deintercalation was recorded for pure $WO_3$ and $WO_3–V_2O_5$ composites and presented in Fig. 1.29. The area of CV curve increased with increasing the $V_2O_5$ content implied that as-prepared composites structures provided an easy way for ions intercalation and charge transfer process.

Also, the CV curve showed typically a quasi-rectangular shape with redox peak which ascribed to electrochemical double-layer capacitance with Pseudocapacitance behavior. Further addition of $V_2O_5$ showed noticeable redox peak increment and it can possibly be attributed to the conversion of charge carriers in valance state of $WO_3$ and $V_2O_5$. Besides, the shape of the CV curve changed at a higher scan rate which represents the instability of as-prepared samples. Also, the specific capacitance of as-synthesized nanocomposites was getting decreased at a higher current density. We have observed a shift in anode peak potential for hybridized nanocomposites which was attributed to increasing the $V_2O_5$ content with $WO_3$.

In general, the specific capacitances of working electrode materials were greatly influenced by electrolyte concentration, the surface area of the electrode material. The maximum specific capacitance of pure $WO_3$ and hybridized $WO_3–V_2O_5$ nanocomposites were recorded as 44 F/g, 100 F/g, 173 F/g, and 102 F/g, respectively. Obviously, the addition of the 4 wt% $V_2O_5$ showed the highest specific capacitance of

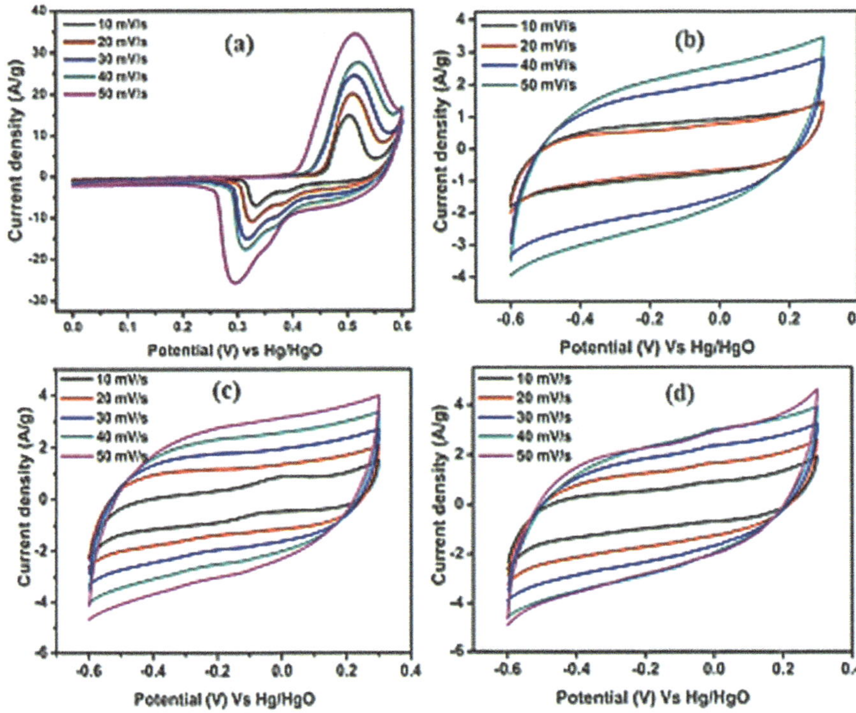


**Fig. 1.28** Cyclic voltammetric curves of $WO_3$ (**a**) and $WO_3$–$V_2O_5$ (**b**, **c**, **d**) nanocomposites in 20% KOH electrolyte

$WO_3$–$V_2O_5$ (c) due to the excellent host matrix provided for the charge transfer process in the intervalence state between $WO_3$ and $V_2O_5$. Further increasing a load of $V_2O_5$ (6 wt%) drastically drops off the specific capacitance of $WO_3$–$V_2O_5$ composites to 102 F/g. The Coulombic efficiency of $WO_3$–$V_2O_5$ composites showed as 100% up to 5000 cycles and a long cycle test showed the capacitance retention as approximately 126% in KOH electrolytes and presented in Fig. 1.30.

The same electrode materials were analyzed in the 1M $H_2SO_4$ electrolyte for comparison. The pure $WO_3$ has shown the highest capacitance of 246 F/g and $WO_3$–$V_2O_5$ composites have shown poor performance such as the maximum capacitance was less than 17 F/g. The CV & GCD curves of as-synthesized composites showed in Figs. 1.31 and 1.32 demonstrated the unsmooth curves which possibly attributed to the corrosion of the electrode material in $H_2SO_4$ electrolyte and high diffusion resistance faced by the ions. These parameters showed the unsatisfactory formation of EDLC. Besides, a high deviation of symmetrical behavior and comparatively less GCD performance of $WO_3$–$V_2O_5$ could be accounted for the electrochemical decomposition of electrolyte and instability of electrode material as an argument by

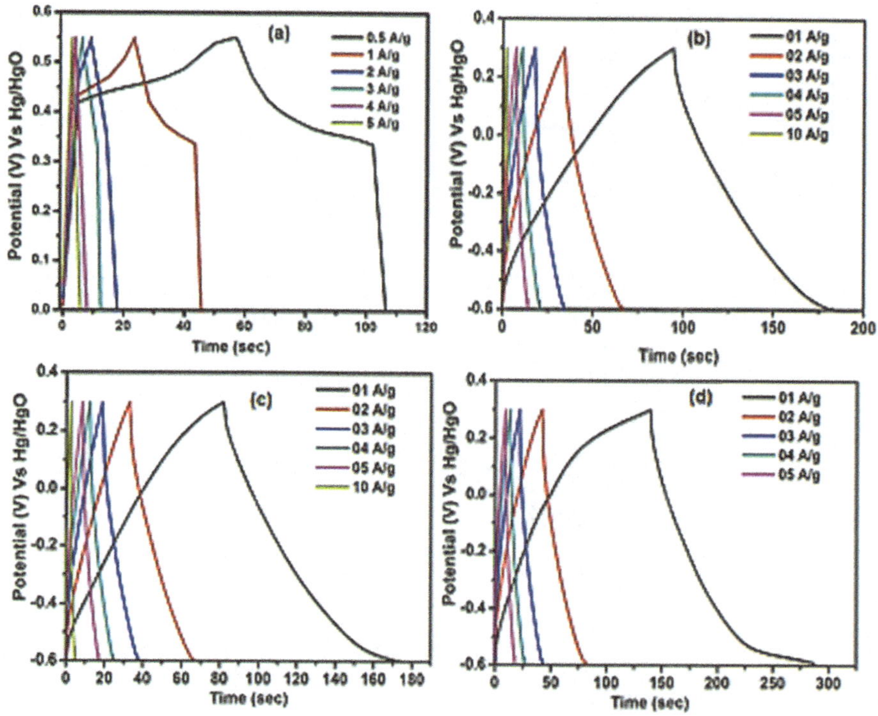


**Fig. 1.29** Galvanostatic charge–discharge curves of $WO_3$ (**a**) and $WO_3–V_2O_5$ (**b**, **c**, **d**) nanocomposites in 20% KOH electrolyte

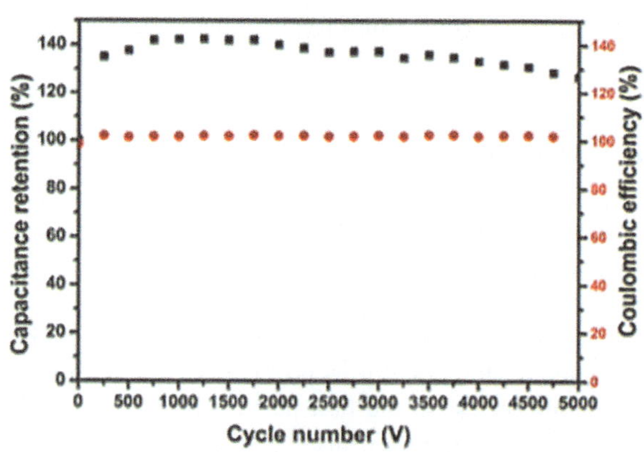


**Fig. 1.30** Cycle performance and Coulombic efficiency of $WO_3–V_2O_5$ (**b**, **c**, **d**) nanocomposites in 20% KOH electrolyte

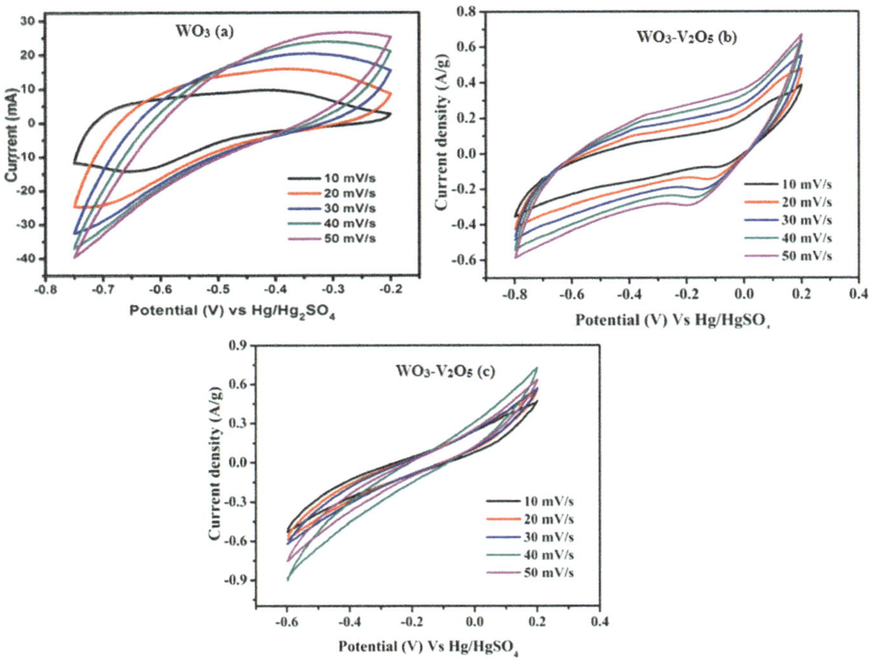


**Fig. 1.31** Cyclic voltammetric curves of $WO_3$ (**a**) and $WO_3$–$V_2O_5$ (**b**, **c**, **d**) nanocomposites in 1M $H_2SO_4$ electrolyte

Mohd Hanappi et al. [69]. Thus, unbalanced ions have gathered on the micropores of electrode surfaces duration the charging-discharging process.

## Conclusion

In this present review, tungsten oxide has been focused on the high performance of the supercapacitor applications. Though several metal oxides have been investigated for the energy storage device, still, $WO_3$ studied for supercapacitor electrode material owing to adjustable characteristics towards high performance. This article was designed systematically to provide the outline of recent development using $WO_3$ nanocomposites. In the first part of the article, we have discussed the major classifications of nanomaterials such as 0D, 1D, 2D, and 3D and followed by that we have paid little attention to components in supercapacitors and challenges encountered by researchers and advantages of a supercapacitor. The high performance of supercapacitor is a relay of electrode materials structure and morphology which is directly affected by synthesizing methods. So some commonly utilized physics and chemical methods also focussed briefly. Finally, electrochemical investigates of

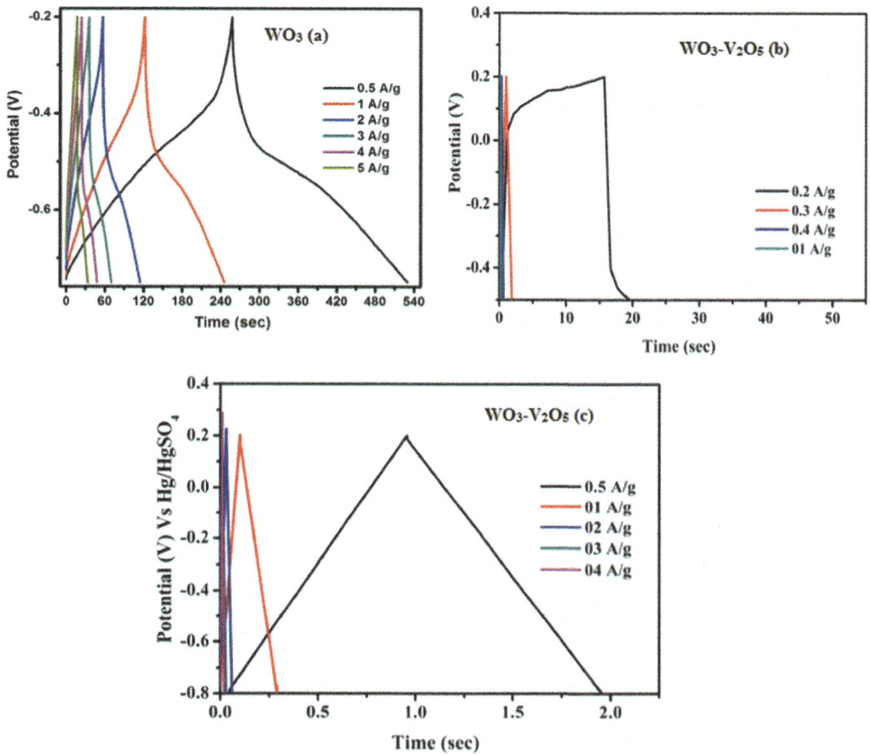


**Fig. 1.32** Galvanostatic charge–discharge curves of $WO_3$ (**a**) and $WO_3$–$V_2O_5$ (**b**, **c**, **d**) nanocomposites in 1M $H_2SO_4$ electrolyte

$WO_3$ based hybrid nanocomposites were explained adequately. However, some challenges remained untouched, particularly, in the synthesizing method and structures of nanocomposites need to be optimized to conquer some unambiguous factors. Hence, intensive research works have been initiated across the globe to fabricate active nanocomposites for supercapacitors.

## References

1. https://www.electricalindia.in/future-perspective-for-renewable-energy-in-india/
2. Pushparaj, V.L., Sreekala, S., Nalamasu, O., Ajayan, P.M.: Flexible energy storage devices using nanomaterials. In: Semiconductor Nanomaterials for Flexible Technologies, pp. 227–245. William Andrew Publishing (2010)
3. Becker, H.I., General Electric Co.: Low voltage electrolytic capacitor. U.S. Patent 2,800,616 (1957)

4. Samantara, A.K., Ratha, S.: Historical background and present status of the supercapacitors. In: Materials Development for Active/Passive Components of a Supercapacitor, pp. 9–10. Springer, Singapore (2018)
5. Sharma, P., Bhatti, T.S.: A review on electrochemical double-layer capacitors. Energy Convers. Manage. **51**(12), 2901–2912 (2010)
6. Lu, Q., Chen, J.G., Xiao, J.Q.: Nanostructured electrodes for high-performance pseudocapacitors. Angew. Chem. Int. Ed. **52**(7), 1882–1889 (2013)
7. Conway, B.E.: Transition from "supercapacitor" to "battery" behavior in electrochemical energy storage. J. Electrochem. Soc. **138**(6), 1539–1548 (1991)
8. Muzaffar, A., Ahamed, M.B., Deshmukh, K., Thirumalai, J.: A review on recent advances in hybrid supercapacitors: design, fabrication and applications. Renew. Sustain. Energy Rev. **101**, 123–145 (2019)
9. Samantara, A.K., Ratha, S.: Components of supercapacitor. In: Materials Development for Active/Passive Components of a Supercapacitor, pp. 11–39. Springer, Singapore (2018)
10. Wu, Z., Zhu, Y., Ji, X., Banks, C.E.: Transition metal oxides as supercapacitor materials. In: Nanomaterials in Advanced Batteries and Supercapacitors, pp. 317–344. Springer, Cham (2016)
11. Han, Y., Dai, L.: Conducting polymers for flexible supercapacitors. Macromol. Chem. Phys. **220**(3), 1800355 (2019)
12. Pan, H., Li, J., Feng, Y.: Carbon nanotubes for supercapacitor. Nanoscale Res. Lett. **5**(3), 654 (2010)
13. Wang, Y., Zeng, J., Li, J., Cui, X., Al-Enizi, A.M., Zhang, L., Zheng, G.: One-dimensional nanostructures for flexible supercapacitors. J. Mater. Chem. A **3**(32), 16382–16392 (2015)
14. Zhong, C., Deng, Y., Hu, W., Qiao, J., Zhang, L., Zhang, J.: A review of electrolyte materials and compositions for electrochemical supercapacitors. Chem. Soc. Rev. **44**(21), 7484–7539 (2015)
15. Gleiter, H.: Nanostructured materials: basic concepts and microstructure. Acta Mater. **48**(1), 1–29 (2000)
16. Skorokhod, V.V., Uvarova, I.V., Ragulya, A.V.: Physico-chemical kinetics in nanostructured systems. Kyiv: Academperiodica **1**, 180–192 (2001)
17. Pokropivny, V.V., Ivanovskii, A.L.: New nanoforms of carbon and boron nitride. Russ. Chem. Rev. **77**(10), 837 (2008)
18. Kim, Y.T., Han, J.H., Hong, B.H., Kwon, Y.U.: Electrochemical deposition of CdSe quantum dot arrays on large-scale graphene electrodes using mesoporous silica thin film templates. Adv. Mater **22**, 515–518 (2010)
19. Hou, J., Chen, H.Y., Zhang, S., Li, G., Yang, Y.: Synthesis, characterization, and photovoltaic properties of a low band gap polymer based on silole-containing polythiophenes and 2, 1, 3-benzothiadiazole. J. Am. Chem. Soc. **130**(48), 16144–16145 (2008)
20. Zheng, L., Chi, Y., Dong, Y., Lin, J., Wang, B.: Electrochemiluminescence of water-soluble carbon nanocrystals released electrochemically from graphite. J. Am. Chem. Soc. **131**(13), 4564–4565 (2009)
21. Nozik, A.J.: Quantum dot solar cells. Phys. E **14**(1–2), 115–120 (2002)
22. Choi, M.K., Yang, J., Hyeon, T., Kim, D.H.: Flexible quantum dot light-emitting diodes for next-generation displays. NPJ Flex. Electron. **2**(1), 10 (2018)
23. Samuel, I.: Colloidal nanocrystals: electrifying quantum dots for lasers. Nat. Mater. **17**(1), 9 (2018)
24. Haug, R.J., Dilger, M., Schmidt, T., Blick, R.H., Klitzing, K.V., Eberl, K.: Single-electron transistors with quantum dots. Phys. B **227**(1–4), 82–86 (1996)
25. Periasamy, P., Krishnakumar, T., Sathish, M., Devarajan, V.P., Siril, P.F., Chavali, M.: Investigation of electrochemical properties of microwave irradiated tungsten oxide ($WO_3$) nanorod structures for supercapacitor electrode in KOH electrolyte. Mater. Res. Exp. **5**(8), 085007 (2018)
26. Liu, X., Sheng, G., Zhong, M., Zhou, X.: Hybrid nanowires and nanoparticles of $WO_3$ in a carbon aerogel for supercapacitor applications. Nanoscale **10**(9), 4209–4217 (2018)

27. Kumar, R.D., Andou, Y., Karuppuchamy, S.: Facile synthesis of Co–$WO_3$/functionalized carbon nanotube nanocomposites for supercapacitor applications. J. Mater. Sci. Mater. Electron. **28**(7), 5425–5434 (2017)
28. Yao, S., Zheng, X., Zhang, X., Xiao, H., Qu, F., Wu, X.: Facile synthesis of flexible $WO_3$ nanofibers as supercapacitor electrodes. Mater. Lett. **186**, 94–97 (2017)
29. Hai, Z., Akbari, M.K., Wei, Z., Xue, C., Xu, H., Hu, J., Zhuiykov, S.: Nano-thickness dependence of supercapacitor performance of the ALD-fabricated two-dimensional $WO_3$. Electrochim. Acta **246**, 625–633 (2017)
30. Ke, Q., Wang, J.: Graphene-based materials for supercapacitor electrodes—a review. J. Mater. **2**(1), 37–54 (2016)
31. Montemor, M.F., Eugénio, S., Tuyen, N., Silva, R.P., Silva, T.M. Carmezim, M.J.: Nanostructured transition metal oxides produced by electrodeposition for application as redox electrodes for supercapacitors. In: Handbook of Nanoelectrochemistry: Electrochemical Synthesis Methods, Properties, and Characterization Techniques, pp. 681–714 (2016)
32. Kumar, K.S., Choudhary, N., Jung, Y., Thomas, J.: Recent advances in two-dimensional nanomaterials for supercapacitor electrode applications. ACS Energy Lett. **3**(2), 482–495 (2018)
33. Tian, J., Lin, J., Xu, M., Wan, S., Lin, J., Wang, Y.: Hexagonal boron nitride catalyst in a fixed-bed reactor for exothermic propane oxidation dehydrogenation. Chem. Eng. Sci. **186**, 142–151 (2018)
34. Back, S., Siahrostami, S.: Noble metal supported hexagonal boron nitride for the oxygen reduction reaction: a DFT study. Nanosc. Adv. **1**(1), 132–139 (2019)
35. Shi, L., Zhao, T.: Recent advances in inorganic 2D materials and their applications in lithium and sodium batteries. J. Mater. Chem. A **5**(8), 3735–3758 (2017)
36. Brar, V.W., Sherrott, M.C., Jariwala, D.: Emerging photonic architectures in two-dimensional opto-electronics. Chem. Soc. Rev. **47**(17), 6824–6844 (2018)
37. Aivazian, G.: Optoelectronic properties of two-dimensional materials. Doctoral dissertation (2015)
38. You, P., Tang, G., Yan, F.: Two-dimensional materials in perovskite solar cells. Mater. Today Energy **11**, 128–158 (2019)
39. Das, S., Pandey, D., Thomas, J., Roy, T.: The role of graphene and other 2D materials in solar photovoltaics. Adv. Mater. **31**(1), 1802722 (2019)
40. Yang, S., Jiang, C., Wei, S.H.: Gas sensing in 2D materials. Appl. Phys. Rev. **4**(2), 021304 (2017)
41. Shi, S., Hu, R., Wu, E., Li, Q., Chen, X., Guo, W., Sun, C., Hu, X., Zhang, D., Liu, J.: Highly-sensitive gas sensor based on two-dimensional material field effect transistor. Nanotechnology **29**(43), 435502 (2018)
42. Lokhande, V., Lokhande, A., Namkoong, G., Kim, J.H., Ji, T.: Charge storage in $WO_3$ polymorphs and their application as supercapacitor electrode material. Res. Phys. **12**, 2012–2020 (2019)
43. Sarkar, D., Mukherjee, S., Pal, S., Sarma, D.D., Shukla, A.: Hexagonal $WO_3$ nanorods as ambipolar electrode material in asymmetric $WO_3$//$WO_3$/$MnO_2$ supercapacitor. J. Electrochem. Soc. **165**(10), A2108–A2114 (2018)
44. https://shellzero.wordpress.com/2012/05/14/inert-gas-condensation-method/
45. Hodges, A.M., Beck, T.W., Johansen, O., Maxwell, I.A., USF Filtration and Separations Group Inc.: Electrochemical method. U.S. Patent 5,942,102 (1999)
46. Kumar, P., Sarswat, P.K., Free, M.L.: Hybridized tungsten oxide nanostructures for food quality assessment: fabrication and performance evaluation. Sci. Rep. **8**(1), 3348 (2018)
47. Ramimoghadam, D., Bagheri, S., Hamid, S.B.A.: Progress in electrochemical synthesis of magnetic iron oxide nanoparticles. J. Magn. Magn. Mater. **368**, 207–229 (2014)
48. Ganesan, R., Gedanken, A.: Synthesis of $WO_3$ nanoparticles using a biopolymer as a template for electrocatalytic hydrogen evolution. Nanotechnology **19**(2), 025702 (2007)
49. Karatas, A., Hilal Algan, A.: Template synthesis of tubular nanostructures for loading biologically active molecules. Curr. Top. Med. Chem. **17**(13), 1555–1563 (2017)

50. Mattevi, C., Kim, H., Chhowalla, M.: A review of chemical vapour deposition of graphene on copper. J. Mater. Chem. **21**(10), 3324–3334 (2011)
51. Reiners, G.: Corrosion books: chemical vapor deposition. By Jong-Hee Park und TS Sudarshan-materials and corrosion 2/2003. Mater. Corros. **54**(2), 127–127 (2003)
52. https://sites.google.com/site/nanomodern/Home/CNT/syncnt/cvd
53. Shi, S., Hwang, J.Y.: Microwave-assisted wet chemical synthesis: advantages, significance, and steps to industrialization. J. Min. Mater. Charact. Eng. **2**(02), 101 (2003)
54. Palanisamy, P., Thangavel, K., Murugesan, S., Marappan, S., Chavali, M., Siril, P.F., Perumal, D.V.: Investigating the synergistic effect of hybridized $WO_3$–ZnS nanocomposite prepared by microwave-assisted wet chemical method for supercapacitor application. J. Electroanal. Chem. **833**, 93–104 (2019)
55. Rao, B.G., Mukherjee, D., Reddy, B.M.: Novel approaches for preparation of nanoparticles. In: Nanostructures for Novel Therapy, pp. 1–36. Elsevier (2017)
56. Chaitoglou, S., Amade, R., Bertran, E.: Evaluation of Graphene/$WO_3$ and Graphene/$CeO_x$ structures as electrodes for supercapacitor applications. Nanoscale Res. Lett. **12**(1), 635 (2017)
57. Zhuzhelskii, D.V., Tolstopjatova, E.G., Eliseeva, S.N., Ivanov, A.V., Miao, S., Kondratiev, V.V.: Electrochemical properties of PEDOT/$WO_3$ composite films for high performance supercapacitor application. Electrochim. Acta **299**, 182–190 (2019)
58. Evans, J.E., Springer, K.W., Zhang, J.Z.: Femtosecond studies of interparticle electron transfer in a coupled CdS–$TiO_2$ colloidal system. J. Chem. Phys. **101**(7), 6222–6225 (1994)
59. Butler, M.A., Ginley, D.S.: Prediction of flatband potentials at semiconductor-electrolyte interfaces from atomic electronegativities. J. Electrochem. Soc. **125**(2), 228–232 (1978)
60. Kim, H., Tak, Y., Senthil, K., Joo, J., Jeon, S., Yong, K.: Novel heterostructure of CdS nanoparticle/$WO_3$ nanowhisker: synthesis and photocatalytic properties. J. Vac. Sci. Technol. B Microelectron. Nanometer Struct. Process. Meas. Phenom. **27**(5), 2182–2186 (2009)
61. Hong, S.J., Lee, S., Jang, J.S., Lee, J.S.: Heterojunction $BiVO_4$/$WO_3$ electrodes for enhanced photoactivity of water oxidation. Energy Environ. Sci. **4**(5), 1781–1787 (2011)
62. Prasannalakshmi, P., Shanmugam, N., Kumar, A.S.: Electrochemistry of $TiO_2$/CdS composite electrodes for supercapacitor applications. J. Appl. Electrochem. **47**(8), 889–903 (2017)
63. Patil, D.S., Pawar, S.A., Shin, J.C.: Core-shell structure of $Co_3O_4$@ CdS for high performance electrochemical supercapacitor. Chem. Eng. J. **335**, 693–702 (2018)
64. Pu, J., Cui, F., Chu, S., Wang, T., Sheng, E., Wang, Z.: Preparation and electrochemical characterization of hollow hexagonal $NiCo_2S_4$ nanoplates as pseudocapacitor materials. ACS Sustain. Chem. Eng. **2**(4), 809–815 (2013)
65. Zhu, J., Chen, S., Zhou, H., Wang, X.: Fabrication of a low defect density graphene-nickel hydroxide nanosheet hybrid with enhanced electrochemical performance. Nano Res. **5**(1), 11–19 (2012)
66. Kumar, R.D., Andou, Y., Sathish, M., Karuppuchamy, S.: Synthesis of nanostructured Cu-$WO_3$ and $CuWO_4$ for supercapacitor applications. J. Mater. Sci. Mater. Electron. **27**(3), 2926–2932 (2016)
67. Qiu, M., Sun, P., Shen, L., Wang, K., Song, S., Yu, X., Tan, S., Zhao, C., Mai, W.: $WO_3$ nanoflowers with excellent pseudo-capacitive performance and the capacitance contribution analysis. J. Mater. Chem. A **4**(19), 7266–7273 (2016)
68. Periasamy, P., Krishnakumar, T., Sandhiya, M., Sathish, M., Chavali, M., Siril, P.F., Devarajan, V.P.: Preparation and comparison of hybridized $WO_3$–$V_2O_5$ nanocomposites electrochemical supercapacitor performance in KOH and $H_2SO_4$ electrolyte. Mater. Lett. **236**, 702–705 (2019b)
69. Mohd Hanappi, M.F.Y., Deraman, M., Suleman, M., Mohd Nor, N.S., Sazali, N.E.S., Hamdan, E., Moh Tajuddin, N.S., Basri, N.H., Mohd Jasni, M.R., Othman, M.A.R.: Influence of aqueous KOH and $H_2SO_4$ electrolytes ionic parameters on the performance of carbon-based supercapacitor electrodes. Funct. Mater. Lett. **10**(03), 1750013 (2017)

# Chapter 2
# Biomaterials: An Introduction to Materials for Biomedical Applications

**Joydeep Biswas and Bandita Datta**

## Introduction

Biomaterials are employed for creating tools to exchange a constituent or a utility of the human body in a very secure, trustworthy, and biologically satisfactory way. Biomaterials are customized as an assembly or tool or material, i.e., applied for restoring a natural tissue or employing, for instance, as an embedded material for the required functions [1]. Biomaterials are moreover fabricated of a natural or artificial genesis. Established based on resources, biomaterials are differentiated into several types. Biomaterials vary in employing their chemical, physical, mechanical, and chiefly biological properties [2]. Currently, diverse progress is made for fetching up the naturally emerged biomaterials as they are simply biodegradable and mainly eco-friendly [3]. Now, these materials are usually used for the treatment of an illness or a wound [4]. Biocompatibility and biodegradability are the primary features of any biomaterial that formulates it appropriate for biomedical applications. Biomaterials should not have any diverse effects when being used on the host body. The numerous devices which are employed include sutures, catheters, man-made tissue, needles, etc. [5]. Materials of biological importance were developed in the few eras which are employed to work beside the natural tissues or an artificial organ in a human body [6]. The frequent materials of biomedical importance which are employed predominantly are bioceramics, metallic biomaterials, biocomposites, and synthetic biopolymers [7]. Individual material of biomedical importance has diverse features that enable them to withstand in the specific applications.

---

J. Biswas (✉)
Department of Chemistry, Sikkim Manipal Institute of Technology, Sikkim Manipal University, East Sikkim, Majitar, Rangpo, Sikkim 737136, India
e-mail: joydeep.biswas@smit.smu.edu.in

B. Datta
Department of Chemistry, Amity University, Kolkata Campus, New Town, Kolkata 700156, India



B. P. Swain (ed.), *Nanostructured Materials and their Applications*, Materials Horizons: From Nature to Nanomaterials, https://doi.org/10.1007/978-981-15-8307-0_2

## Classification of Biomaterials

There are diverse varieties of biomaterials that are fabricated and utilized for various purposes. Biomaterials of different origins are illustrated below (Fig. 2.1). In this chapter, we focused mainly on the biomaterials of synthetic origin.

### *Bioceramics*

Ceramics are usually biocompatible, extremely resistant to decomposition and compression having low thermal and electrical conductivities [8]. Bioceramics have been largely fitted to medical grafts [9]. It is having low toxicity and aids in the greater potency in developing fresh bone tissue. In ceramics, few instances embrace orthophosphate that primarily aids in tissue restoration and furthermore employed aimed at bone repairs. Hydroxyapatites are bioceramics, which are chiefly employed as tooth implants because of their magnificent stiffness properties [10]. Bioceramics contain tinny and non-metallic constituents in their structures. Bioceramics are exceptionally unreactive in characteristics, rigid and brittle, which have low tensile strength but having pronounced compressive strength [11]. Bioceramics are chiefly employed in orthopedic implants and dental uses [12]. Bones and teeth are the examples of rigid natural tissues which contain aggregates of bioceramics and biopolymer. Bioceramics, supported by their bioactivity are employed in totally different biomedical uses [13]. For example, biodegradable materials are employed in medical uses like drug delivery. Ceramics containing nanostructured materials are grown in the present day in an exceedingly nice way. So the appliance of ceramic materials differs consistent with the ultimate applications.

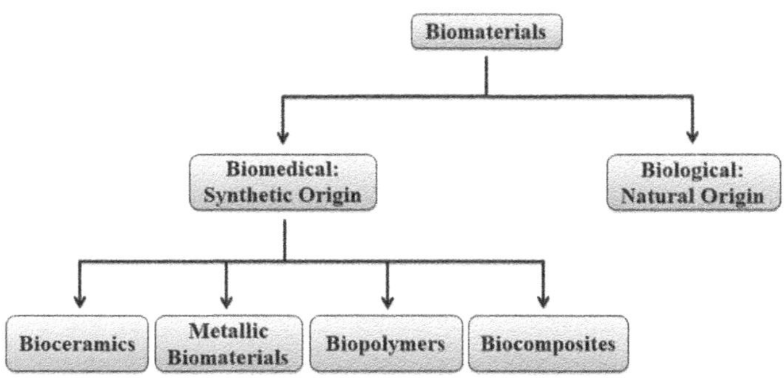


**Fig. 2.1** Flowchart showing different types of biomaterials

**Table 2.1** Primary uses of various metallic biomaterials

| Metallic biomaterials | Primary uses |
|---|---|
| Stainless steel | Surgical instruments, stents, and fracture fixation |
| Titanium alloys | Fracture fixation, bone replacement, pacemaker encapsulation, and dental implants |
| Cobalt and chromium alloy | Bone replacement, heart valves, dental restorations, and dental implants |

## *Metallic Biomaterials*

Metallic biomaterials have been predominantly employed in the uses where the load-bearing characteristic is of importance, viz. applying in knee transplants, bone fracture, and dental implants [14]. Metallic biomaterials have been utilized in numerous biomedical applications because of their outstanding and predominant mechanical characteristics [15]. The most frequently used metallic biomaterials are alloys of iron (e.g., stainless steel), molybdenum, titanium, and cobalt-chromium [16]. The primary uses of various metallic biomaterials are presented in Table 2.1. Although metallic biomaterials have superior mechanical properties, they straightforwardly become corroded in less time after their grafting in the host [15]. It additionally includes a very vital drawback referred to as the toxicity. The above-mentioned characteristics finally begin to affect the living organism wherever they are grafted. These together can terminate the characteristics of the transplant resulting in the weakening of the grafts and lastly leading to the reduction of their compatibility with living organisms [1]. Consequently, it ends up in disaster at few circumstances.

Based on the adjacent circumstance of the host living body, metallic biomaterials must be chosen. Based on the adjacent circumstance of the grafts in the host living body, they must be chosen and employed devoid of inflicting any worse effects on the host living body. Therefore, solely using the preferred starting materials that fit for the graft and the adjacent circumstance of the host living body, the disputes arising from these metallic biomaterials can be escaped by creating improvement of these materials and reduction in the difficulties associated with the chosen material of biomedical importance [15].

## *Biocomposites*

A common discussion regarding the meaning of composite materials between materials scientists and composite experts continues to the present time. Biocomposites denote to the materials generated by combining two or more materials on a macroscopic scale of biological importance, different in physical, compositional, and morphological characteristics [17]. In several instances, subject to the component characteristics, biocomposites have been fabricated having an opinion to produce materials with characteristics designed to achieve precise physical, chemical, and mechanical prerequisites. Hence, in the last forty years, the applications of composites have gradually raised and nowadays composite materials have numerous diverse non-medical appliances, i.e., naval, automotive, aeronautics, etc. [18]. Recently, numerous biocomposites have been examined and verified for biomedical use. Some of the biocomposites are presently available in the consumer markets for their decent benefits than conventional materials. Maximum natural tissues, for example, ligaments, tendons, teeth, bones, skin, etc., are biocomposites fabricated of single components whose shape, amount, distribution, and characteristics govern the concluding activities of the resulting biological tissues [19]. Artificial composites of biomedical importance can, to a certain extent, be employed to fabricate artificial body parts that are capable of copying biological tissues [19].

## *Synthetic Biopolymers*

A synthetic biopolymer is appraised as one of the finest materials for the biomedical ultimate applications. Compared to metallic biomaterials and bioceramics, synthetic biopolymers are broadly employed in varied phases [20]. Man-made polymers are extensively employed in synthetic polymeric drug delivery systems, medical disposable materials, grafts, bandages, prosthetic materials, dental materials, and tissue skilled products, similar to those of metal, and ceramic substituents [21]. The key benefits of the synthetic biopolymers compared to bioceramics or metallic biomaterials are to produce different shapes (latex, film, sheet, fibers, etc.), simple secondary process stages, cheap price, and accessibility with desired physical and mechanical characteristics [22].

Enormous applications employ biopolymers in diverse aspects. Vast signs of progress in the area of synthetic biopolymers are still created that is so thought-about as an associate biological method [23]. The maximum of the biomedical uses employs biopolymers of natural origin rather than utilizing the synthetic ones [24]. Biocompatibility, lighter in weight, flexibility, resistant to biochemical attacks and most importantly biodegradability are the chief properties of synthetic biopolymers [25]. The above-mentioned characteristics build them unique in their applications in the field of biomedical. A diverse range of synthetic biopolymers is employed for treatment as well as in the practice of medicine, clinical inspections, and surgical treatments [26].

Among the polymers developed for numerous medical functions, a definite set of polymers is termed as synthetic biopolymers which they are employed in straight interaction with active cells of the human body. Uses of synthetic biopolymers have turn out to be significantly great in the last few decades [27]. There are varied synthetic biopolymers employed which are stated below.

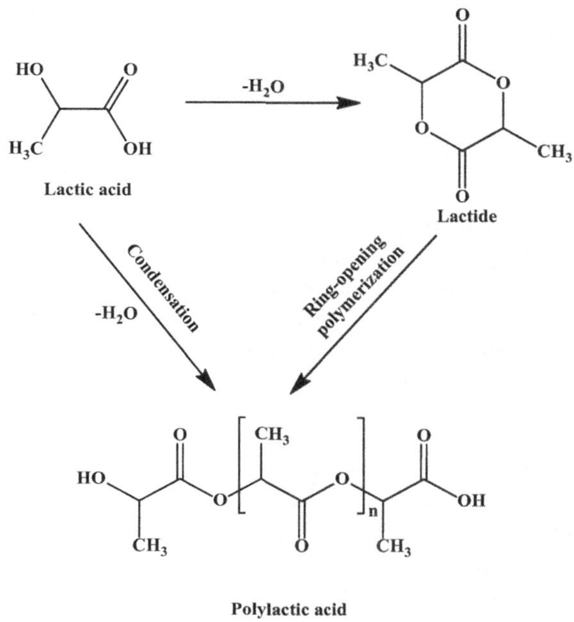


**Fig. 2.2** Schematic of the formation of polylactic acid

## Polylactic Acid

Aliphatic polyester, e.g., polylactic acid (PLA), is most importantly biodegradable in nature and having thermoplastic characteristics. The monomeric unit, i.e., lactic acid, is acquired from natural sources utilizing a fermentation method employing sugarcane or corn. Polylactic acid is then synthesized via either condensation or ring-opening polymerization of the previously obtained monomeric unit, i.e., lactic acid (Fig. 2.2). In a class of degradable polymers, PLA is one of the foremost vital biodegradable and biocompatible synthetic biopolymers [28]. Besides its use in biomedical, automobile, and textile industries, polylactic acid symbolizes an honest contender to create disposable packaging because of its excellent mechanical characteristics, and processability [29]. PLA is an enormously employed biopolymer because of its inherent extraordinary mechanical strength capability for load-bearing uses [30]. PLA is employed in interior fixation parts, like pins, plates, rods, and screws to assist the restoration of fractured bones and to grip them together [31].

## Polyhydroxyalkanoates

Polyhydroxyalkanoates (PHA) or polyhydroxyalkanoate acids (Fig. 2.3) having fairly biocompatible and biodegradable characteristics are categorized as linear polyester [32]. Polyhydroxyalkanoates are fabricated naturally by microorganisms, i.e., bacteria usually cultured on fresh materials obtained from agriculture. This

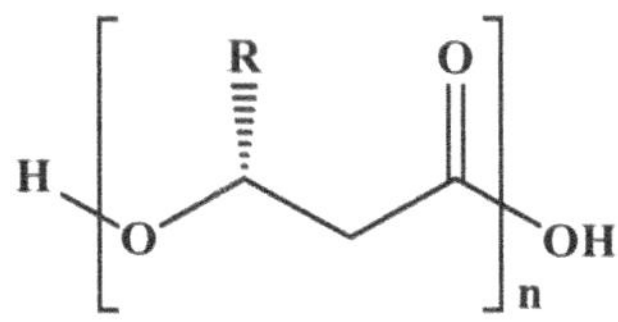

**Fig. 2.3** Chemical structure of polyhydroxyalkanoate

synthetic biopolymer could be managed to generate a diverse range of worthwhile goods, where their ability to degrade in nature and genuineness is fairly advantageous mainly for use in agriculture, and wrapping for single-use type. Polyhydroxyalkanoate possesses favorable characteristics like extraordinary biodegradability in numerous surroundings. Polyhydroxyalkanoates are biosynthesized by living organisms and classified as polyesters which signify a prospective viable alternative for thermoplastics derived from fossil fuel [33]. Maximum available polyhydroxyalkanoates in the markets are acquired with bacterial cultures developed on natural raw material, i.e., glucose in germ-free environments; however, current research emphasizes the application of trashes as growth media for bacteria. Polyhydroxyalkanoates are also acquired from the bacterial cell and subsequently prepared and treated using a technique called extrusion for the fabrication of hard, and malleable material appropriate not only for the foremost evaluated biomedical uses but moreover appraised for uses comprising molded products, films, wrapping, non-woven textiles, glues, and paper coatings. Polyhydroxyalkanoate and its copolymers are broadly employed as clinical graft constituents that comprise cardiovascular repair coverings, bone plates, sutures, and cartilage repair parts. Polyhydroxyalkanoate is employed in drug release uses because of its precise degradability and compatibility with the host living body [34].

## Polyvinyl Alcohol and Its Hydrogel

Polyvinyl alcohol (PVA) is an artificial, biocompatible, and non-toxic synthetic biopolymer which is remarkably well suited for a diverse range of uses. This synthetic biopolymer is compatible and non-toxic with living organisms, water-soluble, and suitable for the reproduction of active tissues. Polyvinyl alcohol possesses decent oxygen permeability due to which it displays no immunogenic effects and attributes magnificent film-forming, remoisturizing, and emulsifying characteristics. Polyvinyl alcohol has been synthesized from vinyl acetate via polymerization followed by treatment with alcoholic NaOH (Fig. 2.4).

Hydrogels are 3D cross-linked hydrophilic polymers having the capacity to hold water by their polymeric network [35]. Hydrogels are those polymeric network structure which swells in the presence of water. PVA hydrogels are considered to be hydrophilic gels which are erstwhile named as mixture gels [36]. Polyvinyl alcohol hydrogels have been categorized as low-, medium-, and high-swelling hydrogels subject to the extent of cross-linking chemically and the quantity of liquid $H_2O$ hold by the polymeric network [37]. Because of their sensitivity toward H-bondings and

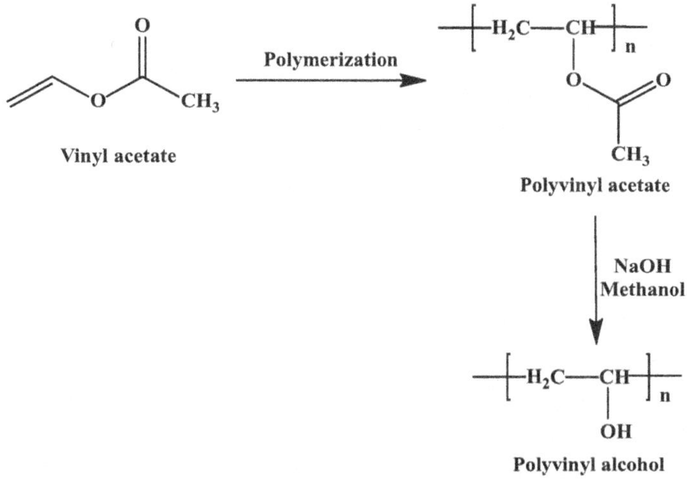


**Fig. 2.4** Schematic of the formation of polyvinyl alcohol

extreme crystallization, they usually suggest very less swelling capability, creating PVA hydrogels nevertheless very appropriate for precise clinical and pharmaceutical uses [38]. Because of their simple structure and exceptional characteristics like strength, swelling, film-forming, adhesiveness, biocompatibility, non-toxic, and non-cancerous, these synthetic biopolymers have found uses in numerous manufacturing units such as textile, paper, adhesives, food, medicine, and pharmaceutical uses specifically [38]. Characteristics like swelling, elastic nature, high water content, and biocompatibility create polyvinyl alcohol hydrogel matrices a prospective contender for the natural tissue regeneration component. PVA hydrogels are investigated for instance artificial cartilages, artificial heart linings, soft contact lens material, pancreas membranes, skin and catheters [37]. In earlier investigations of PVA hydrogels for clinical uses, gel and physical characteristics were examined, and compatibility with human blood was the vital anxiety for the scientists. Due to their drug compatibility, biocompatibility, film-forming, water-solubility, decent mechanical, and swelling characteristics, the polyvinyl alcohol hydrogels are examined for drug release processes [38]. The extent of crystallinity plays a chief part in regulating the release of the drug from hydrogels. Overall, polyvinyl alcohol hydrogels are often fabricated as matrices for drug release stages. Fixing gelation characteristics, solubility and the addition of copolymers have collectively been employed to regulate the drug delivery from polyvinyl alcohol polymeric networks.

#### Polycaprolactone

Synthetic biopolymers having biodegradable and biocompatible characteristics are getting significant in clinical uses. Polycaprolactone is an example of such a synthetic

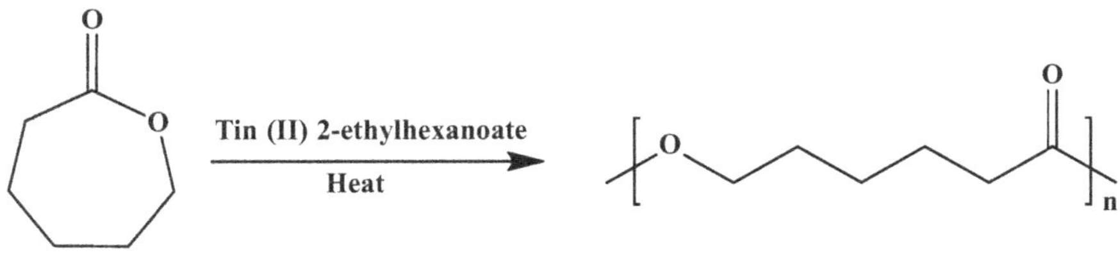


**Fig. 2.5** Synthesis of polycaprolactone via rig-opening polymerization

biopolymer that is biodegradable polyester. For its compatibility with the host living body, polycaprolactone is extremely popular. Polycaprolactone is having predominant applications in targeted drug release [39]. Due to its elongated degradation time, polycaprolactone is considered to be an important synthetic biopolymer in biomedical use. Polycaprolactone has a low melting point which makes it a well-known synthetic biopolymer. Therefore, polycaprolactone can straightforwardly be molded as per the ultimate applications [40]. This synthetic biopolymer is very resistant to organic solvents and even water. Polycaprolactone is employed as graftable tools for drug delivery in an extended duration. Employing hydrolysis reaction, polycaprolactone can be fragmented by the cleavage of ester linkages in physiological environments. Polycaprolactone is employed in vesicle membranes, scaffolds, and joints. Polycaprolactone is being employed to encapsulate drugs for specific drug delivery. Polycaprolactone is additionally employed in the fabrication of polyurethanes which is a polymer composed of monomeric units joined by carbamate linkages. This synthetic biopolymer is synthesized via ring-opening followed by the polymerization reaction of the monomeric unit, i.e., $\varepsilon$-caprolactone (CL). The monomeric unit, i.e., $\varepsilon$-caprolactone is a 7-membered cyclic organic compound which is classified as cyclic ester, i.e., lactone. Ring-opening followed by the polymerization reaction of $\varepsilon$-caprolactone results in the production of the synthetic biopolymer, i.e., polycaprolactone [41]. Ring-opening polymerization reaction is usually catalyzed by metal catalysts, i.e., Tin (II) 2-ethyl hexanoate (Fig. 2.5). It is troublesome to get rid of the metal catalysts once the reaction is completed, and these catalysts are moreover harmful to the host living body. Therefore, an alternative path of employing biocatalysts, i.e., enzymes, has been utilized to diminish the toxic effects. The above-mentioned methodology is cost-effective since the biocatalysts, i.e., enzymes, employed will be restored per se after the reaction and might be recycled.

### Polyglycolide

Biopolymer, viz., polyglycolide is one of the initial decomposable artificial biopolymers employed for clinical uses. The chemical structure of polyglycolide is shown in Fig. 2.6. It has been an extremely crystalline biopolymer and hence demonstrates an excessive tensile modulus along with terribly low solubility in organic media. Despite its little solubility, polyglycolide has been tailored into a diverse range of systems

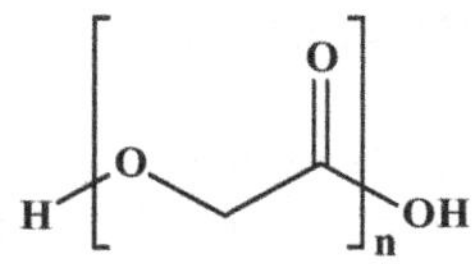

**Fig. 2.6** Chemical structure of polyglycolide

and assemblies. Compression molding, extrusion, and injection are several techniques employed to improve polyglycolide-based biomaterials for clinical uses [42]. Because of its superb fiber-creating capability, this synthetic biopolymer had been originally examined for improving absorbable stitches. Polyglycolide materials are broadly employed as support structures for tissue regeneration because of its magnificent degradability, cell viability on the support structures, and decent preliminary mechanical characteristics [43]. Non-woven polyglycolide material-fibrin glue-fused structure is presently going through clinical trials. It has been studied as an alternative biocompatible meninx because of its magnificent skin-closing capability devoid of the requirement of stitches and its capability to assist renewal of natural tissues of living organisms. Because of its decent preliminary mechanical characteristics, these synthetic biopolymers are examined as bone interior fixation parts. The low solubility, excessive rate of degradability, and degradation in acidic media restrict the biomedical uses for polyglycolide. Thus, many copolymers comprising glycolide monomeric units have been tailored to beat the intrinsic drawbacks of polyglycolide.

## Conclusion

Materials of biomedical importance are tailored in an extensive variety chiefly based on its properties and clinical ultimate applications. It has generated a decent influence in varied biomedical uses. Bioceramics supported their bioactivity are employed in totally different biomedical uses. Metallic biomaterials have been utilized in numerous biomedical applications because of their outstanding and predominant mechanical characteristics. Some of the biocomposites are presently available in the consumer markets for their decent benefits than conventional materials. Biodegradable synthetic biopolymers have been prepared from wide-ranging renewable resources that are environmentally safe. It has some disadvantages; however, it does not affect the business applications more broadly. The benefit of the synthetic biopolymers is that it has higher performance in their medicine property. Wide-ranging progressions are designed in the improvement of the bioceramics, metallic biomaterials, synthetic biopolymers and biocomposites, and the improvement of the characteristics of these bio-based materials in the preferred biomedical ultimate uses.

**Acknowledgements** The author, Joydeep Biswas, wishes to acknowledge the Department of Chemistry, Sikkim Manipal Institute of Technology for help and support. The other author, Bandita Datta, acknowledges the Department of Chemistry, Amity University, Kolkata for help and support.

## References


1. Chena, F.M., Liu, X.: Advancing biomaterials of human origin for tissue engineering. Prog. Polym. Sci. **53**, 86–168 (2016)
2. Mitragotri, S., Lahann, J.: Physical approaches to biomaterial design. Nat. Mater. **8**, 15–23 (2009)
3. Patel, P.N., Parmar, K.G., Nakum, A.N., Patel, M.N., Patel, P.R., Patel, V.R., Sen, D.J.: Biodegradable polymers: an ecofriendly approach in newer millenium. Asian J. Biomed. Pharm. Sci. **1**, 23–39 (2011)
4. Mir, M., Ali, M.N., Barakullah, A., Gulzar, A., Arshad, M., Fatima, S., Asad, M.: Synthetic polymeric biomaterials for wound healing: a review. Prog. Biomater. **7**, 1–21 (2018)
5. Tathe, A., Ghodke, M., Nikalje, A.P.: A brief review: biomaterials and their application. Int. J. Pharm. Pharm. Sci. **2**, 19–23 (2010)
6. Mihov, D., Katerska, B.: Some biocompatible materials used in medical practice. Trakia J. Sci. **8**, 119–125 (2010)
7. Parida, P., Behera, A., Mishra, S.C.: Classification of biomaterials used in medicine. Int. J. Adv. Appl. Sci. **1**, 125–129 (2012)
8. Chevalier, J., Gremillard, L.: Ceramics for medical applications: a picture for the next 20 years. J. Eur. Ceram. Soc. **29**, 1245–1255 (2009)
9. Ginebra, M.P., Espanol, M., Maazouz, Y., Bergez, V., Pastorino, D.: Bioceramics and bone healing. EFORT Open Rev. **3**, 173–183 (2018)
10. Prakasam, M., Locs, J., Salma-Ancane, K., Loca, D., Largeteau, A., Berzina-Cimdina, L.: Biodegradable materials and metallic implants—a review. J. Funct. Biomater. **8**, 44 (2017)
11. Prakasam, M., Locs, J., Salma-Ancane, K., Loca, D., Largeteau, A., Berzina-Cimdina, L.: Fabrication, properties and applications of dense hydroxyapatite: a review. J. Funct. Biomater. **6**, 1099–1140 (2015)
12. Nabiyouni, M., Brückner, T., Zhou, H., Gbureck, U., Bhaduri, S.B.: Magnesium-based bioceramics in orthopedic applications. Acta Biomater. **66**, 23–43 (2018)
13. Thamaraiselvi, T.V., Rajeswari, S.: Biological evaluation of bioceramic materials—a review. Trends Biomater. Artif. Organs **18**, 9–17 (2004)
14. Ige, O.O., Umoru, L.E., Adeoye, M.O., Adetunji, A.R., Olorunniwo, O.E., Akomolafe, I.I.: Monitoring, control and prevention practices of biomaterials corrosion—an overview. Trends Biomater. Artif. Organs **23**, 93–104 (2009)
15. Prasad, K., Bazaka, O., Chua, M., Rochford, M., Fedrick, L., Spoor, J., Symes, R., Tieppo, M., Collins, C., Cao, A., Markwell, D., Ostrikov, K., Bazaka, K.: Metallic biomaterials: current challenges and opportunities. Materials **10**, 884 (2017)
16. Eliaz, N.: Corrosion of metallic biomaterials: a review. Materials **12**, 407 (2019)
17. Fratzl, P., Weinkamer, R.: Nature's hierarchical materials. Prog. Mater. Sci. **52**, 1263–1334 (2007)
18. Thori, P., Sharma, P., Bhargava, M.: An approach of composite materials in industrial machinery: advantages, disadvantages, and applications. Int. J. Res. Eng. Technol. **02**, 350–355 (2013)
19. Patel, N.R., Gohil, P.P.: A review on biomaterials: scope, applications & human anatomy significance. Int. J. Emerg. Technol. Adv. Eng. **2**, 91–101 (2012)
20. Bose, S., Ke, D., Sahasrabudhe, H., Bandyopadhyay, A.: Additive manufacturing of biomaterials. Prog. Mater. Sci. **93**, 45–111 (2018)
21. Ulery, B.D., Nair, L.S., Laurencin, C.T.: Biomedical Applications of Biodegradable Polymers. J. Polym. Sci. B Polym. Phys. **49**, 832–864 (2011)
22. Rao, S.H., Harini, B, Shadamarshan, R.P.K., Balagangadharan. K., Selvamurugan, N.: Natural and synthetic polymers/bioceramics/bioactive compounds-mediated cell signaling in bone tissue engineering. Int. J. Biol. Macromol. **110**, 88–96.
23. Rothemund, S., Teasdale, I.: Preparation of polyphosphazene: a tutorial review. Chem. Soc. Rev. **45**, 5200–5215 (2016)

24. Gunathilake, T.M.S.U., Ching, Y.C., Ching, K.Y., Chuah, C.H., Abdullah, L.C.: Biomedical and microbiological applications of bio-based porous materials: a review. Polymers **9**, 160 (2017)
25. Vroman, I., Tighzert, L.: Biodegradable polymers. Materials **2**, 307–344 (2009)
26. Bressan, E., Favero, V., Gardin, C., Ferroni, L., Iacobellis, L., Favero, L., Vindigni, V., Berengo, M., Sivolella, S., Zavan, B.: Biopolymers for hard and soft engineered tissues: application in odontoiatric and plastic surgery field. Polymers **3**, 509–526 (2011)
27. Mierzwa-Hersztek, M., Gondek, K., Kopeć, M.: Degradation of polyethylene and biocomponent-derived polymer materials: an overview. J. Polym. Environ. **27**, 600–611 (2019)
28. Alizadeh-Osgouei, M., Li, Y., Wen, C.: A comprehensive review of biodegradable synthetic polymer-ceramic composites and their manufacture for biomedical applications. Bioact. Mater. **4**, 22–36 (2019)
29. Gonçalves, C., Gonçalves, I.C., Magalhães, F.D., Pinto, A.M.: Poly(lactic acid) composites containing carbon-based nanomaterials: a review. Polymers **9**, 269 (2017)
30. Nagarajan, V., Mohanty, A.K., Misra, M.: Perspective on polylactic acid (PLA) based sustainable materials for durable applications: focus on toughness and heat resistance. ACS Sustain. Chem. Eng. **4**, 2899–2916 (2016)
31. Uhthoff, H.K., Poitras, P., Backman, D.S.: Internal plate fixation of fractures: a short history and recent developments. J. Orthop. Sci. **11**, 118–126 (2006)
32. Misra, S.K., Valappil, S.P., Roy, I., Boccaccini, A.R.: Polyhydroxyalkanoate (PHA)/inorganic phase composites for tissue engineering applications. Biomacromolecules **7**, 2249–2258 (2006)
33. Bugnicourt, E., Cinelli, P., Lazzeri, A., Alvarez, V.: Polyhydroxyalkanoate (PHA): review of synthesis, characteristics, processing and potential applications in packaging. eXPRESS Polym. Lett. **8**, 791–808 (2014)
34. Gadgil, B.S.T., Killi, N., Rathna, G.V.N.: Polyhydroxyalkanoates as biomaterials. MedChemComm **8**, 1774–1787 (2017)
35. Chai, Q., Jiao, Y., Yu, X.: Hydrogels for biomedical applications: their characteristics and the mechanisms behind them. Gels **3**, 6 (2017)
36. Santos, A.M.N., Moreira, A.P.D., Carvalho, C.W.P., Luchese, R., Ribeiro, E., McGuinness, G.B., Mendes, M.F., Oliveira, R.N.: Physically cross-linked gels of PVA with natural polymers as matrices for Manuka honey release in wound-care applications. Materials **12**, 559 (2019)
37. Reis, E.F.D., Campos, F.S., Lage, A.P., Leite, R.C., Heneine, L.G., Vasconcelos, W.L., Lobato, Z.I.P., Mansur, H.S.: Synthesis and characterization of poly (vinyl alcohol) hydrogels and hybrids for rMPB70 protein adsorption. Mat. Res. **9**, 185–191 (2006)
38. Gajra, B., Pandya, S.S., Vidyasagar, G., Rabari, H., Dedania, R.R., Rao, S.: Polyvinyl alcohol hydrogel and it's pharmaceutical and biomedical applications: a review. Int. J. Pharm. Res. **4**, 20–26 (2012)
39. Li, Z., Tan, B.H.: Towards the development of polycaprolactone based amphiphilic block copolymers: molecular design, self-assembly, and biomedical applications. Mater. Sci. Eng. C **45**, 620–634 (2014)
40. Mochane, M.J., Motsoeneng, T.S., Sadiku, E.R., Mokhena, T.C., Sefadi, J.S.: Morphology and properties of electrospun PCL and its composites for medical applications: a mini-review. Appl. Sci. **9**, 2205 (2019)
41. Rydz, J., Sikorska, W., Kyulavska, M., Christova, D.: Polyester-based (bio)degradable polymers as environmentally friendly materials for sustainable development. Int. J. Mol. Sci. **16**, 564–596 (2015)
42. Fialho, S.L., Cunha, A.D.S.: Manufacturing techniques of biodegradable implants intended for intraocular application. Drug Deliv. **12**, 109–116 (2005)
43. Barnes, C.P., Sell, S.A., Boland, E.D., Simpson, D.G., Bowlin, G.L.: Nanofiber technology: designing the next generation of tissue engineering scaffolds. Adv. Drug Deliv. Rev. **59**, 1413–1433 (2007)

# Chapter 3
# Effect of Surface Roughness on Titanium Medical Implants

**Subhasmita Swain and Tapash Ranjan Rautray**

## Introduction

Implants are the most important revolutions in the field of medicine and dentistry in the past fifty years. These implants are primarily coated with hydroxyapatite (HA) to improve their biocompatibility [1–14]. However, there are two major concerns, namely enhanced susceptibility to bacterial colonization and loss of coating integrity upon using HA-coated implants. Other significant factors that cause implant failures include microbial infection and occlusal trauma [15]. No great difference in microbial colonization on bare Ti surface or HA-coated Ti has been observed [16, 17]. In addition, the composition and sequence of microbial morphotypes in the maturation of subgingival dental plaque have been found to be similar on Ti, HA, and cementum surfaces [18]. From these observations, it can be postulated that enhanced bacterial colonization on HA coatings as compared to that on bare Ti implants/natural teeth could be due to the roughness and hydrophilicity on the HA surface. This further enhances plaque growth predisposing the implants for peri-implantitis [18–20]. In case of implant failure caused due to microbial infection, successful treatment of peri-implant microflora may be complicated by HA surface roughness. Yet another problem arises, that, when the HA-coated surface is exposed to the oral environment, it becomes contaminated and is potentially more difficult for the patient to clean [21].

Some other occurrences of failure of HA-coated implants are also possible, like early dissolution of the coating [22] or fracture of coating-substrate interface leading to further implant mobility and loss [23]. Another major concern of medical implants is coating resorption. In most studies, evaluation of HA coating resorption has been performed only on stable, static, and unloaded implants, which may not be the actual

S. Swain · T. R. Rautray (✉)
Biomaterials and Tissue Regeneration Laboratory, Centre of Excellence in TM Sciences, Siksha O Anusandhan (Deemed to be University), Bhubaneswar 751030, Odisha, India
e-mail: tapash.rautray@gmail.com



B. P. Swain (ed.), *Nanostructured Materials and their Applications*, Materials Horizons: From Nature to Nanomaterials, https://doi.org/10.1007/978-981-15-8307-0_3

condition inside the oral cavity. Actual conditions may warrant the worse performance of the coatings. An early study done in an unstable mechanical situation revealed resorption of 65% of the 50 μm thick HA coating in a 16-week implantation period [24, 25]. The reasons for all of these instances are attributed to the dissolution of HA coatings and the presence of osteoclast-like cells [25]. Although HA coatings help in the deposition of bone on the surface of the prosthesis, there were occasional foci of bone remodeling around the implant, including osteoblast-mediated removal of the HA coating along the adjacent bone [26]. Due to partial or total resorption, there is also a chance of attenuation of initial bone apposition over time [27]. In spite of these adverse happenings, the osteoconductive effect of HA coatings is prolonged with more bone ingrowth after 16 weeks as compared to the uncoated implants under loaded conditions [28]. It could be summarized that the resorption of HA coatings could have damaging effects on the clinical use of these implants [29].

As compared to orthopedic implants, dental implants have been less developed. Common implants used in the dental industry are stainless steel, Co–Cr alloys, Ti alloys, cemented or screwed Au alloys for root-canal treatment [30]. After this, dental porcelain caps can be cemented onto the root post. Titanium implants can be screwed into the bone beneath extracted teeth. Under this condition, the modified surface of the Ti implants gets sufficient osseointegration with bone and can be used as a strong support base where the Ti structure can be fitted, to give a shape to the tooth assembly. Implants after fixation may fail after loosening if osseointegration does not take place immediately. For this reason, rapid osseointegration of the implant is vital for a successful implant procedure. To obtain rapid osseointegration, the implant surface should be fabricated in a way to readily adhere, spread, and differentiate osteoblast cells and the least number of fibroblast cells should survive on that surface that negatively affects bone-to-implant contact [31]. Though the exact role of surface properties on bone growth is incompletely understood, it has been found that topography, chemical composition, purity, and charging of surfaces have the most favorable property to induce bone on implant surface [32]. However, recent research on surface modification of titanium and its alloys has achieved significant success to improve and accelerate osseointegration. Osseointegration of the bare Ti surface can be improved by coating it with HA or bioglass [30].

This review deals with surface treatment of titanium and titanium alloys to accelerate the osseointegration of dental implants and focusing primarily on surface roughness aspects.

## Ti and Ti Alloys for Dental Implants

Titanium and its alloys are regarded as gold standard materials for dental implants. As compared to other alloys such as Au, Ti is a pure element that does not show any physico-chemical reactions or allergies in the body environment [33]. Grade I Ti in combination with amalgam or Au can be used to avoid any chemical reactions in the mouth environment. $TiO_2$ film forms on Ti in the mouth which is electrically neutral. Moreover, the dielectric constant of $TiO_2$ is similar to water for which its taste is also neutral. Again due to its very low thermal conductivity below that of Au-based alloys tilts the balance towards Ti in dental applications [34]. Teeth equipped with Ti are not greatly influenced by the fast thermal changes arising from the intake of hot or cold food or drink intake. Due to its low density, Ti has got large dental restorations like crowns, bridges, implant superstructures, and removable partial dentures, since four-times heavier gold implants can be felt like a foreign body [35].

Commercially, pure titanium (cp-Ti), Ti–6Al–4V, CoCrMo alloys, and ceramics are commonly used in dental implants. Ceramics are used as coatings for both endosseous and subperiosteal implants in the form of HA. Root forms of $Al_2O_3$ (single crystal α-alumina and sapphire) are also used for this purpose [36].

## Osseointegration

Osseointegration refers to both structural and functional aspects since it requires the establishment of a close bone-to-implant contact and functional loads would be transmitted over an extended period without any deleterious effects. As long as these requirements are met and tissues are healthy, osseointegration does not involve the extent of the bone-implant contact. The factors affecting osseointegration are known now, and a successful outcome can be obtained by placing a suitable implant in a specific location [37].

Osseointegration is actually dependent on the biomedical property of an implant. Certain bioactive materials attract biochemical bonding with biomechanical interlocking. Osseointegration takes place at a faster rate while biomechanical bonding takes weeks to develop. Calcium phosphate-coated Ti implants possess both bonding mechanisms because of their roughness [38]. An implant is said to have osseointegrated when there is no micro-motion between the implant and the bone [39, 40].

## Dental Implant Materials

The biocompatibility of materials used for the replacement or augmentation of biological tissues is a very serious concern. In dental implant prosthetic reconstruction of the oral-maxillofacial region, the implant extends from the bone, across the epithelial layers, into the oral cavity. The functional aspects are really complex, including a series of chemical, biochemical, mechanical, and biomechanical challenges, since there is the transfer of force from the occlusal surfaces of teeth through the crown and neck connective region of the implant and into the implant for interfacial transfer to the supporting soft and hard tissues. The disciplines of biomaterials and biomechanics are complementary to the understanding of device-based functions. The physical, chemical, and mechanical properties of the basic material components must be fully evaluated for dental implant applications. Macroscopic implant shape should ensure the effective microscopic transfer of stress and strain along the biomaterial surface to tissue interface [41].

## Surface Roughening of Titanium

Machined titanium implants without any surface modification generally have a smooth surface. But it has been observed that titanium substrates with rough surfaces show faster and higher quality bone-bonding capability [42, 43]. The modified surface of a biomaterial differs in morphology and composition from that of the bulk surface. Differences between the bulk and modified surface can arise in terms of rearrangement of molecules, surface reaction, and contamination. Bioinert materials do not release any biologically active or toxic elements, and the surface properties govern the biological response. Subsequently, the topography of the surface also accounts for the mechanical balance of the tissue-implant interface. Biomaterials placed in the body environment may come into contact with cells in the related tissues for longer times for which they are harmless to the organism [44].

As mentioned earlier, surface topography or roughness plays an important role in the performance of the bioimplants. A straightforward method of producing a roughened surface is to polish it with coarse particles. For metallic surfaces such as Ti, the surface can be blasted with abrasive grits such as $Al_2O_3$, giving rise to a rough surface with pits and peaks distributed in a random fashion. Local deformations are formed due to this process, and the roughness parameter that depends on the particle size is expressed as Ra value. Ti plasma spraying (TPS) is another technique used to produce rough surfaces, and molten Ti is spread onto the material surface which forms protruding structures after getting cooled. Acid etching, a common technique carried out using an aqueous mixture of nitric acid and hydrofluoric acid at a rate of 10:1, results in a surface with a Ra value of $\approx 0.59\ \mu m$. These techniques can also be used in tandem to produce a very rough surface. For example, blasting followed by etching with acid produces a surface with a Ra value of $\approx 4.39\ \mu m$ on the Ti surface. In the following sections, these roughening techniques are discussed in detail [44].

Roughness, which is a topographical property of an implant surface, depends on the surface area which may be classified into three levels such as (i) macro-topographic ($Sa > 10\ \mu m$), (ii) micro-topographic ($1\ \mu m \le Sa < 10\ \mu m$), and (iii) nano-level ($<1\ \mu m$). Macro-topographic all classes of roughness are independent of each other. A surface can behave smoothly for macro-topographic levels while at the same time behave roughly on the micro-topographic level. Four different types of surface roughness are depicted in Fig. 3.1. An implant in a screw form already has a macro-topographical roughness arising out of its peaks and valleys while an untreated implant has a smooth surface at the micro-topographical and nano-sized level [45]. By surface modifying, the implant surface in any of these three levels of roughness increases its osseointegration capability [32]. The proposed mechanism that enhances

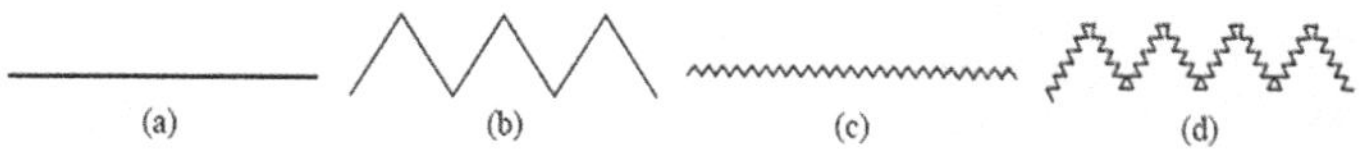


**Fig. 3.1** Different types of surface roughness: **a** Smooth surface; **b** Macrotopographically rough but microtopographically smooth surface; **c** Macrotopographically smooth but microtopographically rough surface; **d** Both macrotopographically and microtopographically rough surface

osseointegration may be a larger surface area which anchor higher number of host cells such as osteoblasts and fibroblasts adhere to implant surface [46]. Osteoblasts survive more on the rough surfaces than fibroblasts [47], which leads to better bone-bonding capacity in the presence of less fibers. Moreover, rough surfaces host high anchorage to the new bone formation that gives the implant long-term stability [46]. The diameter of pores about 150 $\mu$m is similar to the size of lacunae in the human spongy bone that for which this size of pores is ideal for all kinds of tissue ingrowth [48]. But the other two classes of roughness play a significant role in osteoblast attachment [49, 50] because osteoblasts can best fit into the smaller grooves [51] and show proliferation, differentiation, and bone production [52]. Works in literature cite that micro-topographical roughness provides the highest osseointegration and torsion resistance than the others [53, 54]. Peri-implantitis and ionic leakage from the implant surfaces are the two major drawbacks seen in macro-topographical roughness [55]. In general, an increase in micro-topographical roughness on implants is presently the major goal for biomedical researchers that may be obtained by the use of the following methods.

## *Titanium Plasma Spraying*

Roughened titanium surfaces have been successfully employed in orthopedics and especially for endosteal applications. For this reason, dental implant manufacturers have modified the Ti surface by roughening methods mentioned earlier. TPS produced the first rough Ti surface that was introduced into implant dentistry. This process is characterized by high-velocity molten drops of metal being sprayed onto the implant body to a thickness of 10–40 $\mu$m, with the objective of getting a greater surface area for bone attachment. TPS implants demonstrated satisfactory, long-term results in fully and partially edentulous patients [56].

Because of the high temperature of plasma spray, both Ti and HA become thermodynamically unstable. Mandibular symphysis anterior to the mental foramen uses implants coated by the TPS technique. TPS coatings are usually produced in vacuum to avoid the oxidation of metal particles. Babbush et al. [56] provided a sample condition of TPS coating where the specimen was coated at 15,000 °C using an Ar flame by spraying powdered Ti onto a core maintained at a temperature not exceeding 220 °C. A coating thickness of about 0.04–0.05 mm thick was obtained by this method.

**Fig. 3.2** SEM micrographs fracture for **a** TPS after 12 weeks of implantation, **b** HA after 12 weeks of implantation, **c** TPS after one-year loading and **d** HA after one-year loading, C-coatings, B-bone, and Arrows indicate cracks at the coating metal interface

This coating showed a bond strength of 0.5 $kg/mm^2$ and did not impair with fatigue resistance.

Ong et al. [57] carried out the animal studies of the plasma-sprayed HA and Ti implants by taking cp-Ti implants as control (Fig. 3.2). After 12 weeks of implantation and 1 year after loading, interfacial bone-implant strength (IBIS) and bone contact length (BCL) were estimated. Matching was observed in all groups at 12 weeks after implantation and after one-year loading in normal bone.

In a similar fashion, the rougher TPS surface also not corroborate to high pull-out strength because of the weak coating-substrate interface. While BCL for HA implants was found to be much higher than TPS and Ti implants, high BCL could not be correlated to pull out strength because of weak IBIS [57].

## *Grit Blasting*

Grit blasting, sandblasting, or air-crashing are techniques used to produce rough surfaces and are performed with spray gun-like tools that spray a stream of abrasive sand or grit under pressure. The abrasive materials usually get through the surface of the object being blasted to result in surfaces with a dull luster or riddled with holes [58–60].

The blasting is done in an enclosed booth with holes cut in the front to hold long, heavy-duty rubber gloves for manipulating the work under the abrasive spray. The sandblasting booth should be equipped with an extractor fan to remove dust and minimize the amount of dust in the atmosphere. A large window is also required to view the process [60].

Grit or abrasive blasting is done to remove rust, oxide layers, and old coatings from metal surfaces to remove heavy contamination. Blasting is a quick and efficient process and can be controlled easily. It needs the arrangement to contain the blast materials generating dust, and it should be renewed regularly to avoid contamination and provide efficiency. The blasting procedures cannot be performed on delicate surfaces to avoid physical damage. Similarly, some complicated geometries cannot be blasted as well [60].

Oxides of Ti and Al of different particle sizes are used in grit blasting. $Al_2O_3$ with particles of 30 and 60 grit size shows better osseointegration simultaneously increasing shear strength at bone-implant contact when compared to control Ti surface. The grit-blasted surface may have irregular pits and dimensions which may be due to the effect of various parameters like time of blasting, pressure, particle size, the distance between the gun, and substrate, but the desired surface roughness can be controlled too by controlling the parameters. However, there is also a disadvantage in that the particles remain embedded on the implant surface and disrupt osseointegration [61].

Apart from grit blasting, other methods such as wire brushing and abrasion with sandpaper, emery cloth, or metal wool are used to prepare surfaces. All of these methods are highly effective techniques used to remove unwanted particles on the surface of a material. Before the surface is modified, it is cleaned while in some cases, cleaning is also done after mechanical surface preparation, especially if the abrasion process is being used as a pretreatment to a later coating process. The blasted region has to be degreased and neutralized so as to prevent contamination. After abrasion, solid contaminants are removed by dry and clean air and solvent wiping [62]. There are two variants of abrasive blasting, namely dry abrasive blasting and wet abrasive blasting.

#### Dry Abrasive Blasting

In dry abrasive blasting, nonmetallic grits such as SiC, $SiO_2$, $Al_2O_3$, flintstones, or glass beads are used. The size of these particles may vary depending upon the roughness to be made.

Each substrate has an abrasive particle size specificity. The substrate reacts favorably in this particle size range. It has been established that the adhesion strength of the material increases for the roughened surface. The performance of the joint surface is dependent on different types of mechanical abrasion rather than on the abrasion depth. Compared to a simple machined surface, sandblasting provides greater adhesion for the stainless steel and aluminum surfaces. However, pronounced roughening can also leave voids at the interface at the tissue-implant surface which acts as stress raisers that may have an adverse impact on the joint. Thus, the recommended values of abrasive particle sizes for metals and polymers vary from 80–325 to 120–220 grit ranges [63].

### Wet Abrasive Blasting

A wet abrasive blasting process may be adapted instead of conventional dry blasting. $Al_2O_3$ or glass bead slurry with particle size in the range of 20–325 grit size is used as the abrasive medium. The proportion of 3:1 by volume is maintained for water to abrasive in this method. Wet abrasive blasting is versatile in the sense that it can contain a wide range of abrasive materials [64].

It was found from a comparative in vitro study that after culture of MC3T3-E1 mouse osteoblast-like cells, among (i) mirror-polished Ti (smooth Ti), (ii) $Al_2O_3$ grit-blasted Ti (Alumina-Ti), (iii) biphasic calcium phosphate grit-blasted Ti (BCP–Ti), and (iv) a commercially available implant (SLA) (Figs. 3.3 and 3.4), BCP-Ti-blasted surface showed higher average surface roughness (Ra = 2.5 μm) than the other grit-blasted groups. While MC3T3-E1 cells were well-spread, attached, and proliferated on all the surfaces, (BCP–Ti) and SLA groups showed excellent growth after 2 days

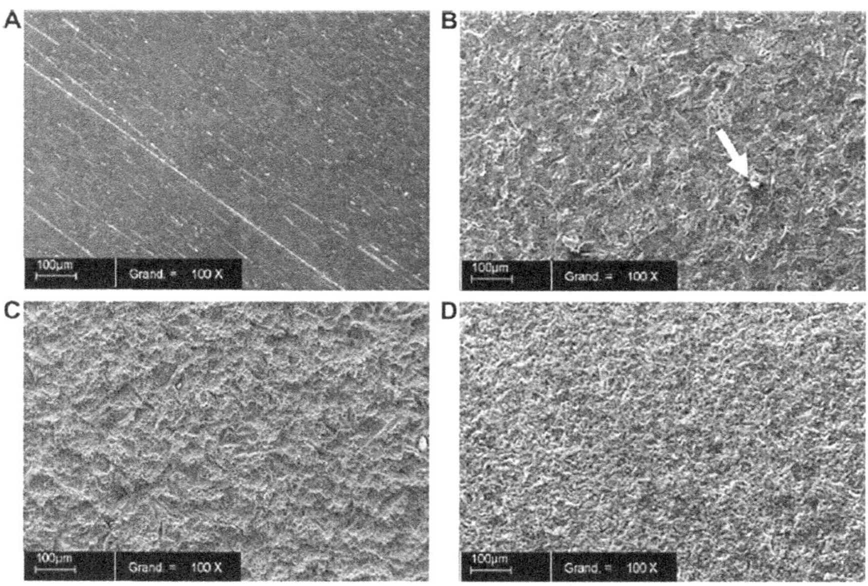


**Fig. 3.3** SEM micrographs (X100) showing the surface roughness of the tested Ti disks **a** smooth Ti, **b** Alumina-Ti, **c** SLA, and **d** BCP-Ti, bar = 100 μm

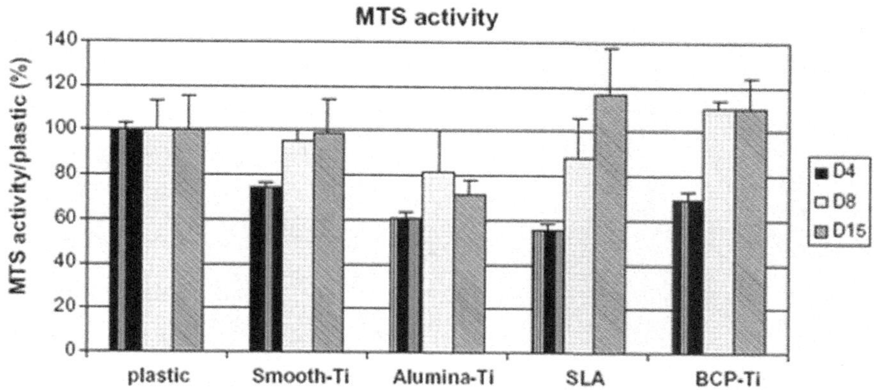


**Fig. 3.4** Cell viability of osteoblastic cells cultured on the Ti surfaces for 4, 8, and 15 days. The MTS activity was measured as optical density in quadrupole and two series of experiments ($n = 8$) and normalized to cell culture plastic (100%) in order to demonstrate differences in cell viability on the Ti substrates. No statistical differences ($p > 0.05$) were observed between groups and delays

of culture. Of the four surfaces, Alumina-Ti disks were the most hydrophilic while SLA was the most hydrophobic. Alumina-Ti had higher cell viability, followed by SLA, BCP-Ti, and then by Smooth-Ti [62].

## *Laser-Induced Surface Roughening*

A laser beam has been used to create rough surfaces. A distinct advantage in this method is that the laser can be targeted precisely and with predetermined angulations (e.g., apical, coronal, or perpendicular to the surface) [65]. It is possible to create regularly oriented micro-retention in contrast to the completely non-oriented surface roughness configurations produced in techniques like TPS coating or sandblasting. In vitro studies have also shown good cell adhesion and proliferation on the laser-treated surfaces [65–67].

Lawrence et al. studied the wettability characteristics of an Nd: YAG laser-treated Ti6Al4V biometal and the effects thereof on the response of osteoblast cells (Figs. 3.5 and 3.6). Wettability characteristics of the laser-treated surface were improved when compared with the untreated Ti6Al4V surface. These improvements were attributed to surface oxygen content, roughness, and elevation in the polar component of the surface energy. When surface roughness was increased, the polar component of surface energy and surface oxygen content promoted cell proliferation. Hence, Nd: YAG laser treatment was thus proved to be more effective for osteoblast cell adhesion than the conventional methods [68].

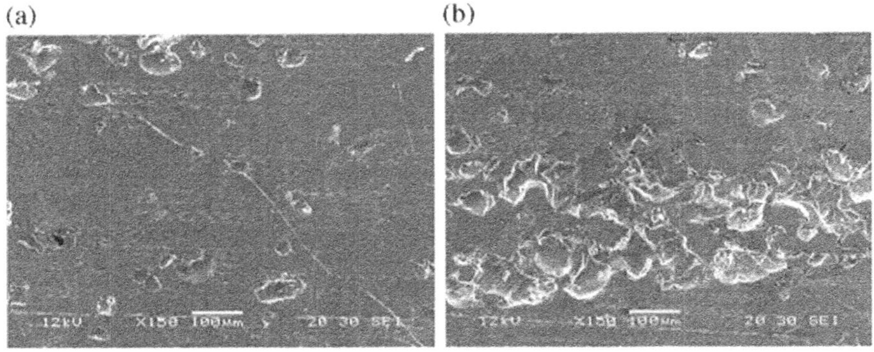


**Fig. 3.5** SEM image of the **a** untreated and **b** Nd: YAG laser-treated Ti6Al4V alloy

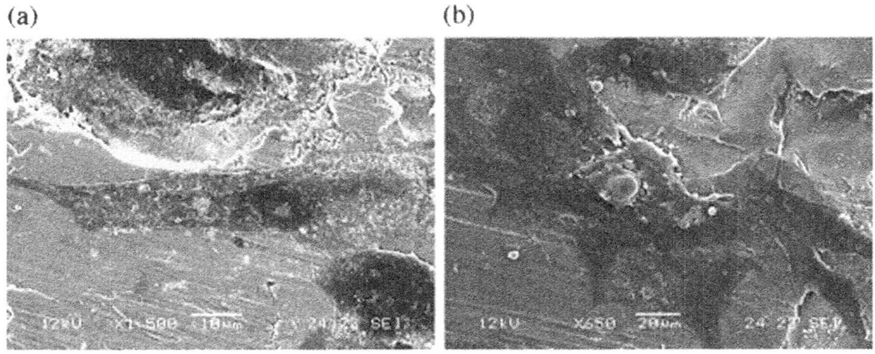


**Fig. 3.6** SEM image of osteoblast cells on Ti6Al4V alloy when **a** untreated and **b** Nd: YAG laser-treated

## *Sandblasting with Large Grit and Acid Etching (SLA)*

It can be suggested that the Sa values are less than 0.5 μm for smooth implants, 0.5–2.0 μm for moderately rough implants and more than 2 μm for rough implants [69]. Sandblasting with large grit and acid etching (SLA) implants maintain a moderately rough surface that favors bone ingrowth on their surfaces. When compared to other surfaces, SLA surfaces showed the highest bone-to-implant contact even though they are not the roughest among all [70].

SLA treatment is a widely used technique to improve the surface of implants for clinical use. In SLA treatment, the titanium surface is initially sandblasted with large-sized grits for creating a roughened texture. Subsequently, acid etching was performed on the blasted surface form micro-pits. The surface thus formed was wavy at the macro-level and rough micro-dimensionally that favors osseointegration [71]. SLA surfaces lead to multi-tier textured topography with multi-layered pores

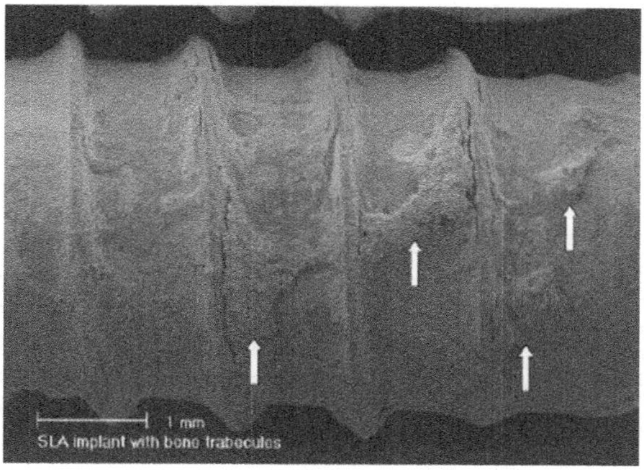


**Fig. 3.7** Surface analysis of an SLA torqued implant

which are demonstrated to enhance bone apposition, healing, and higher removal torque in histomorphometric studies. It was observed that SLA produces fine pits with roughened textures (Fig. 3.7). Dental implants made of titanium were compared after SLA treatment. It was observed that 1.5 times higher osteoblasts, and some fibroblasts were seen to be attached to the SLA implant as compared to that without SLA treatment [47]. 30% stronger in torsion resistance and 5% stronger in pressure resistance were observed in SLA-treated implants as compared to acid-etched implant surface [72].

On the other hand, in an animal model, it was observed that SLA surfaces reduced healing time as compared to the TPS surfaces [73]. A significant amount of bone deposition was observed in SLA titanium implants as shown in Fig. 3.8 where the arrows indicate the fractured bone site (a) and bone ingrowth inside pits (b). Moreover, SLA titanium showed even better osseointegration than hydroxyapatite-coated titanium [74].

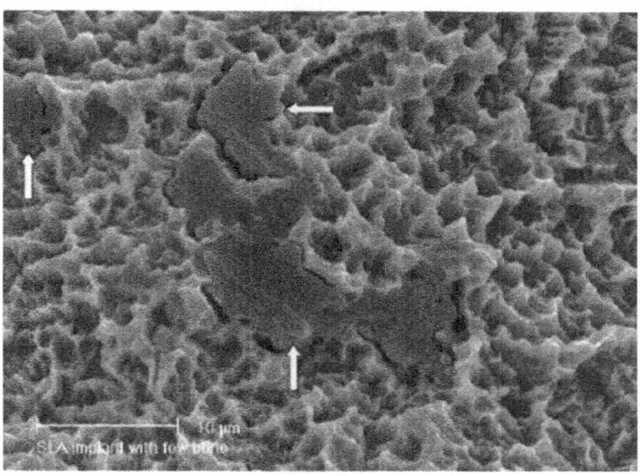


**Fig. 3.8** Details of the attached bone (X6000). Note the bone ingrowth into the pits (arrows)

## *Electron Beam Melting (EBM)*

Electron beam melting (EBM) is a computer-controlled technique that uses an electron gun to melt deposit metal powders in the three-dimensional layer-by-layer structure [75]. EBM has found applications in aerospace and medical and orthopedic industries [76]. The electron beam emerging from a tungsten filament is focused and controlled on the implant by a set of magnetic coils [77]. It is an ultraclean process that produces high strength materials because the whole process is carried out in a vacuum. The modified structure has a porous scaffold structure that helps in osseointegration with the host bone. After the structure is made, it is allowed to cool inside that chamber which is subsequently transferred to a blast chamber where additional processes such as milling, turning, polishing, and grinding was carried out [75]. Ti–6Al–4V implants synthesized by EBM at different locations, angles, and separation lengths showed different sizes of porosity that controlled its hardness and mechanical properties [77]. The porous structure present in the implant highly favors bone ingrowth of the tissue.

Hara et al. compared an EBM titanium implant with a fiber mesh-coated implant in an animal model, and they measured the bone-bonding strength and histological bone ingrowth after 4 and 12 months of implantation, and they found that the EBM implant to be better than the other [78]. Heinl et al. synthesized an EBM Ti–6Al–4V structure with interconnected pore channels that can be used for tissue ingrowth and vascularization. After treating its surface with HCl and NaOH solutions, it induced HA formation in SBF [76].

## *Roughening of Implants by Anodization*

By using potentiostatic or galvanostatic anodization for roughening of the substrate using strong acids such as HF, $HNO_3$, $H_3PO_4$, or $H_2SO_4$ at high current density (200 $A/m^2$) or potential (100 V), highly roughened surfaces can be produced. In higher current densities or potential, the surface oxide layer gets dissolved along current convection lines and thickens in other regions for which the dissolution produces micro- and nanopores on implant surface [79, 80]. Depending upon these process parameters, different morphologies and oxide properties were obtained.

It is also possible to anodize Ti or its alloy sheet or plate or even powder compact. In a recent work [81], nanostructured Ti6Al4V alloy was fabricated by mechanical alloying and roughened by using anodization in phosphoric acid solution (Fig. 3.9). The alloy with a grain size of 35 nm was etched along the grain boundaries that removed the grains by forming pores with an average diameter of 60 nm. Mechanically alloyed samples showed lower resistance to corrosion in Ringer's solution, as compared to the microcrystalline alloy, but anodization etching improved the corrosion resistance. The corrosion resistance of the microcrystalline alloy showed superior properties than nanocrystalline surfaces. Anodic oxidation, as a technique, could

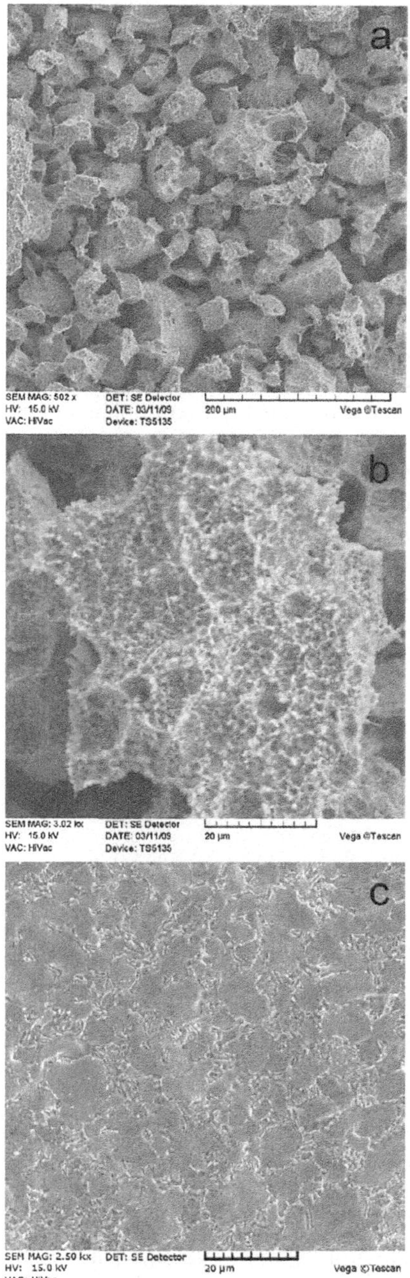


**Fig. 3.9** Mechanically alloyed Ti6Al4V after sintering and electrochemical etching (**a**, **b**); for comparison is shown electrochemically etched bulk microcrystalline sample (**c**)

improve the roughness of the surface and can also create necessary levels of porosity for good bone-bonding and can also improve corrosion resistance in simulated body fluid environment.

## *Chemical Etching*

Etching the surface of metallic implants by acid has been shown to improve osseointegration. It has been shown that the removal torque of acid-etched implants is higher than the machined Ti. Removal torque is a measure of the extent of attachment of body tissues with the implants during in vivo studies and is also an indicator of the extent of bone-bonding with the implant. Higher torque removal values indicate an increase in the bone-to-implant contact [82, 83].

Yang et al. studied the effect of acid etching of Ti screw implants on their osseointegration behavior. Threaded Ti implants with thread separation space of 0.7 mm, a diameter of 3 mm, and length of 8–10 mm were used. The implants were separated into 'machined' and 'roughened' types (Fig. 3.10). The roughened category samples were prepared by a three-level method. First, the samples were polished, roughened by sandblasting SiC at 4 MPa pressure, cleaned with acetone, washed with 75% alcohol and water, and then ultrasonicated. The chemical etching was performed using 0.11 mol HF and 0.09 mol $HNO_3$ for 10 min and then oven-dried at 50 °C for

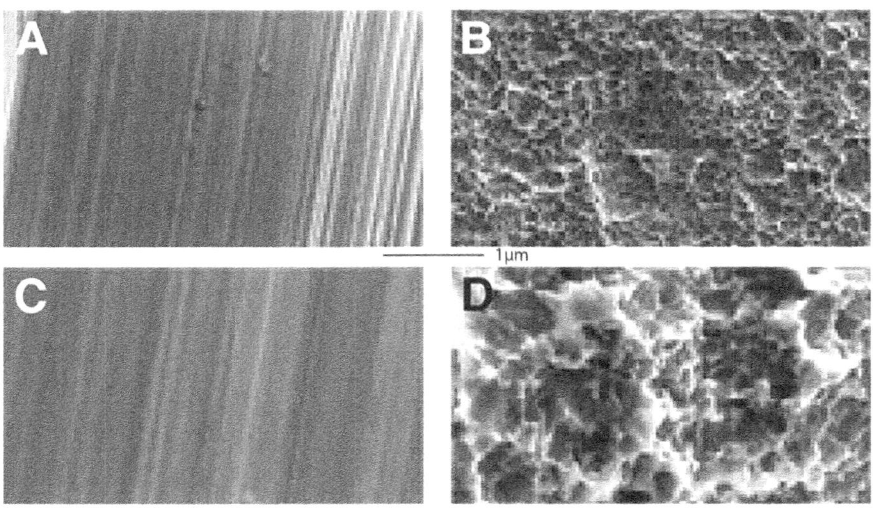


**Fig. 3.10** FESEM micrograph of the machined surface (**A**, **C**) and roughened surfaces (**B**, **D**). The strips were clearly seen on the machined surface. The roughened surface was quite irregular. The porous structures appeared on the roughened surface

24 h. Then the samples were further roughened using 5.80 mol HCl and 8.96 mol $H_2SO_4$ at 80 °C for 30 min and oven-dried at 50 °C for 24 h [82].

Animal testing of these implants showed 66.21%, 89.06%, and 115.00% greater removal torque (RTQ) values, respectively, than did their machined counterparts at 2, 4, and 8 weeks. The in vivo studies showed that the bone-to-implant contact (BIC) and peri-implant formation of bone increased in the case of roughened implants. It was thus stated that the roughened surface produced by acid etching can provide the osteoconductive potential to the surface of implants for better bone integration [82].

## *Cryogenic, Hydrodynamic and Polymeric Blasting*

It is also possible to employ special mechanical blasting systems like cryogenic, hydrodynamic, and polymeric blasting on the surfaces on which grit blasting cannot be performed [84]. Hydrodynamic abrasive-free blasting and cryogenic blasting are also employed for surface preparation [85]. In abrasive blasting, abrasive can contaminate the abraded surface and sometimes recycling of the abrasive is prohibited. Cryogenic and hydrodynamic methods are best-suited in order to overcome these difficulties.

Pelletized $CO_2$ with the temperature at $-73$ °C has been used as fluidized abrasive to remove corrosion or contaminated coatings. Supercritical $CO_2$ has also been used to dissolve some organic or inorganic contaminants [86].

Polymeric blasting uses hard plastic material and is used to remove contaminants and coatings for use in various applications. Different ranges of grits with specific hardness range are available depending on the types of contaminants, coatings, etc. Grits are chosen to remove unwanted particles keeping the base metal and its delicate parts intact [87].

## *Ultraviolet Light Treatment*

When Ti or Ti alloys are treated with ultraviolet (UV) light, some temporary chemical alterations occur within the $TiO_2$ molecules, which consequently results in a photocatalytic chemical reaction. This concept of the photochemical reaction of oxides has been applied to develop pollutant-degrading and anti-bacterial materials [88]. The molecular structure of the $TiO_2$ surface is altered by this phenomenon [89] by the creation of surface oxygen vacancies at bridging sites by converting $Ti^{4+}$ sites to $Ti^{3+}$ sites. These are favorable for dissociative water adsorption. Hence, the better water wetting is the cause of alteration of surface chemistry brought out by $TiO_2$ photocatalytic activity [90, 91].

Liu et al. carried out work on the effect of UV on oxidized titanium specimens. They found that there was 100% cell attachment and 10% cell activity on the surface

[92]. They reported that $TiO_2$ plasma-sprayed coatings (with and without nanostructures) with UV-A treatment and dipping the samples in SBF showed enhanced apatite formation with the nanostructured $TiO_2$ coatings. All these works prove the efficacy of UV-treatment in increasing the biocompatibility of titanium surfaces [93–95].

## Calcium Phosphate Bioceramics Coatings

Calcium phosphate coating improves the bioactivity of the surface drastically. Several phases of calcium phosphate are used in medical applications, including HA, carbonated apatite, octacalcium phosphate (OCP) [96], calcium pyrophosphate dihydrate (CPPD), dicalcium phosphate dihydrate (DCPD) or brushite, β-tricalcium phosphate (β-TCP), and amorphous calcium phosphate (ACP) [97]. The stability of given calcium phosphate in a medical application depends on the crystal phase, crystal size and perfection, the temperature used during processing, etc. At physiological temperature and pH, HA is the stable phase, and it generally takes a long time to resorb via physicochemical dissolution. However, bone cells and other cells called macrophages can initiate cell-mediated resorption by changing the local pH to acidic. pH may change from neutral to acidic when the tissue is damaged [98].

HA coating on titanium implants is performed to achieve rapid osseointegration that leads to faster stabilization of the implant with the surrounding bone that extends the functional life of implants [99]. Many in vivo studies have been performed to predict bone response to HA-coated titanium implants [100–102] where uncoated or untreated or sandblasted titanium was taken as a control group to compare with that of HA-coated titanium implants. It was found from the studies that the HA-coated titanium showed better mechanical properties than its uncoated counterparts in the early stage of healing. Cook et al. reported 5–8 times to increase in the mean interface strength for HA-coated implant with respect to the uncoated but sandblasted titanium [100]. For long-term stability, the HA-coated titanium implants are not well defined [99]. The thickness of HA coating plays a major role where it was suggested that thinner HA coating provides better mechanical strength by preserving the porosity of the HA surface in an effective manner [102]. Moreover, the pure phase of hydroxyapatite is desired on the implant so that other phases of calcium phosphate present in the coating would not hamper its resorption properties by releasing ions from other phases of HA. In the ion release mechanism, density and crystallinity of the coating are drastically affected [99]. The adhesion strength of the coating with the implant is another factor that plays an important role in the long-term stability of the implant. In the case of weak adhesion strength, the coating is separated from the implant causing implant failure [99]. Hence, the HA layer should be strongly bonded to the titanium implant to provide implant integrity by facilitating the required transmission of load from implant to the adjacent bone [102–108].

## *Biomimetic Deposition*

Plasma spraying has long been implemented for HA coating on Ti and Ti alloys. But because of the deposition of coating at very high temperature, implants do not have long-term stability in bio-environment and the stability of the coatings is questioned based on implant loosening and implant failures. Because of the brittle nature of HA, the thick coatings produced by plasma spray technique lead to cracking. These failures result in degradation and poor fatigue performance of these coatings. On the other hand, thin coatings cannot give protection from metal ion release and dissolve faster although they can perform better mechanically [109]. This HA deposition provides a coating with better mechanical strength and gives protection from metal ion release [110, 111].

To induce nucleation of HA, it is necessary to create Ti–OH groups or triple bond Si–O(−) on Ti surfaces. Pretreatment of Ti is done using $H_2O_2$, NaOH, or electrochemical oxidation to produce the desired Ti–OH groups [112]. In this method, HA is precipitated at a very slow rate and the precipitated layer is loose but can be used to form bonding with bone.

Bigi et al. demonstrated a method in which a uniform layer of HA is coated on etched Ti and Ti alloy substrates in a few hours (Fig. 3.11). In spite of the relatively high rate of nucleation and growth, HA crystals have a higher mean dimension and degree of crystallinity than those deposited from SBF solutions. Hence, this method can be successfully applied to obtain quickly formed uniform coatings of nanocrystalline HA on Ti substrates [113].

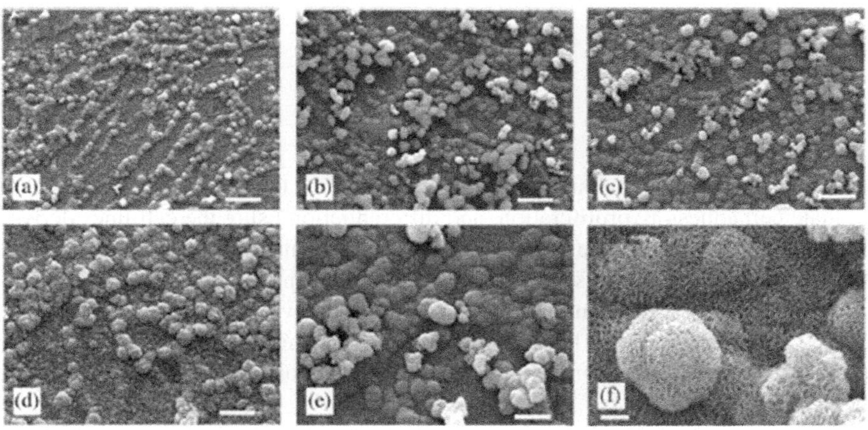


**Fig. 3.11** SEM micrographs of the surface of the etched Ti6Al4V substrates after soaking in Ca/P for 3 h (**a**, **d**), 1 day (**b**) and 3 days (**c**, **e**, **f**). Bars = 10 μm (**a**, **b**, **c**) and 5 μm (**d**, **e**), 1 μm (**f**)

## Properties of the Treated Surface

Each substrate has an abrasive size specificity. The substrate reacts favorably with an abrasive only in a specific size range. As a general observation, BIC increases with an increase in surface roughness. BIC depends mostly on the type of roughness created by sandblasting or machined surface than the depth of roughness. As compared to machined surfaces, sandblasted aluminum and stainless steel show better adhesion of cells. Extreme roughness is not advantageous either, where voids are formed at the interface that can act as stress raisers and are unfavorable to cell adhesion during actual practice. 80–325 abrasive grit size range is recommended for metals, and 120–220 size range is preferred for polymeric substrates [114]. Polyolefins and fluorocarbons, which have low energy surfaces, should be abraded after the application of adhesives because the roughness created on low energy surface creates air pockets in the valleys and crevices. The air pockets present in the implants act as stress raisers and weaken the joint. It is not advisable to use mechanical abrasion or surface roughness if the adhesive does not make less than a 90° water wetting angle.

Buser et al. studied the estimation of BIC for (i) HA-coated, (ii) acid-etched, (iii) TPS, and (iv) sandblasted surfaces in a pig model. While highest BIC was observed for HA-coated implants (60–70%), sandblasting followed by acid etching ($HCl/H_2SO_4$) implants had BIC of (50–60%) that followed by sandblasted and TPS-coated implants (30–40%). But among these coatings, HA-coated implants showed more resorption [114]. In another study, Novaes et al. employed a canine model to show that TPS (48.9%) and HA (57.9%) implants showed higher BIC than machined implants (41.7%). SBM implants (68.5%) showed the highest bone-implant contact [115].

## *In Vitro Studies*

Before performing animal studies, cell culture studies (in vitro) are carried out to evaluate cell adhesion, proliferation, and differentiation using the cell lines derived from animal or human bone tissues. These are preliminary tests carried out before the in vivo animal tests. These tests are less expensive, and the properties of implants such as the effect of implant surface on cells are known in a shorter time by this method. Numerous studies have been performed on the roughened surface of the implants which indicate improvement in cellular behavior with an increase in surface roughness [116–120].

## *In Vivo Studies*

Properties of the roughened implants using animal models have been performed to establish the relationship between histomorphometry and fixation of bone to implant mechanically. Moreover, the percentage of the implant to the new bone formation with healing times and the torque and pushout strengths are also measured. It has not been clearly established whether HA-coated implants or shot-blasted implants are better for bioimplant applications. While implants coated with HA showed better short-term behavior, shot-blasted HA implants show better osseointegration for longer times, i.e., four months to one year after implantation. This behavior is attributed to the implant surface roughness where the morphology of the implant surface changes drastically after shot-blasting which subsequently provides better bone fixation at micron level [121–128].

There are also reports showing faster resorption of the bone tissues in contact with the plasma-sprayed HA implants. This phenomenon is attributed to the degradation of the coating and the further release of HA particles to the surrounding bone [129].

## Summary

Many dental implant materials are available with proven clinical efficacy, but attention is still needed on their cell culture and animal studies. The commercially available implants show overall poor performance, and their studies are not standardized by using various cell lines, implant surfaces, and animal studies. Moreover, the surface morphology, microstructures, topography, and overall surface states on biological activities of titanium systems are poorly understood. Hence, future research should focus on overcoming these drawbacks and develop surfaces with controlled topography and surface chemistry that can provide better osseointegration and mechanical strength of the implants. These therapeutic strategies will be useful in enhancing the osseointegration of bio-implants for their overall success. Even though significant research has been done on fabricating and understanding titanium implant surface treatment techniques, still multi-dimensional exploration in this aspect has to be done to overcome its drawbacks.

## References

1. Rautray, T.R., Narayanan, R., Kim, K.H.: Ion implantation of titanium based biomaterials. Prog. Mater. Sci. **56**, 1137–1177 (2011)
2. Rautray, T.R., Narayanan, R., Kwon, T.Y., Kim, K.H.: Accelerator-based synthesis of hydroxyapatite by MeV ion implantation. Thin Solid Films **518**, 3160–3163 (2010)
3. Lee, K.W., Bae, C.M., Jung, J.Y., Sim, G.B., Rautray, T.R., Lee, H.J., Kwon, T.Y., Kim, K.H.: Surface characteristics and biological studies of hydroxyapatite coating by a new method. J. Biomed. Mater. Res. Part B Appl. Biomater. **98**, 395–401 (2011)

4. Yang, J.H., Kim, K.H., You, C.K., Rautray, T.R., Kwon, T.Y.: Synthesis of spherical hydroxyapatite granules with interconnected pores channels using camphene emulsion. J. Biomed. Mater. Res. Part B Appl. Biomater. **99**, 150–157 (2011)
5. Swain, S., Rautray, T.R.: Silver doped hydroxyapatite coating by sacrificial anode deposition under magnetic field. J. Mater. Sci. Mater. Med. **28**, 160 (2017)
6. Swain, S., Rautray, T.R., Narayanan, R.: Sr, Mg, and Co substituted hydroxyapatite coating on $TiO_2$ nanotubes formed by electrochemical methods. Adv. Sci. Lett. **22**, 482–487 (2016)
7. Rautray, T.R., Swain, S., Kim, K.H.: Formation of anodic $TiO_2$ nanotubes under magnetic field. Adv. Sci. Lett. **20**, 801–803 (2014)
8. Rautray, T.R., Mohapatra, B., Kim, K.H.: Fabrication of strontium-hydroxyapatite scaffolds for biomedical applications. Adv. Sci. Lett. **20**, 879–888 (2014)
9. Rautray, T.R., Narayanan, R., Kwon, T.Y., Kim, K.H.: Surface modification of titanium and titanium alloys by ion implantation. J. Biomed. Mater. Res. **93**, 581–591 (2010)
10. Rautray, T.R., Das, S.L., Rautray, A.C.: In situ analysis of human teeth by external PIXE. Nucl. Instr. Meth. Phys. Res. B. **268**, 2371–2374 (2010)
11. Rautray, T.R., Kim, K.H.: Nanoelectrochemical coating on titanium for bioimplant applications. Mater. Technol. **25**, 143–148 (2010)
12. Rao, K.S., Anupama, P.M., Mahesh, D.G., Rao, R.V., Rautray, T.R., Venkateswarulu, P.: External pixe—a novel approach for determining trace elements in formation process. Int. J. Appl. Biol. Pharm. Technol. **1**, 68–78 (2010)
13. Rautray, T.R., Mishra, S., Patnaik, S.K., Vijayan, V., Panigrahi, S.: Analysis of human bone and teeth. Ind. J. Phys. **81**, 99–102 (2007)
14. Rautray, T.R., Vijayan, V., Panigrahi, S.: Synthesis of hydroxyapatite at low temperature. Ind. J. Phys. **81**, 95–98 (2007)
15. Rosenberg, E.S., Torosian, J.P., Slots, J.: Microbial differences in 2 clinically distinct types of failure of osseointegrated implants. Clin. Oral Impl. Res. **2**, 135–144 (1991)
16. Mombelli, A., Van Oosten, M.A., Schurch, E., Jr., Land, N.P.: The microbiota associated with successful or failing osseointegrated titanium implants. Oral Microbiol. Immunol. **2**, 145–151 (1987)
17. Becker, W., Becker, B.E., Newman, M.G., Nyman, S.: Clinical and microbiologic findings that may contribute to dental implant failure. Int. J. Oral Maxilofac. Impl. **5**, 31–38 (1990)
18. Rams, T.E., Roberts, T.W., Feik, D., Molzan, A.K., Slots, J.: Clinical and microbiological findings on newly inserted hydroxyapatite-coated and pure titanium human dental implants. Clin. Oral Impl. Res. **2**, 121–127 (1991)
19. Gatewood, R., Cobb, C.M., Killoy, W.J.: Microbial colonization on natural tooth structure compared with smooth and plasma-sprayed dental implant surfaces. Clin. Oral Impl. Res. **4**, 53–64 (1993)
20. Johnson, B.W.: HA-coated dental implants: long-term consequences. J. Calif. Dent. Assoc. **20**, 33–41 (1992)
21. Zablotsky, M.H., Diedrich, D.L., Meffert, R.M.: Detoxification of endotoxin-contaminated titanium and hydroxyapatite-coated surfaces utilizing various chemotherapeutic and mechanical modalities. Impl. Dent. **1**, 154–158 (1992)
22. Nancollas, G.H., Tucker, B.E.: Dissolution kinetics characterization of hydroxyapatite coatings on dental implants. J. Oral Implantol. **20**, 221–226 (1994)
23. Cranin, A.N., Mehrali, M., Baraoidan, M.: Hydroxyapatite-coated endosteal implants from the clinicians' perspective. J. Oral Implantol. **20**, 235–239 (1994)
24. Van Blitterswijk, C.A., Grote, J.J., Kuypers, W., Blok van Hoek, C.J., Daems, W.T.: Bioreactions at the tissue/hydroxyapatite interface. Biomaterials **6**, 243–251 (1985)
25. Muller-Mai, C.M., Voigt, C., Gross, U.M.: Incorporation and degradation of hydroxyapatite implants of different surface roughness and surface structure in bone. Scan. Microsc. **4**, 613–624 (1990)
26. Bauer, T.W., Geesink, R.C.T., Zimmerman, R., McMahon, J.T.: Hydroxyapatite-coated femoral stems. Histological analysis of components retrieved at autopsy. J. Bone Joint Surg. **73**, 1439–1452 (1991)

27. Block, M.S., Finger, I.M., Fontenot, M.G., Kent, J.N.: Loaded hydroxylapatite-coated and grit-blasted titanium implants in dogs. Int. J. Oral Maxillofac. Impl. **4**, 219–225 (1989)
28. Soballe, K., Brockstedt-Rasmussen, H., Hansen, E.S., Bunger, C.: Hydroxyapatite coating modifies implant membrane formation. Controlled micro motion studied in dogs. Acta Orthop. Scand. **63**,128–140
29. Ong, J.L., Chan, D.C.N., Bessho, K.: HA coatings on dental implants. In: Wise D.L., Trantolo, D.J., Lewandrowski, K.U., Gresser, J.D., Cattaneo, M.V., Yaszemski, M.J. (eds.) Biomaterials Engineering and Devices: Human Applications, pp. 49–60. Humana Press, NJ (2000)
30. Smallman, R.E., Bishop, R.J. (eds.): Modern Physical Metallurgy and Materials Engineering. Butterworth-Heinemann Publishers, UK (1999)
31. Marcolongo, M., Dinardo, N.J., Hickok, N., Tuan, R., Pourezzaei, K., Beard, R., Brennan, D., Heipp, P., Phan, T.: Osteoblast attachment on biomaterials as a function of surface charge. MRS Symp. Proc. **550**, 121–126 (1999)
32. Le Guéhennec, L., Soueidan, A., Layrolle, P., Amouriq, Y.: Surface treatments of titanium dental implants for rapid osseointegration. Dent. Mater. **23**, 844–854 (2007)
33. Ortrun, E.M.P.: Unalloyed titanium alloys for implants in bone surgery. Injury **31**, 7–13 (2000)
34. Ratner, B.D. (ed.): A perspective on titanium biocompatibility. Titanium in Medicine. In: Engineering materials, pp. 1–12. Springer, Heidelberg (2001)
35. Leyens, C., Peters, M. (eds.): Titanium and Titanium Alloys: Fundamentals and Applications. Wiley-VCH, Chichester (2003)
36. Natiella, J.R., Odontocompatibility, A.V., Recum, E., Jacobi, J.E.: Handbook of Biomaterials Evaluation: Scientific, Technical, and Clinical Testing of Implant Materials, pp. 567–584. CRC Press, Boca Raton (1999)
37. Brånemark, P.I., Adell, R., Albrektsson, T., Lekholm, U., Lundkvist, S., Rockler, B.: Osseointegrated titanium fixtures in the treatment of edentulousness. Biomaterials **4**, 25–28 (1983)
38. Albrektsson, T.: Direct bone anchorage of dental implants. J. Prosthet. Dent. **50**, 255–261 (1983)
39. Oshida, Y. (ed.): Future perspectives, In: Bioscience and Bioengineering of Titanium Materials. Elsevier Publishers, UK (2007)
40. Kim, K.H., Narayanan, R., Rautray, T.R.: Surface Modification of Titanium for Biomaterial Application. Nova Publisher, USA (2010)
41. Doll, B.A., Kukreja, A., Seyedain, A., Braun, T.W.: Dental implants. In: Guelcher, S.A., Hollinger, J.O. (eds.) An Introduction to Biomaterials, pp. 295–416. CRC Press, Boca Raton (2006)
42. Buser, D., Schenk, R.K., Steinemann, S., Fiorellini, J.P., Fox, C.H., Stich, H.: Influence of surface characteristics on bone integration of titanium implants. A histomorphometric study in miniature pigs. J. Biomed. Mater. Res. **25**, 889–902 (1991)
43. Carlsson, L., Röstlund, T., Albrektsson, B., Albrektsson, T.: Removal torques for polished and rough titanium implants. Int. J. Oral Maxillofac. Implants. **3**, 21–24 (1988)
44. Wen, X., Wang, X., Zhang, N.: Microsurface of metallic biomaterials: a literature review. J. Biomed. Mater. Eng. **6**, 173–189 (1996)
45. Klokkevold, P.R., Johnson, P., Dagrostari, S., Caputo, A., Davie, J.E., Nishimura, R.D.: Early endosseous integration enhanced by dual acid etching of titanium: a torque removal study in the rabbit. Clin. Oral Implants Res. **12**, 350–357 (2001)
46. Nishimoto, S.K., Nishimoto, M., Park, S.W., Lee, K.M., Kim, H.S., Koh, J.T., Ong, J.L., Liu, Y., Yang, Y.: The effect of titanium surface roughening on protein absorption, cell attachment, and cell spreading. Int. J. Oral Maxillofac. Implants. **23**, 675–680 (2008)
47. Schuler, M., Owen, G.R., Hamilton, D.W., de Wild, M., Textor, M., Brunette, D.M., Tosatti, S.G.: Biomimetic modification of titanium dental implant model surfaces using the RGDSP-peptide sequence: a cell morphology study. Biomaterials **27**, 4003–4015 (2006)
48. LeGeros, R.Z.: Calcium phosphate materials in restorative dentistry: a review. Adv. Dent. Res. **2**, 164–180 (1988)

49. Bowers, K.T., Keller, J.C., Randolph, B.A., Wick, D.G., Michaels, C.M.: Optimization of surface micromorphology for enhanced osteoblast responses in vitro. Int. J. Oral Maxillofac. Implants. **7**, 302–310 (1992)
50. Brett, P.M., Harle, J., Salih, V., Mihoc, R., Olsen, I., Jones, F.H., Tonetti, M.: Roughness response genes in osteoblasts. Bone **35**, 124–133 (2004)
51. Chehroudi, B., Gould, T.R.L., Brunette, D.M.: Effects of a grooved titanium-coated implant surface on epithelial cell behavior in vitro and in vivo. J. Biomed. Mater. Res. **23**, 1067–1085 (1989)
52. Schwartz, Z., Martin, J.Y., Dean, D.D., Simpson, J., Cochran, D.L., Boyan, B.D.: Effect of titanium surface roughness on chondrocyte proliferation, matrix production, and differentiation depends on the state of cell maturation. J. Biomed. Mater. Res. **30**, 145–155 (1996)
53. Wennerberg, A., Hallgren, C., Johansson, C., Danelli, S.: A histomorphometric evaluation of screw-shaped implants each prepared with two surfacė roughnesses. Clin. Oral Implants Res. **9**, 11–19 (1998)
54. Wennerberg, A., Albrektsson, T., Andersson, B., Krol, J.J.: A histomorphometric and removal torque study of screw-shaped titanium implants with three different surface topographies. Clin. Oral Implants Res. **6**, 24–30 (1995)
55. Becker, W., Becker, B.E., Ricci, A., Bahat, O., Rosenberg, E., Rose, L.F., Handelsman, M., Israelsom, H.: A prospective multicenter clinical trial comparing one- and two-stage titanium screw-shaped fixtures with one-stage plasma-sprayed solid-screw fixtures. Clin. Implant Dent. Relat. Res. **2**, 159–165 (2000)
56. Babbush, C.A., Kent, J.N., Misiek, D.J.: Titanium plasma sprayed (TPS) screw implants for the reconstruction of the edentulous mandible. J. Oral Maxillofac. Surg. **44**, 274–282 (1986)
57. Ong, J.L., Carnes, D.L., Bessho, K.: Evaluation of titanium plasma-sprayed and plasma-sprayed hydroxyapatite implants in vivo. Biomaterials **25**, 4601–4606 (2004)
58. Goldberg, V.M., Stevenson, S., Feighan, J., Davy, D.: Biology of grit-blasted titanium alloy implants. Clin. Orthop. Relat. Res. **319**, 122–129 (1995)
59. Castner, D.G., Ratner, B.D.: Biomedical surface science: foundations to frontiers. Surf. Sci. **500**, 28–60 (2002)
60. Mohammadi, Z., Ziaei-moayyed, A.A., Mesgar, A.S.: Grit blasting of Ti–6Al–4V alloy: optimization and its effect on adhesion strength of plasma-sprayed hydroxyapatite coatings. J. Mater. Process Technol. **194**, 15–23 (2007)
61. Orsini, G., Assenza, B., Scarano, A., Piattelli, M., Piattelli, A.: Surface analysis of machined versus sandblasted and acid-etched titanium implants. Int. J. Oral Maxillofac. Implants. **15**, 779–784 (2000)
62. Le Guehennec, L., Lopez-Heredia, M.A., Enkel, B., Weiss, P., Amouriq, Y., Layrolle, P.: Osteoblastic cell behaviour on different titanium implant surfaces. Acta Biomater. **4**, 535–543 (2008)
63. Pegueroles, M., Gil, F.J., Planell, J.A., Aparicio, C.: The influence of blasting and sterilization on static and time-related wettability and surface-energy properties of titanium surfaces. Surf. Coat. Technol. **202**, 3470–3479 (2008)
64. Gil, J., Planell, J.A., Padrós, A., Aparicio, C.: The effect of shot blasting and heat treatment on the fatigue behavior of titanium for dental implant applications. Dent. Mater. **23**, 486–491 (2007)
65. Gotz, H.E., Muller, M., Emmel, A., Holzwarth, U., Erben, R.G., Stangl, R.: Effect of surface finish on the osseointegration of laser-treated titanium alloy implants. Biomaterials **25**, 4057–4064 (2004)
66. Gaggl, A., Schultes, G., Muller, W.D., Karcher, H.: Scanning electron microscopical analysis of laser-treated titanium implant surfaces—a comparative study. Biomaterials **21**, 1067–1073 (2000)
67. Szabelski, P.: Modeling of surface roughening induced by laser ablation. Appl. Surf. Sci. **219**, 300–310 (2003)
68. Lawrence, J., Hao, L., Chew, H.R.: On The Correlation Between Nd: YAG laser-induced wettability characteristics modification and osteoblast cell bioactivity on a titanium alloy. Surf. Coat. Technol. **200**, 5581–5589 (2006)

69. Wenneberg A., Albrektsson, T.: Implant surfaces beyond micron roughness. experimental and clinical knowledge of surface topography and surface chemistry. Inter. Dent. **8**, 14–18 (2006)
70. Elias C.N. (ed.). Factors affecting the success of dental implants. In: Implant Dentistry—A Rapidly Evolving Practice, Implant Dentistry, pp. 319–364. InTech Open (2011)
71. Wennerberg, A., Albrektsson, T.: Suggested guidelines for the topographic evaluation of implant surfaces. Int. J. Oral Maxillofac. Impl. **15**, 331–344 (2000)
72. Li, D., Ferguson, S.J., Beutler, T., Cochran, D.L., Sittig, C., Hirt, H.P., Buser, D.: Biomechanical comparison of the sandblasted and acid-etched and the machined and acid-etched titanium surface for dental implants. J. Biomed. Mater. Res. **60**, 325–332 (2002)
73. Roccuzzo, M., Bunino, M., Prioglio, F., Bianchi, S.D.: Early loading of sandblasted and acid-etched (SLA) implants: a prospective split-mouth comparative study. Clin. Oral Impl. Res. **12**, 572–578 (2001)
74. Cochran, D.L., Schenk, R.K., Lussi, A.: Bone Response to unloaded and loaded titanium implants with a sandblasted and acid-etched surface: a histometric study in the canine mandible. J. Biomed. Mater. Res. **40**, 1–11 (1998)
75. Nasr, E.A., Ahmari-Al, A., Kamrani, A., Moiduddin, K.: Digital design and fabrication of customized mandible implant. In: World Automation Congress. TSI Press, USA (2014)
76. Heinl, P., Muller, L., Korner, C., Singer, R.F., Muller, F.A.: Cellular Ti-5Al-4V structures with interconnected macro porosity for bone implants fabricated by selective electron beam melting. Acta Biomater. **4**, 1536–1544 (2008)
77. Galarraga, H., Lados, D.A., Dehoff, R.R., Kirka, M.M., Nandwana, P.: Effects of the microstructure and porosity on properties of Ti–6Al–4VELI alloy fabricated by electron beam melting (EBM). Addit. Manuf. **10**, 47–57 (2016)
78. Hara, D., Nakashima, Y., Sato, T., Hirata, M., Kanazawa, M., Kohno, Y., Yoshimoto, K., Yoshihara, Y., Nakamura, A., Nakao, Y., Iwamoto, Y.: Bone bonding strength of diamond-structured porous titanium-alloy implants manufactured using the electron beam-melting technique. Mater. Sci. Eng. C. **59**, 1047–1052 (2016)
79. Sul, Y.T., Johansson, C.B., Jeong, Y., Roser, K., Wennerberg, A., Albrektsson, T.: Oxidized implants and their influence on the bone response. J. Mater. Sci. Mater. Med. **12**, 1025–1031 (2001)
80. Larsson, C., Thomsen, P., Aronsson, B.O., Rodahl, M., Lausmaa, J., Kasemo, B., Ericson, L.E.: Bone response to surface-modified titanium implants: studies on the early tissue response to machined and electropolished implants with different oxide thicknesses. Biomaterials **17**, 605–616 (1996)
81. Jakubowicz, J., Adamek, G.: Preparation and properties of mechanically alloyed and electrochemically etched porous Ti–6Al–4V. Electrochem. Comm. **11**, 1772–1775 (2009)
82. Yang, G.L., He, F.M., Yang, X.F., Wang, X.X., Zhao, S.F.: Bone responses to titanium implants surface-roughened by sandblasted and double etched treatments in a rabbit model. Oral. Surg. Oral Med. Oral Pathol. Oral. Radiol. Endod. **106**, 516–524 (2008)
83. Guo, J., Padilla, R.J., Ambrose, W., De Kok, I.J., Cooper, L.F.: The effect of hydrofluoric acid treatment of $TiO_2$ grit blasted titanium implants on adherent osteoblast gene expression in vitro and in vivo. Biomaterials **28**, 5418–5425 (2007)
84. Hutson, A.L., Niinomi, M., Nicholas, T., Eylon, D.: Effect of various surface conditions on fretting fatigue behavior of Ti–6Al–4V. Int. J. Fatigue. **24**, 1223–1234 (2002)
85. Pei, Y., Shaha, K.P., Chen, C.Q., van der Hulst, R., Turkin, A., Vainshtein, D.I., De Hosson, J.T.M.: Growth of nanocomposite films: from dynamic roughening to dynamic smoothening. Acta Mater. **57**, 5156–5164 (2009)
86. Deligianni, D.D., Katsala, N., Ladas, S., Sotiropoulou, D., Amedee, J., Missirlis, Y.F.: Effect of surface roughness of the titanium alloy Ti–6Al–4V on human bone marrow cell response and on protein adsorption. Biomaterials **22**, 1241–1251 (2001)
87. Harper, C.A., Petrie, E.M.: Plastics Materials and Processes: A Concise Encyclopedia. Wiley, 2003
88. Keleher, J., Bashant, J., Heldt, N., Johnson, L., Li, Y.: Photo-catalytic preparation of silver-coated $TiO_2$ particles for antibacterial applications. World J. Microbiol. Biotechnol. **18**, 133–139 (2002)

89. Wang, R., Hashimoto, K., Fujishima, A., Chikuni, M., Kojima, E., Kitamura, A., Shimohigoshi, M., Watanabe, T.: Light-induced amphiphlic surface. Nature **388**, 431–432 (1997)
90. Zubkov, T., Stahl, D., Thompson, T.L., Panayotov, D., Diwald, O., Yates, J.T.: Ultraviolet light-induced hydrophilicity effect on $TiO_2$ (110) (1 × 1) dominant role of the photooxidation of adsorbed hydrocarbons causing wetting by water droplets. J. Phys. Chem. B. **109**, 15454–15462 (2005)
91. Ogawa, T., Sukotjo, C., Nishimura, I.: Modulated bone matrix related gene expression is associated with differences in interfacial strength of different implant surface roughness. J. Prosthodont. **11**, 241–247 (2002)
92. Liu, X., Zhao, X., Li, B., Cao, C., Dong, Y., Ding, C., Chu, P.K.: UV-Irradiation-induced bioactivity on $TiO_2$ coatings with nanostructural surface. Acta Biomater. **4**, 544–552 (2008)
93. Aita, H., Att, W., Ueno, T., Yamada, M., Hori, N., Iwasa, F., Tsukimura, N., Ogawa, T.: Ultraviolet light-mediated photofunctionalization of titanium to promote human mesenchymal stem cell migration, attachment, proliferation and differentiation. Acta Biomater **5**, 3247–3257 (2009)
94. Stein, G.S., Lian, J.B.: Molecular mechanisms mediating proliferation differentiation interrelationships during progressive development of the osteoblast phenotype. Endocr. Rev. **14**, 424–442 (1993)
95. Sawase, T., Jimbo, R., Baba, K., Shibata, Y., Ikeda, T., Atsuta, M.: Photo induced hydrophylicity enhances initial cell behavior. Clin. Oral Implants Res. **19**, 491–496 (2008)
96. MaThew, M., Takagi, S.: Crystal structures of calcium orthophosphates. In: Chow, L.C., Eanes, E.D. (eds.) Octacalcium Phosphate, pp. 1–16. Karger Publishers, Basel (2001)
97. McCarthy, G.: Calcium pyrophosphate dihydrate, hydroxyapatite and miscellaneous crystals. In: Klippel, J.H., Stone, J.H., Crofford, L.J., White, P.H. (eds.) Primer on The Rheumatic Diseases, pp. 263–270. Springer Publishers, New York (2008)
98. Swain S.: Surface modification of titanium for bioimplant applications. Ph.D. thesis. Siksha O Anusandhan University, Odisha, India (2017) (unpublished)
99. Hench, L.L., Wilson, J. (eds.): An Introduction to Bioceramics. World Scientific, Singapore (1993)
100. Cook, S.D., Kay, J.F., Thomas, K.A., Jarcho, M.: Interface mechanics and histology of titanium and hydroxylapatite-coated titanium for dental implant applications. Int. J. Oral Maxillofac. Impl. **2**, 15–22 (1987)
101. Sergio, A., Jr., Rumpel, E., Kauschke, E., Fanghänel, J., König, B., Jr.: Hydroxyapatite grafting promotes new bone formation and osseointegration of smooth titanium implants. Ann. Anat. **188**, 143–151 (2006)
102. Søballe, K.: Hydroxyapatite ceramic coating for bone implant fixation. Mechanical and histological studies in dogs. Acta Orthop. Scand. Suppl. **64**, 1–58 (1993)
103. Weinlaender, M., Kenney, E.B., Lekovic, V., Beumer, J., III., Moy, P.K., Lewis, S.: Histomorhometry of bone apposition around three types of endosseous dental implants. Int. J. Oral Maxillofac. Impl. **7**, 491–496 (1992)
104. Søballe, K., Hansen, E.S., Brockstedt-Rasmussen, H., Bünger, C.: Hydroxyapatite coating converts fibrous tissue to bone around loaded implants. J. Bone JoInt Surg. **75B**, 270–278 (1993)
105. Bauer, T.W., Taylor, S.K., Jiang, M., Medendorp, S.V.: An indirect comparision of third-body wear in retrieved hydroxyappatite-coated, porous and cemented femoral components. Clin. Orthop. Rel. Res. **298**, 11–18 (1994)
106. Bloebaum, R.D., Beeks, D., Dorr, L.D., Savory, C.G., Dupont, J.A., Hofmann, A.A.: Complications with hydroxyapatite particulate separation in total hip arthoplasty. Clin. Orthop. Rel. Res. **298**, 19–26 (1994)
107. Jones, J.D., Lupori, J., Van Sickels, J.E., Gardner, W.: A 5-year comparison of hydroxyapatite-coated titanium plasma-sprayed cylinder dental implants. Oral Surg. Oral Med. Oral Pathol. Oral Radiol. Endod. **87**, 649–652 (1999)

108. Cook, S.D., Barrack, R.L., Thomas, K.A., Haddad, R.J., Jr.: Quantitative histologic analysis of tissue growth into porous total knee components. J. Arthroplasty. **4**, S33–S43 (1989)
109. Sun, L., Berndt, C.C., Gross, K.A., Kucuk, A.: Material fundamentals and clinical performance of plasma-sprayed hydroxyapatite coatings: a review. J. Biomed. Mater. Res. **58**, 570–592 (2001)
110. Habibovic, P., Barrere, F., Bliterswijk, C.A.V., Groot, K.de., Layrolle, P.: Biomimetic hydroxyapatite coating on metal implants. J. Am. Ceramic. Soc. **85**, 517–522 (2002)
111. Li, F., Feng, Q.L., Cui, F.Z., Li, H.D., Schubert, H.: A simple biomimetic method for calcium phosphate coating. Surf. Coat. Tech. **154**, 88–93 (2002)
112. Jonasova, L., Muller, F.A., Helebrant, A., Strnad, J., Greil, P.: Biomimetic apatite formation on chemically treated titanium. Biomaterials **25**, 1187–1194 (2004)
113. Bigi, A., Boanini, E., Bracci, B., Facchini, A., Panzavolta, S., Segatti, F., Sturba, L.: Nanocrystalline hydroxyapatite coatings on titanium: a new fast biomimetic method. Biomaterials **26**, 4085–4089 (2005)
114. Wennerberg, A.: The role of surface roughness for implant incorporation in bone. Cells and Mater. **9**, 1–19 (1999)
115. Novaes, A.B., Souza, S.L.S., Oliveira, P.T., Souza, A.M.: Histomorphometric analysis of the boneimplant contact obtained with 4 different implant surface treatments placed side by side in the dog mandible. Int. J. Oral Maxillofac. Impl. **17**, 377–383 (2002)
116. Naji, A., Harmand, M.F.: Cytocompatibility of two coating materials, amorphous alumina and silicon carbide, using human differentiated cell cultures. Biomaterials **12**, 690–694 (1991)
117. Puleo, D.A., Holleran, L.A., Doremus, R.H., Bizios, R.: Osteoblast responses to orthopedic implant materials in vitro. J. Biomed. Mater. Res. **25**, 711–723 (1991)
118. Keller, J.C., Stanford, C.M., Wightman, J.P., Draughn, R.A., Zaharias, R.: Characterization of titanium implant surfaces III. J. Biomed. Mater. Res. **28**, 939–946 (1994)
119. Martin, J.Y., Schwartz, Z., Hummert, T.W., Schraub, D.M., Simpson, J., Lankford, J., Jr., Dean, D.D., Cochran, D.L., Boyan, B.D.: Effect of titanium surface roughness on proliferation, differentiation and protein synthesis of human osteoblast-like cells (MG63). J. Biomed. Mater. Res. **29**, 389–401 (1995)
120. Batzer, R., Liu, Y., Cochran, D.L., Szmuckler-Moncler, S., Dean, D.D., Boyan, B.D., Schwartz, Z.: Prostaglandins mediate the effects of titanium surface roughness on MG63 osteoblast-like cells and alter cell responsiveness to 1 alpha, 25-(OH)2D3. J. Biomed. Mater. Res. **41**, 489–496 (1998)
121. Gross, U., Muller-Mai, C.M., Fritz, T., Voigt, C., Knarse, W., Schmitz, H.J.: Implant surface roughness and mode of load transmission influence peri-implant bone structure, In: Heimke, G., Soltesz, U., Lee, A.J.C. (eds.) Clinical Materials, pp. 303–308. Elsevier, Amsterdam (1990)
122. Piattelli, A., Manzon, L., Scarano, A., Paolantonio, M., Piatelli, M.: Histologic and histomorphometric analysis of the bone response to machined and sand-blasted titanium implants: an experimental study in rabbits. Int. J. Oral. Maxill. Impl. **13**, 805–810 (1998)
123. Wong, M., Eulenberger, J., Schenk, R., Hunziker, E.: Effect of surface topology on the osseointegration of implant materials in trabecular bone. J. Biomed. Mater. Res. **29**, 1567–1575 (1995)
124. Feighan, J.E., Goldberg, V.M., Davy, D., Parr, J.A., Stevenson, S.: The influence of surface-blasting on the incorporation of titanium-alloy implants in a rabbit intramedullary model. J. Bone JoInt Surg. Am. **77**, 1380–1395 (1995)
125. Wennerberg, A., Albrektsson, T., Andersson, B.: An animal study of c.p. titanium screws with different surface topographies. J. Mater. Sci. Mater. Med. **6**, 302–309 (1995)
126. Johansson, C.B., Cho, I., Heo, S.J., Sawai, T., Sawase, T., Wennerberg, A., Meredith, N.: Techniques to quantify the incorporation of machined and blasted implants in bone. In: Proceedings of the 13th European Conference on Biomaterials, p. 33. Göteborg, Sweden, 4–7 Sept 1997
127. Dhert, W.J., Thomsen, P., Blomgren, A.K., Esposito, M., Ericson, L.E., Verbout, A.J.: Integration of press-fit implants in cortical bone: a study on interface kinetics. J. Biomed. Mater. Res. **41**, 574–583 (1998)

128. Brånemark, I., Zrab, G.A., Albrektsson, T.: Tissue-Integrated Prostheses: Osseointegration in Clinical Dentistry. QuIntessenz Verlag-GmbH, Berlin (1987)
129. Park, E., Condrate, R.A., Sr., Hoelzer, D.T., Fischman, G.S.: Interfacial characterization of plasma-spray coated calcium phosphate on Ti–6Al–4V. J. Mater. Sci. Med. **9**, 643–649 (1998)

# Chapter 4
# Magnetic Nanomaterials and Their Biomedical Applications

**Papori Seal, Dipraj Saikia, and J. P. Borah**

## Introduction

Nanotechnology has become one of the most stimulating fields in physics, chemistry along with engineering and biology. The fabrication of nanostructured materials in the field of science and technology is an extensive and interdisciplinary area of research and development which has been developing worldwide in recent years. Nanomaterials having unique structural, optical, magnetic, and electrical properties are prospective for the applications in electronics as well as biomedical fields. Scientists and engineers have been developing an extensive variety of ways for the construction of nanostructured materials to take advantage of their improved and altered physical, mechanical, and chemical properties compared to their larger-scale structures [1–3]. The movement of nanotechnology towards biomedical applications involves elimination of the artificial disciplinary limitations and fabrication of the nanostructured materials that are allowed to interact with the body at small scales with high efficiency. Moreover, the development of nanoscience in biomedical field comprises nanomaterials that are applied to improve tissue regeneration, such as restoring cartilage function to overcome arthritis, in vitro engineered organ patches or biomaterials for in situ regeneration of bones, etc. [4]. Thus, it can be speculated from the recent research and developments in the field of nanoscience and technology that it can highly possible for manipulation and fabrication of nanostructured materials in large proficiency. Therefore, numerous scientists and researchers throughout the world recently, are focusing exhaustive research and development of biocompatible and toxin-free nanostructured materials for biomedical applications [5–7].

---

P. Seal · D. Saikia · J. P. Borah (✉)
Department of Physics, National Institute of Technology Nagaland, Dimapur, Nagaland 797103, India
e-mail: jpborah@rediffmail.com



B. P. Swain (ed.), *Nanostructured Materials and their Applications*, Materials Horizons: From Nature to Nanomaterials, https://doi.org/10.1007/978-981-15-8307-0_4

## Magnetic Materials

The development of science and technology in the field of magnetism has become widespread in recent times due to their remarkable properties in the nanoscale size. Considerable developments have been made in synthesizing magnetic materials with desired size, morphology, surface chemistry, and chemical stability, in order to improve their applications in the field of biomedical science and technology. Magnetic materials are basically classified in accordance with their responses under the application of external magnetic field. The responses of the materials under the field are mainly determined by the intrinsic spin associated with the electrons, in addition to the orbital motion of electrons rotating around the nucleus [8]. The value of magnetic moment of a material is a measure of the strength of the magnetism that is present. Depending upon alignment of the magnetic moments and change of magnetic properties with respect to the external applied magnetic field, the magnetic materials are classified mainly into five groups such as diamagnetic, paramagnetic, ferromagnetic, antiferromagnetic, and ferromagnetic.

## Diamagnetic Materials

These are the kind of magnetic materials in which outer-most valence shells are completely filled and no unpaired electrons are present, which results zero net magnetic moment. However, when these materials are experienced by external magnetic field, the magnetic moments will be aligned in the opposite to the direction of the external magnetic field giving rise to the negative magnetic susceptibility. The magnetic susceptibility of these materials is independent of temperature. Diamagnetic materials include quartz, calcites, gold, silver, etc. (Fig. 4.1).

## Paramagnetic Materials

This type of magnetic material has unfilled energy levels in which electrons in their outer-most shells are unpaired which gives rise to a net magnetic moment. In these materials, there are no interactions between the magnetic dipole moments and they are randomly oriented due to thermal agitations (Langevin model) [9]. However, under the application of an external magnetic field the magnetic moments tend to align in the direction of the external magnetic field giving rise to a resultant magnetization which results in a small positive value of magnetic susceptibility. The efficiency of an external magnetic field in aligning the magnetic moments is opposed by the randomizing effects of temperature resulting in a temperature-dependent of susceptibility which is known as the Curie law. Some examples of paramagnetic materials are Mg, Li, Mo, Ta, etc. [10] (Fig. 4.2).

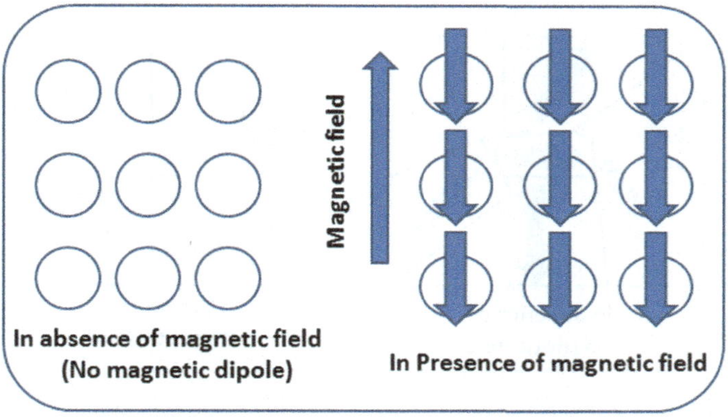


**Fig. 4.1** Orientation of magnetic moments of diamagnetic materials in absence and presence of magnetic field

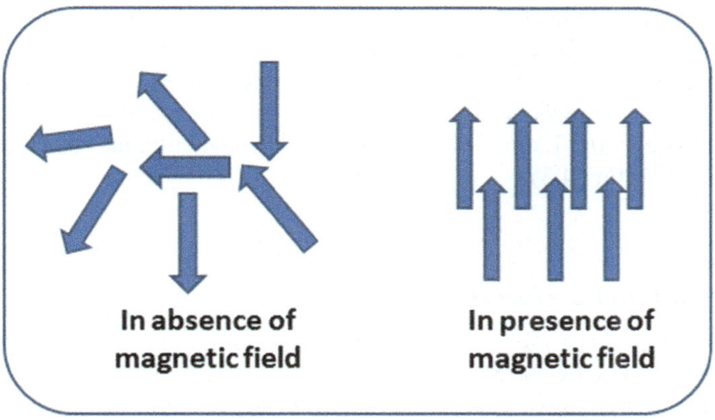


**Fig. 4.2** Orientation of magnetic moments of paramagnetic materials in absence and presence of magnetic field

## *Ferromagnetic Materials*

These types of magnetic materials have the magnetic moments that interact to align parallel to each other. The ferromagnetism in these materials is explained in classical theory postulated by Weiss in 1907 in which it was observed that a molecular field is present within the ferromagnetic material, which is can able to magnetize the material to saturation. In the quantum mechanical point of view, Heisenberg model of ferromagnetism illustrates the presence of exchange interactions between the neighboring individual moments which leads to the parallel alignment of magnetic moments. Weiss discovered the presence of magnetic domains in the ferromagnetic materials that are smaller regions inside the materials in which the individual magnetic

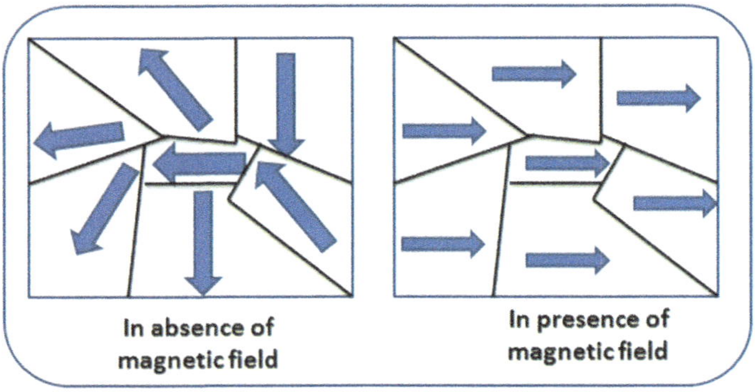


**Fig. 4.3** Orientation of magnetic moments of ferromagnetic materials in absence and presence of magnetic field

moments are aligning into a specific direction. The variation of these magnetic domains under the application of external magnetic field determines the magnetization of the materials. The ferromagnetic properties of these materials are manifested by the Curie temperature ($T_c$) [9]. The Curies temperature is transition temperature below which the materials obey ferromagnetic state and beyond which the materials become paramagnetic. Examples of these type of magnetic materials includes Fe, Co, Ni, etc. [11] (Fig. 4.3).

## *Antiferromagnetic Materials*

This branch of magnetic materials is similar to the ferromagnetic materials, in which the exchange interactions between the magnetic moments of the atoms lead to the antiparallel alignment. These materials are characterized by a transition temperature which is called Neel temperature ($T_N$). Below the Neel temperature the materials possess antiferromagnetic behavior, and above that the materials become paramagnetic. Hematite, chromium, iron manganese (FeMn), nickel oxide (NiO), etc. are some examples of antiferromagnetic materials [11] (Fig. 4.4).

## *Ferrimagnetic Materials*

These types of magnetic materials have complex crystal structure compared to the structure of pure elements. In these materials, some magnetic spins of the atoms are aligned parallel to each other while some of them are aligned antiparallel to each other. In other words, the spins are unequal, resulting a net magnetic moment. The magnetic properties of these materials are similar to that of ferromagnetic material, except the

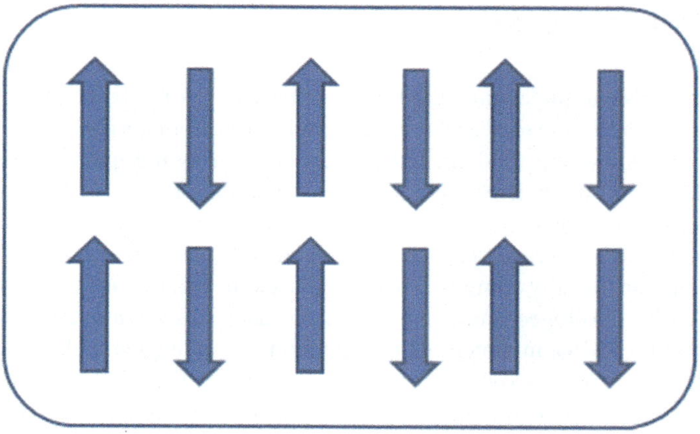

**Fig. 4.4** Orientation of magnetic moments of antiferromagnetic materials

saturation magnetization is smaller than ferromagnetic materials [12]. Ferrimagnetic materials have high resistivity values and highly anisotropic in nature which is very much useful for biomedical applications. Known ferromagnetic materials include Ferrite, magnetic garnets, Al, Co, Ni, Zn, etc. (Fig. 4.5).

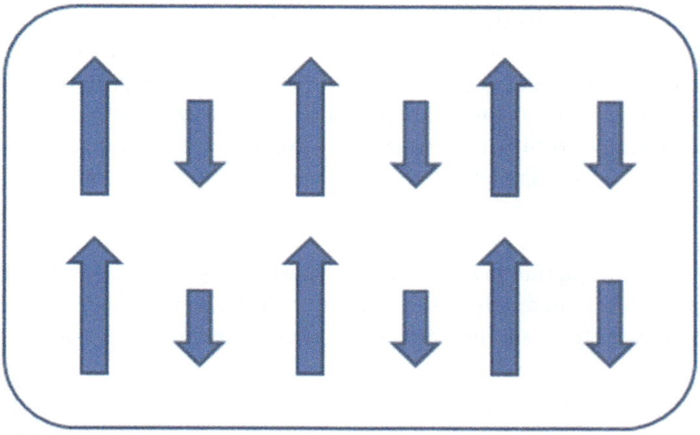

**Fig. 4.5** Orientation of magnetic moments of ferrimagnetic materials

## *Small Particle Magnetism*

It is indispensable to understand the fundamental effects of magnetism of the materials on the matter in the nanoscale range. Similar to nanomaterials, magnetic nanomaterials also show the fundamental properties such as the quantum size effect, surface effect, and the macroscopic quantum tunneling effect [10, 13]. Along with that, properties like superparamagnetism, low Curie temperature, high coercivity, and high magnetic susceptibility can be availed in magnetic nanomaterials [14]. Magnetic nanomaterials are usually composed of magnetic elements such as Fe, Co, Ni, and the Ferrites. A well-developed semiconducting nanomaterial known as diluted magnetic semiconductor (DMS), in which the introduction of the magnetic elements into the semiconductors produces room temperature ferromagnetism. Those nanomaterials utilize the spin polarize nature in order to induce ferromagnetism in the semiconductors, which have technological applications in high-performance spintronic devices [15]. Magnetic nanomaterials offer necessarily high value of saturation magnetization and magnetic susceptibility in addition to the extra anisotropy contributions or shifted loops after field cooling, which enhances the development in biomedical applications [16, 17]. In the nanocrystals, the increased surface to volume ratio of the magnetic nanomaterials consequences the enhancement of surface energy along with the reactivity, which provides a new prospect in the biomedical field. The size of MNPs can also be controlled easily, ranging from few nanometers up to tens of micrometers, comparable to those of cells (10–100 μm), proteins (5–50 nm) and genes (2 nm wide and 10–100 nm long), etc.[18]. As stated, superparamagnetism is an imperative magnetic property of the nanomaterials, which is appeared in the nanoparticles of very small sized. When the size of the nanoparticles is very small, the resultant magnetization randomly flips direction under the effect of temperature. The time between two flips is known as the Néel relaxation time. When the time of measurement of the magnetization of the nanoparticles is much longer compared to Néel relaxation time without the presence of external field, their average magnetization seems to be zero, and they are said to be in superparamagnetic state [19]. Magnetic nanoparticles having superparamagnetic property are considered predominantly as prospective materials in the application of cancer therapy like hyperthermia, due to disappearance of the magnetization after the removal of external field [20] (Fig. 4.6).

Particularly, magnetic NPs are of large interest, having successfully demonstrated their utility in a widespread range of applications, catalysis, mineralogy (such as the selection of ores), informatics (such as data storage), environmental science (such as pollutants concentration), as well as biomedicine due to their attractive properties in physics and chemistry [21–23] In biomedical application, magnetic field based is preferred to electrical field based applications. Since weak interaction of magnetic field with biological molecules allows deep penetration into the body whilst electric fields get strongly attenuated by tissue [24]. Although there were presumptions that small size of magnetic nanomaterial makes them easy to traverse cells, tissues and organelles, it has been recently reported other way [25]. Factually, physico-chemical

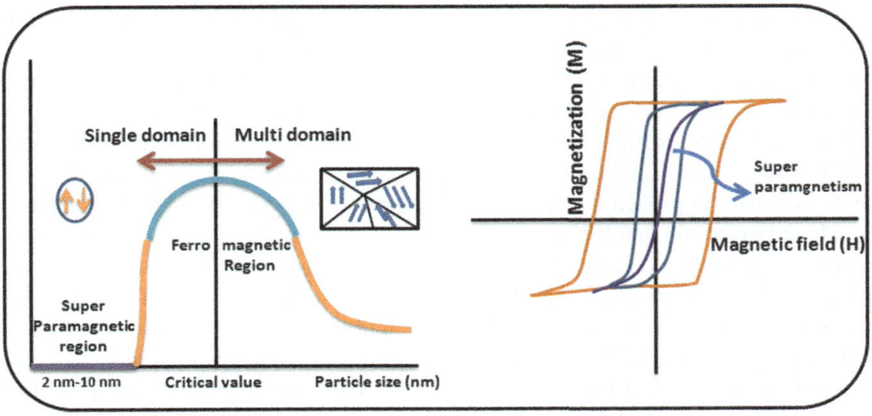


**Fig. 4.6** The qualitative behavior of the size-dependent coercivity of magnetic nanoparticles

properties of nanomaterial plus the identity of the functional molecules added onto their surfaces govern the track of nanomaterial introduced into the body. Thus, clear understanding of their characteristics, such as size uniformity, surface area, adsorption kinetics, biocompatibility, superparamagnetism, and magnetic moment is required to finely tune MNPs during the production process for specific purposes. Moreover, based on requirements magnetic nanomaterials can be modified using polymer organic immobilization, nonpolymer organic immobilization, inorganic molecule immobilization, and targeted ligand modification [26].

## Biomedical Applications

### *Targeted Drug Delivery*

Application of medicine by traditional approach is impaired by low selectivity, poor distribution, limited effectiveness, and released drug interacting with healthy cells or organelles creating side effects. These problems can be terminated by the use of controlled drug delivery [27]. In late 1970, Widder et al. [28] introduced the idea of using magnetic nanomaterials as a carrier of drug to be released on specific targets. This idea proposes that under the application of external magnetic field, magnetic nanomaterial attached to therapeutic agents can be guided to target specific site and drugs be delivered. This method reduces side effects of drugs by targeting specific tissue/organ and also controlled release of drugs prevents underdoing or overdosing problems [27, 29]. Therapeutic nucleic acid molecules, proteins that click with target cells, etc. can be attached to or encapsulated within magnetic nanomaterial core. Improvisation of targeting can be done by tuning magnetic properties of magnetic nanomaterial in use and modifying their surface with high-affinity ligands [30]. The therapeutic agent can be encapsulated to functionalized magnetic material surface [31, 32], linked to magneto-liposome [32], packed in stimuli sensitive hydrogel/polymer configuration. After necessary modification, the magnetic nanomaterial conjugate is implanted or administrated orally into biological system [33, 34]. Under guidance of magnetic field, drug release can be operated internally by click chemistry approach dependent on properties of target cell/organelles, change in pH, and external stimuli like confined increase in temperature [35]. In this application, morphology, charge, surface chemistry are especially important to assure presence of nanoparticles for the required period of time. It has been reported that nanoparticle of size range of 10-100 nm are appropriate for drug delivery application [36]. Nanoparticles below 10 nm size get easily removes by renal clearance and extravasation [37]. Other work suggested, following systematic administration nanoparticle smaller than spleen cut off (200 nm) can permeate into large tumors [38]. To evade retico endothelial system and increase circulation time magnetic nanomaterial can be coated with hydrophilic coating like PEG, monosialoganglioside can, polyvinylpyrrolidone (PVP), etc. [39].

## *Magnetic Bio-separation Technique*

Magnetic bio-separation technique involves extraction of bio-entities from biological environment by attaching them onto the surface of magnetic nanomaterials under application of external magnetic field [40]. Steps of magnetic separation can be completed in a single test tube [41]. Owing to this simplicity of the technique and availability of easy to modify magnetic nanomaterials, it has gained tremendous attention [42]. Magnetic nanomaterials with controlled surface properties and magnetic characteristics, can adsorb the desired product and withdraw it selectively from the solution. As non-toxic nature of $Fe_3O_4$ nanoparticles have been well established, superparamagnetic $Fe_3O_4$ with on and off nature of magnetization is potentially employed in this method [43]. To gain biocompatibility and colloidal stability of magnetic nanomaterial in solution, their surface is generally modified with polymers, organic or inorganic materials with hydrophilic tail [44]. For instance, Khng et al. studied functionalized magnetic particles attached to the carboxylic acid group, Jeong et al. [43] separated red blood cells, lung cancer cells, breast cancer cells and bacteria were using immune specific MNPs [43]. Recently, inorganic coating like silica has gain popularity among researchers due to its water solubility, photostability, and colloidal stability [45, 46]. Siloxy group present at one terminal and free end in another of silica-coated magnetic nanomaterials enable better attachment of biocompatible group. There are various specific affinity ligands that can be attached onto nanomaterials to bind with protein complex. For example, dendrimers [47], peptides content in thiol for gold nanoparticles [48], quantum dots to replace mercaptoacetic acid on the nanoparticle surface [49]. Binding of DNA molecules with MNPs follows the same mechanism as with protein complexes.

## *Magnetic Hyperthermia*

Human death related to cancer reached 9.6 million in 2018 worldwide [50]. Surgery to remove tumor, radiation therapy, chemotherapy, combination of them are conventional choice of treatment currently available, with so many adverse side effects. Research is being carried out widely to develop nanomaterial-based techniques like gene therapy, photo-thermal therapy, photodynamic therapy, immunotherapy, magnetic hyperthermia (MH), etc. Compared to healthy cell cancer cells are more thermosensitive in nature, because of its high glycolic activity and low pH [51]; heat therapy has been traditionally used to destroy or terminate growth of cancer cells. Heat therapy is applied in two ways: one is hyperthermia, in here local or whole body temperature is increased up to 41–46 °C by microwaves, radiofrequency radiation or ultrasounds; [52] another is thermal ablation, here, temperature of specific diseased area is increased above 45 °C to destroy the tissue [53]. In magnetic hyperthermia, properties of magnetic nanoparticles (MNPs) are being used to convert electromagnetic energy (applied alternating magnetic field) to thermal energy, first reported by Gilchrist et al. [54]. Extensive studies of ferrites ($MFe_2O_4$ where M=Fe, Mn, Co) in magnetic hyperthermia can be found in the literature for their chemical stability, magnetic tunability, biocompatibility [18, 55]. The first successful clinical trial of interstitial hyperthermia in the treatment of human cancer using magnetic nanoparticles was attempted by Johannsen et al. [6] in 2005. Heat generation mechanism in MNPs is dominated by hysteresis losses, usually dependent on particle size. To reduce magneto-crystalline energy, above a certain diameter magnetic multi-domains are formed, these NPs are characterized by ferromagnetic behavior. In ferromagnetic MNPs with open hysteresis loop, energy dissipation under applied alternating field is dominated by magnetization reversal. While in the size range of single to multi-domain region, heat generation is dominated by relaxation mechanisms, namely, Brownian relaxation and Neel relaxation. Neel relaxation loss takes place due to orientation of magnetic moment of particle with applied alternating magnetic field and Brownian relaxation loss occurs when there is rotation of MNPs, thus creating friction between MNP surface and the solvent under applied alternating magnetic field [56–58]. Although Brownian and Neel relaxation takes place simultaneously, heat generation is dominated by the one having shorter characteristic time. In multi-domain magnetic nanomaterial, domain wall motion generating heat per unit volume per unit time is given by

$$P = \mu f \int H \mathrm{d}M \tag{4.1}$$

where $f$ is the frequency of the applied field, $H$ the field strength and $M$ is the magnetization [9]. The single-domain nanoparticle possesses a high barrier of magnetization reversal along the easy axis resulting in large coercivities and wide hysteresis loop. It is known that hysteresis loss from a single domain particle is higher than any other losses. While for single domain superparamagnetic nanoparticles, linear response theory explains their substantiation of heating by AC magnetic field [59]. Power dissipation according to this formulation is given by

$$P_1 = \mu_0 \pi \chi'' f H^2 \tag{4.2}$$

where $H$ and $f$ are the amplitude and frequency of the AC magnetic field, $\mu_0$ the permeability of free space, $\chi''$ denotes the out of phase component of the colloidal magnetic fluid AC susceptibility which is again expressed as

$$\chi'' = \frac{\omega\tau}{1 + (\omega\tau)^2} \chi_0 \tag{4.3}$$

And actual susceptibility,

$$\chi_0 = \frac{\mu_0 M_s^2 V}{K_B T} \tag{4.4}$$

where $\omega = 2\pi f$, $\tau$ is the effective relaxation time. The characteristic Neel relaxation time of nanoparticles is given by

$$\tau_N = \tau_0 \exp\left(\frac{K_{\text{eff}} V}{K_B T}\right), \quad \tau_0 \approx 10^{-9}\text{s} \tag{4.5}$$

Here the magnetic energy is converted to thermal energy by rapid oscillation of magnetic moments across anisotropy energy barrier ($K_{\text{eff}} V$). While Brownian relaxation infers physical rotation of MNPs, relaxation time here is dependent on viscosity of the medium. Its corresponding rotational diffusion time is given by

$$\tau_B = \frac{3\eta V_H}{K_B T} \tag{4.6}$$

where $\eta$ is the dynamic viscosity of the medium and is the hydrodynamic diameter of the suspended MNPs. As stated earlier, one with shorter relaxation time dominates heat generation mechanism, the effective relaxation time is given by Shilomis' equation [59]:

$$\tau_{\text{eff}} = \frac{\tau_B \tau_N}{\tau_B + \tau_N} \tag{4.7}$$

To measure and quantify capability of MNPs' heat generation, term Specific Absorption Rate (SAR) has been defined as the amount of heat generated per unit mass of MNPs. Since most of the in vitro setup used is non-adiabatic in nature, SAR is estimated from initial rate of rise in temperature as alternating field is applied to considered nanoparticle system using suitable immune probe like liquid thermometer, fiber optic temperature sensor, infrared thermography [60, 61]. Popularly, SAR of MNPs in polar solvent under non-adiabatic conditions is being estimated which is far from being analogous with in vivo condition. To shed light on relative effect of various physiological factors Chen et al. [62] studied blood sugar (dextrose), viscosity (glycerol), protein (bovine serum albumin, BSA), electrolytes (commercial norm-saline) interacting with MNPs as alternating field is applied. They have found that SAR value reduces significantly in electrolyte environment, due to precipitation and agglomeration with ions of sodium whilst SAR value remains unaffected by blood sugar and protein environment. Till date, most successful result of Magnetic hyperthermia has been reported by Clinic of Radiation Therapy in Berlin, of increased life expectancy to about 8 months of a patient diagnosed with repeated glioblastoma, a form of highly malignant brain tumor [63].

## *MRI Contrast Enhancements*

Magnetic resonance imaging (MRI) invented in the early 1970s [64] deals with computer-assisted imaging of human internal organs with high temporal and spatial resolution. The working principle is based on relaxation properties of proton spins present in the human body under a strong magnetic field excited by radiofrequency (RF) signals. To better differentiate between normal tissue and lesions, which often have small relaxation time differences, MRI contrast agents are helpful. Contrast agents induce difference in signal between adjacent regions, which could be, 'tissue and vessel', 'tissue and tissue' and 'tissue and bone', because of their interaction with neighboring water protons. The objective was to develop an in vivo stable material having influence on proton relaxation properties and non-toxic for the given dose in producing contrast enhancement. Nuclei of protons under the application of strong magnetic field aligns along or against it with a precessing frequency called Larmor frequency. Thus induced net magnetization $M$, has two components of $M_1$ (longitudinal component) and $M_2$ (transverse component), which generates interconnection of spins. When a 90° RF signal in resonating range of Larmor frequency is applied, proton spins gets excited to antiparallel state, thus, $M_1$ falls and $M_2$ grows. Again upon the removal of RF signal, excited nuclei relax to lower energy state; relaxation of $M_1$ to initial state is called $T_1$ relaxation and that of decay of $M_2$ is called $T_2$ relaxation. Due to transfer in energy between spin and lattice, there is change in $M_1$ while change in $M_2$ is due to dephasing of spins that is loss of phase coherence. Inhomogeneous magnetic field in the vicinity, produced by external or internal tissue-inherent forces, can affect transverse relaxation process, thus $T_2$* is described as total relaxation time is given by [18]:

$$\frac{1}{T_2^*} = \frac{1}{T_2} + \gamma B_o \tag{4.8}$$

where $\gamma B_o$ denotes the relaxation by the field inhomogeneity. Contrast agents basically shortens these relaxation time, based on its impact more pronounced on either $T_1$ or $\frac{T_2}{T_2^*}$, they are classified as $T_1$ or $T_2$ contrast agent. Commercially, to enhance

$T_1$ weighted image Gadolinium-based contrast agents are in use [65]. To intensify contrast effect of $T_2$ weighted image, superparamagnetic (SPM) Iron oxide nanoparticles are applicable [18]. $Gd^{3+}$ ion having seven unpaired electrons possess strong hydrogen-proton spin-lattice relaxation effect [65]. Albeit, for vivo applications, detoxification of Gd ion needs to be done, for example by Gd ion complexes with stable chelates like diethylenetnaminepentaacetic acid (DTPA) and ethylenediaminetetraacetic acid (EDTA) [66]. The capability of nanoparticles to have different shape and biodistribution gives the opportunity to go beyond conventional imaging by chemical agents. At certain range of magnetic field, SPM particles used as contrast agent are saturated magnetically, thus producing a local magnetic field that perturbs the dipolar field leading to shortening of $T_2$ relaxation time [67].

### *Theragnosis*

Theragnosis is a relatively new concept of all biomedical applications discussed here. It infers combination of detection and treatment of disease simultaneously [68]. Researcher in this field aims at a very promising future of nanomedicine as multi-functionality of MNPs continues to grow. Santiago Sánchez-Cabezas et al. [67] reported ultra-small superparamagnetic iron oxide nanoparticle to be excellent theranostic material with efficient heat generation for hyperthermia and also enhanced $T_1/T_2$ contrast magnetic resonance imaging. Several factors are needed to be optimized, for instance, in large magnetic moment in large iron oxide nanoparticles shows increase its heating efficiency and enhanced $T_2$ signal. In contrast, shortening of $T_1$ relaxation time is favored for small iron oxide nanoparticle [69]. In Fig. 4.7, probable methodology simultaneous drug delivered to cure and detect along with Hyperthermic effect has been depicted.

## Challenges

Use of nanomaterial in a wide range of application is getting exhaustively popular. Till date, 12 nanomedicines were approved by Food and Drug Administration (FDA) and lot more are in last development stage [70]. Toxicity is one of the most important factors that needs to be well addressed in order to rely on nanomaterial applications. Since toxicity depends on various factors like dose, surface chemistry, chemical composition, solubility, bio-distribution, etc. it has been difficult reporting toxicity accurately. Both in vitro and in vivo cytotoxicity must be conducted extensively with various factors understudy to understand toxicity of nanoparticles [71]. Biomedical application of MNPs always involves external magnetic force mostly using permanent magnet to target specific site and control release [72]. As permanent magnet can only penetrate into 8–12 cm depth of tissue [73, 74], this method is not applicable for

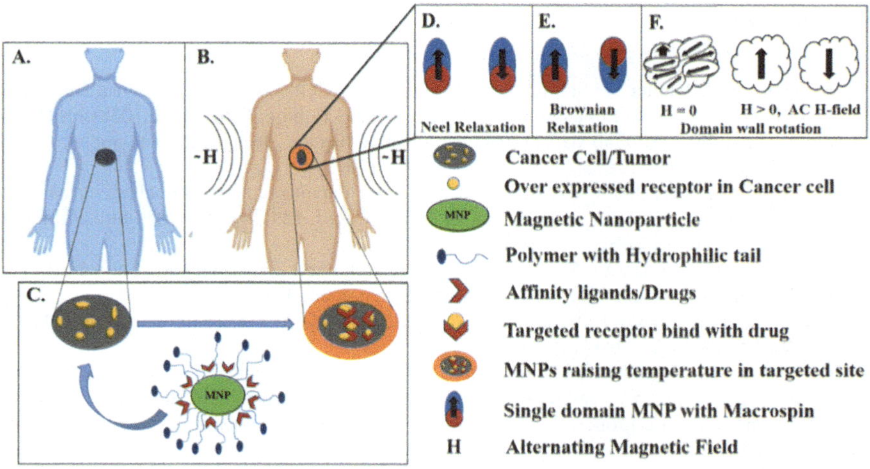


**Fig. 4.7** **a** Tumorous bioenvironment. **b** Targeted application of modified MNP under alternating magnetic field H. **c** MNPs encapsulated with affinity ligand trapped in suitable polymer targeting tumorous cancer cells. **d** Heat generation by Neel relaxation. **e** Brownian relaxation. **f** By domain wall rotation

deep-seated tumors. Modality involving magnetic implants are under development to address these issues [75].

## Conclusion

In this chapter, discussions about fundamental approaches of magnetic nanomaterial to be used in biomedical applications assert that magnetic nanomedicine is promising agents to provide substitution/adjunct to other conventional treatment modalities. Realization of the basic understanding is vital for development of better performing MNPs. Following the recent trend in development of nanotechnology, multimodal therapy that is simultaneous diagnosis, control release of drug and monitoring the interaction process is a far-sighted, yet, an achievable future. Although, only magnetic bio-separation via protein and cell labeling and SPM nanoparticles as MRI contrast enhancement are distributed by market-based technologies. Drug delivery, Magnetic hyperthermia via modified MNPs are at bottleneck to be available in market as present technology is yet to take the leap from laboratory under control to service in use by patients.

## References


1. Stephenson, C., Hubler, A.: Stability and Conductivity of Self Assembled Wires in a Transverse Electric Field. National Publishing Group, pp. 1–9 (2015). https://doi.org/10.1038/srep15044
2. Lyon, D., Hubler, A.: Gap size dependence of the dielectric strength in nano vacuum gaps. IEEE Trans. Dielectr. Electr. Insul. **20,** 1467–1471 (2013)
3. Mathew, D.S., Juang, R.: An overview of the structure and magnetism of spinel ferrite nanoparticles and their synthesis in microemulsions. Chem. Eng. J. **129,** 51–65 (2007). https://doi.org/10.1016/j.cej.2006.11.001
4. Chaudhury, K., Kumar, V., Kandasamy, J., RoyChoudhury, S.: Regenerative nanomedicine: current perspectives and future directions. Int. J. Nanomed. **9**, 4153–4167 (2014)
5. Giustini, A.J., Hoopes, P.J.: Magnetic nanoparticle hyperthermia in cancer treatment, **1**, 17–32 (2010). 10.1142/S1793984410000067
6. Filippousi, M., Angelakeris, M., Katsikini, M., Paloura, E.C., Wang, Y., Zamboulis, D., Van Tendeloo, G., Wang, Y., Zamboulis, D., Van Tendeloo, G.: Surfactant effects on the structural and on the magnetic properties of iron oxide nanoparticles nanoparticles . J. Phys. Chem. (2014). https://doi.org/10.1021/jp5037266
7. Wang, Y.J., Xuan, S., Port, M., Idee, J.: Recent advances in superparamagnetic iron oxide nanoparticles for cellular imaging and targeted therapy research, 6575–6593 (2013)
8. Henner, V.K., Yukalov, V.I., Kharebov, P.V., Yukalova, E.P.: Collective spin dynamics in magnetic nanomaterials. J. Phys. Conf. Ser. **129** (2008). https://doi.org/10.1088/1742-6596/129/1/012015.
9. O'Handley, R.C.: Modern Magnetic Materials. Wiley, New York (2000)
10. Taylor, P.: Annals of science understanding macroscopic quantum phenomena: the history of superfluidity 1941–1955 understanding macroscopic quantum phenomena: the history of superfluidity 1941–1955, 37–41 (1955). https://doi.org/10.1080/00033798800200291
11. Neel, L.: Antiferromagnetism and ferrimagnetism. Proc. Phys. Soc. **65** (1952)
12. Cullity, B.D., Graham, C.D.: Introduction to Magnetic Materials. Wiley, New York (2009)
13. Jaeger, G.: Macroscopic realism and quantum measurement : measurers as a natural kind . Phys. Scripta (2014). https://doi.org/10.1088/0031-8949/2014/T163/014017
14. Frenkel, J., J., Doefman, ©, : Spontaneous and Induced Magnetisation in Ferromagnetic Bodies. Nat. Publ. Group Nat. **126**(1930), 274–275 (1930)
15. Saikia, D., Borah, J.P.: Ferromagnetic ordering inchemically synthesized ZnS: Mn diluted magnetic semiconductor: a density functional theory explanation. Phys. Lett. A. **1**, 2–5 (2017). https://doi.org/10.1016/j.physleta.2017.09.018
16. Grasset, F., Duguet, E.: Magnetic Nanoparticle Design for Medical Diagnosis and Therapy, pp. 2161–2175, (2004)
17. Mosayebi, J., Kiyasatfar, M., Laurent, S.: Synthesis, Functionalization , and Design of Magnetic Nanoparticles for Theranostic Applications (2017). https://doi.org/10.1002/adhm.201700306.
18. Pankhurst, Q.A., Connolly, J., Jones, S.K., Dobson, J.: 11Applications of magnetic nanoparticles in biomedicine. J. Phys. D Appl. Phys. **36**, 167–181 (2003). https://doi.org/10.1088/0022-3727/36/13/201
19. Blanco-andujar, C.: Design of iron oxide-based nanoparticles for MRI and magnetic hyperthermia (2016)
20. Torres, T.E., Mayoral, E.L. Jr., Ibarra, A., Marquina, C., Ibarra, M.R., Goya, G.F., Torres, T.E., Lima, E., Mayoral, A., Ibarra, A., Marquina, C., Ibarra, M.R.: el-Arrhenius model for highly anisotropic $CoxFe_{32}$ x $O_4$ Validity of the N e nanoparticles, 183902 (2015). https://doi.org/10.1063/1.4935146
21. Martins, P., Silva, M.: Determination of the Magnetostrictive Response of Nanoparticles Via Magnetoelectric Measurements, 9457–9461 (2015). https://doi.org/10.1039/c5nr01397f
22. Liu, X., Liu, J., Zhang, S., Nan, Z., Shi, Q.: Structural, Magnetic, and Thermodynamic Evolutions of Zn-Doped $Fe_3O_4$ Nanoparticles Synthesized Using a One-Step Solvothermal Method (2016). https://doi.org/10.1021/acs.jpcc.5b10618

23. Lu, A., Salabas, E.L., Schüth, F.: Magnetic Nanoparticles: Synthesis , Protection, Functionalization , and Application Angewandte, 1222–1244 (2007). https://doi.org/10.1002/anie.200602866
24. Batlle, X.: Am´ılcar Labarta, Finite-size effects in fine particles : magnetic and transport properties. J. Phys. D Appl. Phys. **35**, R15–R42 (2002)
25. Johannsen, M., Gneveckow, U., Eckelt, L., Feussner, A., Waldöfner, N., Scholz, R., Deger, S., Wust, P., Loening, S.A., Jordan, A.: Clinical hyperthermia of prostate cancer using magnetic nanoparticles: Presentation of a new interstitial technique. Int. J. Hyperth. **21**, 637–647 (2005). https://doi.org/10.1080/02656730500158360
26. Chosy, E.J., Nakamura, M., Melnik, K., Comella, K., Lasky, L.C., Zborowski, M., Chalmers, J.J.: Characterization of antibody binding to three cancer-related antigens using flow cytometry and cell tracking velocimetry . Biotech. Bioeng. (2003). https://doi.org/10.1002/bit.10581
27. Assa, F., Jafarizadeh-malmiri, H., Ajamein, H., Vaghari, H., Anarjan, N., Ahmadi, O., Berenjian, A.: Critical Reviews in Biotechnology Chitosan magnetic Nanoparticles for Drug Delivery Systems, 8551 (2016). https://doi.org/10.1080/07388551.2016.1185389.
28. Maeda, M., Kuroda, C.S., Shimura, T., Tada, M., Abe, M., Yamamuro, S., Sumiyama, K., Handa, H., Maeda, M., Kuroda, C.S., Shimura, T., Tada, M., Abe, M.: Magnetic carriers of iron nanoparticles coated with a functional polymer for high throughput bioscreening magnetic carriers of iron nanoparticles coated with a functional polymer for high throughput bioscreening. J. Appl. Phys. **103**, 97–100 (2010). https://doi.org/10.1063/1.2165127
29. Cregg, P.J., Murphy, K., Mardinoglu, A.: Inclusion of interactions in mathematical modelling of implant assisted magnetic drug targeting. Appl. Math. Model. **36**, 1–34 (2012). https://doi.org/10.1016/j.apm.2011.05.036
30. Mou, X., Ali, Z., Li, S., He, N.: Applications of Magnetic Nanoparticles in Targeted Drug Delivery System, 54–62 (2015). https://doi.org/10.1166/jnn.2015.9585
31. Hola, K., Markova, Z., Zoppellaro, G., Tucek, J., Zboril, R.: Tailored functionalization of iron oxide nanoparticles for MRI, drug delivery, magnetic separation and immobilization of biosubstances. Biotechnol. Adv. (2015). https://doi.org/10.1016/j.biotechadv.2015.02.003
32. Tai, L., Tsai, P., Wang, Y., Wang, Y.: Thermosensitive liposomes entrapping iron oxide nanoparticles for controllable, 135101 (2009). https://doi.org/10.1088/0957-4484/20/13/135101
33. Nemati, Z., Salili, S.M., Alonso, J., Ataie, A., Das, R., Phan, M.H., Srikanth, H.: Superparamagnetic iron oxide nanodiscs for hyperthermia therapy: does size matter? J. Alloys Compd. **714**, 709–714 (2017). https://doi.org/10.1016/j.jallcom.2017.04.211
34. Mcbain, S.C.: Magnetic nanoparticles for gene and drug delivery. Int. J. Nanomed. **3**, 169–180 (2008)
35. Arriortua, A.O.K., Insausti, M., Lezama, L., De Muro, I.G., Garaio, E., Mart, J., Fratila, R.M., Morales, M.P., Eceiza, M., Sagartzazu-aizpurua, M., Jesus, M.: RGD-Functionalized $Fe_3O_4$ nanoparticles for magnetic hyperthermia. Colloids Surf. B Biointerfaces (2018). https://doi.org/10.1016/j.colsurfb.2018.02.031
36. Davis, M.E., Chen, Z.G., Shin, D.M.: Nanoparticle therapeutics: an emerging treatment modality for cancer. Coll. Rev. Nat. J. **7**, 771–782 (2008). https://doi.org/10.1038/nrd2614
37. Venturoli, D., Rippe, B.: Ficoll and dextran vs. globular proteins as probes for testing glomerular permselectivity: effects of molecular size, shape, charge , and deformability, 605–613 (2005). https://doi.org/10.1152/ajprenal.00171.2004
38. Hu-lieskovan, S., Heidel, J.D., Bartlett, D.W., Davis, M.E., Triche, T.J.: Sequence-Specific Knockdown of EWS-FLI1 by Targeted, Nonviral Delivery of Small Interfering RNA Inhibits Tumor Growth in a Murine Model of Metastatic Ewing ' s Sarcoma, 8984–8993 (2005). https://doi.org/10.1158/0008-5472.CAN-05-0565
39. Allen, T.M., Hansen, C.: Pharmacokinetics of stealth versus conventional liposomes: effect of dose. Biochem. Biophys. Acta Biomembr. **1068**, 133–141 (1991)
40. Wu, A., Ou, P., Zeng, L.: Biomedical applications of magnetic nanoparticles **5**, 245–270 (2010). https://doi.org/10.1142/S1793292010002165
41. Safarik, I., Safarikova, M.: Magnetic techniques for the isolation and purification of proteins and peptides. BioMagn. Res. Technol. **17**, 1–18 (2004). https://doi.org/10.1186/1477-044X-2-7

42. Nguyen, D.T., Kim, K.: Functionalization of magnetic nanoparticles for biomedical applications. Kor. J. Chem. Eng. **31**, 1289–1305 (2014). https://doi.org/10.1007/s11814-014-0156-6
43. Jeong, B.U., Teng, X., Wang, Y., Yang, H., Xia, Y.: Superparamagnetic Colloids: Controlled Synthesis and Niche Applications, 33–60 (2007). https://doi.org/10.1002/adma.200600674
44. Fatima, H., Kim, K.: Magnetic nanoparticles for bioseparation. Kor. J. Chem. Eng. **32**, 1–11 (2016). https://doi.org/10.1007/s11814-016-0349-2
45. Erathodiyil, N., Ying, J.Y.: Bioimaging Applications **44**, 925–935 (2011). https://doi.org/10.1021/ar2000327
46. Chen, Y., Yin, Q., Ji, X., Zhang, S., Chen, H., Zheng, Y., Sun, Y., Qu, H., Wang, Z., Li, Y., Wang, X., Zhang, K., Zhang, L., Shi, J.: Biomaterials Manganese oxide-based multifunctionalized mesoporous silica nanoparticles for pH-responsive MRI, ultrasonography and circumvention of MDR in cancer cells. Biomaterials **33**, 7126–7137 (2012). https://doi.org/10.1016/j.biomaterials.2012.06.059
47. Wang, Y.A., Li, J.J., Chen, H., Peng, X.: Stabilization of inorganic nanocrystals by organic dendrons. J. Am. Chem. Soc. **124**, 2293–2298 (2002)
48. Hussain, I., Nichols, R.J., Schiffrin, D.J., Brust, M., Fernig, D.G.: Rational and Combinatorial Design of Peptide Capping Ligands for Gold Nanoparticles, 10076–10084 (2004)
49. Gao, X., Chan, W.C.W.: Quantum-dot nanocrystals for ultrasensitive biological labeling and multicolor optical encoding. J. Biomed. Opt. **7**, 532–537 (2002). https://doi.org/10.1117/1.1506706
50. Bray, F., Ferlay, J., Soerjomataram, I., Siegel, R.L., Torre, L.A., Jemal, A., Global cancer statistics, : GLOBOCAN estimates of incidence and mortality worldwide for 36 cancers in 185 countries. CA. Cancer J. Clin. **68**(2018), 394–424 (2018). https://doi.org/10.3322/caac.21492
51. Koltai, T. : Cancer: fundamentals behind pH targeting and the double-edged approach, 6343–6360 (2016)
52. Habash, R.W.Y., Bansal, R., Krewski, D., Alhafid, H.T.: Thermal Therapy, Part 2: Hyperthermia Techniques, vol. 34, pp. 491–542 (2006)
53. Habash, R.W.Y., Bansal, R., Krewski, D., Alhafid, H.T.: Thermal Therapy, Part 1: An Introduction to Thermal Therapy, vol. 34, 459–489 (2006)
54. Gilchrist, R.K., Medal, W.D., Shorey, R.C., Hanselman, J.C., Parrott., Taylor, C.B: Selective inductive heating of lymph nodes. Ann. Surg. **146**, 596–606 (1957). https://www.ncbi.nlm.nih.gov/pubmed/13470751%0Ahttps://www.pubmedcentral.nih.gov/articlerender.fcgi?artid=PMC1450524
55. Mazarío, E., Sánchez-Marcos, J., Menéndez, N., Cañete, M., Mayoral, A., Rivera-Fernández, S., De La Fuente, J.M., Herrasti, P.: High specific absorption rate and transverse relaxivity effects in manganese ferrite nanoparticles obtained by an electrochemical route. J. Phys. Chem. C. **119** (2015). https://doi.org/10.1021/jp510937r
56. Kumar, C.S.S.R., Mohammad, F.: Magnetic nanomaterials for hyperthermia-based therapy and controlled drug delivery. Adv. Drug Deliv. Rev. **63**, 789–808 (2011). https://doi.org/10.1016/j.addr.2011.03.008
57. Hergt, R., Dutz, S., Robert, M., Zeisberger, M.: Magnetic particle hyperthermia: nanoparticle magnetism and materials development for cancer therapy, 2919 (2006). https://doi.org/10.1088/0953-8984/18/38/S26
58. Hergt, R., Andr, W., Ambly, C.G., Hilger, I., Kaiser, W.A., Richter, U., Schmidt, H.: Physical limits of hyperthermia using magnetite fine particles. IEEE Trans Magn. **34**, 3745–3754 (1998)
59. Rosensweig, R.E.: Heating magnetic fluid with alternating magnetic field. J. Magn. Magn. Mater. **252**, 370–374 (2002)
60. Lahiri, B.B., Ranoo, S., Zaibudeen, A.W., Philip, J.: Magnetic hyperthermia in magnetic nanoemulsions: Effects of polydispersity, particle concentration and medium viscosity. J. Magn. Magn. Mater. **441**, 310–327 (2017). https://doi.org/10.1016/j.jmmm.2017.05.076
61. Lahiri, B.B., Muthukumaran, T., Philip, J.: Magnetic hyperthermia in phosphate coated iron oxide nanofluids. J. Magn. Magn. Mater. **407**, 101–113 (2016). https://doi.org/10.1016/j.jmmm.2016.01.044

62. Hsieh, S., Huang, B.Y., Hsieh, S.L., Wu, C.C., Wu, C.H., Lin, P.Y., Huang, Y.S., Chang, C.W.: Green fabrication of agar-conjugated $Fe_3O_4$ magnetic nanoparticles, 445601 (2010). https://doi.org/10.1088/0957-4484/21/44/445601
63. Jordan, A.: Hyperthermia classic commentary: Inductive heating of ferrimagnetic particles and magnetic fluids: physical evaluation of their potential for hyperthermia ' by Andreas Jordan et al. Int J **25**, 512–516 (2009). https://doi.org/10.3109/02656730903183445
64. Lauterbir, P.C.: © 1973 Nature Publishing Group Nature 242, 190 (1973)
65. Hanns-Joachim Weinmann, R.C.B., Press, W.R., Wesbey, G.E.: Characteristics of complex. Ame. J. Roentgenol. **142**, 619 (1984). doi:https://doi.org/10.2214/ajr.142.3.619
66. Hirano, S., Suzuki, K.T.: Exposure metabolism, and toxicity of rare earths and related compounds Environ. Health Perspect. **104**, 85–95 1996
67. Santiago Sánchez-Cabezas, R.M.-M., Roberto, M.-R., Juan, G, Félix, S, Dalton.: Trans. R. Soc. Chem. **48**, 3883–3892 (2019). doi:https://doi.org/10.1039/C8DT04685A
68. Cardoso, V.F., Francesko, A., Ribeiro, C., Bañobre-lópez, M., Martins, P., Lanceros-mendez, S.: Advances in magnetic nanoparticles for biomedical applications. Adv. Healthc. Mater. **1700845**, 1–35 (2017). https://doi.org/10.1002/adhm.201700845
69. Lee, N., Hyeon, T.: Nanomedicine Themed Issue **41** 2575–2589 (2012). 10.1039/c1cs15248c
70. Jones, C.F., Grainger, D.W.: In vitro assessments of nanomaterial toxicity ☆. Adv. Drug Deliv. Rev. **61**, 438–456 (2009). https://doi.org/10.1016/j.addr.2009.03.005
71. Liu, X.F.E.I., Guan, Y.U.N.L.I.N., Yang, D.Z.H.I, Li, Z.H.I., K.D.E. Yao, Antibacterial Action of Chitosan and Carboxymethylated, 1324–1335 (2000)
72. Kettering, M., Winter, J., Zeisberger, M., Bremer-Streck, S., Oehring, H., Bergemann, C., Alexiou, C., Hergt, R., Halbhuber, K.J., Kaiser, W.A., Hilger, I.: Magnetic nanoparticles as bimodal tools in magnetically induced labelling and magnetic heating of tumour cells: an in vitro study. Nanotechnology **18**, 9 (2007). 10.1088/0957-4484/18/17/175101
73. Goodwin, S., Peterson, C., Hoh, C., Bittner, C.: Targeting and retention of magnetic targeted carriers (MTCs ) enhancing intra-arterial chemotherapy, 194, 132–139 (1999)
74. Neuberger, T., Scho, B., Hofmann, M., Von Rechenberg, B.: Superparamagnetic nanoparticles for biomedical applications: possibilities and limitations of a new drug delivery system . J. Magn. **293**, 483–496 (2005). https://doi.org/10.1016/j.jmmm.2005.01.064
75. Price, P.M., Mahmoud, W.E., Al-ghamdi, A.A., Bronstein, L.M.: Magnetic drug delivery: where the field is going. Front. Chem. **6**, 1–7 (2018). https://doi.org/10.3389/fchem.2018.00619

# Chapter 5
# Iron Oxide/Reduced Graphene Oxide Composites for the Sensing of Toxic Chemicals

**Rabina Bhujel, Sadhna Rai, Utpal Deka, Joydeep Biswas, and Bibhu Prasad Swain**

## Introduction

The requirement of a fresh and healthy environment is supreme for all living beings, and at present, its pollution has become a big issue throughout the world. In the near coming future, this problem will take off to an alarming level due to the rapid industrialization and rising population. The pollutants like $NH_3$, $SO_2$, $NO_2$, CO, $H_2O_2$, Pb, As, Cd, Hg, etc. are recognized as the toxic and hazardous chemicals for causing environmental pollution as well as are the cause of many life-threatening diseases. Therefore, the detection of these toxic chemicals becomes extremely important to make a pollution-free environment. In this context, it is crucial and important to develop a cost-effective, reliable and ultrasensitive device for the sensing of toxic chemicals, which can mainly work at room temperature. Metal oxide semiconductors (MOSs) are one of them which are considered as a useful material for sensing, owing to their superior electrical and electrochemical properties. Among the various

R. Bhujel (✉) · S. Rai
Centre for Materials Science and Nanotechnology, Sikkim Manipal Institute of Technology, Sikkim Manipal University, Majhitar, Rangpo 737136, India
e-mail: rabina29@gmail.com

U. Deka
Department of Physics, Sikkim Manipal Institute of Technology, Sikkim Manipal University, Majhitar, Rangpo 737136, India

J. Biswas
Department of Chemistry, Sikkim Manipal Institute of Technology, Sikkim Manipal University, Majhitar, Rangpo 737136, India

B. P. Swain
Department of Physics, National Institute of Technology Manipur, Langol, Imphal 795004, India
e-mail: bibhuprasad.swain@gmail.com



B. P. Swain (ed.), *Nanostructured Materials and their Applications*, Materials Horizons: From Nature to Nanomaterials, https://doi.org/10.1007/978-981-15-8307-0_5

metal oxides, iron oxides are one of them which claim to be a very cheap abundant metal oxide that can be used as the main material for sensors.

## *Iron Oxide and Its Properties*

Iron oxides are the very important inorganic materials which are mainly consisting of iron (Fe) and oxygen. Iron oxides are abundant, cheap and also environment-friendly. Depending upon the oxidation state and structure, iron oxides are of three types (i) hematite ($\alpha$-$Fe_2O_3$), (ii) maghemite ($\gamma$-$Fe_2O_3$) and (iii) magnetite ($Fe_3O_4$).

### Hematite ($\alpha$-$Fe_2O_3$)

Hematite is an *n*-type semiconductor having a bandgap of 2.3 eV in ambient conditions which is black and shiny when in a crystal form and has a red color when it is finely powdered. It is one of the oldest known oxides of iron which is mainly found in rocks and soils. Due to the small bandgap and extremely good electrical properties, these oxides have widespread applications in the fields of pigments [1], supercapacitors [2], various electrical devices [3] and also in sensors [4]. Hematite crystal has a hexagonal packed structure where the $Fe^{3+}$ ions are in octahedral coordination with the oxygen atoms. The Fe atoms of hematite are in a trivalent state ($Fe^{3+}$); therefore, each oxygen atom is allowed to make a bond with only two Fe ions; hence, only two out of three existing oxygen octahedrons are occupied. The structure of hematite is shown in Fig. 5.1.

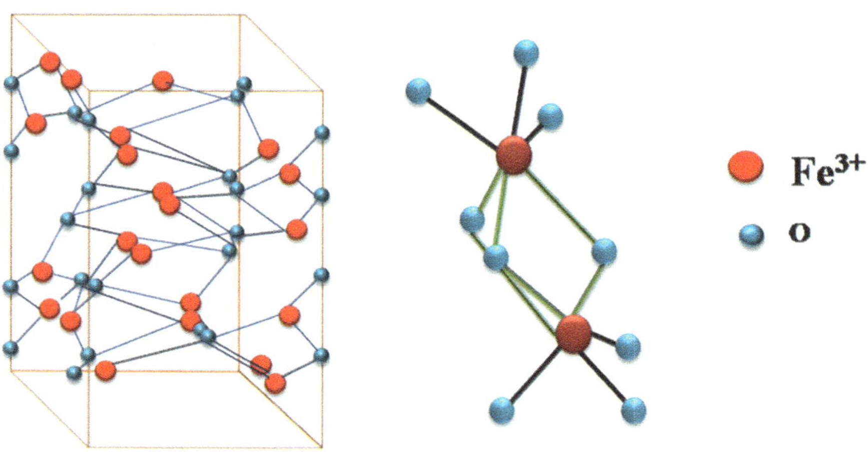


**Fig. 5.1** Structure of hematite ($\alpha$-$Fe_2O_3$)

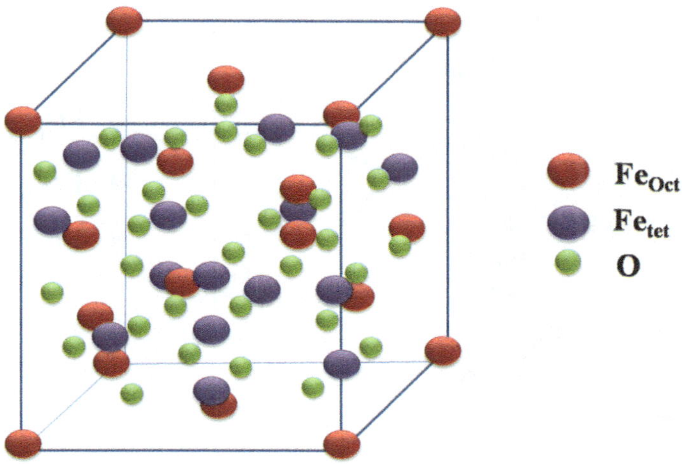


**Fig. 5.2** Structure of maghemite ($\gamma$-$Fe_2O_3$)

### Maghemite ($\gamma$-$Fe_2O_3$)

Maghemite is the second most stable form of iron oxide after hematite which is also found naturally in soils. Maghemite is also an n-type semiconductor material with the bandgap energy of 2.0 eV. Unlike hematite which has antiferromagnetic properties, maghemite is ferrimagnetic. The very good magnetic properties are one of the causes of its application in the field of biomedicine, magnetic pigmentation, etc. In maghemite, the $Fe^{3+}$ cations exist in both octahedral and tetrahedral geometry; hence, the unit cell of maghemite is represented as $(Fe^{3+})8[Fe^{2.5+}]_{16}O_{32}$, where the brackets () and [] are assigned for tetrahedral and octahedral sites, respectively. The crystal of maghemite contains 8/3 vacancies for every 24 sites of Fe in octahedral geometry which is then compensated by the tetrahedral geometry [5]. The structure of maghemite is shown in Fig. 5.2 which helps to understand the unit cell structure of maghemite.

### Magnetite ($Fe_3O_4$)

Magnetite is another important form of iron oxide which is mainly represented as ($FeO.Fe_2O_3$) and is called as ferrous ferrite. Magnetite possesses ferromagnetic character and is one of the strongest magnets found naturally. Magnetite is widely used for molecular imaging, supercapacitors, sensing, magnetic drug targeting, etc. It also falls under the spinal category with the general formula of $A^{2+}B_2{}^{3+}O_4{}^{2-}$, where A and B are divalent, trivalent cations, containing metals like Zn, Mg, Fe, Al, Mn, Ti, Cr and Si. Depending upon the conditions, $Fe^{2+}$ can be replaced by any other divalent cation; hence, it can be either n-type or *p*-type semiconductor depending upon the

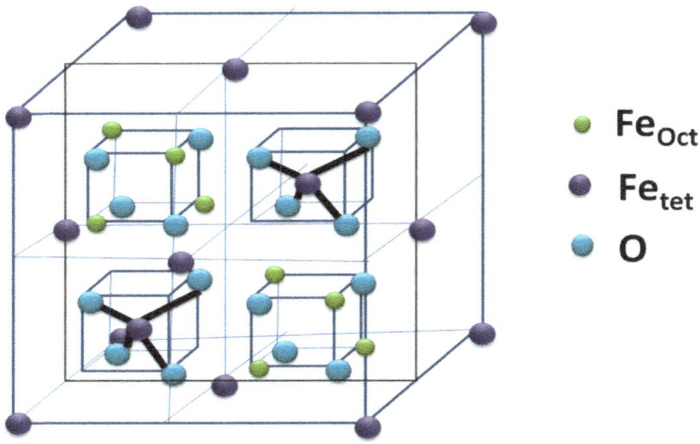


**Fig. 5.3** Structure of magnetite ($Fe_3O_4$)

nature of divalent ions. It has a very small bandgap of 0.1 eV; therefore, magnetite is claimed to have an extremely good electrical conductivity (Fig. 5.3).

## *Reduced Graphene Oxide*

Reduced graphene oxide (rGO) is one of the most extensively used materials recently all over the world due to its widespread applications in the fields of supercapacitors, sensors, photovoltaics, electrical devices, dye degradation, catalysis and many more [6]. rGO is a single sheet of hexagonally arranged $sp^2$ hybridized carbon atoms containing few oxygen functionalities like—hydroxyls, carboxyls, etc. It is also called as chemically obtained graphene and is mainly derived by the reduction of graphene oxide. rGO is black in color and is highly hydrophobic in nature and can be easily separated out in an aqueous solution. Depending upon the presence of different oxygen-containing functional groups in rGO, its properties vary. Hence, the properties of rGO can be easily tuned by varying the oxygen functionality (Fig. 5.4).

## Synthesis of Iron Oxide/rGO Nanocomposites

Various methods have been utilized so far for the preparation of iron oxide/rGO nanocomposites, and their synthesis is broadly categorized into two types, in-situ and ex-situ. In the in-situ process, the metal nanoparticle precursor is seeded onto the surface of GO sheet followed by the reduction of GO with the simultaneous nucleation and growth of the desired nanoparticle. In the ex-situ method, rGO and metal

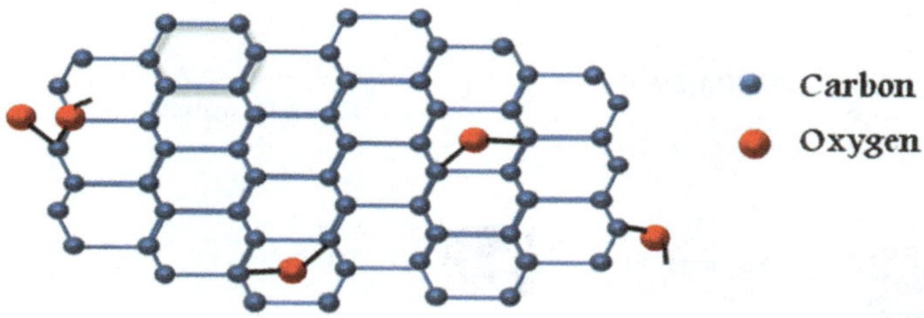


**Fig. 5.4** Sheet-like structure of rGO

nanoparticles are at first prepared separately and then the metal nanoparticles are incorporated in the graphene sheet through some covalent interactions. In-situ preparation is much easier and cost-effective as compared to the ex-situ process hence, in-situ methods are favored over the ex-situ procedures. The detailed explanation of some of the common techniques is given below.

## *Hydrothermal Reduction*

In a typical hydrothermal process at first GO, dispersion is made by ultrasonication (1 mg/ml) GO water solution for an hour. Meanwhile, iron salts like $FeCl_3$, $FeSO_4$ or $Fe(NO_3)_2$ are also dissolved in water and stirred, and 1 ml hydrazine hydrate or any other reducing agent is added. After that, the GO dispersion is added to the previously prepared aqueous solution of iron oxide salts. The resulting mixture is then transferred into an autoclave, and a temperature of ~180 °C is maintained for 10 h. Sometimes ethylene glycol and urea are also added to the mixture as a capping agent in order to prevent the reduction of formed iron oxides [7]. The autoclave is then kept outside and cooled at room temperature. At last, the obtained black residue is filtered cleaned with ethanol and dried at a temperature of 60 °C in an oven. The combined effect of temperature and pressure on the metal salts improve the quality and different features of nanoparticles. The elevated temperature and pressure enhance the rate of dehydration of reactants and also the formation of very fine crystals of nanoparticles. By changing the parameters like temperature, pressure, concentration, pH, and reaction time, the desired shape and size of nanoparticles can be obtained. Zhang et al. followed a one-step hydrothermal procedure for the synthesis of $Fe_3O_4$/rGO composite for its application in $Li^+$-ion batteries. For the synthesis of $Fe_3O_4$ nanoparticle, $FeSO_4$ was used as the precursor material [8]. Hao et al. studied the effect of temperature, precursor concentration and reaction time on the morphology and size of iron oxide nanoparticles [9]. At higher temperature, small-sized nanoparticles were obtained, whereas at lower temperature the nanoparticles have comparatively bigger size. Also, it was observed that the particles size and particle distribution increased proportionally with the precursor concentration. Due

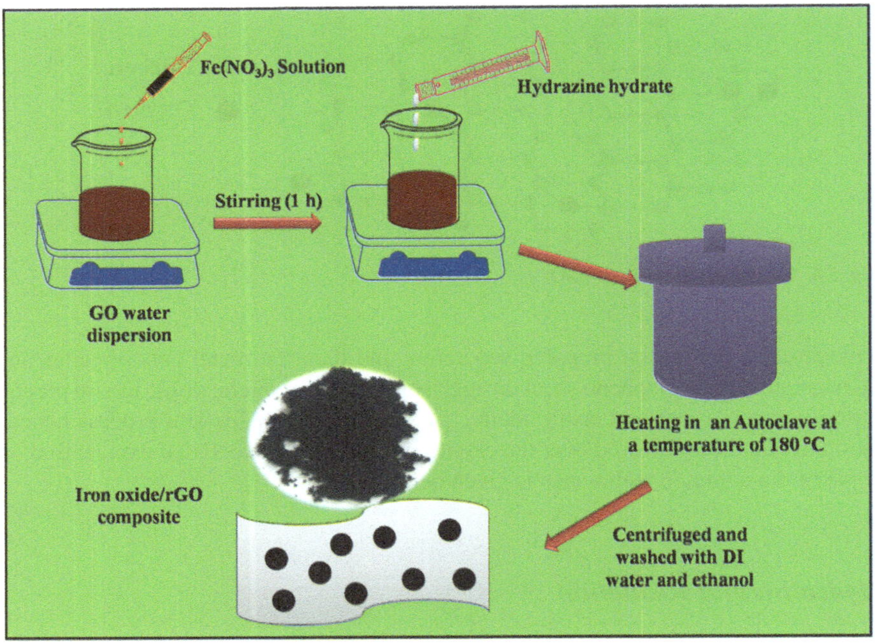


**Fig. 5.5** Hydrothermal synthesis of iron oxide/rGO nanocomposites

to its easy step and environment-friendly nature, this method is one of the most commonly used methods for the synthesis of metal oxide/rGO composites (Fig. 5.5).

## *Thermal Reduction*

Thermal reduction is quite similar to that of hydrothermal reduction process which can be performed in atmospheric pressure and by maintaining a particular temperature. It has been used widely for the synthesis of various nanoparticles including $Fe_2O_3$ and $Fe_3O_4$ as it is easy and time-consuming. Guo et al. synthesized graphene oxide-$Fe_2O_3$ hybrid material by following the thermal reduction procedure in which 0.2M of Fe(NO3)3 was added to the GO dispersion and stirred for 2 h at 60 °C. After that, the GO/Fe(NO3)3 suspension was again aged for 8 h at 60 °C. The formation of graphene oxide-$Fe_2O_3$ was confirmed with the various characterizations like XRD, Raman, FTIR and XPS [10]. Meng et al. synthesized $\alpha$-$Fe_2O_3$/rGO composite by following the simple in-situ thermal reduction process. For the formation of the composite, they used $FeCl_3{\cdot}6H_2O$ as the precursor for $\alpha$-$Fe_2O_3$. Different ratio of $\alpha$-$Fe_2O_3$ by changing the amount of $FeCl_3{\cdot}6H_2O$ added in the GO dispersion. The solution was heated at a temperature of 800 °C for 15 min [11].

### *Chemical Reduction*

Chemical reduction is also one of the most commonly used methods for the synthesis of various metal oxide/rGO composites. This synthesis also starts with the preparation of GO/metal salt dispersion followed by the addition of different reducing agents. The commonly used reducing agents are hydrazine hydrate, ascorbic acid, sodium borohydride, zinc metal, etc. One advantage of following this method is tuning the degree of reduction and different properties by varying the reducing agents. Also, the requirement of low temperature makes it energy efficient. This method has easy synthesis steps but at the same time is considered to be harmful due to the excessive use of different chemicals.

### *Co-precipitation Method*

Co-precipitation is on the most widely used method for the preparation of a variety of metal oxide/rGO composites. In this method, at first GO water dispersion was made by ultrasonication GO and water solution for an hour. After that, the metal salts are added and again stirred for about an hour followed by the addition of hydrazine hydrate. The mixture is then stirred and heated to about 80–90 °C for 2 h. The obtained ppt. is filtered and cleaned with DI water and ethanol. Bhujel et al. prepared rGO/$Fe_2O_3$ nanocomposites by following co-precipitation method by taking $Fe(NO_3)_3$ as the precursor for iron oxide and hydrazine hydrate as the reducing agent. The as-prepared composite was used as an electrode material for the supercapacitors [2]. This method includes the precipitation of $Fe^{2+}$ and $Fe^{3+}$ simultaneously in an aqueous medium. If oxygen is present in solution, then complete precipitation of $Fe_3O_4$ (magnetite) takes place with the $Fe^{3+}:Fe^{2+}$ ratio of 2:1 and gets converted to maghemite. Hence, in order to prepare magnetite one needs to maintain an oxygen-free environment. In this method, at first the metal precursor is converted into metal hydroxide which then gets converted into the desired metal oxide nanoparticles. Hui et al. synthesized hydrophilic $Fe_3O_4$ nanoparticles and observed that by controlling the synthesis parameters, the size of nanoparticles can be tuned within the range of 20–40 nm [12]. This is one of the cheapest and environment-friendly methods for the synthesis of metal oxide/rGO composites (Figs. 5.6 and 5.7).

### *Microwave-Assisted Synthesis*

In this method at first, $Fe(OH)_3$ is prepared by adding $FeCl_3$ in distilled water kept under boiling. GO water dispersion is made by ultrasonication the GO water mixture for an hour. The resulting GO dispersion and $Fe(OH)_3$ sol is then mixed and is stirred

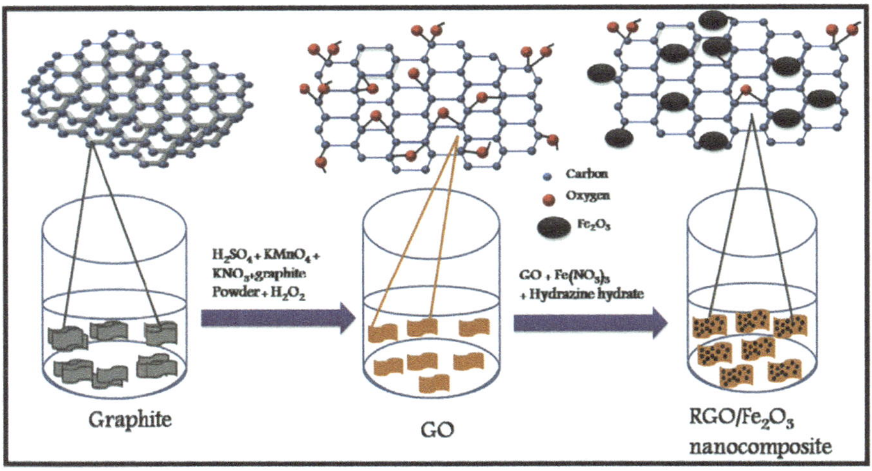


**Fig. 5.6** Synthesis of $Fe_2O_3$/rGO nanocomposites via co-precipitation method [2]

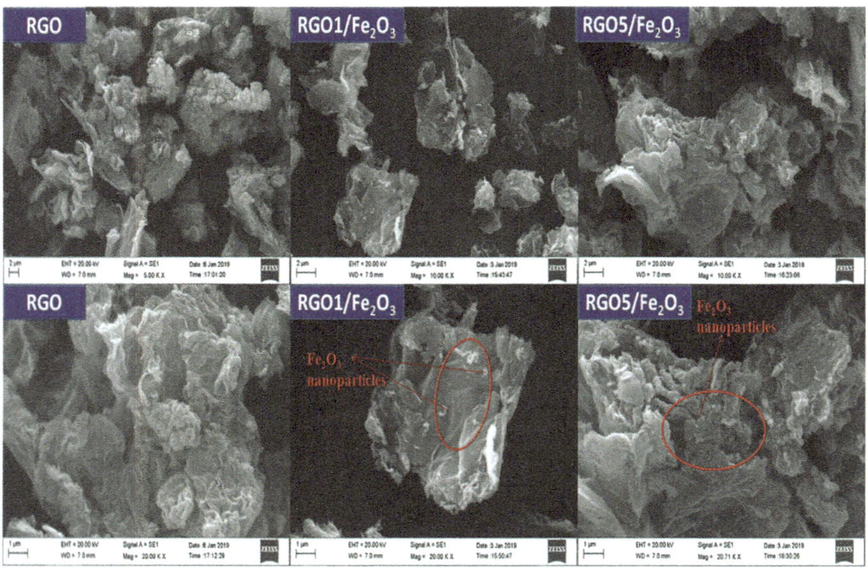


**Fig. 5.7** SEM images of iron oxide/rGO nanocomposites synthesized by co-precipitation method [2]

for an hour with the addition of ~1 ml hydrazine hydrate. The solution was then kept in a microwave reactor under the microwave power of 800 W for about 45 s. The resulting black precipitate is then cleaned several times with DI water and ethanol and dried it in an oven at a temperature of 60 °C. Zhu et al. prepared $\alpha$-$Fe_2O_3$/rGO as anode materials for high-performance lithium-ion batteries [13].

## Synthesis of Graphene Oxide

Graphene oxide is another important material required for the synthesis of metal oxide/rGO composites. Various chemical methods have been used so far for the synthesis of GO by varying the oxidizing agents, acids and also the reaction procedure [14]. The typical synthesis of GO starts with the oxidation of graphite layers followed by its exfoliation with the help of different oxidizing agents. The most commonly used synthesis procedure of GO is shown in the flow chart diagram as shown in Fig. 5.8.

The as-synthesized GO has sheet-like structure with a number of stacking as shown by the SEM image in Fig. 5.9, the crumpling and stacking in GO sheets are expected to occur due to the H-bonding interactions between the oxygen-containing functional groups present in graphene sheet. The extent of oxidation and quality of graphene sheet depends upon the different reaction parameters like oxidizing agents, reaction time, concentration of the GO dispersion, etc.

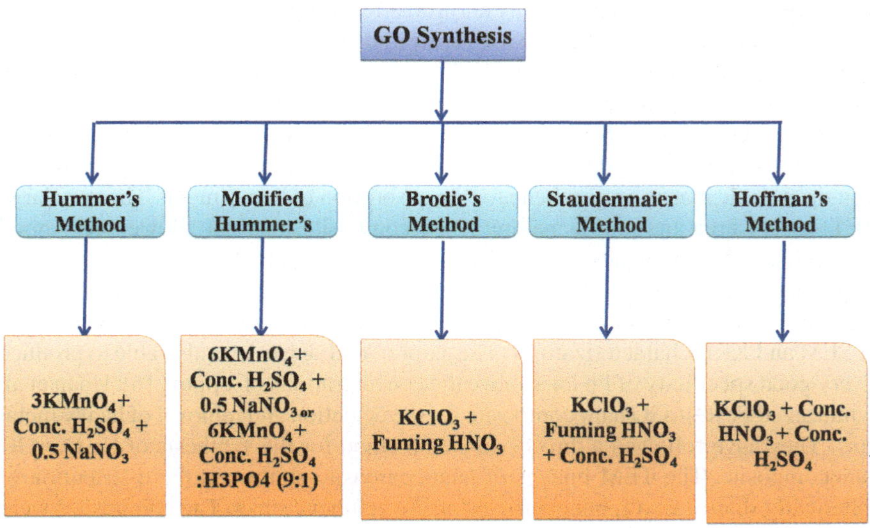


**Fig. 5.8** Synthesis procedures for GO

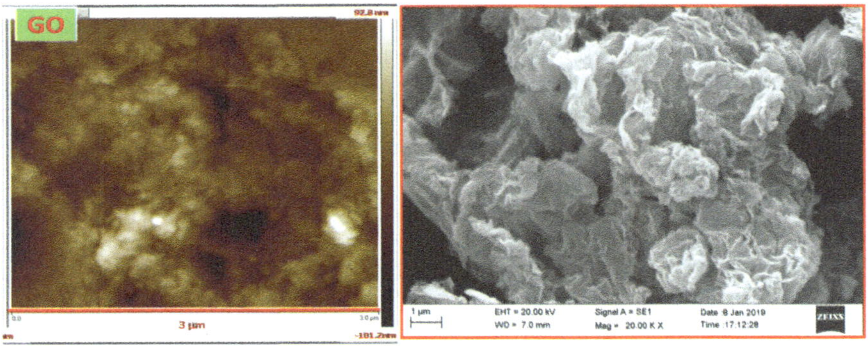


**Fig. 5.9** AFM and SEM images of GO

# Iron Oxide/rGO Nanocomposites for Sensing of Various Toxic Chemicals

## *Sensing of Heavy Metal Ions*

Heavy metal ions such as Pb, Hg, As, Cd and Bi have very harmful effects on human beings as well as on the ecological environments. The waste disposals coming out of different industries have been causing pollution to the water resources which is life-threatening to the marine animals. Also, the contamination of these heavy metals to the drinking water can cause many harmful effects on the human being. This heavy metal poisoning can cause damage to the respiratory tract, kidney, liver, deficiency in hemoglobin, infertility and many more. Therefore, in order to control such pollution, one needs to develop devices that can easily detect and remove these heavy metal ions from the water resources. So far iron oxide/rGO composites have played an important role in the adsorption and detection of these metal ions. Bhunia et al. prepared porous iron–iron oxide matrix on graphene for the adsorption of heavy metal. The very large BET surface area and porous nature of the composite material allowed very good adsorption of heavy metals like As and Hg [15]. Alwahib et al. fabricated rGO/$\gamma$-$Fe_2O_3$ nanocomposite for detecting Pb ions in water using surface plasmon resonance method. The formation of rGO/$\gamma$-$Fe_2O_3$ nanocomposite was confirmed with the help of SEM and XRD characterizations. The as-prepared sensor was also able to produce a very good specificity of Pb ions above the concentrations of 1 ppm [16]. Hoan et al. synthesized $Fe_3O_4$/rGO nanocomposite for the detection and removal of heavy metal ions. They have followed a direct one-step method for the synthesis of $Fe_3O_4$/rGO nanocomposite. The TEM images of nanocomposite confirmed the distribution of spherically shaped $Fe_3O_4$ nanoparticles on the graphene sheet. The composites were able to adsorb As Ni and Pb ions [17].

## *Sensing of Toxic Gases*

Over the last few decades, the production of toxic gases like $NO_2$, $SO_2$, CO, $H_2S$, $NH_3$, etc. has become a crucial issue for causing environmental pollution. These gases are colorless and odorless but are extremely harmful to health concern as it can cause vomiting, nausea, dizziness and breathing problem, and in some case, it can even cause death. Therefore, it becomes necessary to prepare materials that can easily sense and detect these toxic gases. Reduced graphene oxide a 2-D material with a honeycomb structure has attracted the interest of researchers across the world due to its extremely good properties. rGO is capable of detecting very minute changes that occur while the adsorption and desorption of gas molecules. Also, the addition of various metal oxides helps to improve the adsorption property of rGO. Iron oxide is one of them, which are considered as the cheap and eco-friendly metal oxide material that can enhance the sensing property of rGO. Moreover, the sheet-like structure of rGO and homogenous dispersion of metal oxide nanoparticles is one reason for the higher specific surface area. To understand the sensing of different gases, the mechanism for CO sensing is given below

$$O_2\,(\text{gas}) \rightleftharpoons O_2\,(\text{adsorbed})$$
$$O_2\,(\text{adsorbed}) + e^-(\text{from rGo/Fe}_2\text{O}_3) \rightleftharpoons O_2^-$$
$$2CO + O_2^- \rightleftharpoons 2CO_2 + e^-$$

At first, $O_2$ gas gets adsorbed on the surface of the rGO/$Fe_2O_3$ composite which then captures $e^-$ and forms $O_2^-$. When CO is supplied, it gets oxidized to $CO_2$ with the adsorbed oxygen anion [18]. As reported earlier by Dong et al., rGO/$Fe_2O_3$ composites with the spherically shaped iron oxide nanoparticles well dispersed on rGO sheet are also capable of detecting gases like $NO_2$ and acetone [19–21]. The sensing mechanism is similar to that reported earlier for CO detection (Fig. 5.10).

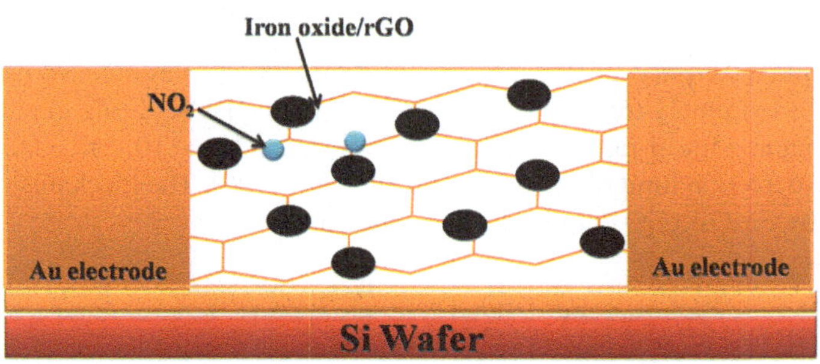


**Fig. 5.10** $NO_2$ sensor fabricated using iron oxide/rGO nanocomposite

## *Sensing of Harmful Chemicals*

Chemicals containing phenol groups are widely used in various industries like drugs, agriculture, dyes and also in various pesticides as an intermediate compound. The common waste coming out from these industries are alkylphenols, cresols, p-nitrophenol, chlorinated or brominated benzenes, etc. which has harmful effects on human health. Therefore, it is important to develop methods for determining the presence of these toxic materials in order to control environmental pollution. In past decades for the detection of these chemicals, people used to follow very expensive and time taking methods like gas chromatography, UV–Visible spectroscopy, fluorescence, etc. But recently, the development of electrochemical sensors has become center of attraction for the entire scientist worldwide due to its extremely high sensitivity and simple working methods. Therefore, various materials have been prepared up to now in order to detect these harmful chemicals. Graphene/metal oxide composites are one of them. The large surface area of graphene sheet with the distribution of active metal oxide nanoparticles provides excellent properties to these composite materials. Iron oxide/rGO composite is one of them having good sensing properties due to its high surface area, high electrical conductivity and very good magnetic properties. Iron oxide/rGO composites were used earlier for the detection of chemicals like p-nitrophenol [22] which is highly toxic. The sensing of para-nitrophenol takes place through the following mechanism as shown below. At first, p-nitrophenol gets reduced to p-aminophenol through an irreversible reduction process which then again undergoes reversible oxidation to p-nitrophenol and vice versa (Fig. 5.11).

Next is diphenylamine (DPA) which is mainly used in fruit harvesting. The residue of DPA in fruits is carcinogenic and hence poses harmful effects on human health. To control the excess use of DPA, it's monitoring and detection are essential. Iron oxide/rGO composites with molecular imprinting polymers are an important material that is capable of detecting DPA [23]. Again Rhodamine B (RhB) is another chemical used commonly in textile and dyes industry. The contamination of RhB in water is needed to be controlled, or else this could be one of the most important chemicals for causing water pollution. $Fe_3O_4$/rGO composite has played an important role in the detection and adsorption of RhB [24]. $Fe_3O_4$/.rGO composites are also used for the detection of extremely toxic inorganic chemicals like $H_2O_2$. Xiong et al. observed that the obtained linear range concentration was from $5.0 \times 10^{-7}$ to $3.0 \times 10^{-3}$ M having the detection limit of $1.8 \times 10^{-7}$ M (S/N = 3) [25]. The mechanism of $H_2O_2$ detection by $Fe_3O_4$/.rGO composites is as shown below

$$Fe^{3+} \underset{-e^-}{\overset{+e^-}{\rightleftharpoons}} Fe^{2+} \xrightarrow[H^+]{H_2O_2} Fe^{3+}$$

The reaction is a simple electron transfer process between $Fe^{3+}$ and $Fe^{2+}$ on the $Fe_3O_4$ catalysts with a chemical reduction reaction of $H_2O_2$ by $Fe^{2+}$. Furthermore by adding different functionalities to these iron oxide/rGO nanocomposites, it can be used for the detection and adsorption of various toxic chemicals (Table 5.1).

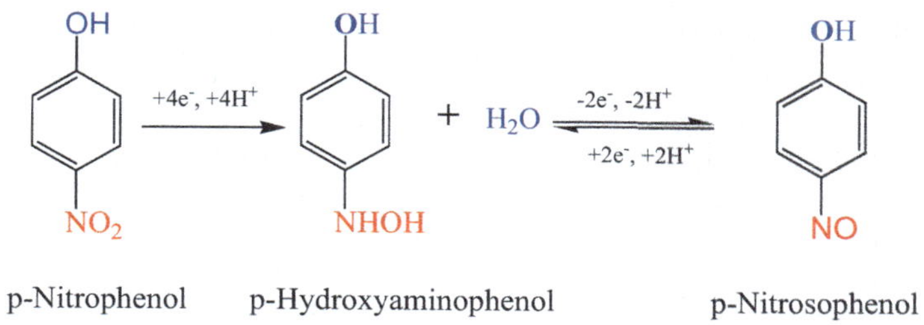


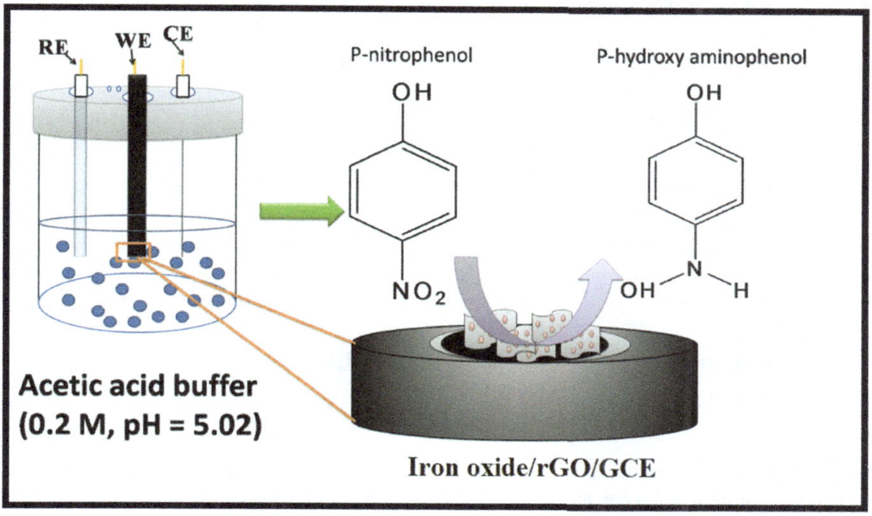


**Fig. 5.11** Electrochemical sensing of p-nitrophenol using iron oxide/rGO/GCE

## Conclusion

Iron oxide/rGO composites have been widely studied in the past decades due to its easy availability and extremely good properties. The high active surface area and synergistic effect between the iron oxide nanoparticles and rGO sheet facilitate a very good sensing property to different toxic chemicals. Iron oxide/rGO nanocomposites have been prepared by following a variety of synthesis techniques, including hydrothermal method, co-precipitation, thermal reduction, chemical reduction, microwave-assisted synthesis, etc. Reaction parameters like reaction temperature, precursor concentration, reaction time, pH, etc. have a great impact on the size of iron oxide nanoparticles and its distribution on the rGO sheet. Therefore, by varying the reaction parameters, one can tune the properties of this composite. Finally, the application of iron oxide/rGO composites in sensing of various toxic chemicals has been achieved with proper mechanisms.

**Table 5.1** Sensing of various toxic chemicals using iron oxide/rGO composites

| S. No. | Type of composite | Synthesis procedure | Detected chemical | References |
|---|---|---|---|---|
| 1 | rGO-Fe(0)-$Fe_3O_4$ | Thermal reduction | As(V), Hg(II), Cd(II), Pb(II), Cr(III) | [15] |
| 2 | rGO/$\gamma$-$Fe_2O_3$ | Simple reduction | Pb(II) | [16] |
| 3 | $Fe_3O_4$/rGO | Ex-situ method | As(V), Ni(II), and Pb(II) | [17] |
| 4 | rGO/$\alpha$-$Fe_2O_3$ | Solvothermal method | CO gas | [18] |
| 5 | rGO/$\alpha$-$Fe_2O_3$ | Hydrothermal method | $NO_2$ gas | [19] |
| 6 | rGO/$\alpha$-$Fe_2O_3$ | Electrospinning method | Acetone gas | [20] |
| 7 | rGO/$Fe_3O_4$ | Co-precipitation Method | p-nitrophenol | [22] |
| 8 | RGO/$Fe_3O_4$-IL-MIP | Solvothermal method | Diphenylamine | [23] |
| 9 | $Fe_3O_4$/rGO | Co-precipitation Method | Rhodamine B | [24] |
| 10 | $Fe_3O_4$/rGO | Chemical reduction | $H_2O_2$ | [25] |

**Acknowledgements** This work was financially supported by Dr. Ramdas Pai and Mrs. Vasanthi Pai endowment fund as acknowledged by one of the author's Rabina Bhujel.

## References

1. Opuchovic, O., Kareiva, A.: Historical hematite pigment: synthesis by an aqueous sol-gel method, characterization, and application for the coloration of ceramic glazes. Ceram. Int. **41**, 4504–4513 (2015)
2. Bhujel, R., Rai, S., Deka, U., Swain, B.P.: Electrochemical, bonding network and electrical properties of reduced graphene oxide-$Fe_2O_3$ nanocomposite for supercapacitor electrodes applications. JALCOM **792**, 250–259 (2019)
3. Mansour, S.A., Ibrahim, M.M.: Electrical investigation of nanostructured $Fe_2O_3$/p-Si heterojunction diode fabricated using the sol-gel technique. J. Electron. Mater. **46**, 6502–6507 (2017)
4. Leonardi, S.G., Mirzaei, A., Bonavita, A., Santangelo, S., Frontera, P., Pantò, F., Antonucci, P.L., Neri, G.: A comparison of the ethanol sensing properties of -iron oxide nanostructures prepared via the sol-gel and electrospinning techniques. Nanotechnology **27**, 075502 (2016)
5. Movlaee, K., Ganjali, M., Norouzi, P., Neri, G.: Iron-based nanomaterials/graphene composites for advanced electrochemical sensors. Nanomater **7**, 406 (2017)
6. Saha, V.C., Sabuj, M.M.A., Shams, P., Rahman, S., Qadir, M.R., Islam, M.R., Gulshan, F.: Synthesis and characterization of reduced graphene oxide reinforced polymer matrix composite. IOP Conf. Ser. Materi. Sci. Eng. **438**, 012008 (2018)

7. Cao, W., Ma, Y., Zhou, W., Guo, L.: One-pot hydrothermal synthesis of RGO-$Fe_3O_4$ hybrid nanocomposite for removal of Pb(II) via magnetic separation. Chem. Res. Chin. U. **31**, 508–513 (2015)
8. Zhang, X., Jiang, B., Xie, Y., Du, F.: One-pot hydrothermal synthesis of $Fe_3O_4$/rGO nanocomposite for enhanced lithium storage, Indian. J. Chem. **53** A, 265–273 (2014)
9. Hao, Y., Teja, A.S.: Continuous hydrothermal crystallization of –$Fe_2O_3$ and $Co_3O_4$ Nanoparticles. J. Mater. Res. **18,** 415–422 (2011)
10. Guo, S., Zhang, G., Guo, Y., Yu, J.C.: Graphene oxide–$Fe_2O_3$ hybrid material as highly efficient heterogeneous catalyst for degradation of organic contaminants. Carbon **60**, 437–444 (2013)
11. Meng, F., Li, J., Cushing, S.K., Bright, J., Zhi, M., Rowley, J.D., Wu, N.: Photocatalytic water oxidation by hematite/reduced graphene oxide composites. ACS Catal. **3**, 746–751 (2013)
12. Hui, C., Shen, C., Yang, T., Bao, L., Tian, J., Ding, H., Li, C., Gao, H.J.: Large-scale $Fe_3O_4$ nanoparticles soluble in water synthesized by a facile method. J. Phys. Chem. C **112**, 11336–11339 (2008)
13. Zhu, S., Chen, M., Ren, W., Yang, J., Qu, S., Li, Z., Diao, G.: Microwave assisted synthesis of $\alpha$-$Fe_2O_3$/reduced graphene oxide as anode material for high performance lithium ion batteries. New J. Chem. **39**, 7923–7931 (2015)
14. Singh, R.K., Kumar, R., Singh, D.P.: Graphene oxide: strategies for synthesis, reduction and frontier applications. RSC Adv. **6**, 64993–65011 (2016)
15. Bhunia, P., Kim, G., Baik, C., Lee, H.: A strategically designed porous iron-iron oxide matrix on graphene for heavy metal adsorption. Chem. Commun. **48**, 9888–9890 (2012)
16. Alwahib, A.A.A., Mustapha Kamil, Y., Abu Bakar, M.H., Noor, A.S.M., Yaacob, M.H., Lim, H.N., Mahdi, M.A.: Reduced graphene oxide/maghemite nanocomposite for detection of lead ions in water using surface plasmon resonance. IEEE Photon. J. **10**, 1–10 (2018)
17. Vuong Hoan, N.T., Anh Thu, N.T., Duc, H.V., Cuong, N.D., Quang Khieu, D., Vo, V.: $Fe_3O_4$/reduced graphene oxide nanocomposite: synthesis and its application for toxic metal ion removal. J. Chem. **2016**, 1–10 (2016)
18. Basu, A.K., Chauhan, P.S., Awasthi, M., Bhattacharya, S.: $\alpha$-$Fe_2O_3$ loaded rGO nanosheets based fast response/recovery CO gas sensor at room temperature. Appl. Surf. Sci. **465**, 56–66 (2019)
19. Dong, Y., Zhang, X., Cheng, X., Xu, Y., Gao, S., Zhao, H., Huo, L.: Highly selective $NO_2$ sensor at room temperature based on nanocomposites of hierarchical nanosphere-like $\alpha$-$Fe_2O_3$ and reduced graphene oxide. RSC Adv. **4**, 57493–57500 (2014)
20. Guo, L., Kou, X., Ding, M., Wang, C., Dong, L., Zhang, H., Lu, G.: Reduced graphene oxide/$\alpha$-$Fe_2O_3$ composite nanofibers for application in gas sensors. Sens. Actuat. B-Chem. **244**, 233–242 (2017)
21. Xu, S.P., Sun, F.Q., Pan, Z.Z., Huang, C.W., Yang, S.M., Long, J.F., Chen, Y.: Reduced graphene oxide-based ordered macroporous films on a curved surface: general fabrication and application in gas sensors. ACS Appl. Mater. Interfaces **8**, 3428–3437 (2016)
22. Cheng, Y.: A sensor for detection of 4-nitrophenol based on a glassy carbon electrode modified with a reduced graphene oxide/$Fe_3O_4$ nanoparticle composite. Int. J. Electrochem. Sci. **12**, 7754–7764 (2017)
23. Liu, L., Zhu, X., Zeng, Y., Wang, H., Lu, Y., Zhang, J., Li, L.: An electrochemical sensor for diphenylamine detection based on reduced graphene oxide/$Fe_3O_4$-molecularly imprinted polymer with 1,4-Butanediyl-3,3'-bis-l-vinylimidazolium Dihexafluorophosphate ionic liquid as cross-linker. Polymers **10**, 1329 (2018)
24. Long, M., Qin, Y., Tan, B., Zhou, B.: Rhb adsorption performance of magnetic adsorbent $Fe_3O_4$/RGO composite and its regeneration through a fenton-like reaction. Nano-Micro Lett. **6**, 125–135 (2014)
25. Xiong, L., Zheng, L., Xu, J., Liu, W., Kang, X., Wang, Y., Xia, J.: A non-enzyme hydrogen peroxide biosensor based on $Fe_3O_4$/RGO nanocomposite material. ECS Electrochem. Lett. **3**, B26–B29 (2014)

# Chapter 6
# Reduced Graphene Oxide for Advanced Energy Applications

**Sadhna Rai, Rabina Bhujel, Joydeep Biswas, and Bibhu Prasad Swain**

## Introduction

With the advancement of human civilization, the need for energy generation and its conservation has become a major concern. The energy consumption has increased alarmingly, and day by day, the conventional sources of energy are depleting which possess a great threat to the civilization itself. In this regard, researches are working interminably to meet the energy demands of the society. There is a shift from fossil fuels to renewable source of energy mainly solar, wind, hydel and geothermal energies. But only the generation of energy is not enough to meet up with the energy demands. Therefore, development of energy storage devices in large scale is also important.

Since graphene was discovered in 2004, it has been studied comprehensively for its applications in various fields. Graphene is a single sheet of carbon hexagons which can be used as the building foundations for other forms of carbon; for instance, it can be folded to give 0D fullerenes, rolled to form 1D carbon nanotubes and piled to present 3D graphene [1]. Graphene exhibits exceptional properties like excellent optical transparency, zero bandgap, high carrier mobility, mechanical strength and electrical conductivity. The reported values of charge mobility and electrical

S. Rai (✉) · R. Bhujel
Centre for Materials Science and Nanotechnology, Sikkim Manipal Institute of Technology, Sikkim Manipal University, Majhitar, Rangpo, Sikkim 737136, India
e-mail: sadhnarai15@gmail.com

J. Biswas
Department of Chemistry, Sikkim Manipal Institute of Technology, Sikkim Manipal University, Majhitar, Sikkim 737136, India

B. P. Swain
Department of Physics, National Institute of Technology, Manipur, Langol Rd, Imphal, Manipur 795004, India
e-mail: bibhuprasad.swain@gmail.com; bpswain@nitmanipur.ac.in



B. P. Swain (ed.), *Nanostructured Materials and their Applications*, Materials Horizons: From Nature to Nanomaterials, https://doi.org/10.1007/978-981-15-8307-0_6

conductivity graphene is about $2 \times 10^5$ cm$^2$ (Vs)$^{-1}$ and $1 \times 10^8$ Sm$^{-1}$, respectively, which enables it to be used in different electronic devices [2]. It also possesses high specific area (2360 m$^2$g$^{-1}$) which is typically valuable for energy storage devices, viz. lithium-ion batteries and supercapacitors [3]. Also, possessing excellent thermal conductivity (~5000 W/mK) makes it suitable for use in both energy-generating and storing devices [4]. The most studied graphene-based derivatives for these purposes are graphene oxide (GO) and reduced graphene oxide (rGO).

Graphene oxide possesses 2-D structure and various oxygen functionalities such as hydroxyl, epoxy, carbonyl and carboxyl groups on its basal plane and on the edges. These functionalities make GO dispersible in water. GO is inexpensive, can be synthesized without any hassle and is easy to handle. The electrical conductivity of GO is low, but being optically transparent, it can be used as a good current collector in solar cells. The conductivity of GO is practically improved by the removal of oxygen functionalities which is attained by the reduction of graphene oxide. For large-scale application in energy devices, rGO can be used as a suitable alternative because of the ease of its synthesis and mass production compared to pristine graphene.

## Synthesis Procedures

The synthesis procedures for graphene oxide and reduced graphene oxide are presented below.

### *Graphene Oxide*

The synthesis procedures for graphene oxide include Brodie's method, Staudenmaier method, Hummer's method and Tour's method (Fig. 6.1.). These methods are discussed below in chronological order.

#### Brodie's Method

B.C. Brodie first synthesized GO in 1859 while investigating the chemistry of graphite [5]. He took a mixture of graphite with three times potassium chlorate ($KClO_3$) and fuming nitric acid ($HNO_3$). It was left at 60 °C for three to four days in a water bath. It was washed with water so as to make it acids and salts free and dried in a water bath. The oxidation procedure was repeated until no further changes were observed generally four times. The product was left to dry at 100 °C. The crystals were extremely small, thin and had imperfect structure due to which the goniometry reflective measurement was failed. The elemental investigation of the product affirmed the molecular formula of the product was $C_{11}H_4O_5$. However, a

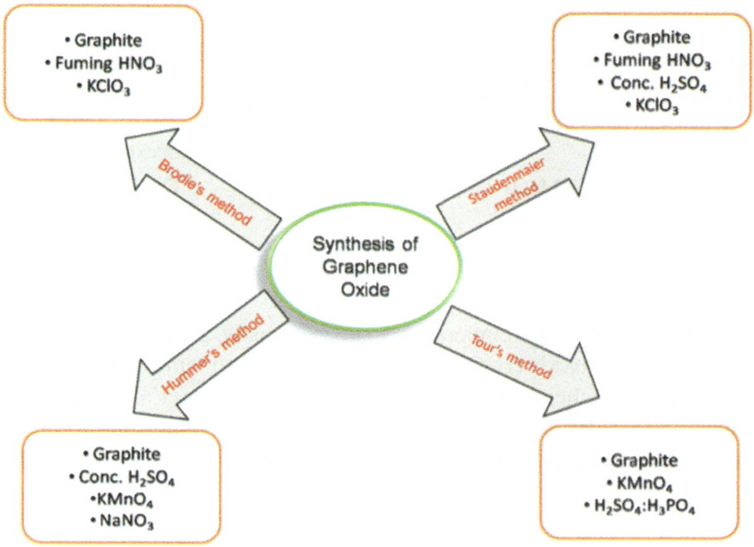


**Fig. 6.1** Different synthesis procedures for graphene oxide

limited study was done by him owing to inadequate techniques accessible at that time which leaves a huge room for further study and analysis.

### Staudenmaier's Method

Later, in 1898 Staudenmaier made some improvement to Brodie's work [6]. He added concentrated sulfuric acid ($H_2SO_4$) replacing the two-third amount of $HNO_3$. $KClO_3$ was gradually added to the reaction mixture in batches, and the reaction flask was kept in ice bath. The procedure followed by Staudenmaier was simpler, and GO produced was highly oxidized. However, the reaction produced chlorine dioxide ($ClO_2$) gas which was removed by an inert gas which otherwise would lead to explosions. Thus, this method also required further modifications.

### Hummer's Method

A rapid and relatively safer process for synthesis of graphene oxide was introduced by chemists W. S. Hummers, Jr., and R. E. Offeman in the year 1958 [7]. A mixture of graphite, sodium nitrate ($NaNO_3$) and potassium permanganate ($KMnO_4$) was taken in the ratio 1:0.5:3 with concentrated sulfuric acid keeping the temperature below 20 °C. The ice bath was removed and the temperature of the reaction mixture was raised to 35 °C keeping at this temperature for 30 min. Water was then added

**Table 6.1** Comparison of synthesis procedures for graphene oxide

| Year | Method | Oxidizing agent | Solvent | Disadvantages |
|---|---|---|---|---|
| 1859 | Brodie | $KClO_3$ | $HNO_3$ | Limited study |
| 1898 | Staudenmaier | $KClO_3$ | $HNO_3 + H_2SO_4$ | Time-consuming, hazard of explosion |
| 1958 | Hummer | $KMnO_4 + NaNO_3$ | $H_2SO_4$ | Evolution of toxic gases like $NO_2$ and $N_2O_4$ |
| 2010 | Tour | $KMnO_4$ | $H_2SO_4 + H_3PO_4$ | – |

slowly which led to a rise in temperature to 98 °C. Lastly, the dispersion was treated with 3% hydrogen peroxide ($H_2O_2$) whose purpose was to remove the residual $Mn^{+2}$ ions. They completed their whole reaction within 2 h. Hummer's product resulted in higher degree of oxidation compared to Staudenmaier's. According to them, well-oxidized GO has bright yellow color and poorly oxidized ones have green or black tint.

#### Tour's Method

In 2010, a group of researchers from Rice University, Texas, presented a superior method for the synthesis of GO [8]. In this method, $NaNO_3$ was eliminated from the reaction, and the quantity of $KMnO_4$ was raised to six equivalents. The reaction was performed taking a mixture of concentrated $H_2SO_4/H_3PO_4$ in the ratio 9:1 as the reaction medium. They found that the method was more efficient than Hummer's and produced more oxidized graphene oxide. $H_3PO_4$ contributed towards formation of GO with regular structure and less ruptured basal plane. This is possible as $H_3PO_4$ forms a five-membered ring with the formation of two vicinal diols on the basal plane of graphite after oxidation [9]. This protocol is considered environmental friendly because it avoids creation of toxic gases like $NO_2$ and $N_2O_4$ (Table 6.1).

## *Reduced Graphene Oxide*

GO can be reduced to rGO using several reduction strategies, viz. thermal, chemical, photo and solvothermal reduction as shown in Fig. 6.2.

#### Thermal Reduction

The reduction of GO by application of heat is known as thermal annealing reduction. The thermal reduction is a simple, safe and efficient mode for reduction of GO. The mechanism involves removal of $H_2O$ and oxygen functionalities of graphene oxide

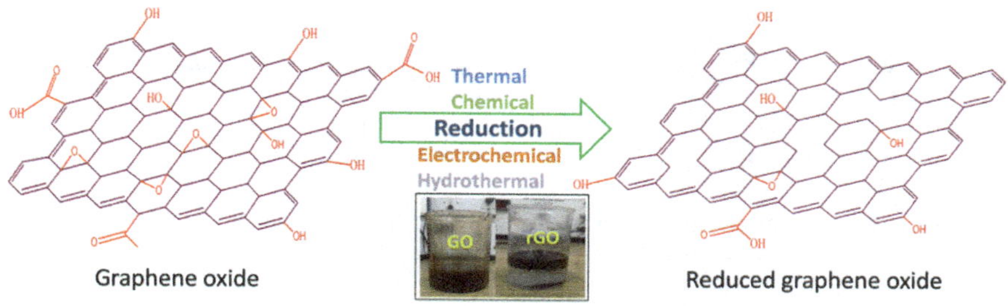


**Fig. 6.2** Reduction of graphene oxide to reduced graphene oxide

such as hydroxyl, epoxy and carboxyl on application of heat. These oxide groups are removed in multiple steps in the form of gases like $CO_2$ and CO. The gases evolved create pressure between the graphene sheets which consecutively help in the exfoliation process. McAllister et al. reported a pressure of 40 and 130 MPa is generated at 300 and 1000 °C, respectively. They evaluated Hamaker coefficient and found only 25 MPa of pressure which is required to separate two GO sheets [10]. Another significant aspect for thermal annealing reduction of GO is the annealing atmosphere. The annealing should be performed in vacuum [11] or in an inert [12] or reducing atmosphere [13, 14]. With the elimination of oxygen functionalities, the conductivity of rGO increases which can be achieved at higher temperatures. However, this process possesses some drawbacks.

1. The heating at high temperatures involves more power consumption.
2. GO must be heated slowly in order to reduce defects created in the structure, but this makes the processing time-consuming.

### Chemical Reduction

Chemical reduction involves reduction by chemical reaction of GO with chemical reagents. It is considered more efficient than thermal reduction method in terms that it does not require the equipment and the environment for annealing. The reaction can be fairly conducted at room temperature or by slight heating. It is a cheap and easy path for mass production of graphene. Hydrazine and its derivatives such as hydrazine monohydrate and dimethylhydrazine have widely been acknowledged as an efficient reducing agent for the synthesis of rGO [15–18]. The reduction can be obtained by adding the reagent to aqueous dispersion of GO and heating the reaction mixture. The mechanism for the reduction of epoxide group present in GO is shown in Scheme 1 as proposed by Stankovich et al. [19]. Sodium borohydride is also used for the reduction of GO. Shin et al. revealed that $NaBH_4$ showed more efficiency than hydrazine for the reduction of GO. $NaBH_4$ slowly hydrolyzes in water mainly used for dispersion and exfoliation of GO by freshly prepared solutions which can effectively reduce GO. $NaBH_4$ effectively reduces C=O groups but shows less efficiency towards reduction of epoxy and carboxylic groups which results in residual alcohol groups

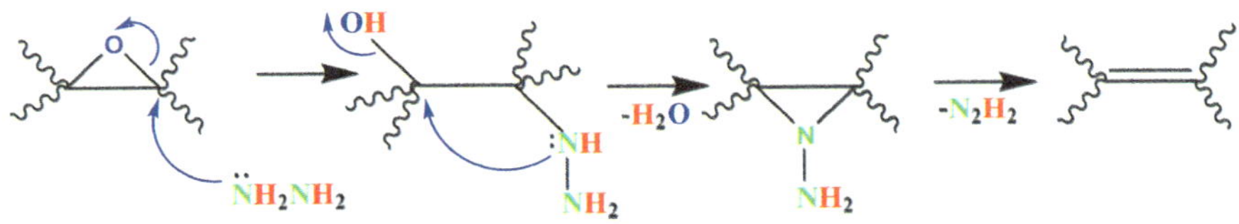


**Scheme 6.1** Reaction mechanism for the reduction of the epoxide group of GO using hydrazine

after reduction. To overcome it, Gao et al. improved the reduction process by using concentrated $H_2SO_4$ at 180 °C after reduction with $NaBH_4$. Recently, a lot of work has been done using ascorbic acid, and it is considered that it can replace hydrazine for reduction of GO [20]. Unlike hydrazine, ascorbic acid is non-toxic and is chemically more stable in water than $NaBH_4$ (Scheme 6.1).

## Electrochemical Reduction

Moreover, the electrochemical reduction is a promising method for the reduction of GO. It is performed in an electrochemical cell involving an electrolyte at room temperature. The commonly used electrolytes are ionic liquids, buffer solution, aqueous, acidic or basic solvents [21–25]. The reduction occurs due to exchange of electrons between the electrode and GO. Since it is performed at room temperature and excludes any harsh chemical reducing agents, it rules out the hazards of high temperature and toxic chemicals. Hence, it opens up a new way for low-cost, time-saving, low temperature and environmental-friendly reduction of GO compared to the conventional thermal and chemical reduction methods. The reduction mainly occurs due to following redox reactions [26].

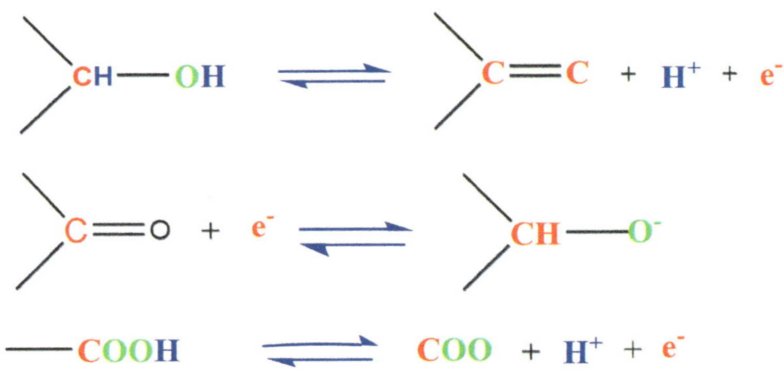

### Hydrothermal Reduction

Another promising approach for the reduction of GO is solvothermal reduction method. The word solvothermal can be explained as "solvo" meaning solvent and "thermal" meaning heat. In this method, chemical reactions are performed in sealed vessels at the supercritical temperature of solvents which is achieved by the increase in pressure. If the solvent is replaced by water, the process is called hydrothermal. As the temperature and pressure have increased, the solubility and reactivity of the reagents are thus enhanced resulting in reactions which are not possible otherwise [27]. The main advantages of using water are its environmental-friendly nature and cost-effectiveness. It has advantages of being non-hazardous and nonflammable. Also, being volatile, it can simply be removed from the products. Zheng et al. anticipated a mechanism for the reduction of GO by hydrothermal method [28]. In hydrothermal reduction of GO, hydroxyl groups are formed after ring opening of epoxy groups which follows $H^+$ catalyzed dehydration at moderate temperature. The ring opening of epoxy groups is catalyzed by both acidic and alkaline conditions. In case of $H^+$ catalyzed dehydration, intramolecular or intermolecular dehydration can take place. In alkaline condition, it is usually intramolecular reduction. The source of $H^+$ ions for protonation of hydroxyl groups is the supercritical water. The hydroxyl groups along with nearby hydrogen atoms are eliminated which result in the recovery of $\pi$-bonding and thus the $sp^2$ carbons. In acidic conditions, oxygen groups are protonated and reduction occurs by intermolecular dehydration. An ether structure is formed when water molecule is eliminated by combination of two hydroxyl groups situated on the edges of two neighboring GO sheets. In weak acidic conditions, both intra- and intermolecular dehydration may occur.

## Advanced Energy Applications

The challenge of generating energy for meeting up the demands of the people is increasing at a rapid rate. Lately, the derivatives of graphene particularly rGO have been studied extensively for the application in energy generating and storing devices. Its unique optical and electrical properties make it suitable for applications in the field of energy and hence widely used in solar cells and supercapacitor devices.

### *Solar Cells*

Solar power is most abundantly available, a harmless and clean energy source for sustaining energy. An efficient approach of harnessing solar energy is its conversion to electrical energy using solar cells. rGO is used in different types of solar cells such as dye-sensitized solar cells (DSSC) and organic solar cells (OSC).

## Organic Solar Cells (OSC)

The conventional OSCs are usually made up of a cathode, an electron transport layer, an active layer, a hole transport layer and an anode. In OSCs, rGO can perform the role of electrode, hole transport layer or electron transport layer (Fig. 6.3).

### GO as an Electrode

Indium tin oxide (ITO) is a usually used electrode in OSCs owing to their high transparency and conductivity. ITO has drawbacks like that it is sensitive to chemicals, expensive and brittle in nature [29]. So the need for its substitute arises which can overcome these drawbacks. The high transparency, flexibility, specific area, conductivity and most importantly low cost of rGO make it a suitable candidate to replace ITO [30]. It can be adopted as a transparent flexible anode or cathode in OSC showing large power conversion efficiency (PCE) of about 0.78 and 0.31%, respectively [31, 32].

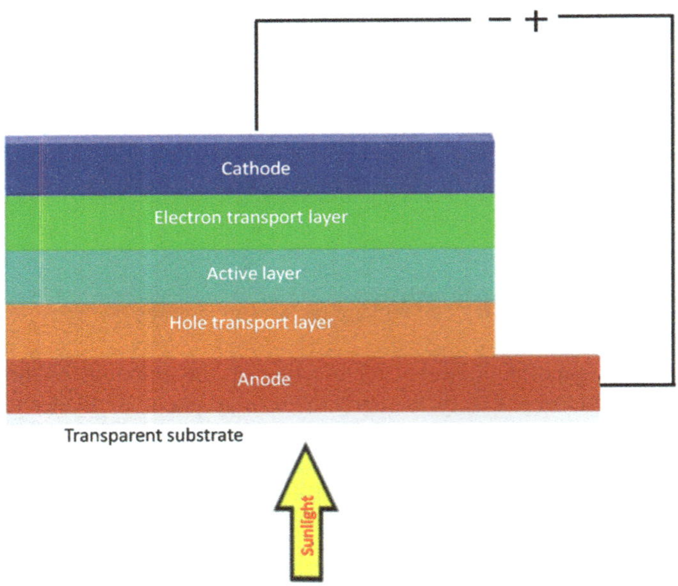


**Fig. 6.3** Schematic of an organic solar cell

GO as Hole Transport Layer (HTL)

The function of HTL is to improve the efficiency of the anode electrode to collect extract positive carriers [33]. HTL helps to prevent direct contact between active layer and anode minimizing of current leakage which reduced recombination [34]. The most frequently used material as HTL in OSCs is organic poly(3,4-ethylenedioxythiophene poly(styrene sulfonic acid) (PEDOT:PSS). But it is acidic which may corrode the ITO. Also, being hygroscopic it can easily absorb water and degrade the device lifetime. rGO is found to replace PEDOT:PSS giving PCE which is analogous with PEDOT:PSS devices [35]. The optical properties, electrical conductivity and work function of rGO can be tuned by reduction or by doping. Jeon et al. fabricated OSC using vanadyl phosphate decorated rGO as HTL which showed higher work function and an increased PCE value of 8.32% [36]. Cheng et al. functionalized rGO with fluorine and used it as an HTL [34]. The fluorinated rGO showed an enhanced conductivity and work function value. The device attained a PCE value of 8.6% with higher stability compared to PEDOT:PSS-based devices.

GO as Electron Transport Layer (ETL)

The role of ETL is to improve the efficiency of the cathode electrode to collect and extract negative charge carriers [33]. Some of the examples of ETL are alkali metals like lithium fluoride (LiF) [37], metal oxides, viz. titanium suboxide ($TiO_x$) [38], zinc oxide (ZnO) [39], calcium oxide (CaO) [40] and organic polymers such as bathocuproine (BCP), 2,9-dimethyl-4,7-diphenyl-1, and 10-phenanthroline [41]. Metal oxide incorporated rGO can serve as an efficient ETL due to enhanced charge transport property of rGO in comparison with metal oxide alone. The metal oxide/rGO provides better electron extraction developing the performance of the solar cell. The solar cell using metal oxide/rGO nanohybrid as ETL fabricated by Jayawardena et al. showed a PCE value in the range of 7.4–7.5% [42]. Lee et al. utilized ZnO/rGO as ETL in ITO/ZnO:rGO/P3HT:PCBM/PEDOT:PSS/Ag solar cell and found the incorporation of rGO led to the enhanced device performance [43]. This was observed due to increased electron conductivity and reduced charge transfer resistance which led to substantial increase in Jsc and fill factor boosting up the PCE value to 4.1%.

**Dye-Sensitized Solar Cell (DSSC)**

The key components of a dye-sensitized solar cell are working electrode, electrolyte, dye and counter electrode as shown in Fig. 6.4. The working electrode (FTO) in DSSC is dipped in dye and the counter electrode (Pt) in an electrolyte ($I^-$) which are then fixed together to prevent electrolyte from leaking [44]. Pt is mainly utilized as the counter electrode because of its high conductivity and noncorrosive nature to

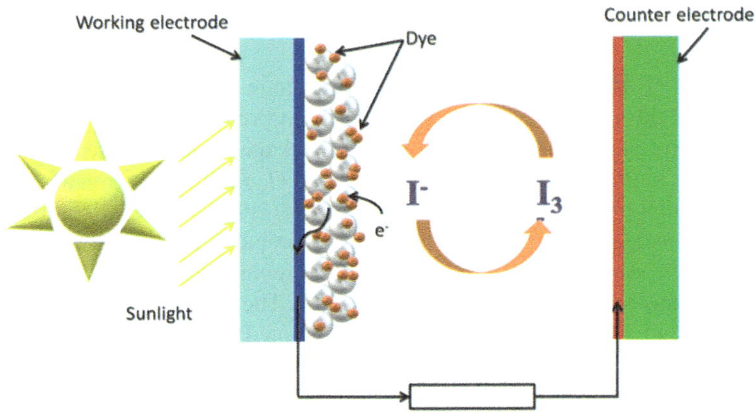


**Fig. 6.4** Schematic of a DSSC

electrolyte. However, Pt is highly expensive which may increase the fabrication cost of solar cell.

rGO is generally found to replace Pt in DSSCs owing to its small manufacturing charge and high conductivity. The oxygen functionalities present in rGO provide sites for catalytic activity which supports the redox reaction of the electrolyte [45]. Zheng et al. fabricated a DSSC using rGO film as the counter electrode [46]. They found the efficiency of solar cell is based on the thickness of rGO film. A high PCE value of 7.19% was obtained when the thickness of rGO was 15 μm. Qiu et al. used hydrazine hydrate to treat rGO electrodes [47]. The addition of hydrazine hydrate resulted in large surface area, higher catalytic activity and lower sheet resistance. With the increase in amount of hydrazine hydrate, the PCE value was found to increase to 2.622%. The thermally treated GO may show an increase in efficiency of DSSC. GO was treated at different temperatures 150, 250 and 350 °C among which GO reduced at 350 °C showed highest PCE of 3.60% [48]. The electrochemical reduction method was followed to fabricate PDDA@ERGO for replacing Pt electrode in DSSC, and an extremely high PCE value of 9.14% was obtained [49].

## *Supercapacitors*

Supercapacitors are devices that store energy and have larger power and energy density than batteries. In a supercapacitor, the electrolyte is sandwiched between the two electrodes. Different types of electrolytes such as aqueous, organic or ionic electrolytes may be used. Aqueous electrolytes (aqueous solution of $Na_2SO4$, $H_2SO_4$, KOH, etc.) possess many advantages like high ionic mobility, high-conductivity, low-cost, non-toxic, nonflammable and are easy to handle. On the contrary, organic electrolytes (acetonitrile, propylene carbonate, tetraethylammonium tetrafluoroborate)

**Table 6.2** Photovoltaic characteristics of DSSCs using rGO as the counter electrode

| Reducing agent | $V_{oc}$ (V) | $J_{sc}$ (mA/cm$^2$) | FF (%) | PCE (%) | References |
|---|---|---|---|---|---|
| Zn metal | 0.78 | 12.82 | 0.72 | 7.19 | [46] |
| Hydrazine hydrate | 0.684 | 9.475 | 0.405 | 2.622 | [47] |
| Thermal (350 °C) | 0.66 | 16.35 | 33.33 | 3.60 | [48] |
| Electrochemical | 0.692 | 18.77 | 74 | 9.14 | [49] |
| Thermal (200 °C) | 0.58 | 1.38 | 44 | 0.97 | [50] |

have low conductivity, expensive, more toxic, inflammable and difficult to handle. However, organic electrolytes possess larger potential window compared to aqueous electrolytes [51]. Ionic liquids can operate in wider potential window than aqueous and organic electrolytes [52]. Reduced graphene oxide possesses high electrical conductivity and large specific surface area which enables it to be used as electrodes for application in supercapacitors. Researchers have made use of different electrolytes for studying the application of rGO in supercapacitor application as shown in Table 6.2. For example, Johra et al. studied the supercapacitive behavior of hydrothermally reduced rGO in $H_2SO_4$ and found, the specific capacitance value to be 367 F/g at a current density of 1 A/g [53]. Ambrosi et al. synthesized electrochemically exfoliated graphene and found the specific capacitance value as 106 and 78 F/g in $Na_2SO_4$ and $H_2SO_4$, respectively [54]. Zhao et al. found that at a current density of 0.4 A/g the specific capacitance value was 260.5 F/g for thermally reduced GO in KOH [55] (Table 6.3).

rGO follows the mechanism of electrochemical double-layer capacitors (EDLC). EDLCs consist of an electrolyte, two electrodes and a separator. On application of voltage, there is accumulation of charge on the electrodes, and a difference in potential is created which results in attraction of opposite charges on the electrodes. A double layer of charge is produced on the interface between electrode and electrolyte which leads to diffusion of ions of the electrolyte as shown in Fig. 6.5 [61]. pH also possesses noteworthy role in the supercapacitance behavior of rGO. Ha et al. showed rGO at pH 10 possessed larger specific surface area of 216 m$^2$/g compared to that of rGO at pH 2 with 155 m$^2$/g [62]. The rGO at higher pH exhibited higher specific conductance (182 F/g) than that at lower pH (143 F/g). Bai et al. deliberated the supercapacitive behavior of rGO at basic, neutral and acidic conditions [63]. rGO at basic medium showed a specific capacitance value of 185 F/g with EDLC behavior. Moreover, rGO at neutral and acidic medium presented EDLC along with pseudocapacitive behavior showing specific capacitance value of 225 F/g and 230 F/g, respectively.

## Conclusion

Reduced graphene oxide is reduced form of graphene oxide containing a lower number of oxygen functionalities. Various techniques such as thermal, chemical,

**Table 6.3** Specific capacitance and charge retention of rGO in different electrolytes

| Electrolyte | Current density (A/g) | Specific capacitance (F/g) | No of cycles | Capacitance retention (%) | References |
|---|---|---|---|---|---|
| $H_2SO_4$ | 367 | 1 | 1000 | 107.7 | [53] |
| $H_2SO_4$ | 253 | 0.2 | 3000 | 102.5 | [56] |
| $H_2SO_4$ | 348 | 0.2 | 3000 | 120 | [57] |
| $Na_2SO_4$ | 106 | 0.1 | – | – | [54] |
| KOH | 260.5 | 0.4 | – | – | [55] |
| KOH | 225 | 0.5 | 1200 | 93 | [58] |
| Tetraethyl ammonium tetrafluoroborate in acetonitrile (TEA BF4/AN) | 120 | 10 | 2000 | 95 | [59] |
| 1-butyl-3 methyl-imidazolium tetrafluoroborate in acetonitrile BMIM ($BF_4$/AN) | 169.06 | 1.6 | 3000 | 96.6 | [60] |
| 1-butyl-3 methyl-imidazolium hexafluorophosphate (BMIP$F_6$) | 158 | 0.2 | – | – | [57] |

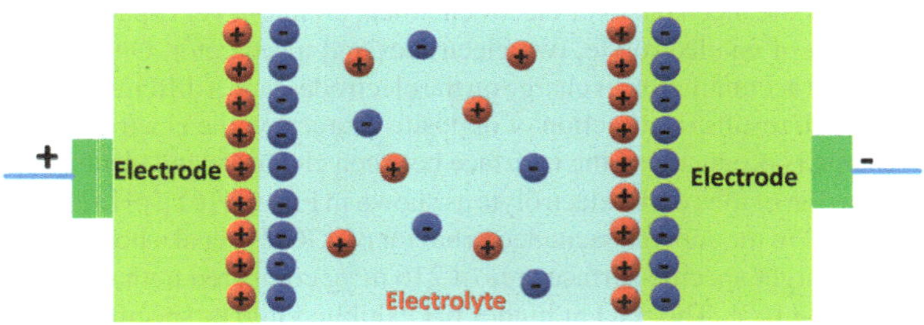


**Fig. 6.5** Schematic of the electrochemical double-layer capacitor

electrochemical and hydrothermal are used to reduce GO. Then, rGO synthesized via these various techniques could be utilized for applications in energy conversion and storage systems. Nevertheless, synthesis of high-quality rGO using environmental-friendly techniques still remains a challenge. rGO serves as electrode, HTL as well as ETL in OSCs. It is also used to substitute the Pt counter electrode in DSSCs owing to its higher conductivity and low-cost production. The use of rGO as electrode for supercapacitors has highly gained attention. Supercapacitors with high specific capacitance value and large cyclic stability can be fabricated using rGO as the electrode material. Thus, rGO acts as a wonder material which can be used for the advancement of energy applications.

**Acknowledgements** The authors thank the UGC-DAE-CRS Indore (Project Ref: CRC-IC-MSR-07/CRS-215/2017-18/1296) and Dr. Ramdas Pai and Vasanthi Pai Endowment (Project Ref: SMU/ENDOW/2016-17/292/002) for providing Grant-in-Aid for Junior Research Fellowship to Ms. Sadhna Rai and Ms. Rabina Bhujel, respectively.

## References

1. Yang, Z., Ren, J., Zhang, Z., Chen, X., Guan, G., Qiu, L., Zhang, Y., Peng, H.: Recent advancement of nanostructured carbon for energy applications. Chem. Rev. **115**, 5159–5223 (2015)
2. Bolotin, K.I., Sikes, K.J., Jiang, Z., Klima, M., Fudenberg, G., Hone, J., Kim, P., Stormer, H.L.: Ultrahigh electron mobility in suspended graphene. Solid State Commun. **146**, 351–355 (2008)
3. Stoller, M.D., Park, S., Yanwu, Z., An, J., Ruoff, R.S.: Graphene-based ultracapacitors. Nano Lett. **8**, 3498–3502 (2008)
4. Balandin, A.A., Ghosh, S., Bao, W., Calizo, I., Teweldebrhan, D., Miao, F., Lau, C.N.: Superior thermal conductivity of single-layer graphene. Nano Lett. **8**, 902–907 (2008)
5. Brodie, B.C.: On the atomic weight of graphite. Philos. Trans. R. Soc. London. **149**, 249–259 (1859)
6. Staudenmaier, L.: Verfahren zur Darstellung der Graphitsäure. Ber. Der Dtsch. Chem. Gesellschaft. **32**, 1394–1399 (1899)
7. Hummers, W.S., Offeman, R.E.: Preparation of graphitic oxide. J. Am. Chem. Soc. **80**, 1339–1339 (1958)
8. Marcano, D.C., Kosynkin, D.V., Berlin, J.M., Sinitskii, A., Sun, Z., Slesarev, A., Alemany, L.B., Lu, W., Tour, J.M.: Improved synthesis of graphene oxide. ACS Nano **4**, 4806–4814 (2010)
9. Higginbotham, A.L., Kosynkin, D.V., Sinitskii, A., Sun, Z., Tour, J.M.: Lower-defect graphene oxide nanoribbons from multiwalled carbon nanotubes. ACS Nano **4**, 2059–2069 (2010)
10. McAllister, M.J., Li, J.-L., Adamson, D.H., Schniepp, H.C., Abdala, A.A., Liu, J., Herrera-Alonso, M., Milius, D.L., Car, R., Prud'homme, R.K., Aksay, I.A.: Single sheet functionalized graphene by oxidation and thermal expansion of graphite. Chem. Mater. **19**, 4396–4404 (2007)
11. Becerril, H.A., Mao, J., Liu, Z., Stoltenberg, R.M., Bao, Z., Chen, Y.: Evaluation of solution-processed reduced graphene oxide films as transparent conductors. ACS Nano **2**, 463–470 (2008)
12. Wang, X., Zhi, L., Müllen, K.: Transparent, conductive graphene electrodes for dye-sensitized solar cells. Nano Lett. **8**, 323–327 (2008)
13. Wu, Z.S., Ren, W., Gao, L., Liu, B., Jiang, C., Cheng, H.M.: Synthesis of high-quality graphene with a pre-determined number of layers. Carbon N. Y. **47**, 493–499 (2009)
14. Li, X., Wang, H., Robinson, J.T., Sanchez, H., Diankov, G., Dai, H.: Simultaneous nitrogen doping and reduction of graphene oxide. J. Am. Chem. Soc. **131**, 15939–15944 (2009)
15. He, Q., Wu, S., Gao, S., Cao, X., Yin, Z., Li, H., Chen, P., Zhang, H.: Transparent, flexible, all-reduced graphene oxide thin film transistors. ACS Nano **5**, 5038–5044 (2011)
16. Mattevi, C., Eda, G., Agnoli, S., Miller, S., Mkhoyan, K.A., Celik, O., Mastrogiovanni, D., Granozzi, G., Garfunkel, E., Chhowalla, M.: Evolution of electrical, chemical, and structural properties of transparent and conducting chemically derived graphene thin films. Adv. Funct. Mater. **19**, 2577–2583 (2009)

17. Gómez-Navarro, C., Weitz, R.T., Bittner, A.M., Scolari, M., Mews, A., Burghard, M., Kern, K.: Electronic transport properties of individual chemically reduced graphene oxide sheets. Nano Lett. **7**, 3499–3503 (2007)
18. Stankovich, S., Dikin, D.A., Piner, R.D., Kohlhaas, K.A., Kleinhammes, A., Jia, Y., Wu, Y., Nguyen, S.T., Ruoff, R.S.: Synthesis of graphene-based nanosheets via chemical reduction of exfoliated graphite oxide. Carbon **45**, 1558–1565 (2007)
19. Stankovich, S., Dikin, D.A., Dommett, G.H.B., Kohlhaas, K.M., Zimney, E.J., Stach, E.A., Piner, R.D., Nguyen, S.B.T., Ruoff, R.S.: Graphene-based composite materials. Nature **442**, 282–286 (2006)
20. Fernández-Merino, M.J., Guardia, L., Paredes, J.I., Villar-Rodil, S., Solís-Fernández, P., Martínez-Alonso, A., Tascón, J.M.D.: Vitamin C is an ideal substitute for hydrazine in the reduction of graphene oxide suspensions. J. Phys. Chem. C. **114**, 6426–6432 (2010)
21. Chen, L., Tang, Y., Wang, K., Liu, C., Luo, S.: Direct electrodeposition of reduced graphene oxide on glassy carbon electrode and its electrochemical application. Electrochem. Commun. **13**, 133–137 (2011)
22. Tong, H., Zhu, J., Chen, J., Han, Y., Yang, S., Ding, B., Zhang, X.: Electrochemical reduction of graphene oxide and its electrochemical capacitive performance. J. Solid State Electrochem. **17**, 2857–2863 (2013)
23. Kar, T., Devivaraprasad, R., Singh, R.K., Bera, B., Neergat, M.: Reduction of graphene oxide – a comprehensive electrochemical investigation in alkaline and acidic electrolytes. RSC Adv. **4**, 57781–57790 (2014)
24. Ramesha, G.K., Sampath, S.: Electrochemical reduction of oriented graphene oxide films: an in situ Raman spectroelectrochemical study. J. Phys. Chem. C. **113**, 7985–7989 (2009)
25. Fu, C., Kuang, Y., Huang, Z., Wang, X., Du, N., Chen, J., Zhou, H.: Electrochemical co-reduction synthesis of graphene/Au nanocomposites in ionic liquid and their electrochemical activity. Chem. Phys. Lett. **499**, 250–253 (2010)
26. Zhang, X., Zhang, D., Chen, Y., Sun, X., Ma, Y.: Electrochemical reduction of graphene oxide films: Preparation, characterization and their electrochemical properties. Chin. Sci. Bull. **57**, 3045–3050 (2012)
27. Modeshia, D.R., Walton, R.I.: Solvothermal synthesis of perovskites and pyrochlores: crystallisation of functional oxides under mild conditions. Chem. Soc. Rev. **39**, 4303 (2010)
28. Zheng, X., Peng, Y., Yang, Y., Chen, J., Tian, H., Cui, X., Zheng, W.: Hydrothermal reduction of graphene oxide; effect on surface-enhanced Raman scattering. J. Raman Spectrosc. **48**, 97–103 (2017)
29. Park, H., Chang, S., Zhou, X., Kong, J., Palacios, T., Gradecak, S.: Flexible graphene electrode-based organic photovoltaics with record-high efficiency. ECS Trans. **69**, 77–82 (2015)
30. Manzano-Ramírez, A., López-Naranjo, E.J., Soboyejo, W., Meas-Vong, Y., Vilquin, B.: A review on the efficiency of graphene-based BHJ organic solar cells. J. Nanomater. (2015). https://doi.org/10.1155/2015/406597
31. Yin, Z., Sun, S., Salim, T., Wu, S., Huang, X., He, Q., Lam, Y.M., Zhang, H.: Organic photovoltaic devices using highly flexible reduced graphene oxide films as transparent electrodes. ACS Nano **4**, 5263–5268 (2010)
32. Yin, Z., Wu, S., Zhou, X., Huang, X., Zhang, Q., Boey, F., Zhang, H.: Electrochemical deposition of ZnO nanorods on transparent reduced graphene oxide electrodes for hybrid solar cells. Small **6**, 307–312 (2010)
33. Po, R., Carbonera, C., Bernardi, A., Camaioni, N.: The role of buffer layers in polymer solar cells. Energy Environ. Sci. **4**, 285–310 (2011)
34. Cheng, X., Long, J., Wu, R., Huang, L., Tan, L., Chen, L., Chen, Y.: Fluorinated reduced graphene oxide as an efficient hole-transport layer for efficient and stable polymer solar cells. ACS Omega. **2**, 2010–2016 (2017)
35. Yun, J.M., Yeo, J.S., Kim, J., Jeong, H.G., Kim, D.Y., Noh, Y.J., Kim, S.S., Ku, B.C., Na, S.I.: Solution-processable reduced graphene oxide as a novel alternative to PEDOT:PSS hole transport layers for highly efficient and stable polymer solar cells. Adv. Mater. **23**, 4923–4928 (2011)

36. Jeon, Y.J., Yun, J.M., Kang, M., Lee, S., Jung, Y.S., Hwang, K., Heo, Y.J., Kim, J.E., Kang, R., Kim, D.Y.: 2D/2D vanadyl phosphate (VP) on reduced graphene oxide as a hole transporting layer for efficient organic solar cells. Org. Electron. Phys. Mater. Appl. **59**, 92–98 (2018)
37. Brabec, C.J., Shaheen, S.E., Winder, C., Sariciftci, N.S., Denk, P.: Effect of LiF/metal electrodes on the performance of plastic solar cells. Appl. Phys. Lett. **80**, 1288–1290 (2002)
38. Kim, J.Y., Kim, S.H., Lee, H.H., Lee, K., Ma, W., Gong, X., Heeger, A.J.: New architecture for high-efficiency polymer photovoltaic cells using solution-based titanium oxide as an optical spacer. Adv. Mater. **18**, 572–576 (2006)
39. White, M.S., Olson, D.C., Shaheen, S.E., Kopidakis, N., Ginley, D.S.: Inverted bulk-heterojunction organic photovoltaic device using a solution-derived ZnO underlayer. Appl. Phys. Lett. **89**, 143517 (2006)
40. Zhao, Y., Xie, Z., Qu, Y., Geng, Y., Wang, L.: Effects of thermal annealing on polymer photovoltaic cells with buffer layers and in situ formation of interfacial layer for enhancing power conversion efficiency. Synth. Met. **158**, 908–911 (2008)
41. Mahmoudi, T., Wang, Y., Hahn, Y.B.: Graphene and its derivatives for solar cells application. Nano Energy. **47**, 51–65 (2018)
42. Jayawardena, K.D.G.I., Rhodes, R., Gandhi, K.K., Prabhath, M.R.R., Dabera, G.D.M.R., Beliatis, M.J., Rozanski, L.J., Henley, S.J., Silva, S.R.P.: Solution processed reduced graphene oxide/metal oxide hybrid electron transport layers for highly efficient polymer solar cells. J. Mater. Chem. A. **1**, 9922 (2013)
43. Woo Lee, H., Young, Oh., J., Il Lee, T., Soon Jang, W., Bum Yoo, Y., Sang Chae, S., Ho Park, J., Min Myoung, J., Moon Song, K., Koo Baik, H. : Highly efficient inverted polymer solar cells with reduced graphene-oxide-zinc-oxide nanocomposites buffer layer. Appl. Phys. Lett. **102**, 193903 (2013)
44. Sharma, K., Sharma, V., Sharma, S.S.: Dye-sensitized solar cells: fundamentals and current status. Nanoscale Res. Lett. **13**, 1–46 (2018)
45. Yeh, M.H., Lin, L.Y., Chang, L.Y., Leu, Y.A., Cheng, W.Y., Lin, J.J., Ho, K.C.: Dye-sensitized solar cells with reduced graphene oxide as the counter electrode prepared by a green photothermal reduction process. ChemPhysChem **15**, 1175–1181 (2014)
46. Zheng, H., Neo, C.Y., Mei, X., Qiu, J., Ouyang, J.: Reduced graphene oxide films fabricated by gel coating and their application as platinum-free counter electrodes of highly efficient iodide/triiodide dye-sensitized solar cells. J. Mater. Chem. **22**, 14465 (2012)
47. Qiu, L., Zhang, H., Wang, W., Chen, Y., Wang, R.: Effects of hydrazine hydrate treatment on the performance of reduced graphene oxide film as counter electrode in dye-sensitized solar cells. Appl. Surf. Sci. **319**, 339–343 (2014)
48. Jang, H.S., Yun, J.M., Kim, D.Y., Park, D.W., Na, S.I., Kim, S.S.: Moderately reduced graphene oxide as transparent counter electrodes for dye-sensitized solar cells. Electrochim. Acta. **81**, 301–307 (2012)
49. Xu, X., Huang, D., Cao, K., Wang, M., Zakeeruddin, S.M., Grätzel, M.: Electrochemically reduced graphene oxide multilayer films as efficient counter electrode for dye-sensitized solar cells. Sci. Rep. **3**, 1489 (2013)
50. Yuliasari, F., Aprilia, A., Syakir, N., Safriani, L., Saragi, T., Risdiana, Hidayat, S., Bahtiar, A., Siregar, R., Fitrilawati: Characteristics of thermally reduced graphene oxide thin film as DSSC counter electrode. In: IOP Conference Series: Materials Science and Engineering, vol. 196, 012049 (2017)
51. Zhao, C., Zheng, W.: A review for aqueous electrochemical supercapacitors. Front. Energy Res. **3** (2015). https://doi.org/https://doi.org/10.3389/fenrg.2015.00023
52. Salanne, M.: Ionic liquids for supercapacitor applications. Top. Curr. Chem. **375**, 63 (2017)
53. Johra, F.T., Jung, W.G.: Hydrothermally reduced graphene oxide as a supercapacitor. Appl. Surf. Sci. **357**, 1911–1914 (2015)
54. Ambrosi, A., Pumera, M.: Electrochemically exfoliated graphene and graphene oxide for energy storage and electrochemistry applications. Chem. Eur. J. **22**, 153–159 (2016)
55. Zhao, B., Liu, P., Jiang, Y., Pan, D., Tao, H., Song, J., Fang, T., Xu, W.: Supercapacitor performances of thermally reduced graphene oxide. J. Power Sourc. **198**, 423–427 (2012)

56. Rajagopalan, B., Chung, J.S.: Reduced chemically modified graphene oxide for supercapacitor electrode. Nanoscale Res. Lett. **9**, 535 (2014)
57. Chen, Y., Zhang, X., Zhang, D., Yu, P., Ma, Y.: High performance supercapacitors based on reduced graphene oxide in aqueous and ionic liquid electrolytes. Carbon N. Y. **49**, 573–580 (2011)
58. Lei, Z., Lu, L., Zhao, X.S.: The electrocapacitive properties of graphene oxide reduced by urea. Energy Environ. Sci. **5**, 6391–6399 (2012)
59. Zhang, L.L., Zhao, X., Stoller, M.D., Zhu, Y., Ji, H., Murali, S., Wu, Y., Perales, S., Clevenger, B., Ruoff, R.S.: Highly conductive and porous activated reduced graphene oxide films for high-power supercapacitors. Nano Lett. **12**, 1806–1812 (2012)
60. Sahu, V., Shekhar, S., Sharma, R.K., Singh, G.: Ultrahigh performance supercapacitor from lacey reduced graphene oxide nanoribbons. ACS Appl. Mater. Interfaces. **7**, 3110–3116 (2015)
61. Iro, S.Z.: A brief review on electrode materials for supercapacitor. Int. J. Electrochem. Sci. **11**, 10628–10643 (2016)
62. Ha, T., Kim, S.K., Choi, J.W., Chang, H., Jang, H.D.: pH controlled synthesis of porous graphene sphere and application to supercapacitors. Adv. Powder Technol. **30**, 18–22 (2019)
63. Bai, Y., Rakhi, R.B., Chen, W., Alshareef, H.N.: Effect of pH-induced chemical modification of hydrothermally reduced graphene oxide on supercapacitor performance. J. Power Sourc. **233**, 313–319 (2013)

# Chapter 7
# Nanowires/Graphene Nanocomposites for Photovoltaic Applications

**Sumitra Nongthombam and Bibhu Prasad Swain**

## Introduction

Sustainable energy production has been a global concern to resolve energy crisis issues, as well as environmental pollution problems that arise due to the rapid exhaustion of fossil fuels. Continuous deterioration caused by the harmful gases that are released by conventional energy sources is of serious threat. Therefore, probing for new energy sources that are renewable and eco-friendly benign as a replacement is of immediate need. Photovoltaic has attracted tremendous attraction for new-generation energy sources as the technology offers an inexhaustible, renewable and clean energy production. It generates energy from light via photovoltaic effect through the use of semiconducting materials. The low-cost and pollution-free technology make it significant for applications in numerous purposes—current generation, solar water pumping, transport, space satellites, solar power calculators, etc. [1].

Silicon is the primary semiconducting material in the development of solar technology capturing largely (>90%) in the solar trade as silicon-based photovoltaic are cost-effective and long-lasting [2]. This, however, is limited by the unbalanced cost with efficiency providing only 8–19% of the available solar energy [3, 4]. As such, recent interest is in the search for cost-effective and better efficient materials for making photovoltaic cells.

One important characteristic of PV material for increasing solar cell efficiency is the need for a suitable bandgap that matches the solar spectrum to improve the optoelectronic properties. Another important step to enhance efficiency is to improve

S. Nongthombam (✉) · B. P. Swain
Department of Physics, National Institute of Technology Manipur, Langol Rd, Lamphelpat, Imphal, Manipur 795004, India
e-mail: sumitranongs@gmail.com

B. P. Swain
e-mail: bibhuprasad.swain@gmail.com; bpswain@nitmanipur.ac.in



B. P. Swain (ed.), *Nanostructured Materials and their Applications*, Materials Horizons: From Nature to Nanomaterials, https://doi.org/10.1007/978-981-15-8307-0_7

charge collection method. Therefore, choosing the best absorbing material is essential as the bandgap energy of the material determines how much of the sun's spectrum can be absorbed.

Graphene is rising as a potential PV material that can replace the expensive crystalline silicon. The material offers low-cost and improved efficiency of solar cells delivering double output power as that of silicon [5]. Reduced graphene oxide (rGO) is widely studied as an alternative to graphene for its low-cost cost and scalability where its properties can be tuned for desired results. Graphene being a zero bandgap material is highly efficient for making photoanodes. Also, it possesses excellent thermal, electrical and good mechanical and optical transparency henceforth making it an encouraging material for energy storage applications.

## Photovoltaic Effect

The process by which electrical energy is derived from sunlight's radiation through the use of semiconducting materials is called the photovoltaic effect. When a photovoltaic cell is exposed to sunlight which consists of photons of different wavelengths or frequencies, the light may be absorbed, transmitted or reflected. If a material has a bandgap lower to that of the energy of incident light, the light will be absorbed by the material while the light will pass through if the material has a bandgap greater than incident energy. The PV cell is imparted with certain energy by sunlight to free the electrons which correspond to the photoelectric current. The amount of the current produced depends on the efficiency of the PV device used. Silicon is the most widely used cell material because of its efficiency and largely available on the Earth. Efficiency of a solar cell is an important characteristic which indicates the performance of the cell. It is defined as the amount of energy harnessed from a given solar irradiance. It is expressed as

$$\eta = \frac{P_{\mathrm{m}}}{\varepsilon . A} \tag{7.1}$$

where $\eta$ is the maximum achievable efficiency, $P_{\mathrm{m}}$ is the power output, $E$ is the radiation flux and $A$ denotes solar panel area.

Light is composed of photons which are energy packets of electromagnetic energy. When the light of a suitable wavelength is incident on the PV cell, energy is transferred from photon to the atom in the semiconducting material. This energy imparts the electron in the material creating some free electrons. This, in turn, leads to the formation of holes in the vacancy. In this manner, electron-hole pairs are created. When the external circuit is connected electric current is generated.

## *Working of Photovoltaic Cell*

Consider silicon as a model for exhibiting the photovoltaic effect. Silicon belongs to group 14 elements in the periodic table where each atom consists of 4 valence electrons. Silicon is bonded to four other silicon atoms to be structurally stable, and so many other Si atoms are bonded covalently to form a lattice. Silicon forms the basis of a PV cell. A typical silicon PV cell is shown in Fig. 7.1

In a typical PV cell, the *p-n* semiconductor is formed. When silicon atoms are doped with group 13 elements, there is a deficiency of an electron to form an octet in their bonding, thus creating a vacancy represented as holes resulting in the formation of *p*-region. On the other hand, an n-type region is created by doping Si atoms with group 15 elements. Here, there is an extra unshared electron which increases with the number of bonds formed. Electrons in the n region tend to fill the vacancies in the p-region leading to the diffusion of the two kinds between the *p* and *n* region. Consequently, at the junction, an electric field develops which further restricts the movement of electrons and holes across them. This region around the junction which is devoid of free charge carriers is called the potential barrier.

When the light of certain wavelength falls on the solar panel, some free electrons are imparted from the *n* region. However, these free electrons are restricted to move into the *p* region by the potential barrier. Subsequently, when an external circuit is

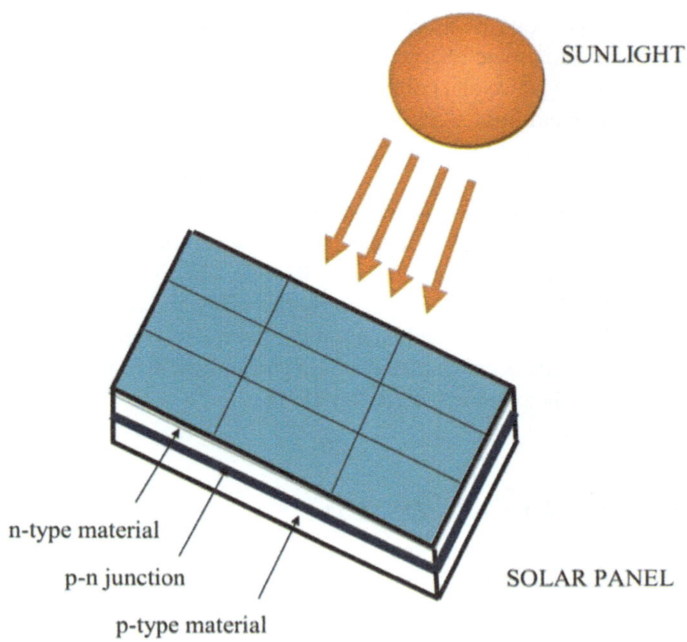


**Fig. 7.1** A schematic view of silicon solar cell

connected with PV cell, electrons flow continuously from the *n* side which is again accelerated by the barrier and holes to the other side resulting in the photocurrent generation.

## Different Kinds of Solar Cells

With the successful silicon solar cell technology, various other solar cells have been investigated in parallel. Solar cells are of three kinds of generation [6, 7] as follows:

1. Silicon-based solar cells,
2. Thin-film solar cells, and
3. Organic/polymer solar cells (OSCs), perovskite solar cells (PSCs), dye-sensitized solar cells (DSSCs) and quantum dot-sensitized solar cells (QDSc) include the third generation.

Recently, in the last decade, third-generation solar cells are widely studied.

### *Organic/polymer Solar Cells (OSCs)*

Organic polymer solar cells are those photovoltaic which uses organic molecules for absorbing photons and create charge carriers to generate electricity via the photoelectric effect. Most organic solar cells use polymer molecules, and they are solution-processable, flexible and light-wearable materials. They have a high optical absorption coefficient which facilitates absorbing a large amount of energy with a small number of materials. The difference between silicon-based solar cells and OSCs is that *p-n* semiconducting junction is formed in silicon-based solar cells while separate materials are used for donating an electron and accepting electron in fabricating OSCs. Although organic polymers are light, flexible and cost-effective production, their advantages are limited by their low carrier mobility and short-lived. Thereby, a good number of researches are conducted on the hybrid solar cells of organic materials with inorganic structures which have high carrier mobility and effective charge transport to enhance the resultant efficiency [8, 9].

### *Dye-Sensitized Solar Cells (DSSCs)*

DSSCs owing to its simplicity in fabrication and low cost are immensely studied for electricity conversion from solar energy with the utilization of photosensitized nanomaterials. Metal oxides and nanowires have been recently studied for tailoring photoanodes in solar cells for increased charge transport and efficiency [10].

However, since they have poor crystallinity and poor transport properties, investigation on modifications of photoanodes with hybrid nanocomposites is on the recent trend [11]. Such composite-based semiconductors exhibit enhanced optical and electrical properties for DSSC applications. Nanocomposites formed by two or more materials with desired performances are mainly studied because of the synergetic effect shown by them.

## *Quantum Dot Dye-Sensitized Solar Cells*

QDSCs are third-generation solar cells which have been considered to be very promising candidates. They are derived from DSCs with quantum dots (QDs) replacing organic molecules for absorbing photons. QDSCs facilitate the use of QDs by exciting multiple electrons with a single photon of light contrary to organic dyes where a single electron is excited per photon. Also, this new-generation PV cells offer solution processability, low-cost production and facile fabrication. The basic components of QDSCs are (1) metal oxide electron transporting/acceptor materials (ETMs), (2) QD sensitizers, (3) counter electrodes and (4) electrolyte or host transporting materials (HTMs) [12].

QDSCs work with a photon of light being incident on a QD particle which causes some of these inside the material to get excited. These photo excited $e^{-}$s are extracted by a photoanode and carry onto the conductive electrode layer leaving behind holes in the QD material. These holes are then filled with $e^{-}$s from the electrolyte which is again supplied by the counter electrode. The process continues to generate the electrical current.

### ETMs

ETMs act as a substrate of photoanode. Necessary/desired properties for ETMs include:

1. Appropriate conduction band edge,
2. High electron mobility,
3. Abundant surface area.

### QD Sensitizers

QD plays the role of extracting solar energy and generate excitons. Characteristics of an ideal QD are:

1. Narrow bandgap,
2. Non-toxic and facile preparation,
3. Stable exposure to heat and light.

QD sensitizers may be deposited over metal oxide semiconductor substrate by in situ or ex situ process.

#### Current Electrodes

In QDSCs, counter electrodes play the role of catalyst and supplied electrons for the reaction in the electrolyte. Characteristics of a good CE are:

1. Highly conductive,
2. High catalytic activity,
3. Stable chemical and mechanical properties.

#### Electrolytes

Electrolytes are core elements of QDSCs that can influence the performance of a photovoltaic. Characteristics of a good electrolyte are:

1. Good chemical stability,
2. Less corrosive towards QD material,
3. Fast regeneration rate,
4. Highly conductive.

Electrolytes are of liquid electrolytes, quasi solid-state electrolytes.

## Transparent Conductive Materials

Transparent conductive materials play an important role in using electrodes in photovoltaics and other electronic devices such as LED, touchscreen, etc. Indium tin oxide (ITO) is the most widely used material. However, due to the limited availability of indium, high cost and brittleness of ITO [13, 14], alternative materials such as graphene, CNTs, nanowires and polymers have been explored [15]. Whatsoever, competing for ITO for better transparent and lower sheet resistance remains a challenge. Nanostructures metal oxides such as zinc oxide and titania are a potential choice for buffer material due to their transparency in the visible and good charge transport properties. However, since they are electron transporters, they must be in an inverted configuration with a transparent electrode acting as the cathode. A study by Dong Seok Leem and his team found that Ag NWs electrode-based solar cells with buffered layers have better power electrical efficiency than without a buffer layer.

## Graphene as Electrode Material in Photovoltaics

Electrode materials play a crucial part in photovoltaics. Indium tin oxide (ITO) based films are the most widely studied electrode materials for photovoltaics because of its excellent optical transparency and good electrical properties. In recent years, graphene has been popularly studied for numerous applications including photovoltaics because of its noble properties. Among its notable properties, its high transparency, high conductivity, good chemical and mechanical strength and largely abundant make it a suitable candidate to be explored as a transparent conducting electrode material for replacement of ITO in photovoltaics. QDSCs have attracted recent interest because of its certain advantages over conventional solar cells in terms of bandgaps tunability, high extinction coefficient and multiple creations of electron-hole pairs in comparison with dyes. However, its benefits are limited by the charge recombination at the interface of the photoanode layer and electrolyte. This can be improved by providing high electrical electrode materials. Thus, varying the photoanodes with graphene that has high electron mobility can increase the charge transport.

Graphene is a 2D material which is considered to be one of the most encouraging materials finding enormous applications in optoelectronics and energy storage. Reduced graphene oxide rGO is the chemically produced graphene providing low-cost and large-scale availability. The reduction of GO plays an important part in the quality of the material, and its physical properties can be tuned by controlling the reduction process which is not possible in the case of graphene. Besides, graphene being almost transparent, giant intrinsic mobility and excellent strength offer for an efficient cost-effective solar cell material providing double output in less time as that obtained by silicon solar material.

## Overview of NW/Graphene Nanocomposites for Photovoltaics Applications

Low-dimensional materials have been studied recently because of their vision in future technologies to produce portable size, flexible and wearable materials. 1D material such as nanowires and nanotubes is believed to produce enhanced charge collection efficiency owing to their short charge transfer length. However, these low-dimensional materials are limited by their low bandgaps and low carrier mobilities. On the other hand, graphene materials find a drawback in the stacking of its atomic sheets, thus decreasing its surface area and pore volume. So, hybridization of graphene with nanowires to construct 3D architectures has the potential ability to exhibit unique structural, chemical and physical properties. Hence, semiconductor NWs/graphene composites have attained much attraction in optoelectronic applications producing high-efficiency and low-cost device.

Silicon is the most widely used solar cell material. However, the bulk material is limited in emitting visible light as it is an indirect bandgap material. This does not happen when going down to the nanoregime and can emit visible light in Si NWs. Recently, Si NWs have been investigated for its optical properties and the resulting electrical efficiency in photovoltaic applications [16, 17]. Studies on the material reveal that nanowires have certain advantages of short collection length for excited carriers resulting in their carrier collection efficiency. Besides, Si NWs arrays are usable as an effective antireflection coating for Si solar cells using this array. However, the limitation lies in the lower electrical efficiency of the NWs as compared to those without NWs. One interesting effect on the study of optical absorbance of Si NWs is that optical absorption properties increase in the high-frequency region and are less in the lower-frequency region in comparison with thin-film Si. The reason may be asserted due to the small reflection and zero transmittance in the high-frequency region.

While various alternatives for ITO were explored, greater transparency and conductivity have been a challenge. Graphene, a 2D material that is almost transparent and high conductivity, is however difficult to compete ITO owing to its high resistance grain boundaries. A new approach is described recently to overcome the lower resistance of graphene by percolation doping with metallic nanowires [18]. Through this, the no. of electronic pathways is increased to bridge the percolation bottleneck, resulting in the enhanced conduction of the graphene. Ag NWs doping on polycrystalline graphene can achieve less resistance ($<10\ \Omega$/sq) maintaining the high transmittance of graphene (>90%). A successful percolation of single-layer graphene with Ag NWs is reported by Chen et al. [19] achieving transparency of 88% with 22 $\Omega$/sq sheet resistance stabilizing to 13 $\Omega$/sq (after 4 months) at the wavelength of 550 nm. The work demonstrates the effective co-percolation network theory where bottleneck conduction of graphene is prevented by Ag NWs and vice versa leading to lesser resistance of grain boundaries. The hybrid structure ensures excellent stability to mechanical pressure, bending and over time yielding a cost-effective, scalable and flexible transparent conducting electrode for photovoltaic applications.

While the growth of semiconducting nanowires over graphene provides a new scope for producing high-performance optoelectronic devices, preserving the structural and electrical properties of graphene has been a challenge. A hydrothermal grown of ZnO NWs on graphene is studied by Park et al. for solar cell applications with modifications using two photoactive materials PbS quantum dots and polymer P3HT to provide hole transport layer. The interfacial polymer growth enhances the conformal, smooth wetting of the ZnO seed layer and subsequent ordered ZnO NW array. The method employs the hybrid structure using PbS quantum dots and polymer P3HT providing PCE of 4.2% and 0.5%, respectively, preserving the structural and electrical properties of graphene [19].

Copper NWs/graphene oxide thin film obtained by Thekkedar [20] and his team present a 43% enhancement in optical properties of the composite in comparison with graphene oxide with 93% optical transmittance and 10 $\Omega$/sq. Its limitation being low carrier mobility is fixed by developing hybrid materials. Another disadvantage of copper material is the oxidative nature of the material. By making composites with graphene, the heterogeneous system can act as an oxidation protective coating.

Since nanostructures possess unique physical and chemical properties differing from that of bulk as a result of quantum confinement effect, a variety of low-dimensional nanostructures has been greatly investigated for their potential applications in optoelectronics [16].

In contrary to silicon, the alloy composition and bandgap of III-V materials can be tuned for a better match of the solar spectrum, thus offering the highest energy conversion efficiency in photovoltaic devices [21].

Nanowires offer wide applications in optoelectronics and energy storage applications because of their large surface to volume ratio and alignment in 1D enhances the charge collection efficiency. Therefore, research has been explored recently on the semiconducting nanowires decorated graphene nanocomposites as photoanodes in photovoltaics applications. Low-dimensional materials are advantageous in making nanocomposites with graphene for photovoltaic applications because of its large surface to volume ratio and fast charge transport.

## Fabrication of Photovoltaic Devices Based on Nanowire/Graphene Transparent Conductive Electrodes

### *Plasma Enhanced Chemical Vapor Deposition (PECVD) Synthesized Cu NWs/Graphene Electrode*

Initially, Cu NWs are synthesized using $CuCl_2.2H_2O$, hexadecyl amine (HMA) and glucose as precursors in a simple way by chemical method followed by a thermal method. Cu NWs are then placed in the PE-CVD chamber. The process temperature of the chamber is maintained at 400 °C. A hydrogen plasma (RF-50W) is applied to the vacuum chamber for a few minutes so that the oxide layer or any other contaminants present on the Cu NWs can be removed. Copper NW/Graphene electrode is prepared by the PECVD process on the glass substrate by supplying reaction gas mixtures of methane and hydrogen (40:1) with methane as the source for graphene. For the fabrication of solar cells, the prepared electrode is spin-coated with poly (3, 4-ethylene dioxythiophene) polystyrene sulfonate (PEDOT: PSS). Here, PEDOT: PSS provides a hole transport layer in the cell. Then poly[[4,8-bis[(2-Ethylhexyl)oxy]benzo[1,2-b:4,5-b′]dithiophene-2,6-diyl][3-fluoro-2-[(2-ethylhexy) carbonyl]thieno [3, 4-b] thiophenediyl]] [6, 6]-phenyl-C71-butyric acid methyl ester (PTB7): ($PC_{71}MB$) is spin-coated on top of the hole transport layer. Finally, LiF and Al are deposited on top of the architecture through thermal evaporation [22].

### *Solution-Processed Ag NWs/Graphene Transparent Electrode for Fabricating Organic Solar Cells (OSCs)*

Photovoltaic device based on Ag NWs/Graphene is fabricated using (PTB7): ($PC_{71}MB$) as photoactive material. Initially, graphene is prepared from graphite through electrochemical exfoliation using ammonium sulfate electrolyte (0.1M aq). Ag NWs dispersed solution is spray-coated over Si substrate. The as-synthesized graphene is made dispersion in *N*, *N*-dimethylformamide (DMF) and sprays coated over the Ag NWs network. Organic solar cells (OSCs) are fabricated based on the hybrid film by uniformly spraying coating (PEDOT: PSS) above it. PEDOT: PSS acts as an interfacial conductive layer (hole transport layer). Barium and aluminum are deposited over PTB7: $PC_{71}MB$ as the cathode. This solution-processed electrode overcomes the Ag NWs limitations reducing its sheet resistance to 83% and its surface roughness by 11.8 nm. The method provides the Ag NWs/graphene-based OSCs devices PCE of 6.6% [23].

## Photovoltaic Measurements

The performance of a PV cell is analyzed by studying the relationship between current and voltage typically employed on the device. Figure 7.2 shows the graphical representation of current versus voltage also known as the *I-V* characteristic curve.

From this study, the efficiency of a given PV cell can be summarized, i.e., the maximum energy output that can be obtained in a given solar cell from the amount of solar energy striking the cell.

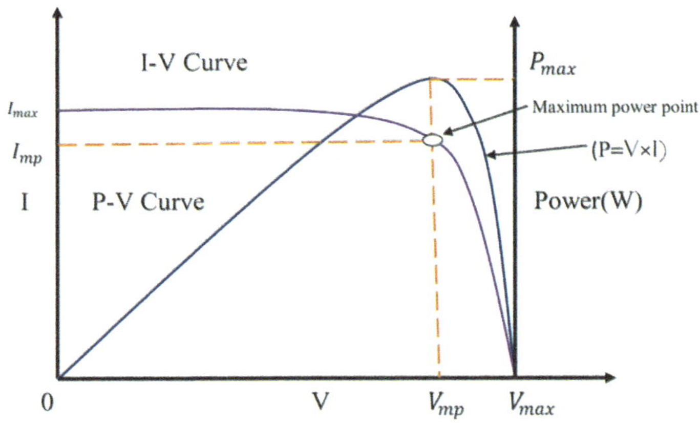


**Fig. 7.2** Current–voltage characteristic curve with the power–current relationship of a solar cell

Theoretically, maximum power is the product of maximum current and maximum voltage given by

$$P_{\mathrm{m}} = I_{\mathrm{m}} \times V_{\mathrm{m}} \tag{7.2}$$

The maximum current is the short circuit current denoted by $I_{sc}$ when the circuit is shorted and so no voltage flows across the cell and the total resistance across it is zero. A maximum voltage is obtained when no external load is connected to the circuit. Here current reaches a minimum and the maximum voltage is known as the open-circuit voltage ($V_{oc}$). However, as seen from the I-V curve, no power is generated at $I_{sc}$ and $V_{oc}$, but at a certain point, power attains a maximum for a given combination of current and voltage denoted by $I_{mp}$ and $V_{mp}$.

$$P_{\mathrm{mp}} = I_{\mathrm{mp}} \times V_{\mathrm{mp}} \tag{7.3}$$

Another important characteristic that can be evaluated from the solar cell is the fill factor defined by the ratio of the maximum power that can be achieved from the solar cell to the theoretical power of the cell. It describes the quality of the solar array.

$$FF = \frac{P_{\mathrm{mp}}}{P_{\mathrm{m}}} \tag{7.4}$$

## Conclusions

In summary, semiconducting nanowires/Graphene nanocomposites obtain enhanced efficiency and greater optical transparency compared to the individual materials. Copper NWs/graphene oxide thin film obtained presents a 43% enhancement in optical properties of the composite in comparison with graphene oxide with 93% optical transmittance and 10 Ω/sq. Percolation of single-layer graphene with Ag NWs can achieve transparency of 88% with 22 Ω/sq sheet resistance stabilizing to 13 Ω/sq (after 4 months) at a wavelength of 550 nm. Thus, the hybrid nanostructure is a promising candidate for high-performance photovoltaic.

## References

1. Fraas, L., Partain, L.: Solar Cells and Their Applications, 2nd edn. Wiley, Hoboken (2010)
2. Zhang, Y., Liu, H.: Nanowires for high-efficiency low-cost solar photovoltaics. J. Cryst. **9**, 87 (2019)
3. Almansouri, I., Ho-Baillie, A., Bremner, S.P., Green, M.A.: Supercharging silicon solar cell performance employing the multijunction concept. IEEE J. Photovolt. **5**, 968–976 (2015)

4. Essig, S., Allebé, C., Remo, T., Geisz, J.F., Steiner, M.A., Horowitz, K., Barraud, L., Ward, J.S., Schnabel, M., Descoeudres, A., Young, L.D., Woodhouse, M., Despeisse, M., Ballif, C., Tamboli, A.: Raising the one-sun conversion efficiency of III–V/Si solar cells to 32.8% for two junctions and 35.9% for three junctions. Nat. Energy **2**, 17144 (2017)
5. Kumar, T.N.: Solar cell made using graphene. IOSE JMCE. **11**, 71–81 (2014)
6. Wu, J.Z., Lan, J., Lin, J.M., Huang, Y., Fan, L., Luo, G.: Electrolytes in dye-sensitized solar cells. Chem. Rev. **115**, 2136–2173 (2015)
7. Wu, J., Lan, Z., Lin, J., Huang, M., Huang, Y., Fan, L., Luo, G., Lin, Y., Xie, Y., Wei, Y.: Counter electrodes in dye-sensitized solar cells. Chem. Soc. Rev. **46**, 5975–6023 (2017)
8. Adikaari, A.D.T., Dissanayake, D.N.M., Silva, S.R.P.: Organic-inorganic solar cells: recent developments and outlook. IEEE J. Sel. Top. Quantum Electron. **16**, 1595–1606 (2010)
9. Yan, L., You, W.: Real function of semiconducting polymer in GaAs/polymer planar heterojunction solar cells. ACS Nano **7**, 6619–6626 (2013)
10. Effendi, N.A.S., Samsi, N.S., Zawawi, S.A., Hassan, O.H., Zakaria, R., Yahya, M.Z.A., Ali, A.M.M.: Studies on Graphene zinc-oxide nanocomposites photoanodes for high-efficient dye-sensitized solar cells. In: AIP Proceedings, vol. 1877 (2017)
11. Maçaira, J., Andrade, L., Mendes, A.: Review on nanostructured photoelectrodes for next-generation dye-sensitized solar cells. Renew. Sustain. Energy Rev. **27**, 334–349 (2013)
12. Pan, J., Rao, H., Mora-Sero, I., Bisquert, J., Zhong, X.: Quantum dot sensitized solar cells. Chem. Soc. Rev. **47**, 7659–7702 (2018)
13. Jeong, C., Nair, P., Khan, M., Lundstrom, M., Alam, M.A.: Prospects for nanowire-doped polycrystalline graphene films for ultratransparent highly conductive electrodes. . Nano Lett. **11**, 5020–5025 (2011)
14. Hu, Y.H., Wang, H., Hu, B.: Thinnest two-dimensional nanomaterial-graphene for solar energy. Chemsuschem **3**, 782–796 (2010)
15. Leem, D.S., Edwards, A., Faist, M., Nelson, J., Bradley, D.D.C., de Mello, J.C.: Efficient organic solar cells with solution-processed silver nanowire electrodes. Adv. Mater. **23**, 4371–4375 (2011)
16. Peng, K., Xu, Y., Wu, Y., Yan, Y., Lee, S.T., Zhu, J.: Aligned single-crystalline si nanowire arrays for photovoltaic applications. Small **1**, 1062–1067 (2005)
17. Hu, L., Chen, G.: Analysis of optical absorption in silicon nanowire arrays for photovoltaic applications. Nano Lett. **7**, 3249–3252 (2007)
18. Chen, R., Das, S.R., Jeong, C., Khan, M.R., Janes, D.B., Alam, M.A.: Co-percolating graphene-wrapped silver nanowire network for high-performance, highly stable transparent conducting electrodes. Adv. Funct. Mater. **23** 5150–5158 (2013)
19. Park, H., Chang, S., Jean, J., Cheng, J.J., Araujo, P.T., Wang, M., Bawendi, M.G., Dresselhaus, M.S., Bulovic, V., Kong, J., Gradecak, S.: Graphene cathode-based ZnO nanowire hybrid solar cells. Nano Lett. **13**, 233–239 (2013)
20. Thekkekkara, L.V., Jason, N.N., Cheng, W., Gu, M.: Stable copper nanowire-graphene oxide thin films for non-linear photonics. OSA Continuum. **2**, 1455–1467 (2019)
21. Czaban, J.A., Thomson, D.A., LaPierre, R.R.: GaAs core-shell nanowires for photovoltaic applications. Nano Lett. **9**, 148–154 (2009)
22. Ahn, Y., Jeong, Y., Lee, D., Lee, Y.: Copper nanowire-graphene core-shell nanostructure for highly stable transparent conducting electrodes. ACS Nano **9**, 3125–3133 (2015)
23. Ricciardulli, A.G., Yang, S., Wetzalaer, G.-J.a.H., Feng, X., Blom, P.W.M.: Hybrid silver nanowire and graphene-based solution-processed transparent electrode for organic optoelectronics. Adv. Funct. Mater. **28**, 1706010 (2018)

# Chapter 8
# Physics of Ion Beam Synthesis of Nanomaterials

**Biswajit Mallick**

## Introduction

Ion Beam Technique is widely spread and very powerful technique for material analysis, material modification and new material synthesis. Again, this is the only technique where both physical and chemical modification is possible simultaneously. This is why ion beam technology is so important and has created interest among research community. Ion beam implantation has been proven to be very successful in modifying the physical–chemical properties of materials. Its ability to introduce any element into any host substrate at concentrations well in excess of solid-solubility limits enables the formation of a compounds which is not readily achievable with conventional doping methods such as solid-state reaction. The ion implantation process is non-equilibrium in nature. In summary, Ion beam implantation and irradiation are versatile tools for synthesis of various nanocrystals, nanocomposite materials, materials properties modifications at nanoscale by introducing foreign element near the surface region of a host matrix, and fabricating new materials and structures.

**What is the Physics ofNanoparticles?**

Electronic states of an atom are typically characterized by discrete energy levels that are often separated by electron volts. The spatial distribution of these states is highly localized. At the nanoscale, the dimension of energy states resides between these limits. An electron confined within a one-dimension (1-D) box of size '$L$' is considered. The lowest energy level of this system is given by $E = (h^2/8\ m\ L^2)$, where '$h$' is the Planck's constant and '$m$' is the mass of the particle (electron). The above relation ($E \propto 1/\ L^2$), if size of the box (particle diameter) is reduced, then $E$ increases, hence a drastic variation of its bulk properties. In the nanoscale

B. Mallick (✉)
Ion Beam Laboratory, Institute of Physics, Bhubaneswar 751005, India
e-mail: bmallick@iopb.res.in



B. P. Swain (ed.), *Nanostructured Materials and their Applications*, Materials Horizons: From Nature to Nanomaterials, https://doi.org/10.1007/978-981-15-8307-0_8

phenomena, the energy level spacing of electronic states of atom increases with reduction in dimensionality of particle and is called Quantum Confinement (QC) [1]. The above reduction in dimensionality approaches to a critical quantum size call the Bohr's exciton radius. Also, the QC phenomena can be readily understood from the well-known Heisenberg's uncertainty principle, which states that: $\Delta P.\Delta x = \hbar/2$, so $\Delta P = \hbar/2.\Delta x$. Considering the energy of a free electron with momentum $P$, i.e., $E = P^2/2m$. Since the uncertainty in momentum cannot exceed the momentum itself $P$, i.e., $P > \Delta P$. So, the energy can be defined as $E < \hbar^2/8m\ L^2$. If one tries to localize the position of an electron by reducing the box size ($L$), its energy must increases and diverges as the confining region vanishes. The QC can be experimentally observed in nanostructures quantum dots of Si and CdSe. Typically a quantum dot is a few nanometers to tens of nanometer size. The smallest dots contain hundreds of thousands of atoms. However, nanocrystalline materials are single-phase or multiphase polycrystals, the crystal size of the order of 1–100 nm in at least 1-D. Depending on the dimensions in which the length scale is nanometers, they can be classified into: (a) nanoparticles (0-D), (b) lamellar structure (1-D), (c) filamentary structure (2-D), and (d) bulk nanostructured (3-D) materials.

**Why Ion beam Implantation/Irradiation is suitable for synthesis of Nanomaterials?**

In physics point of view, the fundamentals of irradiation of a material system for re-establishment of thermodynamic equilibrium are conveniently divided into three categories, which involve different time scales and somewhat different modes of thought [2]. Those three different stages are: Physical stage, Physico-Chemical stage, and Chemical stage.

Physical stage is mainly covers the interaction of high-energy radiation with matter. In this stage, the period of energy transfer (dissipation and energy absorption) by the radiation is an interval of less than $10^{-15}$ s. From the material point of view, the absorption mechanism of radiation in atoms of solids is described in three different ways, *i.e.*, (a) electron displacement, (b) atomic displacement, and (c) generation of an impurity atom (result from nuclear reactions).

Physico-chemical stage is mainly the establishment of thermal equilibrium within the solid due to the transport and disposition of radiation energy. In this stage, the period of happenings is less than $10^{-12}$ to $10^{-9}$ s. There are three alternatives: (a) energy can be emitted as luminance, (b) energy can be dissipated as heat, or (c) energy can be stored as potential energy in the form of reactive species.

Chemical stage is the stage, where the potential energy stored in reactive species is dissipated to establish chemical equilibrium. In this process, the diffusion and reaction rates extend the duration of the chemical stage from a minimum of about $10^{-8}$ s to months or, in some rigid solids it may be years.

**What are the advantages of ion implantation technique?**

Ionizing radiation provides a practical and easy way for modifying the properties or synthesizing the MATERIALS. High-charge state (>3) of Atom is possible using an accelerator, which is very difficult in Chemical, Physical or Solid-State Reaction

Technique. In Ion Beam Synthesis process both physical and chemical modification is possible at the same time.

## Theory

In general, the evolution of nanostructures using ion beam synthesis can be determined by the combined ballistic and thermodynamic effects. Interaction of ion with matter creates opportunity for modification of material, synthesis of new material phases, synthesis of nanomaterials and analysis of materials.

**What will happen if ion beam interact with materials?**

Ions when move through matter, lose energy through collisional interactions with atoms. The above energy loss takes place by mainly two different processes, i.e., (i) elastic Coulomb interaction between the screened nuclear charges of ion and target atom, and (ii) inelastic interaction of ions with bound or free target electrons. This beam energy loss can be calculated using well defined SRIM code [3]. If '$E$' is the energy of the incident ion and '$x$' is the distance covered, the differential function $\frac{\mathrm{d}E}{\mathrm{d}x}$ is called the stopping power or specific energy loss [4–7]. Mathematically, the total energy loss can be derived as,

$$\left(\frac{\mathrm{d}E}{\mathrm{d}x}\right) = \left(\frac{\mathrm{d}E}{\mathrm{d}x}\right)_{\mathrm{nucl}} + \left(\frac{\mathrm{d}E}{\mathrm{d}x}\right)_{\mathrm{elec}} \tag{8.1}$$

It is well reported that the specific energy loss first increases with increasing energy of the incident ions and then starts decreasing. Nuclear stopping is more predominant at low energies, while electronic stopping is more dominant at higher energies. The stopping power ($S$) for a charged particle in a given absorber is simply defined as the differential energy loss per corresponding differential path length, i.e.

$$S = -\frac{\mathrm{d}E}{\mathrm{d}x} \tag{8.2}$$

The value of $-\frac{\mathrm{d}E}{\mathrm{d}x}$ along a projectile particle track is also called rate of energy loss or differential stopping power. The stopping cross-section can be defined as

$$S = -\frac{1}{N}\frac{\mathrm{d}E}{\mathrm{d}x}\quad\left(\frac{eV}{A^{\circ}}\right) \tag{8.3}$$

where $N$ is the number of stopping atoms per unit volume.

## *Electronic Stopping*

If the velocity, $v$, of the incident particle is greater than the critical velocity.

$v_c = v_o Z_1^{2/3}$, where $v_o = e^2/h$ and $Z_1$ is atomic number of the incident ion, the electronic stopping is more prominent [7]. Electronic stopping may be prominent even at particle velocity below the critical velocity $v_c$ for lighter particles incident on heavy element. The classical calculation of energy loss during electronic stopping was derived by Bohr [8] and is given by

$$\left(-\frac{\mathrm{d}E}{\mathrm{d}x}\right)_{\text{elec}} = \frac{4\pi\ Z_1^2 e^4 N\ Z_2}{m_e v^2} \ln\left(\frac{P_{\max}}{P_{\min}}\right) \tag{8.4}$$

where $v$ and $Z_1$ are the velocity and atomic number, respectively, of the projectile particle; $N$ and $Z_2$ are the number density and atomic number, respectively, of the absorber atoms; and $m_e$ and $e$ are, respectively, the electron rest mass and the electronic charge. Maximum and minimum impact parameters, as chosen by Bohr for head-on collision are $P_{\min} = \frac{2\,Z_1\,e^2}{m_e\,v^2}$ and $P_{\max} = \frac{v}{\omega_j}$, respectively, where $\omega_j$ is the characteristic frequency of the $j$th electronic orbit.

For the charged particle of a given charge state, the $S$ increases as the particle velocity decreases. The classical treatment for the specific energy loss was given by Bethe [4, 5, 9] using the Born approximation, who obtained

$$\left(-\frac{\mathrm{d}E}{\mathrm{d}x}\right)_{\text{elec}} = \frac{4\pi\ Z_1^2 e^4 N\ Z_2}{m_e v^2} \left[\ln\frac{2m_e v^2}{I} - \ln\left(1 - \frac{v^2}{c^2}\right) - \frac{v^2}{c^2}\right] \tag{8.5}$$

The experimental parameter $I$ represents the average excitation energy of the electron in the absorber.

Equation (8.5) is valid for different types of charged particles provided their velocity remains large compared with the velocity of the orbital electron or critical velocity in the absorbing atoms. For non-relativistic charged particles ($v \ll c$), the expression (8.5) becomes

$$\left(-\frac{\mathrm{d}E}{\mathrm{d}x}\right)_{\text{elec}} = \frac{4\pi\ Z_1^2 e^4 N\ Z_2}{m_e v^2} \left[\ln\frac{2m_e v^2}{I}\right] \tag{8.6}$$

It is clear from the above expression that $\frac{\mathrm{d}E}{\mathrm{d}x}$ for non-relativistic particles varies as $1/v^2$, or inversely with particle energy ($E$), and the particle with the greatest charge ($Z_1\ e$) will have the largest specific energy loss. That the factor $\frac{\mathrm{d}E}{\mathrm{d}x}$ varies directly with the product $NZ_2$ indicates that high atomic number ($Z_2$) and high-density materials will result in the greatest linear stopping power. Bloch [10], on the basis of the Thomas–Fermi, model has shown that experimental parameter $I = K\ Z_2$, where $K = 11$ eV. Finally Walske [11] and Bichsel [12] added some shell correction, after which the Eq. (6) was modified as

$$\left(-\frac{\mathrm{d}E}{\mathrm{d}x}\right)_{\mathrm{elec}} = \frac{4\pi\, Z_1^2 e^4 N\, Z_2}{m_e v^2}\left[\ln\frac{2m_e v^2}{I}\right] - C_\mathrm{K} - C_\mathrm{L} \tag{8.7}$$

where $C_\mathrm{K}$ and $C_\mathrm{L}$ are contribution due to the stopping process of $K$ and $L$ shell electrons.

## *Nuclear Stopping*

The specific energy loss here is due to collisions of ions and target atoms, and can be derived by considering these independent elastic two-body interactions. The incident ion transfer energy $E^T$ to the target atom can be expressed as a function of $E$ (energy of the projectile particle) and $p$ (impact parameter in two-body collision) and can take any value between zero (for $p = \infty$) and $E^T_{\max}$, where $E^T_{\max}$ is the maximum energy transfer at head-on collision and is given by:

$$E^\mathrm{T}_{\max} = \frac{4M_1 M_2}{(M_1 + M_2)^2}E \tag{8.8}$$

where $M_1$ and $M_2$ are the mass of incident and target atom respectively.

If energy '$E$' of the incident particle be referred to center of mass coordinates, it assumes the value $E_\mathrm{cm}$, which is given by: $E_\mathrm{cm} = E.\frac{\mu}{M_1}$, where $\mu = \frac{M_1 M_2}{M_1+M_2}$ and is known as reduced mass.

The nuclear stopping process assumes more importance for heavy ions at low energy impinging on the absorber. The specific energy loss is given by

$$\left(-\frac{\mathrm{d}E}{\mathrm{d}x}\right)_{\mathrm{nucl}} = \frac{1}{4}\pi\, b^2\, N\, E^T_{\max} \int\limits_{E^T_{\min}}^{E^T_{\max}} \frac{\mathrm{d}E^T}{E^T} \tag{8.9}$$

where '$b$' is the distance of closest approach in head-on collision ($= Z_1\, Z_2\, e^2/E_\mathrm{cm}$).

Hence, the nuclear stopping is defined as

$$\left(-\frac{\mathrm{d}E}{\mathrm{d}x}\right)_{\mathrm{nucl}} = \frac{1}{4}\pi\, b^2\, N\, E^\mathrm{T}_{\max} \ln\left(\frac{E^T_{\max}}{E^T_{\min}}\right) \tag{8.10}$$

where $E^T_{\min}$ is the minimum effective energy transfer corresponding to inter-atomic spacing in the target.

## *Projectile Range*

The total path length covered by the projectile particle in the target is called the range '$R$' of the projectile, while the total path length in the original direction is called projected range. In other words, the penetrating length [5] of the projectile ion particle of the energy $E$ in matter is the range($R$) of the particle and can defined as

$$R = \int_0^R \mathrm{d}x = \int_{E_0}^0 \left(\frac{\mathrm{d}E}{\mathrm{d}x}\right)^{-1} \mathrm{d}E \tag{8.11}$$

The total number of collisions suffered by each ion may be different and also the total path length may be different. Thus there will be distribution of stopping positions. The distribution of stopping positions is usually assumed to be Gaussian. The Gaussian [13, 14] distribution can be written as

$$N(x) = N_p \in^{-\frac{1}{2}X^2} \tag{8.12}$$

where $N(x)$ = concentration of implanted atoms at a depth $x$, $X = \frac{x-R_\mathrm{p}}{\Delta R_\mathrm{p}}$, $R_\mathrm{p}$ = projected range, $\Delta R_\mathrm{p}$ = standard deviation = $\frac{\mathrm{FWHM}}{2.355}$, $N_\mathrm{p}$ = peak concentration of implanted atoms, at a distance $x = R_\mathrm{p} \pm \Delta R_\mathrm{p}$, $N_s$ being the number of implanted ions per unit area, the concentration falls to $\frac{1}{\sqrt{\in}} N_\mathrm{p}$. The number of implanted ions per unit area is given by

$$N_\mathrm{s} = \int_{-\infty}^{+\infty} N(x)\mathrm{d}x \tag{8.13}$$

As, $X = \frac{x-R_\mathrm{p}}{\Delta R_\mathrm{p}} \Rightarrow \Delta R_\mathrm{p}\mathrm{d}X = \mathrm{d}x$, so Eq. (8.13) may be expressed as

$$N_\mathrm{s} = \int_{-\infty}^{+\infty} N_\mathrm{p} \in^{-\frac{X^2}{2}} \Delta R_\mathrm{p}\mathrm{d}X \tag{8.14}$$

$$= 2\Delta R_\mathrm{p} N_\mathrm{p} \int_0^{\infty} \in^{-\frac{X^2}{2}} \mathrm{d}X \tag{8.15}$$

$$= \Delta R_\mathrm{p} N_\mathrm{p} \sqrt{2\pi} \left\{ \sqrt{\frac{2}{\pi}} \int_0^{\infty} \in^{-\frac{X^2}{2}} \mathrm{d}X \right\} \tag{8.16}$$

As, $X \to \infty$, the integral inside the bracket tends to unity, so

$$N_p = \frac{N_s}{\left\{\sqrt{2\pi}\Delta R_p\right\}} \tag{8.17}$$

$$\Rightarrow N_p = \frac{0.4 N_s}{\Delta R_p} \tag{8.18}$$

Putting the value of Eq. (8.18) in Eq. (8.12) we obtain

$$N(x) = \frac{0.4 N_s}{\Delta R_p} \exp\left[-\frac{1}{2}\left[\frac{x - R_p}{\Delta R_p}\right]^2\right] \tag{8.19}$$

$$N(x) = N_p \exp\left[-\frac{1}{2}\left[\frac{x - R_p}{\Delta R_p}\right]^2\right] \tag{8.20}$$

Thus, by knowing the calculated value of average projectile range and standard deviation, the distribution of implanted ions can be easily estimated.

## *Stopping Time*

The stopping time [5], i.e., the time required to stop a charged particle in an absorber, can be deduced from its range and average velocity. For a non-relativistic particle of mass $m$ and energy $E$, the stopping time $t$ can be calculated from its range as

$$t = \frac{R}{\langle v \rangle} = \frac{R}{kc}\sqrt{\frac{mc^2}{2E}}$$

$$t = \frac{R}{k(3 \times 10^8\,\mathrm{m/s})}\sqrt{\frac{931\,\mathrm{MeV/amu}}{2} \times \frac{m_p}{E}} \tag{8.21}$$

where $m_p$ is the particle mass in amu, $R$ is the range of projectile ion, and $\langle v \rangle = k\,v$ ($=v/2$, for uniformly decelerating particle of initial velocity $v$) is the average velocity of the projectile. However, projectile ion particle generally loses energy at a greater rate near the end of its range, and the value of $k$ should be a somewhat higher fraction (assuming $k = 0.60$ for a light charged particle like proton). So

$$t = 1.2 \times 10^{-7} R\sqrt{\frac{m_p}{E}} \tag{8.22}$$

where $t$ is in second, $R$ is in meters, $m_p$ is in amu and $E$ is in MeV.

## *Radiation Dose*

The dose of ion can be expressed [15, 16] in terms of specific energy loss $\frac{dE}{dx}$ and required ion fluence $F$ as

$$D_{\text{ion}} = \frac{\mathrm{d}E}{\mathrm{d}x} F \tag{8.23}$$

However, Experimentally the ion dose can be calculated as

$$N = \frac{F\ A\ q\ e}{S} \tag{8.24}$$

where $F$ (ions/cm$^2$) is the ion fluence, $N$ is the counts, $q$ (Coulomb) is the charge, $e$ is the electronic charge ($1.602 \times 10^{-19}$ coulomb) and $S$ (μC/Count) is the scale factor.

**Conversion of Radiation Dose**

1. International unit—Gray (Gy)
   1 Gy = 1 J/kg = 1 W s/kg; 1 kGy = 1 kJ/kg
2. Equivalent/Effective dose-Sievert (Sv)
   1 Sv = 100 rem (1 rem = 1 rad × Quality Factor; 1 Gy = 100 rad)
3. Energy (physical) unit- Roentgen equivalent physical (rep)

   1 Rad = 1.19 Rep = $6.25 \times 10^{13}$ eV/g

**How to Estimate the Required Fluence/dose for the Material Modification/synthesis and Why?**

The effect of ion irradiation on solids may be classified in a general way according to three characteristic fluence ranges, in which different types of modification occur [17]. A more general parameter to attain these characteristic fluencies is the energy density, $D_e = F.\ S_e$ (eV/cm$^3$), where $F$ is the ion fluence (ions/cm$^2$) and $S_e$ (eV/cm) is the differential value of energy loss [18]. The ion-induced changes in solid polymer taking place at different energy density values are discussed below.

1. Low energy density ($D_e \sim 10^{23}$ eV/cm$^3$): At this energy density regime, new bonds (crosslinks) or chain scissions formation occurs in the polymer. This result in modification in its solubility and molecular weight distribution, change in transport properties, change in polymer morphology, occurrence of unsaturation and occurrence of desorption in simple molecule etc.
2. Medium energy density ($D_e \sim 10^{24}$ eV/cm$^3$): At this energy density regime, the continuous bond arrangement induces a complete change of the polymer structure resulting in change of electronic structure and physical properties, etc.
3. High energy density ($D_e \sim 10^{25}$ eV/cm$^3$): At this energy density regime, the polymer loses memory of its original structure and stoichiometry etc.

## Program Code

A number of well-known scientific software such as SRIM, TRIM, TRYDIN, EnvizION etc. for ion particles; software like SRDN for neutron; and SREM, ESTAR etc. for electrons are available to analyze various radiation-interaction parameters such as energy, specific energy loss, range of projectile ion, straggling, stopping time etc. very accurately. Few examples of above parameters of ion beam interaction using SRIM available at *www.srim.org* are given in Tables 8.1, 8.2 and 8.3. Similarly, parameters viz. stopping, CSDA range of electron interaction with matter using ESTAR (https://physics.nist.gov/PhysRefData/Star/) are shown in Table 8.4. The Monte-Carlo routine analysis is usually carried out using above programs. Most of the above programs were so designed to carry out the simulation methods for implantation work. The abbreviation of the above software are given below.

*SRIM- **S**topping and **R**ange of **I**on in **M**atter
*TRIM- **T**ransport and **R**ange of **I**on in **M**atter
*SREM- **S**topping and **R**ange of **E**lectron in **M**atter
*SRDN- **S**topping **R**ange and **D**amage by **N**eutron

## Instrumentation

The instrument used to generate heavy ion of energy in MeV range for implantation, a nuclear accelerator or high voltage machine is required. For the implantation work, we have used a 3 Million Volt Tandem Pelletron Accelerator available at Institute of Physics (IOP), Bhubaneswar, India. The IOP Pelletron is shown in Fig. 8.1. The major components used for the production of heavy ion for implantation are: ion source, Injector magnet, Accelerator, Electrostatic Quadrupole Triplet, Analyser magnet, Scanner, Implantation chamber, and Current integrator etc. Few major components are defined below to give some idea about the beam tuning and generation of ion beam.

**Why a MV accelerator is required for heavy ion implantation?**

To generate energetic positive ion particle such as proton or heavy ions a Million Voltage accelerator is required. Since, mass of proton or heavy ions are very high in the order of few $10^3$ times more as compared to mass of electrons. Hence, to energizes and accelerate the heavy ions, usually a MV accelerator is required. However, to accelerator an electron or a negative ion particle few, KV potential field is enough.

**Fig. 8.1** 3 MV Tandem Pelletron Laboratory of IOP Bhubaneswar

## *Ion Source*

National Electrostatic Corporation (NEC), USA make negative ion source is used for production of heavy ion from solid samples. The present negative ion source exactly known as Multi-cathode Source of Negative Ion by Cesium Sputtering or MC-SNICS and provision to hold 40 samples at a time. So, 40 or more number of different ion particle can be expected in a single (using compound instead of single element as source materials). The working principle of the source is based on the sputtering technique. Major advantage of this source is rapid cathode change and precise, repeatable positioning without cathode exposure to air. The cathode wheel is controlled via a pneumatic system both for cathode indexing and cathode disk retraction for replacement of the cathode disk.

### *Injector Magnet*

The injector magnet of IOP Pelletron system is a DANFYSIK (DF-54) dipole magnet. The deflection angle of the injector is 90°. The value of magnetic rigidity, $ME/q^2 = 15$ and $r = 0.457$ m for the present injector magnet, which select mass of the ion particle.

### *Accelerator*

A 3MV Tandem Pelletron 9SDH-2 (NEC Make) horizontal machine is available at Ion Beam Laboratory, IOP Bhubaneswar, India. Maximum potential up to 3 million Volts can be achieved in this machine. It is capable of producing 9 MeV double plus ions (symbol: 9-max. energy of double plus ion, *S*- section, *D*- double means tandem, *H*- horizontal machine and 2 number of chains in machine) and very suitable for Implantation purpose.

### *Electrostatic Quadruple Triplet*

The Electrostatic Quadruple Triplet (EQT) is mainly responsible to select the required charge state of the ion beam for the implantation. This is based on the principle of electric field lens. By adjusting EQT one can get maximum beam current and well shaping of the profile. The potential $\varphi$ (in KV) required for the EQT setting is approximately given as $\varphi = (A^2\ K^2)/qV_{\mathrm{T}} = 43.56/qV_{\mathrm{T}} = 21.78$ KV ( let $q = +2$ and $V_{\mathrm{T}} = 1$ MV), where $K$ is a parameter that characterize the lens strength($=D/F \approx 0.3$), $q$ is the charge state of the particle, $V_{\mathrm{T}}$ is the absolute value of the voltage on the singlet elements(=Terminal voltage in MV), $A$ is the aperture radius(22 mm), $\varphi$ is the potential drop through which the incoming particle has been accelerated (KV). Accordingly, the desired charge state can be selected.

### *Analyzer Magnet*

Heart of an Accelerator system is its Analyzer magnet. Precession in Ion beam energy/momentum can be achieved by this device. Analyzing magnet control current appears on both High Energy slit (HE Slit) and Low Energy Slit (LE Slit) meters. The value of magnetic rigidity, $ME/q^2 = 151$ for the present 90°-analyzing magnet, which select energy and momentum of the ion particle.

Calibrate the energy by finding the Nuclear Magnetic Resonance (NMR) field in Gauss i.e., $E = (Bqr/1438.5)^2/M$, so $(B^2) = E\ M \times (1438.5/qr)^2$ in Gauss, where $E$ in MeV, $r = 1.28$ m, $q$ = charge state and $M$ = mass in amu.

Setting of the analyzer magnetic field can be done as follows:

$$B_1 = \frac{\sqrt{M_1\ E_1}}{q_1}$$

$$B_2 = \frac{\sqrt{M_2\ E_2}}{q_2}$$

$$\frac{B_1}{B_2} = \frac{\sqrt{M_1\ E_1}}{q_1} \times \frac{q_2}{\sqrt{M_2\ E_2}}$$

## *Scanner*

Another major components for ion implantation is DEFLECTOR/SCANNER. The scanner is so adjusted to swift the ion particle over certain area (sq. cm). The setting voltage (KV) for certain charges state and energy. Ex. 3 MeV $C^{+2}$, $V$ (KV) = 20.6E (MeV)/$q$ = 20.6 × 3/2 = 30.9 kV (say). The beam deflection is at very small angle nearly 3° to the target. Proper energy and charge state can be selected for implantation.

## *Implantation Chamber*

The implantation chamber is simple a vacuum chamber equipped with all gadgets like vacuum gauge and controller, view ports to see the sample alignment, angular adjustment for ladder, signal ports, and a ladder with special types of attachment as shown in Fig. 8.2. To achieve various physical conditions during experiments or after experiment as defined in later Sect. 'In-Situ Condition', a special type of ladder is required for implantation.

## *Current Integrator*

Main function of the current integrator is to record the beam current or ion dose implanted on the sample. Again, this is also help to measure the counts, beam

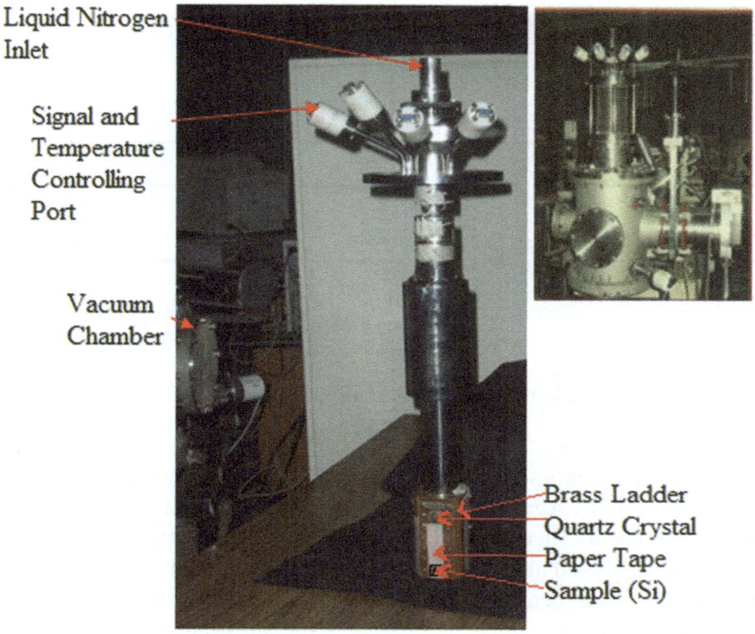


**Fig. 8.2** Implantation chamber ladder with different components used during experiment

current, current rate etc. At present, IOP-Implantation beam line equipped with a DANFYSIK-554 current integrator for the purpose.

## Experimental

### *Synthesis*

Ion beam synthesis (IBS) broadly modeled into three category such as: Collisional mixing (CM) [19–21], Monte-Carlo(MC) [22–24] and Thermodynamic mixing (TM) [25, 26]. Development of each model based on certain principle to calculate the various information about the ion particle. The IBS classification scheme is shown in Fig. 8.3. The collisional mixing or CM model is based on factors viz. electron-molecule Collision and kinds of collision. The electron-molecule collision probability can be calculated using the simulation relation derived by Nanbu [20] Similarly, the Monte-Carlo or MC model is based on three important parameters viz. Ballistic effect, Thermodynamic evolution and Gibbs–Thomson Relation. The transition probability can be find out applying thermodynamic evolution [24]. Also the

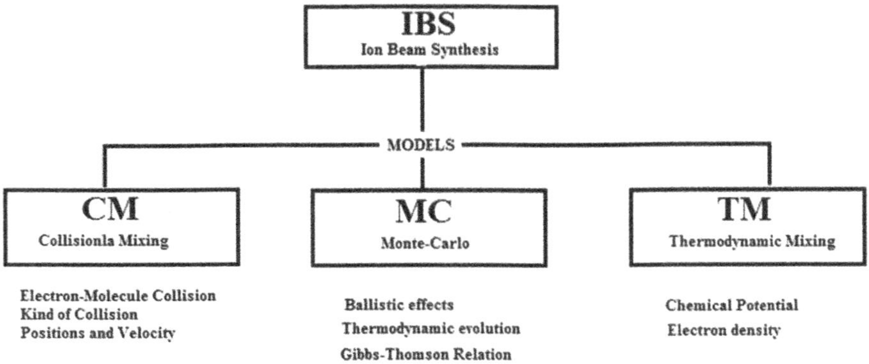


**Fig. 8.3** Classifications of Ion Beam Synthesis Models

nanocluster radius, monomer concentration etc. can be estimated applying Gibbs–Thomson relation [24]. Again, The thermodynamic mixing or TM model based on Chemical potential and Electron density. The chemical potential, Heat of formation of solids can be analyzed applying the relation Miedema et al. [26].

The synthesis of Ion beam prompted nanomaterials is mainly pass through three different processes viz. cleaning of surface of the substrate in vacuum, Implantation of ion particles, and Thermal or laser based annealing. The above processes usually passed through a number of phases viz. Supersaturation, Nucleation, Growth, Ostwald ripening in case of Ion implantation and phases viz. Ostwald ripening, Coalescence, Burred layer in the case of Annealing process. Again, the above processes are depends on various parameters such as material density $\rho$, Charge state of ion $q$, ion energy $E$, ion fluence $F$, ion flux or beam current $I$, substrate temperature $T$. Formation of a well ordered nanoclusters desired a well précised control over above parameters (Fig. 8.4).

**Why annealing is required after high-energy ion beam implantation?**

Annealing (400 °C–900 °C/30 min) after implantation under gas environment (Ar, $N_2$) or vacuum repair crystal damage by restoring the lattice of the substrate atom to its perfect crystalline state. Limitation is that the damage less than a critical value can be repaired, but for damage above the critical value can cause defects leads to stable dislocation. Again, dopant ($10^{12}$–$10^{18}$ dopants/cm$^2$) activation requires annealing, which can cause dopant diffusion, restore the electron and hole mobility, and loss of implanted depth.

## *Material System*

Ion beam implantation technique is capable to implant variety of energetic ion beam on the surface of substrate atom. So, different material system can be synthesized.

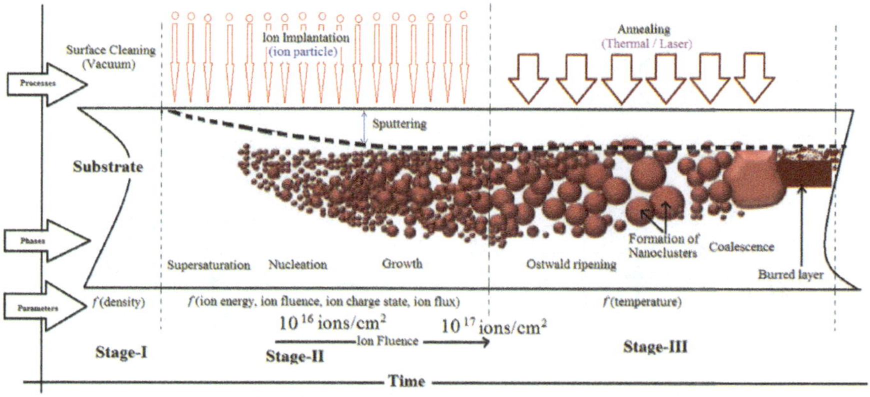


**Fig. 8.4** Process involved in the Synthesis of nanomaterials

Also, a number of crystallographic phases can be formed depends on the appropriate stoichiometry. Few simple model viz. A-A Monetary, A-B Binary and A-B-C Ternary system are discussed here. Examples: (i) Si–Si Monetary system can be formed by the Implantation of 50 keV Silicon ion on Silicon substrate and can penetrate about 713 Å depth (Table 8.1); (ii) Si-Pt Binary system can be formed by the Implantation of 2 MeV Platinum ion on Silicon can penetrate about 4697 Å (Table 8.2); and (iii) Si–C-Pt Ternary system can be formed by the Implantation of 1.5 MeV Carbon ion ($q = +1$) and 9 MeV Platinum ion ($q = +1$) on Silicon and both can penetrate about 2 μm depth on the silicon surface (Tables 8.2 and 8.3) (Fig. 8.5).

## *In-Situ Condition*

A number of physical technique and conditions are required to synthesize new material phases in general and nanomaterial in particular. In-situ specific physical conditions such as: heating, cooling, gas flash, angular orientation etc. required in a well-controlled manner to developed a particular material phase. The In-situ conditions are:

1. Heating or in-situ annealing
2. Cooling at $LN_2$ or controlled cooling
3. Angular orientation of sample with respect to the beam direction
4. Specific gas flash required some time during irradiation or after irradiation.

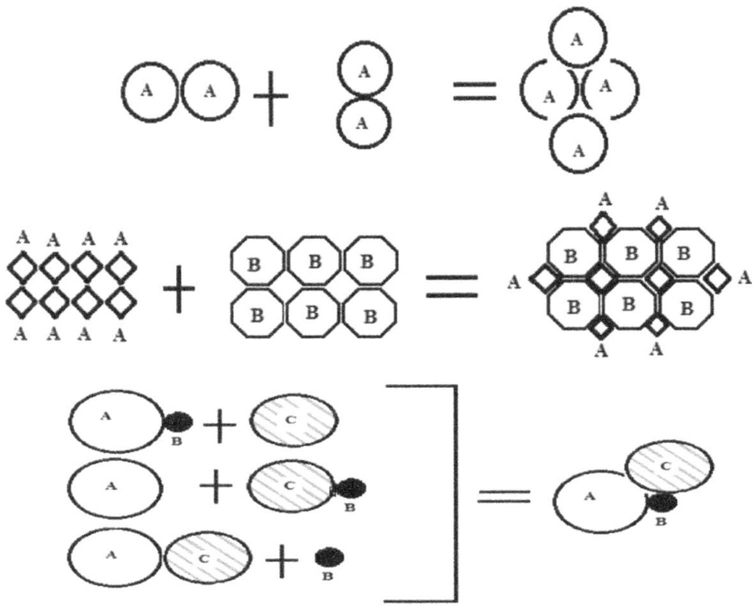


**Fig. 8.5** Synthesized material systems. (i) A-A Monetary system. (ii) A-B binary system, and (iii) A-B-C ternary system

## Applications

A number of nanoparticles synthesized using ion beam technique for different application has been reported by various researchers. Noble metal nanoparticles embedded in transparent matrices have stimulated interest for their optical properties and quantum size effect. In an investigation by Mishra et al. [27], ion-induced synthesis of Au-nanoparticle has been reported. The study of Ni and Co nanoparticles are important for its magnetic properties. Kumar et al. [28] studied the ion beam synthesis of Ni-nanoparticles. Ion beam synthesis of Co-nanoparticle have been reported [29, 30]. The Ge-quantum dots embedded in a transparent matrix exhibit strong size dependence of optical properties and attributed quantum confinement effect. Ion beam synthesis and characterization of Ge-nanoparticle system were reported by Desnica et al. [31]. In Gonźalez-Varona et al., the semiconductor nanoparticles applicable for Si based optoelectronic devices were synthesized using ion beam technique [32]. Applying ion implantation technique, the Ga-nanodots were synthesized and reported by Buckmaster et al. [33]. Nanostructuring by Energetic ion beams has been reported by Avasthi [34]. The nanomaterials structure by ion beam was described by Schmidt [35]. A nice book on Nanofabrication by ion beam sputtering edited by Som and Kanjilal has already been published [36]. The above references related to ion beam prompted nanomaterial synthesis are good source for readers.

## Conclusion

The Synthesis of nanomaterials using ion beam implantation technique is one of the advanced method in the present scenario. Ion beam technique provides a practical and essay way for synthesizing and modifying material of diverse nature. High-charge state (>3) of Atom, which are impossible in chemical, physical or solid-state reaction is only possible by using accelerators. In Ion Beam Synthesis process both physical and chemical modification simultaneously. The synthesis process of ion beam prompted nanomaterials is highly depends on parameters viz. $\rho$, $q$, $E$, $F$, $I$, $T$ etc. The synthesis process usually passed through a number of phases such as supersaturation, Nucleation, Growth, Ostwald ripening, Coalescence, Burred layer before the final result. A number of simulation programs like SRIM, TRIM, TRYDIN, EnvizION etc. for ion particles based on Monte-Carlo are very useful to analyze various beam interaction parameters such as energy, specific energy loss, range of projectile ion, straggling, stopping time etc. accurately. Looking to the importance, a large variety of material system (Monetary, Binary, Ternary) can be synthesized with a number of crystallographic phases depending on ion beam type, fluence, substrate atom, physical conditions applied and appropriate stoichiometry.

**Research Problems**

*Problem-1*

**How to select energy of a Molecular beam for implantation?**

*Ans.:* The energy of a molecular beam can be defined as $E_{m_1} = V_T\left(\frac{m_1}{m_1+m_2} + 1\right) q$,
For example, the energy of the $CN^-$ (nitrogen ion) molecular beam.
$E_{\text{Nitrogen}} = \left(\frac{14}{14+12} + 1\right) q\ V_{\text{T}}$.

$$E_{\text{Carbon}} = \left(\frac{12}{14+12} + 1\right) q\ V_{\text{T}}$$

*Problem-2*

**A researcher interested to irradiate his/her sample (Si of 1 cm²) with a 3 MeV proton at a fluence $1 \times 10^{15}$ protons/cm². If the stable beam current available is 50 nA with a knob setting for scale factor S is 0.1 μC/Count, find the following things:**

(a) **The ion dose in counts $N$ required for irradiation at** $1 \times 10^{15}$ protons/cm²
(b) **Time required to irradiate the sample**
(c) **With same beam current if the researcher interested to irradiate his/her sample at a fluence** $1 \times 10^{17}$ protons/cm², how much time it will take to complete the sample
(d) **What steps he/she should take to reduce the irradiation time?**

(e) **Instead of H⁺beam, if the researcher is interested to irradiated his/her sample with Au⁺³beam of same parameters, how much count of dose required?**

**[Note: Let us say the knob setting value of the DANFYSIK 554 scale is 0.1 (μC/Count) and total division of the rate meter scale is 100 parts]**

*Ans.:* (a) The data given, Fluence, $F = 1 \times 10^{15}$ ions/cm$^2$).
Scale factor $S = (0.1/100) \times 50 = 0.05$ μC/Count.
Area of Sample $A = 1$ cm$^2$.
Charge state of ion (proton) $q = +1$.
Current rate $I = 50$ nA (=50 × $10^{-9}$ C/s).
Working formula for ion dose in counts can be expressed as $N = (FqeA/S)$,
Hence,

$$N = \frac{(1 \times 10^{15}\ \text{ions/cm}^2)\ \times (+1) \times (1.602 \times 10^{-19} C) \times 1\ \text{cm}^2}{0.05\ \times 10^{-6}\ \text{C/Count}} = 3204\ \text{Counts}.$$

The total charge implanted on the sample is.

$$\begin{aligned} Q &= \text{Total Counts } N\ \times \text{ knob setting Value} \\ &= 3204\ \times\ 0.1\ =\ 320.4\,\mu\text{C} \end{aligned}$$

(b) Time required to complete the experiment is $T = Q/I$

$$T = \frac{320.4 \times 10^{-6}\ \text{C}}{50 \times 10^{-9}\ \text{C/s}} = 6408\ \text{s} = 1\ \text{h}\ 47\ \text{min}$$

$$N = \frac{(1 \times 10^{17}\ \text{ions/cm}^2)\ \times (+1) \times (1.602 \times 10^{-19}\ \text{C}) \times 1\ \text{cm}^2}{0.05\ \times 10^{-6}\ \text{C/Count}} = 320,400\ \text{Counts}$$

$$T = \frac{32040 \times 10^{-6}\ \text{C}}{50 \times 10^{-9}\ \text{C/s}} = 640,800\ \text{s} = 178\ \text{h}$$

To complete the implantation work for the single dose, about 7.5 days required. Which is very difficult for the researcher as well as for the laboratory staff. So, one has to reduce this time period.

(d) To reduce the above implantation time (178 h), one has to either increase the beam current or decrease the sample.

Case I, Let beam current increase to 100 nA, then

$$T = \frac{32040 \times 10^{-6}\ \text{C}}{100 \times 10^{-9}\ \text{C/s}} = 320,400\ \text{s} = 89\ \text{h}$$

Case II, Let sample (irradiation area) size reduced to 0.5 × 0.5 cm$^2$

$$N = \frac{(1 \times 10^{17}\ \text{ions/cm}^2)\ \times (+1) \times (1.602 \times 10^{-19} C) \times 0.25\ \text{cm}^2}{0.05\ \times 10^{-6}\ \text{C/Count}}$$

$$= 80,100\ \text{Counts}$$

$$T = \frac{8010 \times 10^{-6}\ \text{C}}{50 \times 10^{-9}\ \text{C/s}} = 160,200\ s = 44.5\ \text{h}$$

(e) Counts required when, $q = +3$, $Au^{+3}$ beam of fluence $F = 1 \times 10^{15}$ ions/cm$^2$

$$N = \frac{(1 \times 10^{15}\ \text{ions/cm}^2) \times (+3) \times (1.602 \times 10^{-19}\text{C}) \times 1\ \text{cm}^2}{0.05 \times 10^{-6}\ \text{C/Counts}} = 9612\ \text{Counts}$$

$$T = \frac{961.2 \times 10^{-6}\ \text{C}}{50 \times 10^{-9}\ \text{C/s}} = 19,224\ \text{s} = 5\ \text{h}\ 34\ \text{min}$$

*Problem-3*

**How to set the Energy of the charge particle in a 3MV Tandem Pelletron Accelerator?**

*Ans.:* The maximum terminal voltage $V_T$ is 3 Million Volts (MV). Generally, the energy of the beam can be calculated as $E = V_T(q + 1)$,or $V_T = E/(q + 1)$. Taking pre-acceleration energy $J$ (= Cathode + Bias + Extractor = 45 keV) into consideration (say = 45KeV), corrected beam energy can be calculated as $E - J = V_T(q + 1)$,so $V_T = (E - J)/(q + 1)$.

Terminal Voltage, $V_T = \frac{E-J}{q+1} = \frac{E-0.045}{q+1}$.

Hence, the corrected energy of the ion beam is

$$E = V_T\ (q + 1)\ +\ J$$

*Problem-4*

**What is the maximum range of highest-energetic proton beam generated using a 3MV Tandem Pelletron accelerator (6 MeV) in Air, Silicon, and Mylar ?**

*Ans.:* The proton beam of maximum energy generated by a 3MV Tandem is.

$E = V_T(q + 1) = 3\ (1 + 1) = 6$ MeV.

The proton range is assumed to be represented by an empirical range relation and power law holds good for the range of proton in an element (Z) or material. So, the range, $R = a\ E^b$ , where "*a*" is the coefficient depend on the material, and the power "b" only slightly depends on the "Z". For proton of energy $E_p$ (MeV) ranging from few MeV to 200 MeV, the range (meter) in air can be calculated as

$$R_{\text{air}} = \left(\frac{1}{9.3}\right)^{1.8} (E)^{1.8}$$

$$= \left(\frac{1}{9.3}\right)^{1.8} (6)^{1.8}$$

$$= 0.018 \times 25.158$$
$$= 0.4543 \text{ m}$$

Simplifying the Bragg–Kleeman rule, the range of proton on silicon (Z = 14, $M$ = 28 and $\rho = 2.33$ g/cm$^3$) can calculated as

$$R_{Si} = 3.2 \times 10^{-4} \frac{\sqrt{M}}{\rho} R_{\text{air}}$$

$$= 3.2 \times 10^{-4} \times \frac{\sqrt{28}}{2.33} \times 45.43$$
$$= 330.15 \times 10^{-4} \text{ cm}$$
$$= 330.15 \, \mu\text{m}$$

For mixture of compound like Mylar, the effective atomic mass can be used instead of atomic mass. The effective atomic mass can be calculated as

$$\sqrt{M_{\text{eff}}} = \sum_i \gamma_i \sqrt{M_i}$$

$$\frac{1}{\sqrt{M_{\text{eff}}}} = \sum_i \frac{\omega_i}{\sqrt{M_i}}$$

where $\gamma_i$ and $\omega_i$ are the atomic fraction and weight fractions of the $i$th element in the compound. Again, using Ziegler's SRIM program one get those range easily. Few examples on range of proton in materials are given in **Appendix**.

**Table 8.1** . TRIM program for Si ion on Si-substrate

```
==============================================================
           SRIM version ---> SRIM-2013.00
           Calc. date   ---> August 30, 2019

==============================================================

 Disk File Name = SRIM Outputs\Silicon in Silicon.txt

 Ion = Silicon [14] , Mass = 27.977 amu

 Target Density =  2.3212E+00 g/cm3 = 4.9770E+22 atoms/cm3
 ======= Target  Composition ========
   Atom   Atom   Atomic    Mass
   Name   Numb   Percent   Percent
   ----   ----   -------   -------
    Si    14    100.00    100.00
 ====================================
 Bragg Correction = 0.00%
 Stopping Units =  keV / micron
 See bottom of Table for other Stopping units

     Ion        dE/dx      dE/dx     Projected  Longitudinal   Lateral
    Energy      Elec.      Nuclear    Range      Straggling   Straggling
  -----------  ---------- ---------- ----------  ----------  ----------

  50.00 keV   2.383E+02  3.438E+02     713 A       287 A       208 A
  60.00 keV   2.652E+02  3.282E+02     848 A       330 A       240 A
  70.00 keV   2.952E+02  3.140E+02     981 A       369 A       272 A
  80.00 keV   3.201E+02  3.010E+02    1113 A       407 A       303 A
  90.00 keV   3.400E+02  2.891E+02    1244 A       442 A       333 A
 100.00 keV   3.562E+02  2.783E+02    1375 A       476 A       363 A

-----------------------------------------------------------
 Multiply Stopping by        for Stopping Units
 -------------------        ------------------
  1.0000E-01                 eV / Angstrom
  1.0000E+00                 keV / micron
  1.0000E+00                 MeV / mm
  4.3083E-03                keV / (ug/cm2)
  4.3083E-03                MeV / (mg/cm2)
  4.3083E+00                 keV / (mg/cm2)
  2.0093E-01                 eV / (1E15 atoms/cm2)
  8.2748E-04                L.S.S. reduced units

==============================================================
 (C) 1984,1989,1992,1998,2008 by J.P. Biersack and J.F. Ziegler
```

**Table 8.2** TRIM program for Pt- ion on Si-substrate

```
=============================================================
          SRIM version ---> SRIM-2013.00
          Calc. date   ---> August 30, 2019

=============================================================

 Disk File Name = SRIM Outputs\Platinum in Silicon.txt

 Ion = Platinum [78] , Mass = 194.965 amu

 Target Density =  2.3212E+00 g/cm3 = 4.9770E+22 atoms/cm3
 ======= Target  Composition ========
   Atom   Atom   Atomic    Mass
   Name   Numb   Percent   Percent
   ----   ----   -------   -------
    Si    14    100.00    100.00
 ====================================
 Bragg Correction = 0.00%
 Stopping Units =  keV / micron
 See bottom of Table for other Stopping units

     Ion       dE/dx      dE/dx     Projected  Longitudinal   Lateral
    Energy      Elec.     Nuclear     Range     Straggling   Straggling
  ----------- ---------- ---------- ----------  ----------  ----------

   1.00 MeV   1.515E+03  2.882E+03   2494 A       393 A       347 A
   2.00 MeV   1.847E+03  2.439E+03   4697 A       651 A       589 A
   2.25 MeV   1.918E+03  2.348E+03   5259 A       713 A       647 A
   5.00 MeV   2.659E+03  1.695E+03   1.15 um     1287 A      1241 A
   8.00 MeV   3.312E+03  1.330E+03   1.80 um     1763 A      1787 A
   9.00 MeV   3.492E+03  1.245E+03   2.01 um     1899 A      1949 A
  10.00 MeV   3.656E+03  1.172E+03   2.22 um     2024 A      2102 A
-----------------------------------------------------------
 Multiply Stopping by        for Stopping Units
 -------------------        ------------------
  1.0000E-01                 eV / Angstrom
  1.0000E+00                 keV / micron
  1.0000E+00                 MeV / mm
  4.3083E-03                 keV / (ug/cm2)
  4.3083E-03                 MeV / (mg/cm2)
  4.3083E+00                 keV / (mg/cm2)
  2.0093E-01                 eV / (1E15 atoms/cm2)
  1.2204E-04                 L.S.S. reduced units

=============================================================
 (C) 1984,1989,1992,1998,2008 by J.P. Biersack and J.F. Ziegler
```

**Table 8.3** TRIM program for C- ion on Si-substrate

```
==============================================================
          SRIM version ---> SRIM-2013.00
          Calc. date   ---> August 30, 2019

==============================================================

 Disk File Name = SRIM Outputs\Carbon in Silicon.txt

 Ion = Carbon [6] , Mass = 12 amu

 Target Density =  2.3212E+00 g/cm3 = 4.9770E+22 atoms/cm3
 ======= Target  Composition ========
    Atom   Atom   Atomic    Mass
    Name   Numb   Percent   Percent
    ----   ----   -------   -------
     Si     14    100.00    100.00
 ======================================
 Bragg Correction = 0.00%
 Stopping Units =  keV / micron
 See bottom of Table for other Stopping units

     Ion       dE/dx      dE/dx    Projected  Longitudinal   Lateral
    Energy     Elec.     Nuclear    Range     Straggling   Straggling
  -----------  ---------- ---------- ----------  ----------  ----------
   1.00 MeV  9.931E+02  8.677E+00    1.57 um      1691 A      2077 A
   1.50 MeV  1.116E+03  6.322E+00    2.03 um      1797 A      2282 A
   2.00 MeV  1.167E+03  5.028E+00    2.46 um      1882 A      2432 A
   5.00 MeV  1.158E+03  2.376E+00    4.99 um      2418 A      3036 A
   8.00 MeV  1.092E+03  1.602E+00    7.65 um      3132 A      3516 A
   9.00 MeV  1.071E+03  1.450E+00    8.57 um      3431 A      3669 A
  10.00 MeV  1.051E+03  1.326E+00    9.51 um      3717 A      3820 A
-----------------------------------------------------------
 Multiply Stopping by        for Stopping Units
 -------------------        -------------------
  1.0000E-01                 eV / Angstrom
  1.0000E+00                 keV / micron
  1.0000E+00                 MeV / mm
  4.3083E-03                keV / (ug/cm2)
  4.3083E-03                MeV / (mg/cm2)
  4.3083E+00                 keV / (mg/cm2)
  2.0093E-01                 eV / (1E15 atoms/cm2)
  2.8503E-03                L.S.S. reduced units

==============================================================
 (C) 1984,1989,1992,1998,2008 by J.P. Biersack and J.F. Ziegler
```

**Table 8.4** ESTAR: Stopping Powers and Ranges for Electrons in Silicon

NIST National Institute of Standards and Technology Physical Meas. Laboratory

estar stopping-power and range tables for electrons

The ESTAR program calculates stopping power, density effect parameters, range, and radiation yield tables for electrons in various materials. Select a material and enter the desired energies specified in MeV, and must be in the range from 0.001 MeV to 10000 MeV.

Help Text version Material composition data

Select a common material: 14: Silicon
or enter a unique material

- Graph stopping power:
  - Total Stopping Power
  - Collision Stopping Power
  - Radiative Stopping Power
- Graph density effect parameter
- Graph CSDA range
- Graph radiation yield
- No graph

Additional Energies (optional):

Use energies from a file* Browse... No file selected.

or

Use energies entered below (one per line)

Include default energies

*Note*: Only stopping powers and the density effect parameter will be calculated if additional energies are used.

Submit Reset

* Your browser must be file-upload compatible.

## SILICON

To download data in spreadsheet (array) form, choose a delimiter and use the checkboxes in the table heading. After downloading, save the output by using your browser's Save As feature.

**Delimiter:**

- space
- | (vertical bar)
- tab (some browsers may use spaces instead)
- newline

Download data Reset

| (required) Kinetic Energy (MeV) | Stopping Power (MeV cm²/g) | | | CSDA Range (g/cm²) | Radiation Yield | Density Effect Parameter |
|---|---|---|---|---|---|---|
| | Collision | Radiative | Total | | | |
| 1.000E-02 | 1.689E+01 | 7.255E-03 | 1.690E+01 | 3.461E-04 | 2.289E-04 | 1.037E-03 |
| 1.250E-02 | 1.432E+01 | 7.431E-03 | 1.433E+01 | 5.074E-04 | 2.780E-04 | 1.332E-03 |
| 1.500E-02 | 1.251E+01 | 7.555E-03 | 1.252E+01 | 6.946E-04 | 3.252E-04 | 1.641E-03 |
| 1.750E-02 | 1.115E+01 | 7.648E-03 | 1.116E+01 | 9.065E-04 | 3.709E-04 | 1.963E-03 |
| 2.000E-02 | 1.010E+01 | 7.720E-03 | 1.011E+01 | 1.142E-03 | 4.151E-04 | 2.298E-03 |
| 2.500E-02 | 8.556E+00 | 7.822E-03 | 8.564E+00 | 1.682E-03 | 5.000E-04 | 3.007E-03 |
| 3.000E-02 | 7.480E+00 | 7.892E-03 | 7.487E+00 | 2.308E-03 | 5.807E-04 | 3.766E-03 |
| 3.500E-02 | 6.682E+00 | 7.946E-03 | 6.690E+00 | 3.016E-03 | 6.579E-04 | 4.572E-03 |

*CSDA- Continuous Slowing Down Approximation

back contents

# Appendix: SRIM Calculation for Penetration Depth of Proton in (a) Mylar, (B) Silicon and (C) Air

```
=================================================================
(A)     SRIM version ---> SRIM-2013.00
        Calc. date  ---> September 06, 2019
=================================================================
Disk File Name = SRIM Outputs\Hydrogen in Mylar (ICRU-222).txt
Ion = Hydrogen [1] , Mass = 1.008 amu
Target Density =  1.3970E+00 g/cm3 = 9.6311E+22 atoms/cm3
======= Target  Composition ========
  Atom  Atom  Atomic   Mass
  Name  Numb  Percent  Percent
  ----  ----  -------  -------
   H     1    036.36   004.20
   C     6    045.45   062.50
   O     8    018.18   033.30
=====================================
Bragg Correction = -4.30%;  Stopping Units =  MeV / (mg/cm2)
See bottom of Table for other Stopping units
    Ion       dE/dx      dE/dx     Projected  Longitudinal  Lateral
   Energy     Elec.     Nuclear     Range     Straggling   Straggling
 ----------- ---------- ---------- ----------  ----------  ----------
  1.00 MeV  2.306E-01  1.733E-04   19.92 um      8648 A      7459 A
  2.00 MeV  1.502E-01  9.442E-05   58.97 um      2.29 um     1.95 um
  3.00 MeV  1.111E-01  6.597E-05  114.93 um      4.74 um     3.58 um
  4.00 MeV  8.908E-02  5.109E-05  187.21 um      7.19 um     5.64 um
  5.00 MeV  7.486E-02  4.187E-05  274.99 um     11.58 um     8.11 um
  6.00 MeV  6.484E-02  3.557E-05  377.77 um     15.75 um    10.96 um
  7.00 MeV  5.738E-02  3.098E-05  495.11 um     19.95 um    14.19 um
  8.00 MeV  5.158E-02  2.748E-05  626.48 um     27.56 um    17.78 um
  9.00 MeV  4.693E-02  2.472E-05  771.75 um     34.70 um    21.73 um
 10.00 MeV  4.311E-02  2.248E-05  930.64 um     41.69 um    26.02 um
-----------------------------------------------------------
Multiply Stopping by        for Stopping Units
-------------------        -------------------
 1.3970E+01                  eV / Angstrom
 1.3970E+02                 keV / micron
 1.3970E+02                 MeV / mm
 1.0000E+00                 keV / (ug/cm2)
 1.0000E+00                 MeV / (mg/cm2)
 1.0000E+03                 keV / (mg/cm2)
 1.4505E+01                  eV / (1E15 atoms/cm2)
 7.0529E+00                 L.S.S. reduced units
=================================================================
(C) 1984,1989,1992,1998,2008 by J.P. Biersack and J.F. Ziegler
```

**(B)** SRIM version ---> SRIM-2013.00
Calc. date ---> September 06, 2019

Disk File Name = SRIM Outputs\Hydrogen in **Silicon**.txt

Ion = Hydrogen [1] , Mass = 1.008 amu

Target Density = 2.3212E+00 g/cm3 = 4.9770E+22 atoms/cm3

======= Target Composition ========

| Atom Name | Atom Numb | Atomic Percent | Mass Percent |
|---|---|---|---|
| Si | 14 | 100.00 | 100.00 |

Bragg Correction = 0.00%
Stopping Units = MeV / (mg/cm2)
See bottom of Table for other Stopping units

| Ion Energy | dE/dx Elec. | dE/dx Nuclear | Projected Range | Longitudinal Straggling | Lateral Straggling |
|---|---|---|---|---|---|
| 1.00 MeV | 1.755E-01 | 1.328E-04 | 16.33 um | 7871 A | 9972 A |
| 2.00 MeV | 1.124E-01 | 7.331E-05 | 47.69 um | 2.04 um | 2.56 um |
| 3.00 MeV | 8.486E-02 | 5.156E-05 | 92.05 um | 4.06 um | 4.66 um |
| 4.00 MeV | 6.906E-02 | 4.010E-05 | 148.36 um | 6.11 um | 7.24 um |
| 5.00 MeV | 5.866E-02 | 3.297E-05 | 215.93 um | 9.54 um | 10.29 um |
| 6.00 MeV | 5.124E-02 | 2.807E-05 | 294.32 um | 12.84 um | 13.76 um |
| 7.00 MeV | 4.564E-02 | 2.450E-05 | 383.14 um | 16.17 um | 17.66 um |
| 8.00 MeV | 4.125E-02 | 2.177E-05 | 482.01 um | 21.92 um | 21.95 um |
| 9.00 MeV | 3.771E-02 | 1.961E-05 | 590.78 um | 27.36 um | 26.64 um |
| 10.00 MeV | 3.479E-02 | 1.786E-05 | 709.23 um | 32.70 um | 31.70 um |

| Multiply Stopping by | for Stopping Units |
|---|---|
| 2.3211E+01 | eV / Angstrom |
| 2.3211E+02 | keV / micron |
| 2.3211E+02 | MeV / mm |
| 1.0000E+00 | keV / (ug/cm2) |
| 1.0000E+00 | MeV / (mg/cm2) |
| 1.0000E+03 | keV / (mg/cm2) |
| 4.6637E+01 | eV / (1E15 atoms/cm2) |
| 2.9650E+01 | L.S.S. reduced units |

(C) 1984,1989,1992,1998,2008 by J.P. Biersack and J.F. Ziegler

```
=====================================================================
(C)     SRIM version ---> SRIM-2013.00
        Calc. date   ---> September 06, 2019
=====================================================================
Disk File Name = SRIM Outputs\Hydrogen in Air (gas).txt
Ion = Hydrogen [1] , Mass = 1.008 amu
Target Density =  1.1400E-03 g/cm3 = 3.8697E+19 atoms/cm3
Target is a GAS
======= Target Composition ========
  Atom  Atom  Atomic   Mass
  Name  Numb  Percent  Percent
  ----  ----  -------  -------
   H     1    025.00   001.42
   N     7    025.00   019.74
   O     8    025.00   022.55
   Ar   18    025.00   056.29
====================================
Bragg Correction = 0.00%
Stopping Units =  MeV / (mg/cm2)
See bottom of Table for other Stopping units

   Ion       dE/dx      dE/dx     Projected  Longitudinal   Lateral
  Energy     Elec.     Nuclear     Range     Straggling   Straggling
 ----------- ---------- ---------- ---------- ---------- ----------
  1.00 MeV  1.856E-01  1.309E-04   30.09 mm     1.42 mm     1.62 mm
  2.00 MeV  1.169E-01  7.193E-05   91.20 mm     3.82 mm     4.40 mm
  3.00 MeV  8.771E-02  5.048E-05  178.44 mm     7.79 mm     8.20 mm
  4.00 MeV  7.109E-02  3.921E-05  289.73 mm    11.77 mm    12.92 mm
  5.00 MeV  6.020E-02  3.220E-05  423.70 mm    18.56 mm    18.51 mm
  6.00 MeV  5.245E-02  2.740E-05  579.53 mm    25.07 mm    24.91 mm
  7.00 MeV  4.663E-02  2.390E-05  756.50 mm    31.64 mm    32.11 mm
  8.00 MeV  4.207E-02  2.122E-05  953.85 mm    43.12 mm    40.07 mm
  9.00 MeV  3.840E-02  1.911E-05    1.17 m     53.95 mm    48.77 mm
 10.00 MeV  3.538E-02  1.739E-05    1.41 m     64.59 mm    58.20 mm
-----------------------------------------------------------
Multiply Stopping by      for Stopping Units
-------------------       ------------------
 1.1400E-02                eV / Angstrom
 1.1400E-01               keV / micron
 1.1400E-01               MeV / mm
 1.0000E+00               keV / (ug/cm2)
 1.0000E+00               MeV / (mg/cm2)
 1.0000E+03               keV / (mg/cm2)
 2.9459E+01                eV / (1E15 atoms/cm2)
 1.7313E+01               L.S.S. reduced units
=====================================================================
(C) 1984,1989,1992,1998,2008 by J.P. Biersack and J.F. Ziegler
```

## References


1. Norris, D.J., Bawendi, M.G.: Measurement and assignment of the size-dependent optical spectrum in CdSe quantum dots. Phys. Rev. B. **53**(24), 16338–16346 (1996)
2. Alger, R.S.: Radiation effects in polymers. In: Fax, D., Labes, M.M., Weissberger, A. (eds.) Physics and Chemistry of the Organic Solid State, pp. 769–849. Vol-II. Wiley, New York (1965)
3. Biersack J.P., Ziegler J.F.: SRIM Programme, www.srim.org (2008)
4. Aliev, A.I., Drynkin, V.I., Leipunskaya, D.I., Kasatkin, V.A.: Handbook of Nuclear data for Neutron Activation Analysis, Israel Program for Scientific Translations, Jerusalem, 22–23 (1970)
5. Leo, W.R.: Technique for Nuclear and Particle Physics Experiments: A How to Approach, pp. 1–109. Springer, Berlin (1987)
6. Carter, G., Grant, W.A.: Ion implantation of semiconductors (Contemporary electrical engineering). In: Beck, A.H.W., Lamb. J (eds.), Edward Arnold, London (1976).
7. Dearnaley, G., Freeman, J.H., Nelson, R.S., Stephen, J.: Ion Implantation. North Holland Publishing Company, Amsterdam (1973)
8. Bohr, N.: On the constitution of atoms and molecules. Phil. Mag. **26** (1913)
9. Bethe, H.A.: Zur Theorie des Durchgangs schneller Korpuskularstrahlen durch Materie. Ann. Phys. **397**, 325–400 (1930)
10. Bloch, F.: Zur Bremsung rasch bewegter Teilchen beim Durchgang durch Materie. Ann. Phys. **16** (1933)
11. Walske, M.C.: The stopping power of K-electrons. Phys. Rev. **88**, 1283–1289 (1952)
12. Bichsel, H.: Hand Book of Physics. McGraw-Hill, New York (1963)
13. Pawar, P.S.: Study of $Al_2O_3$ layers formed by ion implantation and plasma anodization. Ph. D. thesis 23–49 (1991)
14. Mukhi, K.N.: Experimental Nuclear Physics. Vol. **I.** MIR Publications Moscow (1987)
15. Mazzie, R., Tadey, D., Smolko, E., Rocco, C.: Radiation grafting of different monomers onto PP foils irradiated with a 25 MeV proton beam. Nucl. Instrum. Methods Phys. Res. B **208**, 411–415 (2003)
16. Mckenna, C.M.: Faraday cup designs for ion implantation. In: Ryssel, H., Glawischnig, H. (eds.) Ion Implantation Technique, pp. 73–103. Springer, New York (1982)
17. Calcagno, L.: Ion-chains interaction in polymers. Nucl. Instrum. Methods Phys. Res. B **105**, 63–70 (1995)
18. Calcagno, L., Compagnini, G., Foti, G.: Structural modification of polymer films by ion irradiation. Nucl. Instrum. Methods Phys. Res. B **65**, 413–422 (1992)
19. Borgnakke, C., Larsen, P.S.: Statistical collision model for Monte Carlo simulation of polyatomic gas mixture. J. Comput. Phys. **18**, 405–420 (1975)
20. Nanbu, K.: Simple method to determine collisional event in Monte Carlo simulation of electron-molecule collision. Jpn. J. Appl. Phys. **33**, 4752–4753 (1994)
21. Chae, K.H., Choi, J.M., Jung, S.M., Joo, J.H., Kim, J.K., Kang, H.J., Whang, C.N.: Atomic transport in collisional mixing in bilayer structure. Nucl. Instr. Meth. Phys. Res. B **88**, 387–393 (1994)
22. Martin, G., Bellon, P.: Driven alloys. Solid State Phys. **50**, 189–331 (1997)
23. Jaraiz, M., Gilmer, G. H., Poate, J. M., Diaz de la Rubia, T.: Atomistic calculations of ion implantation in Si: point defect and transient enhanced diffusion phenomena. Appl. Phys. Lett. **68**, 409–501 (1996)
24. Strobel, M., Heinig, H., Moller, W.: Understanding ion beam synthesis of nanostructures: Modeling and atomistic simulations. Mat. Res. Soc. Symp. Proc. **647** O2.3.1-O2.3.13 (2001)
25. Miedema, A.R.: On the heat of formation of solid alloys II. J. Less-Common Metals **46**, 67–83 (1976)
26. Miedema, A.R., Boom, R., De Boer, F.R.: On the heat of formation of solid alloys. J. Less-Common Metals **41**, 283–298 (1975)

27. Mishra, Y.K., Singh, F., Avasthi, D.K., Pivin, J.C., Malinovska, D., Pippel, E.: Synthesis of elongated Au nanoparticles in silica matrix by ion implantation. Appl. Phys. Letts. **91** (2007) 063103-1- 063103-3
28. Kumar, P., Kumar, R., Kanjilal, D., Knobel, M., Thakur, P., Chae, K.H.: Ion beam synthesis of Ni nanoparticles embedded in quartz. J. Vac. Sci. Technol. B **26**(4), L36 (2008)
29. Cerruti, C., Stoquert, J.P., Orleans, C., Estournes, C., Grob, J.J., Guille, J.L., Haas, F., Muller, D., Richard-Plouet, M.: Ion beam synthesis of Co nanoparticles in SiO2: Monte Carlo simulation. Nucl. Instrum. Methods Phys. Res. B **216**, 329–333 (2004)
30. Sprouster, D.J., Ridgway, M.C.: Ion beam formation and modification of Cobalt nanoparticles. Appl. Sci. **2**, 396–442 (2012)
31. Buljan, M., Dubcek, P., Siketic, Z., Bogdanovic Radovic, I., Bernstorff, S., Serincan, U., Turan, R.: Ion beam synthesis and characterisation of Ge nanoparticles in $SiO_2$,m Nucl. Instrum. Methods Phys. Res. B **249**, 843–846 (2006)
32. Gonzalez-Varona, O., Perez-Rodriguez, A., Garrido, B., Bonafos, C., Lopez, M., Morante, J.R., Montserrat, J., Rodriguez, R.: Ion beam synthesis of semiconductor nanoparticles for Si based optoelectronic devices. Methods Phys. Res. B **161–163**, 904–908 (2000)
33. Buckmaster, R., Hanada, T., Kawazoe, Y., Cho, M., Yao, T., Urushihara, N., Yamamoto, A.: Novel method for site-controlled surface nanodot fabrication by ion beam synthesis. Nano Letts. **5**(4), 771–776 (2005)
34. Avasti, D.K.: Nanostructuring by energetic ion beams. In: Proceedings of Workshop on Ion Beam Studies of Nanomaterails: Synthesis, Modification and Characterisation. SMR/1758-12 (2006)
35. Schmidt B.: Nanostructures by ion beams. In: Proceedings of Workshop on Ion Beam Studies of Nanomaterails: Synthesis, Modification and Characterisation. **SMR/1758–15** (2006)
36. Som, T., Kanjilal, D.: Nanofabrication by ion beam sputtering: Fundamentals and Applications. Pan Stanford Publishing, CRC Press, Taylor & Francis Group, Boca Raton (2013)

# Chapter 9
# Spectroscopic Characterization of Gallium Nitride Nanowires

**Umesh Rizal and Bibhu Prasad Swain**

## Introduction

Nanometer-scale structures represent an emerging and rapidly expanding area of research due to photon, phonon, and electron confinement. The large surface to volume ratio, constrain at the molecular level and change in the phonon density of states have great impact on the process related properties of material. In this regard, GaN-NWs with wide direct band gaps of 3.4 eV, considered as novel material for fabrication of blue/green light emitting diodes, developing nanoscale emitters, detectors, biological/chemical sensors, laser diodes, and high-temperature devices [1]. The selection of catalyst is important parameter in the growth process of GaN-NWs. The catalyst influences the nucleation at the initial stage. The following important factors decides the formation of GaN-NWs (a) eutectic temperature of metal and GaN alloys, (b) surface energy of catalysts during dissociation, (c) adatom vapor pressure inside the reactor, (d) adatom vapor pressure inside catalyst particles. Lee et al. have synthesized GaN nanorings, nanoribbons and smooth nanowires on MgO substrate using Ag catalyst and observed that the diameter of GaN-NWs varies from 10–20 nm and length extended up to 500 $\mu$m [2]. Maliakkal et al. successfully synthesized Ni catalyzed GaN-NWs of diameter 20–130 nm and length extended up to 6 $\mu$m using MOCVD [3]. Kuykendall et al. synthesized vertical GaN-NWs of diameter 15–200 nm using Fe catalyst [4]. Chen et al. deposited GaN-NWs via conventional APCVD process using In catalyst and found that the diameter of NWs varies from 10

U. Rizal (✉)
Department of Chemistry, Sikkim Government College, Upper Kamrang, Namchi South Sikkim, Sikkim 737126, India
e-mail: umeshrizal680@gmail.com

B. P. Swain (✉)
Department of Physics, National Institute of Technology Manipur, Langol, Imphal, Manipur 795004, India
e-mail: bpswain@nitmanipur.ac.in; bibhuprasad.swain@gmail.com



B. P. Swain (ed.), *Nanostructured Materials and their Applications*, Materials Horizons: From Nature to Nanomaterials, https://doi.org/10.1007/978-981-15-8307-0_9

to 50 nm and length extended up to several μm [5]. Chen et al. reported that the diameter of Au catalyst grown GaN-NWs deposited via CVD and MBE process varies from 20 to 160 nm and 50–200 nm [6]. Mata et al. have correlated surface optical phonon modes in GaN-NWs with nanowire density and diameter [7]. The different $H_2$ flow rate have major impact on process related properties of GaN-NWs (a) passivates the dangling bonds at the growing surface, (b) increase the surface mobility of ad-atoms, (c) reduced the film deposition rate but it increase surface roughness. Dong et al. have revealed that $H_2$ induce step bunching on the surface of GaN thin film under Ga-rich conditions [8]. $H_2$ passivation reduces defect densities and increases the sticking coefficient of NWs [9]. The role of $H_2$ is suggested to thin boundary layer and increase entry length and lower diffusion coefficient for the precursor molecules in the gas phase [10]. Despite of the large quantum of research that has been documented so far in the field of synthesis and characterization of GaN-NWs, the variation of metal catalyst and $H_2$ flow rate that influence on microstructural, structural, vibration, optical and electronic properties has not yet been understood qualitatively. The present paper aimed at investigation of following:

(i) To investigate structural network and phonon lifetime ($\tau$) of catalyst grown GaN-NWs with different $H_2$ flow rate.
(ii) To investigate chemical composition and electronic environments of catalyst grown GaN-NWs with different $H_2$ flow rate.
(iii) To investigate optical properties of catalyst grown GaN-NWs with different excitation energy and $H_2$ flow rate.

## Experimental Details

GaN-NWs were synthesized by using GaN powder as a GaN source and $N_2$, $H_2$ as precursor gases in a horizontal thermal CVD reactor. The p-type Si (100) (resistivity 3–10 Ω-cm) substrate is placed on the downstream to the Ga source with the distance of 4 cm. The Si (100) substrates were cleaned by standard RCA cleaning procedure followed metallic coating of silver (Ag), iron (Fe), indium (In), and nickel (Ni) on c-Si were used as substrate. Then the source and substrate were placed inside the reactor and pumped to sub-atmospheric pressure. The temperature of the furnace was increased to the growth temperature of 1050 °C at a rate of at 5 °C min. $N_2$ and $H_2$ flow rate of 120 sccm and 40–120 sccm, respectively, is used as carrier and reactant gases. The growth duration was fixed for 4 h. After deposition, furnace was cooled down gradually at a rate of 3 °C/min. The microstructure of GaN-NWs was observed by a FE-SEM, JEOL-JEM-3000F. The Raman spectrum was recorded with a Horiba Jobin Yvon at room temperature using the 488 nm line of an $Ar^+$ laser as an excitation source. Structural characterization was done with the help of XRD using (Rigaku Saturn 724 + Cu (Kα) Single crystal X-ray diffractometer). The FTIR spectra were collected with a Perkin Elmer spectrometer (Model: Spectrum 2) operating in transmission mode. The spectra were collected in the ranges from 450 to 800 cm-1 with a resolution of 1 $cm^{-1}$. Room temperature Photoluminescence spectra of GaN-NWs were taken using a florescence spectrophotometer with a Xe discharge lamp. The chemical composition and bonding networks of the as synthesized GaN-NWs were characterized with the help of XPS spectroscopy using (VG ESCA-LABMK II system) equipped with an Mg K$\alpha$ X-ray (1253.6 eV) source.

## Result and Discussion

### *Microstructures of GaN-NWs*

The NWs lengthwise growth results from the difference in accommodation coefficient of eutectic liquid droplet and NWs. The diameter of NWs is controlled by three major parameters (i) size of catalyst define size of eutectic droplet, (ii) size of droplet further determined by growth temperature which controls solubility of reactants, (iii) the diameter of NWs increases if vapor pressure of reactant is above critical value. Kinetic theory can determines how fast the NWs grow and dependency of growth rate. The growth rate dependence can be described by a Gibbs Thomson effect:

$$\delta\mu = \delta\mu_o - \frac{4\Omega\sigma_s}{d} \tag{9.1}$$

where, $\delta\mu$ is difference in chemical potential between gas and solid phase of nanowire material, $\delta\mu_o$ is for the bulk case, $\sigma_s$ is the surface free energy of NWs, $\Omega$ is the atomic volume of NWs. GaN-NWs was grown using Ag, Fe, Ni and In metal catalysts on Si (100) substrates. Figure 9.1(a) shows hair like morphology of GaN-NWs grown using Ag catalyst. These FESEM image reveals, wire like morphology of GaN-NWs which is distributed haphazardly over the entire Si substrate. The diameter of observed NWs is in the range of 40–80 nm with length of several tens of micrometers after 2 h of deposition. Figure 9.1b reveals the nodules like microstructure of GaN-NWs synthesized by using Fe catalyst. In addition to straight NWs, lots of twisted NWs were observed and are randomly distributed on the Si substrates. The diameter of GaN-NWs is in the range of 60–100 nm using Fe catalyst. Figure 9.1c shows the microstructure of In assisted GaN-NWs. The diameter of NWs varies from 20 to 80 nm. The observed GaN-NWs are thinner to Fe catalyst grown GaN-NWs but very similar to Ag catalyst. Some GaN-NWs are bent and coiled. Figure 9.1d shows a FESEM image of GaN-NWs grown by using Ni catalyst. The NWs are straight, long and entangled in a dense microstructural network of GaN of diameter varies from 60 to 120 nm.

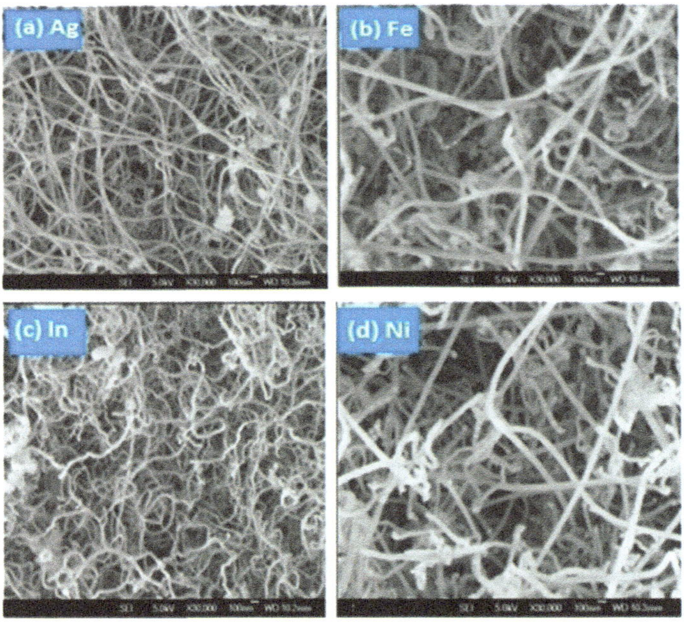


**Fig. 9.1** FESEM image of GaN-NWs synthesized using different catalysts at 80 sccm $H_2$ flow rate [11]

## *Structural Analysis*

XRD spectroscopy is based on the elastic scattering of X-rays from the electron clouds of the individual atoms in the system. XRD data gives us information about (a) crystal structure, (b) phase transition, (c) crystallite size, (d) degree of crystallinity, and (e) residual stress of materials. The diffraction peaks are satisfied by Bragg's equation

$$2d \sin\theta = n\lambda \tag{9.2}$$

(a) Crystallite sizes of GaN-NWs were determined using Scherer's formula.

$$\text{Crystallite Size} = \frac{0.94\lambda}{\beta\text{COS}\theta} \tag{9.3}$$

(b) The peak position shifts to higher diffraction angle indicates tensile stress whereas diffraction peak shifts to lower diffraction angle indicates compression stress in the film.

Figure 9.2a shows the XRD spectra of GaN-NWs grown using Ag catalyst. The diffraction patterns of GaN-NWs are indexed at $2\theta = 32.05°, 34.26°, 36.41°, 47.69°,$

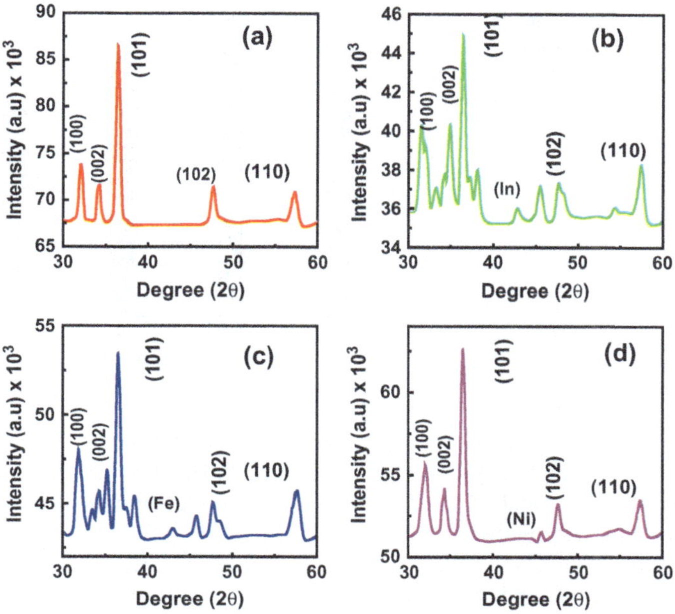


**Fig. 9.2** XRD spectrum of GaN-NWs with **a** Ag, **b** In, **c** Fe and **d** Ni catalysts

57.31° corresponding to (100), (002), (101), (102) planes of crystalline GaN phase. The peak at 36.41° corresponding to (101) plane of GaN shows highest intensity with respect to other peaks. Therefore, we have calculated the crystallite size and stress in GaN-NWs thin film from XRD data using Scherer equation. Figure 9.2b shows six XRD peaks for GaN-NWs grown with In catalyst. These planes are indexed at 31.15°, 34.66°, 36.16°, 42.39°, 47.89°, and 57.12° corresponds to (100), (002), (101), In plane, (102) and (110) planes of crystalline GaN phase. Figure 9.2c shows the GaN-NWs grown with Fe catalyst, the diffraction peak appeared at 31.58°, 35.08°, 36.07°, 45.81°, 47.80°, and 57.54° corresponds to (100), (002), (101), Fe (101), (102) and (110) planes of crystalline GaN phase. Figure 9.2d shows the diffraction patterns of GaN-NWs using Ni catalyst at 31.91°, 34.15°, 36.40°, 45.39°, 47.37°, and 57.12° corresponds to (100), (002), (101), (102) and (110) planes indicates crystalline GaN phase. The XRD patterns of GaN-NWs indicate that it has a wurtzite structure with lattice constants of a = 0.317 nm and c = 0.517 nm (JCPDS number 06–0416 for GaN). The strong intensity of the (101) peak indicates that the GaN-NWs have a preferential growth direction along (101) orientation. Li et al. have observed (100), (101) plane of GaN-NWs grown with Ni catalyst at $32^0$, $35^0$, respectively [12]. However, we have observed the (100) and (101) plane of GaN-NWs with Ni catalyst at 31.91° and 36.40°, respectively. The (101) plane of GaN-NWs grown with Ni catalyst shift to higher diffraction angle indicates tensile stress in the film. Johnson et al. observed (100), (002), (101) planes of Fe catalyst grown GaN-NWs at 32, 35,

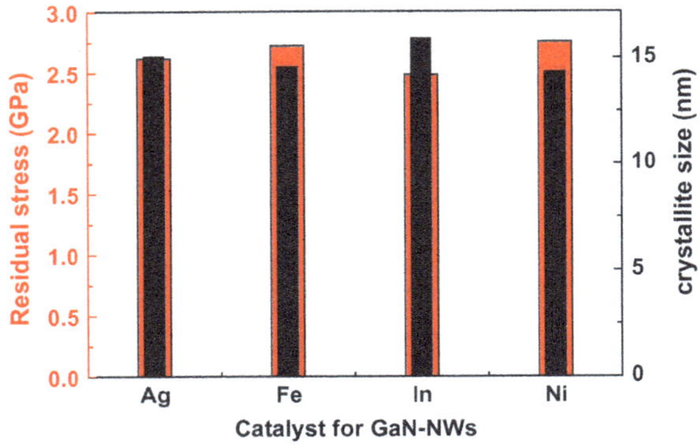


**Fig. 9.3** Residual stress and crystallite size of GaN-NWs grown using Ag, Fe, Ni and In catalysts

and 37°, respectively [13]. However we have observed (100), (002), (101), planes of GaN-NWs with Fe catalyst at 31.58°, 35.08°, 36.07°. The (101) plane of GaN-NWs grown with Fe catalyst shift to lower diffraction angle indicates compressive stress in the film. Zhang et al. have found (100), (002), (101), (102), (110), (103), (112), (201) planes of GaN-NWs grown with in catalyst at ~32°, 34.5°, 37°, 48°, 58°, 64°, 31.9°, 36.4°, 69° and 70°, respectively [14]. We have observed various planes (100), (002), (101), (102) and (110) of GaN-NWs with In catalyst at 31.15°, 34.66°, 36.16°, 47.89°, and 57.12° respectively. The shifting of diffraction plane to lower angle with In catalyst indicates compressive stress in the film. Figure 9.3 depicts the estimated crystallite size and residual stress of GaN-NWs grown using Ag, Fe, Ni and In catalyst. The crystallite size of GaN-NWs grown with Ag, Fe, In and Ni catalyst are 15.06, 14.56, 15.89, and 14.33 nm, respectively. The crystallite size of GaN-NWs grown with Ni catalyst is considerably smaller than the other catalyst. In a similar trend, stress of GaN-NWs grown using Ag, Fe, In and Ni catalyst are observed at 2.618, 2.72, 2.482, and 2.752 GPa respectively. GaN-NWs grown with Ni catalyst experience maximum stress whereas minimum stress is observed for In catalyst grown GaN-NWs.

## *FTIR Analysis*

The vibrational spectra of nanostructure materials are generally different from the spectra of their bulk counterpart due to its large surface to volume ratio. The small size of nanomaterial will cause the breakdown of vibrational selection rule that result in broadening and shifting of various vibrational signatures.

The significant classification for analysis of FTIR data in the GaN–NWs are

(a) Peak intensity of FTIR spectra indicates the amount of concentration or density present in a solid as described by the Lambert Beer Law. Repetition of the same functional group leads to a larger and intense peak.

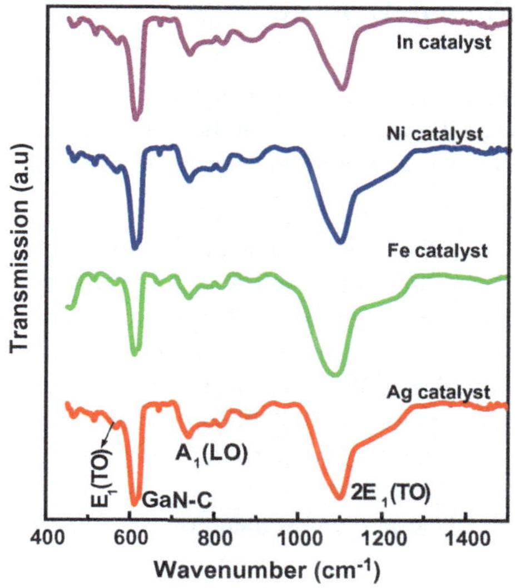


**Fig. 9.4** FTIR spectra of GaN-NWs grown in presence of different catalyst

$$I = I_O e^{-\alpha t} \quad (9.4)$$

(b) Peak shift of FTIR in the GaN-NWs indicates the polarity of the molecule or backing of higher electronegative atoms attached with chemical bonding.
(c) Asymmetric peak in the FTIR spectra signifies another chemical compound with similar bonding strength overlap with each other or relative movement of the atoms connected by the bond.
(d) The FWHM of FTIR spectra indicates (i) coordination number of elements, and (ii) dispersion of chemical bonds or defect states attached to it.

The chemical bonding of GaN-NWs using FTIR spectroscopy was less studied. In the present research we have studied the FTIR spectra of GaN-NWs grown with different metal catalysts. Figure 9.4 shows the transmission FTIR spectra of GaN-NWs with different of catalysts in the range of 400–1400 cm$^{-1}$. The bonding signature of GaN-NWs shows four distinguished vibrational bands at 536, 613, 737, and 1103 cm$^{-1}$ corresponds to $E_1$(TO), GaN-C, $A_1$(LO) and $2E_1$(TO), respectively [15].

Some important observations in FTIR spectra of GaN-NWs with Ag, Fe, Ni and In catalyst are:

(a) The $2E_1$(TO) peak of GaN-NWs grown with Ag, Fe, Ni catalyst shows asymmetry to higher wavenumber region indicates another chemical compound of Ga=N or Ga–N–O with similar bonding strength overlap with each other. However we have not observed asymmetry with In catalyst.
(b) The FWHM of $A_1$(LO) phonon with Ag, Fe, Ni, and In catalyst are 42, 38, 39, and 44 cm$^{-1}$. The maximum value of FWHM is observed with in catalyst indicates increasing of the coordination number whereas minimum value is observed with Fe catalyst.

### *Raman Analysis*

To analysis the experimental data of GaN-NWs the following parameters are considered. It has been shown theoretically as well as experimentally that Raman spectra data yield information about structural aspects of GaN-NWs in the following way.

(a) The increasing of Raman intensity leads to increasing of polarizability of GaN-NWs. The bond polarizability model allows us to calculate Raman spectra out of the Eigen modes obtained with the spring and-mass model.

(b) The increasing of FWHM of Raman spectra leads to decreasing crystallinity in the materials. It may due to increasing of doping or impurity content in the GaN-NWs. Moreover, it can be due to dispersion of nanowire diameter.

(c) The entire GaN-NWs Raman peak is caused by a breathing mode with low frequency proportional to $1/d$, where $d$ is the diameter which is predicted by elasticity theory.
One-phonon Raman process will exhibit a Lorenzian line shape described by

$$I(\omega) = I_o \frac{2A}{\pi} \cdot \frac{\Gamma}{4(\omega - \omega_c)^2 + \Gamma^2} \tag{9.5}$$

Here $I(\omega)$ is the intensity at energy $\omega$, $I_0$ a constant shift and $\omega_c$, $A$, $\Gamma$ the center, area and full width at half maximum (FWHM) of the peak, respectively.

(d) The intensity ratios of $A_1$(TO) to $E_1$(TO) indicate structural anisotropy due to difference in bond polarizability in parallel and perpendicular direction.
The energies of quasi TO modes lie between the energies of $A_1$(TO) and $E_1$(TO) mode and fulfilled the following equations:

$$\omega^2(\theta) = \left[\omega_{E1(TO)} \mathrm{Cos}\theta\right]^2 + \left[\omega_{A1(TO)} \mathrm{Sin}\theta\right]^2 \tag{9.6}$$

(e) The full width of half maxima (FWHM) of Raman spectra represents the inversely related to phonon lifetime ($\tau$) considering from Heisenberg uncertainty principle

$$\frac{1}{\tau} = \frac{\mathrm{FWHM}}{\hbar} \tag{9.7}$$

To simplify above equation

$$\tau = \frac{5.309 \times 10^{-12}}{\mathrm{FWHM}\left(\mathrm{cm}^{-1}\right)} \mathrm{s} \tag{9.8}$$

(f) The dispersion relation $\omega_{(SO)}$ for an infinite long cylindrical wire, in the limit where the phonon wave vector $q \gg c$, can be written as

$$\omega_{\mathrm{SO}}^2 = \omega_{\mathrm{TO}}^2 \frac{\varepsilon_0 - \rho_{nx}\varepsilon_m}{\varepsilon_\infty - \rho_{nx}\varepsilon_m} \tag{9.9}$$

where $\omega_{TO}$ is the frequency of the TO phonon, $\varepsilon_0$ and $\varepsilon_\infty$ are the static and high frequency dielectric constant of the material, respectively, and $\varepsilon_m$ is the dielectric constant of the medium.

Figure 9.5 shows Raman spectra of GaN-NWs grown with Ag catalyst varying $H_2$ flow rates from 40 to 120 sccm. The various Raman active phonon modes appeared at 515.82, 526.25, 562.53, 611.63, 658, 692, and 720 $cm^{-1}$ corresponds to $A_1$(TO), $E_1$(TO), $E_2$(high), DI phonon, SO($A_1$), SO($E_1$), and $A_1$(LO), respectively.

Some important observation in Raman spectra of GaN-NWs with Ag catalyst are address as follows:

(a) The FWHM of $A_1$(TO), $E_1$ (TO), $E_2$(high), DI, SO($A_1$) phonon increases from 18.7 to 26.68, 36.18 to 37.14, 31 to 35.70, 54.11 to 59.56, 54.49 to 55.68 $cm^{-1}$ with increasing $H_2$ flow rate indicates decreasing crystallinity in the materials.

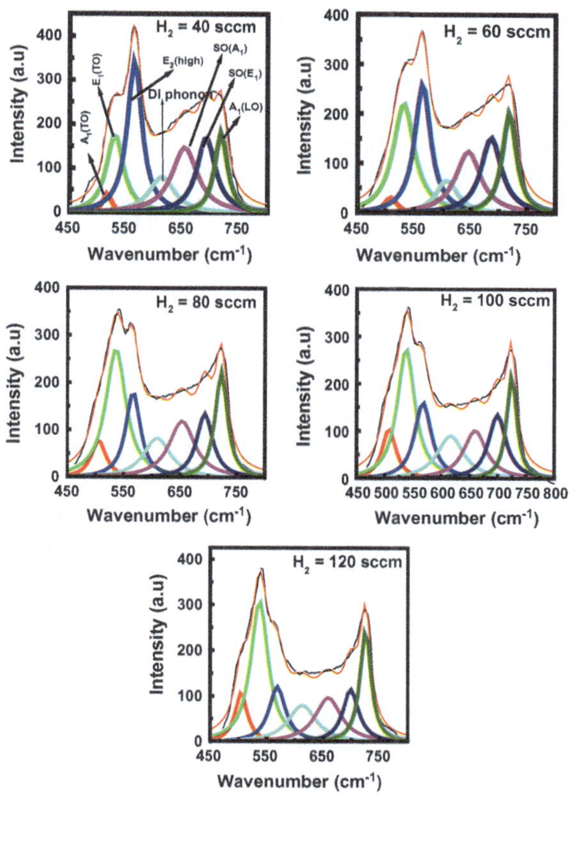


**Fig. 9.5** Raman spectra of GaN-NWs using Ag catalyst with different $H_2$ flow rate

(b) The intensity of $A_1$(TO) phonon peak with Ag catalyst increases with increasing $H_2$ flow rate indicates increasing of polarizability of GaN-NWs due to change in dielectric constant of surrounding chemical environment.
(c) The FWHM of SO($E_1$) and $A_1$(LO) phonon decreases from 40.06 to 34.44 and 26.52 to 23.29 indicates increasing crystallinity of GaN-NWs film.

Figure 9.6 shows Raman spectra of GaN-NWs with Ni catalyst varying $H_2$ flow rates from 40 to 120 sccm. The various Raman active phonon modes appeared at 521.14, 545.55, 559.19, 598.23, 638.98, 691.44, 722.82 $cm^{-1}$ corresponds to A1(TO), E1(TO), E2(high), DI phonon, SO(A1), SO(E1), and A1(LO), respectively.

Some important observation in Raman spectra of GaN-NWs with Ni catalyst are:

(a) The FWHM OF $A_1$(TO), SO($A_1$), SO($E_1$) phonon decreases from 37.59 to 28.77, 75.15 to 68.86, 36.29 to 23.05 $cm^{-1}$ indicates increasing of crystallinity.
(b) The FWHM of $E_1$(TO), $E_2$(high), DI phonon increases from 17.87 to 38.12, 17.77 to 31.53, 32.25 to 64.97 $cm^{-1}$ indicates decreasing of crystallinity.

Figure 9.7 shows Raman spectra of GaN-NWs grown with Fe catalyst varying $H_2$ flow rates from 40 to 120 sccm. The various Raman active phonon modes appeared at 520, 547, 560, 617, 664, 699 and 720 $cm^{-1}$ corresponds to $A_1$(TO), $E_1$(TO), $E_2$(high), DI phonon, SO($A_1$), SO($E_1$), and $A_1$(LO), respectively.

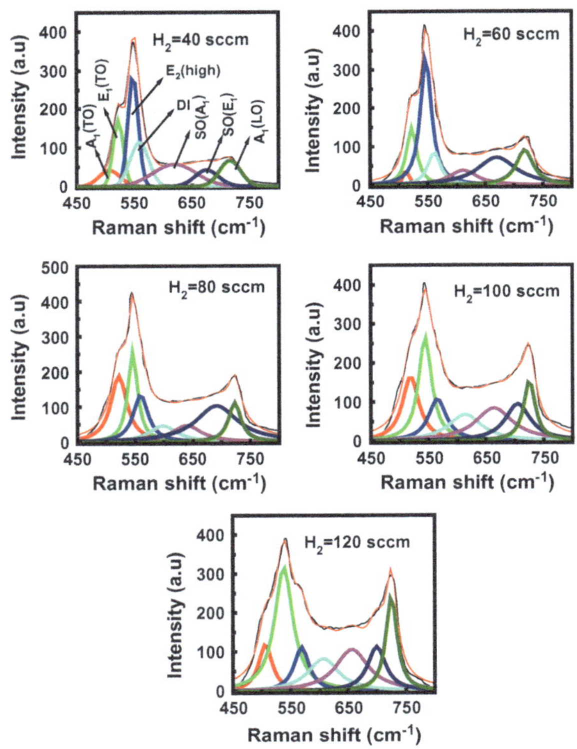


**Fig. 9.6** Raman spectra of GaN-NWs grown using Ni catalyst with different $H_2$ flowrate

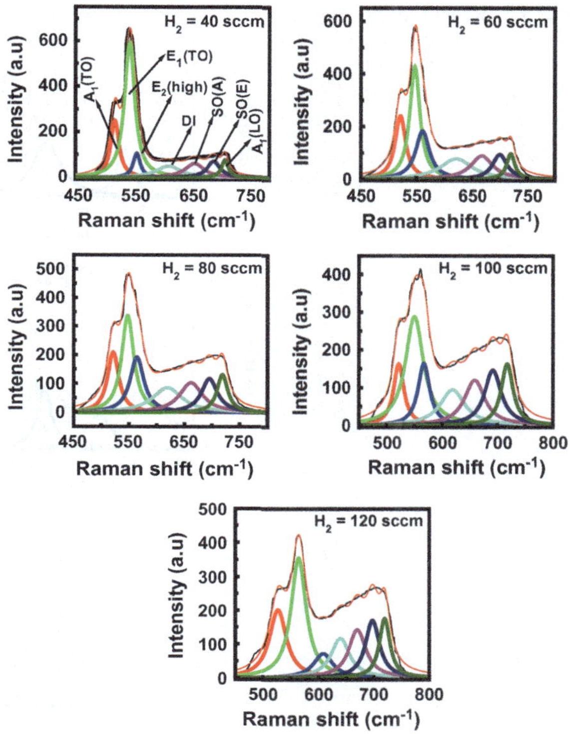


**Fig. 9.7** Raman spectra of GaN-NWs using Fe catalyst with different $H_2$ flow rate

Some important observations were made as follows:

(a) The FWHM of $A_1$(TO), $E_1$(TO), $E_2$(high), SO($E_1$) and $A_1$(LO) phonon increases from 18.08 to 36.72, 22.19 to 32.82, 15.45 to 42.19, 28.87 to 30.67, 14.53 to 23.10 cm$^{-1}$ with increasing $H_2$ flow rate indicates decreasing crystallinity in the materials.

(b) The FWHM of DI and SO($A_1$) phonon decreases from 55.16 to 41.84 and 41.47 to 38.37 indicates increasing crystallinity of film.

Figure 9.8 shows Raman spectra of GaN-NWs grown with In catalyst varying $H_2$ flow rates from 40 to 120 sccm. The various vibration phonons appear at 506, 537, 567, 603, 651, 702, and 725 cm$^{-1}$ corresponds to $A_1$(TO), $E_1$(TO), $E_2$(high), DI phonon, SO($A_1$), SO($E_1$), and $A_1$(LO), respectively.

Some major observations are address as follows:

(a) The FWHM of $A_1$(TO), $E_1$(TO), DI, and $A_1$(LO) phonon increases from 27.66 to 41.48, 36.20 to 36.41, 57.77 to 60.71, and 23.12 to 26.84 cm$^{-1}$ with increasing $H_2$ flow rate indicates decreasing crystallinity in the materials.

(b) The FWHM of $E_2$(high), SO($A_1$), and SO($E_1$) phonon decreases from 31.60 to 30.64, 78.72 to 59.66, and 56.97 to 47.52 cm$^{-1}$ indicates increasing crystallinity of film.

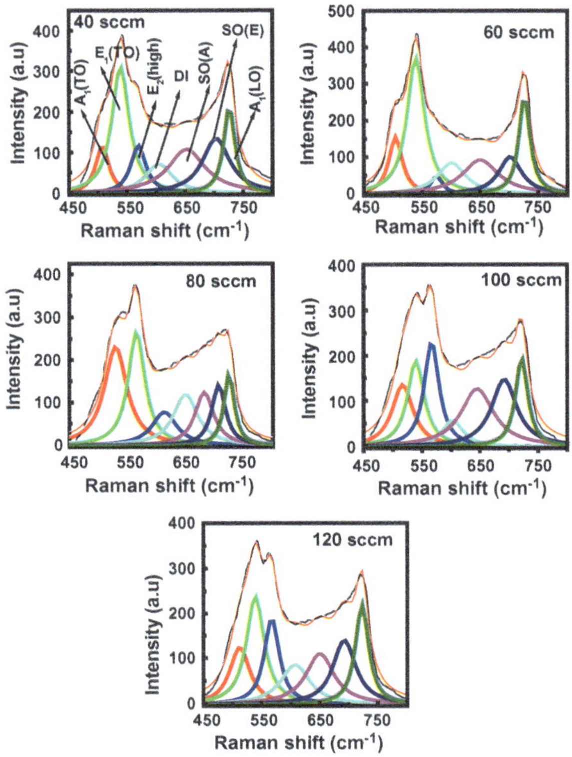


**Fig. 9.8** Raman spectra of GaN-NWs grown using In catalyst with different $H_2$ flowrate

The spatial confinement of optical phonons shows that the Raman spectra shifted in peak position and broadens due to the relaxation of the *q*-vector selection rule in finite size material. This finite size effect give rise to phonon confinement, causing an uncertainty in the phonon wave vector, which typically produces a frequency shift and a line shape broadening and leads to the asymmetrical shape of the Raman spectrum. Such an asymmetrical shape of the Raman modes is associated with the phonon confinement.

Figure 9.9 shows peak position of $A_1$(TO), $E_1$(TO), $E_2$(High), DI Phonon, SO($A_1$), SO($E_1$) and $A_1$(LO) with increasing $H_2$ flow rate from 40 to 120 sccm using different catalysts. The $A_1$(TO) phonon peak of GaN-NWs grown by using Ag, Fe, Ni and In catalyst shifts from 505.86 to 516.49, 520.86 to 526.3, 506.17 to 521.14 and 506.66 to 516.25 $cm^{-1}$. In a similar manner, $E_1$(TO) phonon of GaN-NWs grown by using Ag, Fe, Ni and In catalyst shifts from 529.8 to 537.37, 547.31 to 564.07, 521.71 to 545.55, 537.85 to 564.42 $cm^{-1}$, respectively. The $E_2$(high) phonon of GaN-NWs grown using Ag, Fe, Ni and In catalyst shifted from 564.2 to 568.56, 560.43 to 608.07, 546.88 to 568.29, and 565.43 to 612.01 $cm^{-1}$. Similarly, the peak position of DI phonon of GaN-NWs grown using Ag, Fe, Ni and In catalyst shifts from 608.54 to 623.51, 617.96 to 639.39, 558.52 to 613.16, and 599.19 to 648.66 $cm^{-1}$.

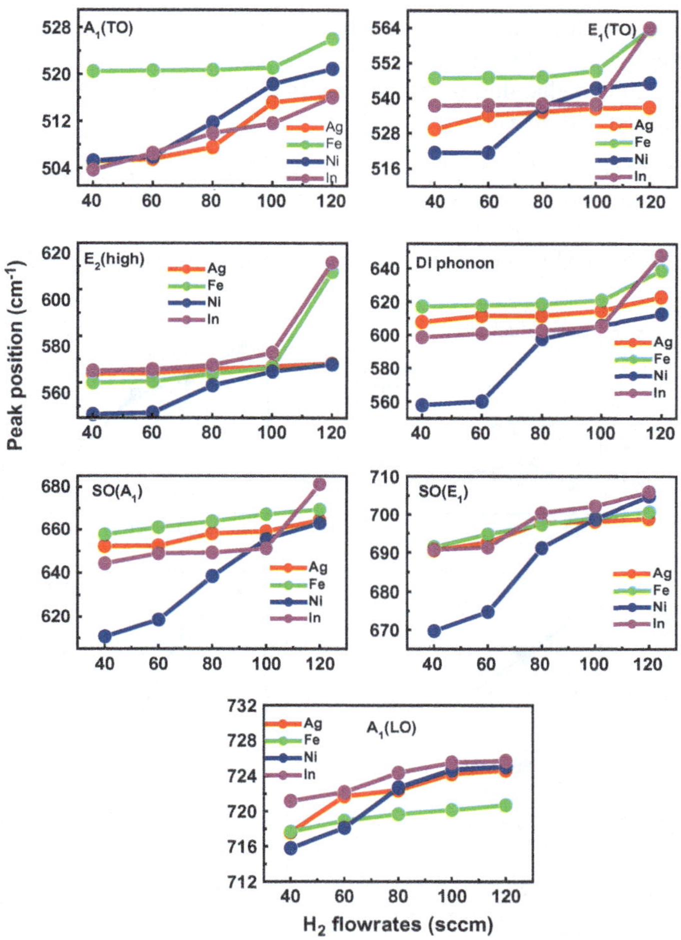


**Fig. 9.9** Peak position of various phonon modes of GaN-NWs $A_1$(TO), $E_1$(TO), $E_2$(High), DI Phonon, SO($A_1$), SO($E_1$) and $A_1$(LO) with different $H_2$ flow rate using Ag, Fe, Ni and In catalysts

The SO(A) phonon of GaN-NWs grown using Ag, Fe, Ni, In catalyst shifts from 652.57 to 664.81, 658.19 to 670.07, 611.1 to 663.39 and 644.57 to 681.64 $cm^{-1}$. In a similar sideline, the SO(E) phonon of GaN-NWs grown using Ag, Fe, Ni, In catalyst shifts from 690.83 to 699.07, 691.78 to 700.84, 669.95 to 704.98, 691.01 to 706.17 $cm^{-1}$, respectively. The $A_1$(LO) phonon of GaN-NWs grown with Ag, Fe, Ni, In catalyst shifts from 717.7 to 724.81, 717.8 to 720.88, 715.89 to 725.2 and 721.28 to 725.82 $cm^{-1}$, respectively.

Some important observations are address as follows:

(a) The peak position of $A_1$(TO), $E_1$(TO), $E_2$(High), DI,SO($A_1$), SO($E_1$) and $A_1$(LO) phonon of GaN-NWs grown with Ag, Ni, Fe and In catalysts shows blue shift with increasing $H_2$ flow rate indicates increase of stress and more structural constrain the in film.

(b) The maximum blue shift is observed for In catalyst whereas minimum blue shift is observed with Fe catalyst.

Figure 9.10 shows phonon lifetime ($\tau$) of $A_1$(TO), $E_1$(TO), $E_2$(high), DI Phonon, SO($A_1$), SO($E_1$) and $A_1$(LO) modes of GaN-NWs grown with different $H_2$ flow rate and catalysts. The phonon lifetime of $A_1$(TO) phonon peak of GaN-NWs grown by using Ag, Fe, Ni and In catalyst increase from 0.2839 to 0.19893, 0.29353 to 0.14458,

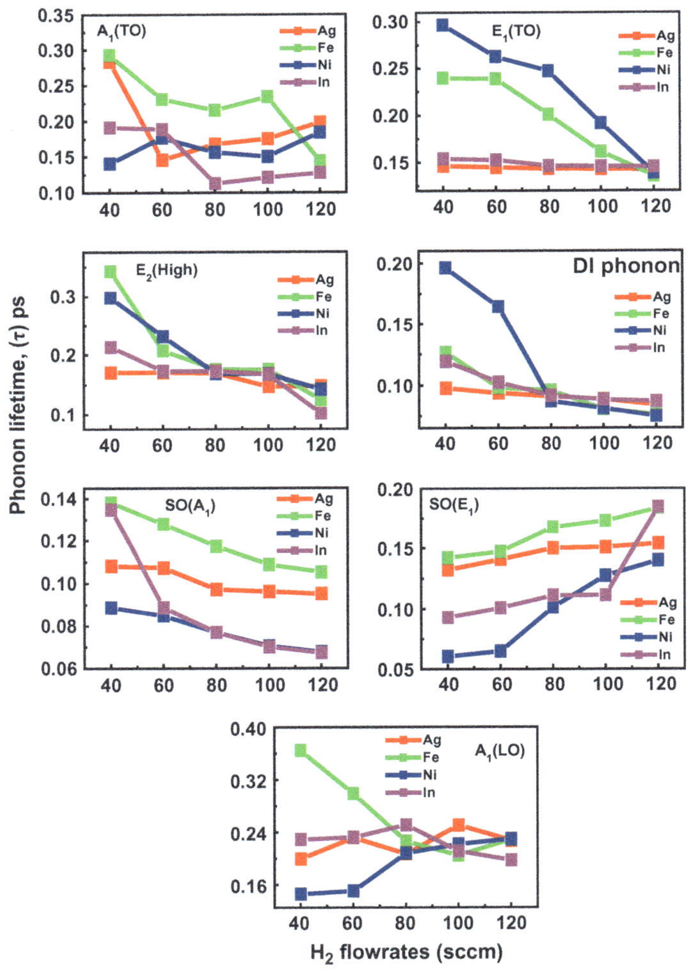


**Fig. 9.10** Phonon lifetime ($\tau$) of $A_1$(TO), $E_1$(TO), $E_2$(high), DI Phonon, SO($A_1$), SO($E_1$) and $A_1$(LO) modes of GaN-NWs grown using Ag, Fe, Ni and In with different $H_2$ flow rate

**Table 9.1** Comparison of phonon lifetime of GaN-NWs grown using different catalysts with literatures [16]

| Phonon mode | $\tau$ = (Bulk ~ 1–10 ps) (literature) | Ag (ps) | Fe (ps) | Ni (ps) | In (ps) |
|---|---|---|---|---|---|
| $A_1$(TO) | 0.28 | 0.19 | 0.22 | 0.16 | 0.15 |
| $E_1$(TO) | 0.15 | 0.14 | 0.19 | 0.22 | 0.15 |
| $E_2$(high) | 0.36 | 0.16 | 0.20 | 0.20 | 0.17 |
| SO($A_1$) | 0.13 | 0.10 | 0.12 | 0.07 | 0.08 |
| SO($E_1$) | 0.24 | 0.14 | 0.16 | 0.09 | 0.12 |
| $A_1$(LO) | 0.35 | 0.22 | 0.27 | 0.19 | 0.22 |
| DI phonon | – | 0.09 | 0.10 | 0.12 | 0.10 |

0.19188 to 0.12797, and 0.14121 to 0.18448 pico second (ps) with increasing $H_2$ flow rate from 40 to 120 sccm. On the other hand, phonon lifetime of $E_1$(TO) phonon of GaN-NWs grown by using Ag, Fe, Ni and In catalyst decrease from 0.14672 to 0.14292, 0.24005 to 0.13625, 0.29704 to 0.13925, 0.15418 to 0.14578 ps, respectively. The phonon lifetime of $E_2$(high) phonon of GaN-NWs grown using Ag, Fe, Ni and In catalyst decrease from 0.17122 to 0.14869, 0.34358 to 0.12581, 0.29876 to 0.14302, 0.21382 to 0.10229 ps. Similarly, the phonon lifetime of DI phonon of GaN-NWs grown using Ag, Fe, Ni and In catalyst decrease from 0.09811 to 0.08461, 0.12689 to 0.07651, 0.19635 to 0.07513, and 0.11928 to 0.08744 ps, respectively. The phonon lifetime of SO(A) phonon varies from 0.10838 to 0.09535, 0.13834 to 0.10547, 0.08887 to 0.06784, 0.1349 to 0.06744 ps. On the other hand, the phonon lifetime of SO(E) phonon of GaN-NWs grown using Ag, Fe, Ni, In catalyst increase from 0.13251 to 0.15413, 0.14244 to 0.18388, 0.06072 to 0.14015, and 0.09319 to 0.18477 ps, respectively. The phonon lifetime of $A_1$ (LO) phonon of GaN-NWs grown with Ag, Fe, Ni, In catalyst varies from 0.20013 to 0.22788, 0.36531 to 0.22983, 0.14627 to 0.23032, and 0.22959 to 0.19774 ps. Table 9.1 shows the comparative study of phonon lifetime of GaN-NWs in present research with the literature.

(a) The shorter phonon lifetime in NWs is due to decay of optical phonon into acoustic phonon. When the phonon lifetime of semiconductor is longer, than the phonon population that interacts with carrier will be the hotter so that net energy relaxation time is longer [17]. Therefore the efficiency of device degrades with longer phonon lifetime of material.
(b) Phonon lifetime can be contributed according to matthiessen rule:

$$\frac{1}{\tau} = \frac{1}{\tau_u} + \frac{1}{\tau_b} + \frac{1}{\tau_m} + \frac{1}{\tau_{e-ph}} \tag{9.10}$$

where $\tau_u$ represents Umklapp effect, $\tau_b$ is due to boundary effect, $\tau_m$ is due to mass difference effect and $\tau_{e-ph}$ represents electron phonon scattering effect.

(c) The shorter phonon lifetime is observed with Ni catalyst whereas longer phonon lifetime is observed with Fe catalyst. Therefore, therefore the growth of GaN-NWs with Ni catalyst is favored whereas Fe catalyst degrades the efficiency of device.

## *PL Analysis*

In the present research, we have studied the optical properties of GaN-NWs grown with different metal catalyst. To study the effect of different metal catalyst on the optical properties of the GaN-NWs, PL spectra were recorded using Xenon discharge lamp as an excitation source.

The significant classifications for analysis of PL spectra of GaN–NWs are.

(a) The peak position of PL spectra shifts due to vacancies, impurities, dangling bonds, strain and structural defects.
(b) FWHM of PL spectra signifies the distribution of defect states, band gap renormalization, and impurity band formation.
(c) Asymmetric peak in the PL spectra indicates electronic transition among defect states.
(d) Peak intensity of PL spectra indicates the concentration of defect states present in a material.

$$I_{\mathrm{PL}} = \frac{1}{\text{Defect states}} \quad (9.11)$$

In the present scenario, GaN-NWs sample were excited with excitation energy of 3.87 eV. The band gap energy of GaN is 3.44 eV. As we know, when the excitation energy is higher than the band gap energy of semiconductor, then they absorb incident energy at the surface states. Hence in our case, the characteristic PL spectrum gives us idea about surface behavior of GaN-NWs. The metal catalyst can be placed according to its valance states. Hence, the PL shows radiative recombination of the surface states. Figure 9.11a shows a broad range emission band of Ag catalyst grown GaN-NWs between 2.4 and 3.2 eV with a strong peak centered at 2.97 eV corresponds to Nitrogen vacancy ($V_{\mathrm{N}}$) in the wire. However, broadband also contains some fine emission peaks centered at 2.58, 2.73, and 2.83 eV corresponds to Ag interstial defect, Nitrogen interstial and Ga vacancy ($V_{\mathrm{Ga}}$) [18].

This result reveals that the surfaces states of GaN-NWs is inherently imperfect and confirms the presence of defects. However, it was observed that the shape of PL spectra gets changed with different catalyst. Figure 9.11b shows broad range PL emission band of GaN-NWs grown using Fe catalyst. This broad band spread from 2.4 and 3.2 eV containing four emission band centered at 2.70, 2.88, 3.02, and 3.10 eV corresponds to Oxygen vacancy ($V_{\mathrm{O}}$), $V_{\mathrm{Ga}}$, $V_{\mathrm{Ga}}$, $V_{\mathrm{N}}$ complexes and Fe interstial defects. In a similar sideline, Fig. 9.11c shows broad range PL emission

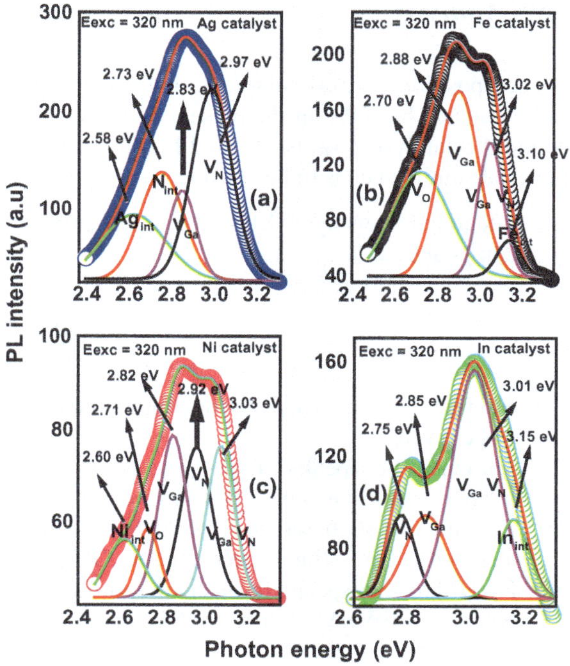


**Fig. 9.11** PL spectra of GaN-NWs grown using **a** Ag, **b** Fe, **c** Ni, and **d** In metal catalysts

band of GaN-NWs grown using Ni catalyst. This broad band spread from 2.4 and 3.2 eV containing five emission band centered at 2.60, 2.71, 2.82, 2.92 and 3.03 eV corresponds to Nickel point defect, $V_{\mathrm{O}}$, $V_{\mathrm{Ga}}$, $V_{\mathrm{N}}$ and $V_{\mathrm{Ga}}$, $V_{\mathrm{N}}$ complexes, respectively. Figure 9.11d shows broad range PL emission band of GaN-NWs grown using In catalyst. This broad band spread from 2.4 and 3.2 eV containing four emission band centered at 2.75, 2.85, 3.01, and 3.15 eV corresponds to $V_{\mathrm{N}}$, $V_{\mathrm{Ga}}$, $V_{\mathrm{Ga}}$, $V_{\mathrm{N}}$ and Indium defect. As it can be seen from these figures, all spectra show strong blue luminescence band at 2.96–3.10 eV. It is well evident that GaN-NWs have the large surface area with unsaturated dangling bonds, which are saturated by bonding with neighboring catalyst and $H_2$ atom. In addition to the blue band, another defect related emission such as green band appears at 2.7 eV [19]. Moreover, incorporation of $H_2$ gas increases the vacancy concentration in the NWs. Here, we further noticed the appreciable reduction of band gap energy of GaN-NWs. Such a reduction of band gap energy could be ascribed to the superposition of several effects such as strain, Stokes shift, defect and impurity states experienced by the GaN-NWs.

## *XPS Analysis*

For composition, core orbital spectra and binding energy of various chemical bonding are calculated by XPS. In the present scenario we are focus on core orbital of Ga, N and the catalysts (Ag, Ni, Fe and In). To analysis the XPS spectra the following physical significance is to be consider.

(a) Integrated intensity of XPS spectra represent amount of element content in the compound (at.%)

$$X_i \text{ at.\%} = \left(\frac{A_i}{S_i}\right) \Big/ \sum\left(\frac{A_i}{S_i}\right) \quad (9.12)$$

where $A_i$ is the area under core orbital spectra, $S_i$ is the atomic sensitivity factor, respectively.

(b) The core orbital peak shifts to higher binding energy in the XPS spectra represent stronger electronegative atom attached to it and vice versa.

(c) Asymmetric shoulder peak of XPS spectra indicates new functional group attached to core orbital spectra in the XPS.

(d) FWHM of XPS spectra inversely proportional to hole life time in the core orbital where $\beta$ is FWHM of core orbital spectra.

$$\tau = (2\pi\beta)^{-1} \quad (9.13)$$

Figure 9.12a shows Ga, N, C and O core orbital spectra of GaN-NWs with 40 sccm $H_2$ flow rate. The binding energy of Ga (3d), Ga (3p), Ga (3s), C (1s), N (1s) were observed at 20.3, 110.4, 162, 276, 398.3 eV, respectively. The binding energy of Ag catalyst was found out at 367.89 eV. The signature of C(1s) and O(1s) were found out due to additional contaminations present in it during deposition of GaN-NWs. Figure 9.12b shows deconvolution of Ag core orbital into four Gaussian peaks according to probability of their co-ordinations number, and electronegative of elements. The binding energy at 367.26 eV corresponds to the Ag bonding with Ag, (Ag–Ag) whereas the binding energy at 369.9 eV corresponds to Ag bonding with Ga, (Ag–Ga). Similarly, the binding energy of 374.87 eV corresponds to Ag bonding with N, (Ag–N) and the binding energy at 376.77 eV corresponding to Ag bonding with Ag pi–bond with N, (Ag=N). The FWHM of Ag–Ag, Ag–Ga, Ag–N, and Ag=N are 4.61, 3.69, 3.37, and 3.21 eV, respectively. The elemental contribution ratios are found as 1: 1.54: 2.45: 1.36 for Ag–Ga, Ag–N, Ag–Ag, Ag=N, respectively.

Figure 9.12c shows Ga, C, N and O core orbital spectra of GaN-NWs with 40 sccm $H_2$ flow rate. The binding energy of Ga(3d), Ga(3p), C(1s), N(1s) were observed at 20.3, 110.4, 286.7, 401.2 eV, respectively. The binding energy of In catalyst was found out at 443.7 eV. The signature of C(1s) and O(1s) were found out due to additional contaminations present in it during deposition of GaN-NWs. Figure 9.12d shows

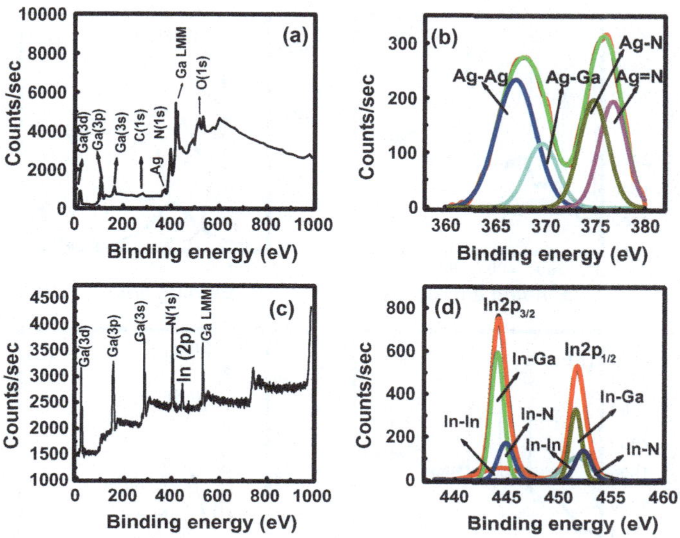


**Fig. 9.12** **a** Broad XPS scan of GaN-NWs using Ag catalyst **b** deconvolution of Ag catalyst core orbital spectra **c** broad XPS scan of GaN-NWs using In catalyst and **d** deconvolution of In catalyst core orbital spectra

deconvolution of In core orbital into Six Gaussian peaks according to probability of their co-ordinations number, and electronegative of elements. The binding energy of In–In, In–Ga, In–N for In($2p_{3/2}$) were observed at 444, 444.4, 444.8 eV and binding energy of In–In, In–Ga, In–N for In($2p_{1/2}$) were observed at 451.5, 451.8, 452.2 eV. The FWHM of In–In, In–Ga, In–N for In($2p_{3/2}$) and In($2p_{1/2}$) are 1.44, 4.57, and 1.78 and 2.9, 1.4, 1.8 eV, respectively. The elemental contribution ratio of In–Ga, In–N, In–In are 1: 1.20: 3.23 for In($2p_{3/2}$) and In–N, In–Ga, In–In are 1: 1.71: 1.24 for In($2p_{1/2}$), respectively.

Figure 9.13a shows Ga, N, C and O core orbital spectra of GaN-NWs with 40 sccm $H_2$ flow rate. The binding energy of Ga(3d), Ga(3p), Ga(3s), N(1s) were observed at 20.2, 154.2, 282.2, 401.7 eV, respectively. The binding energy of Fe catalyst was found out at 709.7 eV. The signature of O(1s) was found out at 739.7 eV due to additional contaminations present in it during deposition of GaN-NWs. Figure 9.13b shows deconvolution of Fe core orbital into Six Gaussian peaks three for $Fe_{(3/2)}$ and three for Fe($2p_{1/2}$) orbital according to probability of their co-ordinations number, and electronegative of elements. The binding energy at 710.5 eV corresponds to the Fe bonding with Fe, (Fe–Fe) whereas the binding energy at 712.1 eV corresponds to Fe bonding with Ga, (Fe–Ga). Similarly, the binding energy of 716.2 eV corresponds to Fe bonding with N, (Fe–N) for Fe($2p_{3/2}$). Similarly for Fe($2p_{1/2}$), Fe–Fe, Fe–Ga and Fe–N peaks are obtained at 724.19, 727.6, 730.49, respectively. The FWHM of Fe–Fe, Fe–Ga, and Fe–N for Fe($2p_{3/2}$) are found at 1.8, 2.77 and 5.95 eV. Similarly for Fe($2p_{1/2}$), the FWHM of Fe–Fe, Fe-Ga and Fe–N peaks are obtained

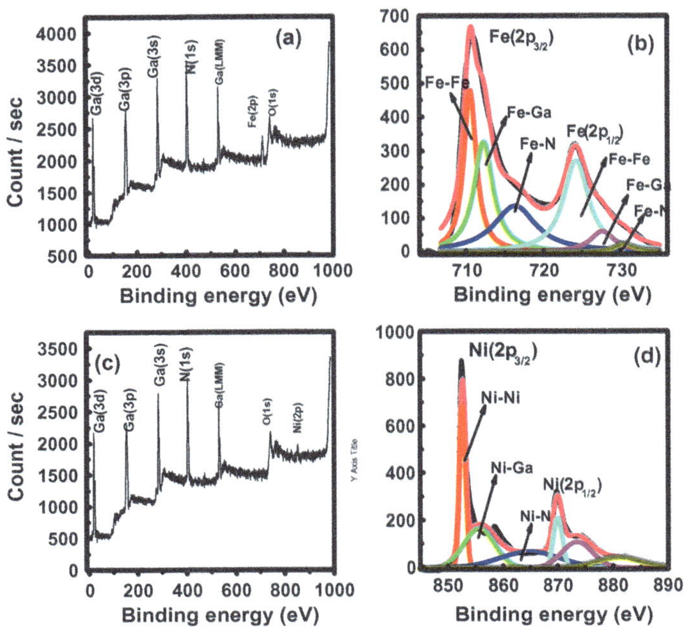


**Fig. 9.13** **a** Broad scan of GaN-NWs using Fe catalyst, **b** deconvolution of Fe catalyst core orbital spectra, **c** broad scan of GaN-NWs using Ni catalyst, and **d** deconvolution of Ni catalyst core orbital spectra

at 3.67, 3.47, 2.6 eV, respectively and elemental contribution ratios for Fe–N, Fe-Ga, Fe–Fe for $Fe2p_{(3/2)}$ and Fe–N, Fe–Ga, Fe–Fe for $Fe(2p_{1/2})$ are 1:1.14:1.12 and 1:3.25:14.55, respectively. Figure 9.13c shows Ga, N, C and O core orbital spectra of GaN-NWs with 40 sccm $H_2$ flow rate. The binding energy of Ga (3d), Ga (3p), Ga (3s), N (1s) were observed at 20.2, 154.2, 282.2, 401.2 eV, respectively. The binding energy of Ni catalyst was found out at 851.7 eV. The signature of O (1s) was found out at 739.7 eV due to additional contaminations present in it during deposition of GaN-NWs. Figure 9.13d shows deconvolution of Ni core orbital into Six Gaussian peaks three for $Ni_{(3/2)}$ and three for $Ni2p_{(1/2)}$ orbital according to probability of their co-ordinations number, electronegative of elements. The binding energy at 852.7 eV corresponds to the Ni bonding with Ni, (Ni–Ni) whereas the binding energy at 853.9 eV corresponds to Ni bonding with Ga, (Ni–Ga). Similarly, the binding energy of 865 eV corresponds to Ni bonding with N, (Ni–N) for $Ni(2p_{3/2})$. Similarly for $Ni(2p_{1/2})$ Ni–Ni, Ni–Ga and Ni–N peaks are obtained at 869.77, 873.4, 881.59 eV, respectively. The FWHM of (Ni–Ni), (Ni–Ga) and (Ni–N) are observed at 1.28, 6.58, and 12.73 eV. Similarly for $Ni(2p_{1/2})$, the FWHM of Ni–Ni, Ni–Ga and Ni–N peaks are obtained at 1.81, 6.59, and 9.30 eV, respectively. The elemental contribution ratios of Ni–N, Ni–Ni, Ni–Ga for $Ni(2p_{3/2})$ are 1:1.26:1.07 and Ni–N, Ni–Ni, Ni–Ga for $Ni(2p_{1/2})$ are 1:1.83:1.10, respectively. Figure 9.14a shows the

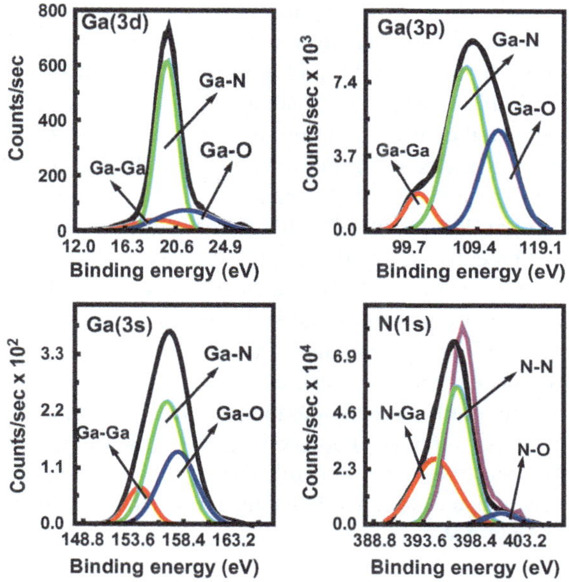


**Fig. 9.14** **a** Deconvolution of Ga(3d) core orbital peak **b** Deconvolution of Ga(3p) core orbital spectra, **c** deconvolution of Ga(3s) core orbital spectra, and **d** deconvolution core orbital spectra of N(1s) binding energy

deconvolution of Ga (3d) core orbital spectra in the GaN-NWs with 40 sccm $H_2$ flow rate. The Ga(3d) core orbital spectra can be deconvoluted in three individual Gaussian peaks, (a) Ga–Ga, Ga orbital influence by Ga orbital, (b) Ga–N, Ga orbital influenced by nitrogen orbital and (c) Ga–O, Ga orbital influence by higher electronegative (O) orbital. Therefore, the binding energy at 19.017, 19.852, 21.450 eV attributed to Ga–Ga, Ga–N and Ga–O, respectively. The FWHM of Ga–Ga, Ga–N and Ga–O are 4.1503, 1.8755 and 4.9059 eV, respectively. If we assume the X-ray cross section of Ga–Ga, Ga–N and Ga–O are same, the contribution of the elemental contribution ratio of orbital are 1.0:6.5:2.1, respectively.

Figure 9.14b shows the deconvolution of Ga (3p) core orbital spectra in the GaN-NWs with 40 sccm $H_2$ flow rate. The Ga(3p) core orbital spectra can be deconvoluted in three individual Gaussian peaks, (a) Ga–Ga, Ga orbital influence by Ga orbital, (b) Ga–N, Ga orbital influenced by nitrogen orbital and (c) Ga–O, Ga orbital influence by outer higher electronegativity O orbital. Therefore, the binding energy at 100.82, 107.77, 112.54 eV attributed to Ga–Ga, Ga–N and Ga–O, respectively, in the Ga (3p) core orbital. The full width at half maxima (fwhm) of Ga–Ga, Ga–N and Ga–O are 3.8939, 5.5701 and 5.1347 eV, respectively. If we assume the X-ray cross section of Ga-Ga, Ga-N and Ga-O are same in the Ga (3p), the contribution of the elemental contribution ratio of orbital are 1.0:6.5: 2.1, respectively. Figure 9.14c shows the deconvolution of Ga (3s) core orbital spectra in the GaN-NWs with 40 sccm $H_2$ flow rate. The Ga (3s) core orbital spectra can be deconvoluted in three individual Gaussian peaks, (a) Ga–Ga, Ga orbital influence by Ga orbital, (b) Ga–N, Ga orbital influenced by nitrogen orbital and (c) Ga–O, Ga orbital influence by outer higher electronegative (O) orbital. Therefore, the binding energy at 154.22,

156.79, 157.82 eV attributed to Ga–Ga, Ga–N and Ga–O, respectively in the Ga (3p) core orbital. The full width at half maxima (fwhm) of Ga–Ga, Ga–N and Ga–O are 2.4581, 3.3758 and 3.5943 eV, respectively. If we assume the X-ray cross section of Ga–Ga, Ga–N and Ga–O are same in the Ga (3p), the contribution of the elemental contribution ratio of orbital are 1.0:4.46: 2.82, respectively. Figure 9.14d shows the deconvolution of core orbital spectra of N (1s) in the GaN-NWs. In a similar trend, the N (1s) peak was deconvoluted into three peaks N-Ga, N–N and N–O at 394.72, 396.77, 401.00 eV, respectively. The FWHM of N-Ga, N–N and N–O are 4.1127, 2.8016, and 3.3517 eV, respectively. If we assume the X-ray cross section of N–Ga, N–N and N–O are same in the N (1s), the contribution of the elemental contribution ratio of orbital are 1.0:1.41:0.15, respectively. Figure 9.15 shows atomic percentage of Ga, N, and Ag, Fe, Ni, In catalysts in GaN-NWs. The atomic percentage is calculated using Eq. (9.13). The atomic percentage of Ni catalyst is minimum whereas maximum value is observed with In catalyst. The presence of higher amount of catalyst degrades the optoelectronic properties of GaN-NWs. Therefore we can minimize the catalyst contribution using Ni catalyst whereas higher percentage of retained In catalyst in GaN-NWs deteriorates the device performance of GaN-NWs.

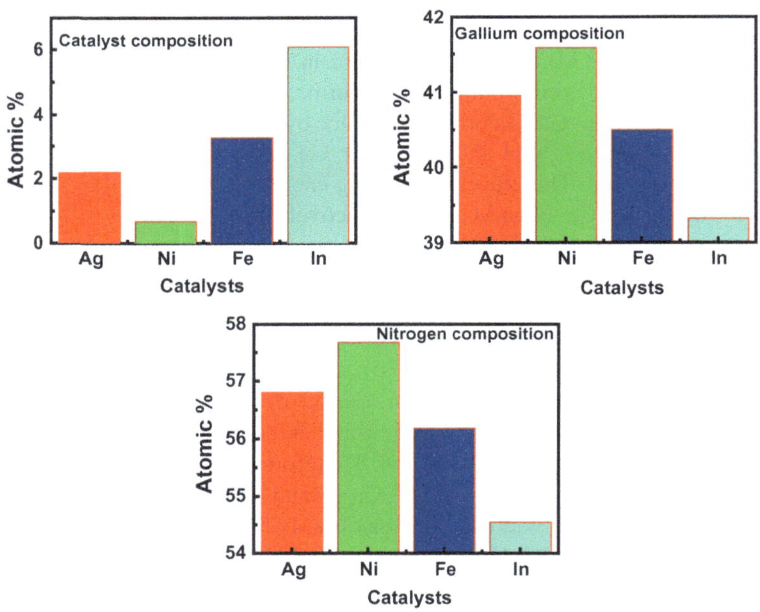


**Fig. 9.15** Atomic % v/s different catalysts in GaN-NWs

## Conclusion

In summary, we have prepared GaN-NWs by Thermal CVD process with different $H_2$ flow rates in the presence of four different catalysts (Ag, Fe, In, Ni). The larger diameter GaN-NWs were formed with Ni and Fe catalyst whereas smaller diameter NWs were obtained with In and Ag catalyst. The (101) plane of GaN-NWs grown with Ni catalyst shift to higher diffraction angle indicates tensile stress whereas with Fe catalyst it shifts to lower diffraction angle indicates compressive stress. The crystallite size and residual stress of GaN-NWs grown with Ag, Fe, In and Ni catalyst are 15.06, 14.56, 15.89, and 14.33 nm and 2.618, 2.72, 2.482, and 2.752 GPa, respectively. The $2E_1$(TO) FTIR peak of GaN-NWs grown with Ag, Fe, Ni catalyst shows asymmetry to higher wavenumber region indicates another chemical compound of Ga=N or Ga–N–O with similar bonding strength overlap with each other. However we have not observed asymmetry with In catalyst. The FWHM of $A_1$(LO) phonon with Ag, Fe, Ni, and In catalyst are 42, 38, 39, and 44 $cm^{-1}$. The maximum value of FWHM is observed with In catalyst indicates increasing of the coordination number whereas minimum value is observed with Fe catalyst. The FWHM of $A_1$(TO), $E_1$(TO), $E_2$(high), DI, SO($A_1$) phonon of GaN-NWs with Ag catalyst increases whereas FWHM of SO($E_1$) and $A_1$(LO) phonon decreases. The intensity of $A_1$(TO) phonon peak with Ag catalyst increases with increasing $H_2$ flow rate indicates increasing polarizability of GaN-NWs due to change in dielectric constant of surrounding chemical environment. We have calculated the phonon lifetime of GaN-NWs grown with Ag, Ni, Fe, and In catalyst from Raman spectra. The result confirms that growth of GaN-NWs with Ni catalyst is favored due to its shorter phonon lifetime whereas Fe catalyst having longer phonon lifetime degrades the efficiency of device. PL spectrum of GaN-NWs grown with Ag catalyst reveals broad range emission band between 2.4 and 3.2 eV with a strong peak centered at 2.97 eV corresponds to Nitrogen vacancy ($V_N$) in the wire. However, broadband also contains some fine emission peaks centered at 2.58, 2.73, and 2.83 eV corresponds to Ag interstial defect, Nitrogen interstial and Ga vacancy ($V_{Ga}$). The optical properties of the GaN-NWs reveal the strong blue shift which indicates the dense micro structural network of nanowires with incorporated hydrogen atom and catalyst. The incorporated $H_2$ atom passivates the various defect states. Therefore the passivated defect substantially enhance the PL emission so that such defect derived PL emissions are bright which provides ample opportunity for this material to be used in much needed optoelectronic industry. Further the catalyst also plays a vital role in modifying the band gap energy of GaN-NWs which further improves the optical property of the nanowires. Finally, the various chemical networks present in the GaN-NWs were confirmed by XPS spectroscopy.

**Acknowledgements** This study was supported by the Department of Science and Technology, Govt. of India (Project no.: SB/FTP/ETA-295/2011). Umesh Rizal acknowledged financial support from the Department of Biotechnology, Government of India (Project No: BCIL/NER-BPMC/2012/650).

## References


1. Stach, E.A., Pauzauskie, P.J., Kuykendall, T., Goldberger, J.R., He, R., Yang, P.: Watching Gallium nitride nanowires grow. Nano Lett. **3**, 867–869 (2003)
2. Lee, Z.J., Chen, X.L., Li, H.J., Tu, Q.Y., Yang, Z., Xu, Y.P., Hu, B.Q.: Synthesis and Raman scattering of GaN nanorings, nanoribbons and nanowires. Appl. Phys. A **72**, 629–632 (2001)
3. Maliakkal, C.B., Hatui, N., Bapat, R.D., Chalke, B.A., Rahman, A.A., Battacharya, A.: The mechanism of Ni assisted GaN nanowire growth. Nano Lett. **16**, 7632–7638 (2016)
4. Kuykendall, T., Pauzauskie, P., Lee, S., Zhang, Y., Goldberger, J., Yang, P.: Metalorganic chemical vapour deposition route to GaN-NWs with triangular cross sections. Nano Lett. **3**, 1063–1066 (2003)
5. Chen, C.C., Yeh, C.C., Chen, C.H., Yu, M.Y., Liu, H.L., Wu, J.J., Chen, K.H., Chen, L.C., Peng, J.Y., Chen, Y.F.: Catalytic growth and characterization of GaN-NWs. J. Am. Chem. Soc. **123**, 2791–2798 (2001)
6. Chen, R.S., Tsai, H.Y., Chan, C.H., Huang, Y.S., Chen, Y.T., Chen, K.H., Chen, L.C.: Comparison of CVD and MBE-grown GaN nanowires: crystallinity, photoluminescence, and photoconductivity. J. Electron. Mater. **44**, 177–187 (2015)
7. Mata, R., Hestroffer, K., Budagosky, J., Cros, A., Bougerol, C., Renevier, H., Daudin, B.: Nucleation of GaN nanowires grown by plasma-assisted molecular beam epitaxy: the effect of temperature. J. Cryst. Growth **334**, 177–181 (2011)
8. Dong, Y., Feenstra, R.M., Greeve, D.W., Moore, J.C., Sievert, M.D., Baski, A.A.: Effect of hydrogen on the morphology and electrical properties of GaN grown by Plasma assisted MBE. Appl. Phys. Lett. **86**, 1219141–1219143 (2005)
9. Fukuta, F., Oshima, T., Fujimura, S., Haneda, H., Murakami, K.: Hydrogen passivation of donors and hydrogen states on heavily doped n-type silicon. Jpn. J. Appl. Phys. **35**, 3937–3941 (1996)
10. Van, D., Walle, C.G., Johnson, N.M.: Hydrogen in III–V nitrides. Semicond. Semimetal, **57**, 157–183 (1999)
11. Rizal, U., Swain, B.P., Raman characterization of GaN-NWs deposited by chemical vapor deposition. In: Garg, A., et al. (eds.), Advances in power system and energy. Lecture notes in electrical engineering, vol. 436, pp. 47–61 (2018)
12. Li, P., Liu, Y., Meng, X.: GaN-NWs and nanotube growth by CVD method at different $NH_3$ flow rate. Rev. Mex. Fis. **62**, 219–222 (2016)
13. Johnson, J.L., Choi, Y., Ural, A.: GaN-NWs and nanoribbons growth from ion implanted iron catalyst. J. Vac. Sci. Technol. B **26**, 1841–1847 (2008)
14. Zhang, J., Zhang, L.: Growth of semiconductor gallium nitride NWs with different catalysts. J. Vac. Sci. Technol., B **2**, 2415–2419 (2003)
15. Guo, D.J., Abdulagatov, A.I., Rourke, D.M., Bertness, K.A., George, S.M., Lee, Y.C., Tan, W.: GaN NWs functionalized with atomic layer deposition techniques for enhanced immobilization of biomolecules. Langmuir **26**, 18382–18391 (2010)
16. Sahoo, P., Dhara, S., Dash, S., Tyagi, A.K., Raj, B.: Surface optical modes in GaN NWs. Int. J. Nanotechnol. **7**, 823–827 (2010)
17. Ridley, B.K.: Large band gap semiconductors. Tr. J. Phys. **23**, 577–582 (1999)
18. Reshchikov, M.A., Morkov, H.: Luminescence properties of defect in GaN. J. Appl. Phys. **97**, 1–95 (2005)
19. Chang, K.W., Wu, J.J.: Low temperature catalytic synthesis of gallium nitride NWs. J. Phys. Chem. B **106**, 7796–7799 (2002)

# Chapter 10
# Investigation of PANI/Graphene for Gas Sensor Applications

**Sayantan Sinha and Bibhu Prasad Swain**

## Introduction

Graphene is a two-dimensional sheet of carbon atoms with $sp^2$-hybridized bondings having honeycomb-like lattice structure. It has superior thermal conductivity ($\sim$5000 $Wm^{-1}$ $K^{-1}$), high carrier mobility (200,000 $cm^2/(V\ s)$), and high specific surface area (2630 $m^2g^{-1}$) [1]. For these reasons, graphene became very popular in many applications like supercapacitor, solar cell, photo catalysis, biomedicine, electrochemical, gas sensors, etc. since the first successful mechanical exfoliation of graphene layer from graphite by Geim et al. in 2004. In the same year, Novoselov et al. studied mechanically exfoliated graphene for sensing $NO_2$, $H_2O$, $NH_3$, and $CO_2$ gasses and the detection was successful even at a very low concentration of those gasses (~1 ppm) [2]. But the mechanical exfoliation was not a practically yielding technique and that is why it leads the researchers to find out some other methods for exfoliating the graphite such as chemical vapour deposition (CVD), chemically modification of graphite by introducing oxygen atoms into the graphitic planes. In chemically modified graphene (CMG), the introduced oxygen atoms disrupts the weak Vander wall bonds between the graphite planes. But CMG also suffers a drawback of high electrical resistivity. To decrease the resistance, CMG is reduced, i.e., the oxygen atoms are removed from the material.

On the other hand, conducting polymers such as polyaniline (PANI) and polypyrrole are also used as sensing material. The electronic structure of these polymers is made up of conjugated π-bonds which go through some changes during the chemical reactions (either redox or acid–base type) between polymer and the species

S. Sinha (✉) · B. P. Swain
Department of Physics, National Institute of Technology Manipur, Langol, Imphal 795004, India
e-mail: syntan.sinha@gmail.com

B. P. Swain
e-mail: bibhuprasad.swain@gmail.com



B. P. Swain (ed.), *Nanostructured Materials and their Applications*, Materials Horizons: From Nature to Nanomaterials, https://doi.org/10.1007/978-981-15-8307-0_10

adsorbed in the surface of the polymer. PANI is usually used to detect humidity, ammonia ($NH_3$), $NO_2$, etc. [1]. The basic reasons that PANI became so important sensing material are the easy synthesis process, can be used in room temperature, high sensitivity, fast response time, etc.

Both the materials graphene and PANI show the sensing property but this can be further enhanced significantly by introducing the graphene into the polymer matrices of PANI. The PANI/Graphene composite has a large surface-to-volume ratio, enhanced electrochemical and mechanical properties [3, 4]. Surface-to-volume ratio plays an important role in the improvement of sensitivity of the material.

## Preparation of Graphene via Chemical Exfoliation of Graphite

Ruolff's group first demonstrated the method of deriving a single-layer graphene via a solution-based process in 2006. A water dispersible graphite oxide (GO) is first prepared by Hummer's method as will be discussed in this section and then the oxygenated groups will be removed by adding hydrazine hydrate. GO is the graphene layers stacked with AB stacking (in Fig. 10.1 AA and AB stacking are illustrated).

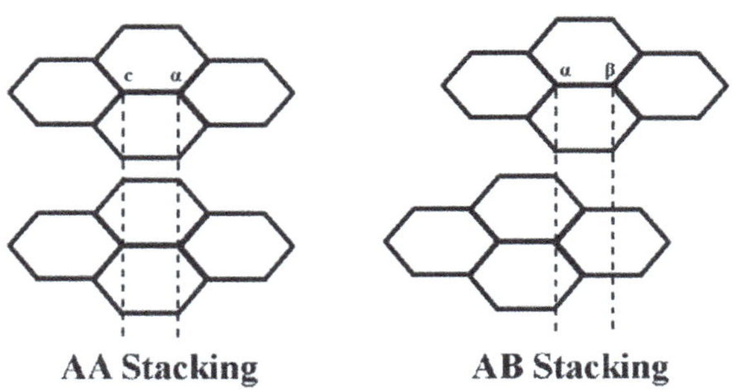


**Fig. 10.1** Schematic shows the AA and AB stacking of graphene layers. In AA stacking structure, the carbon atoms are aligned with same lateral coordinates in both the adjacent layers while in AB stacking, one layer is shifted with reference to another by the vector equal to the edge of the hexagon

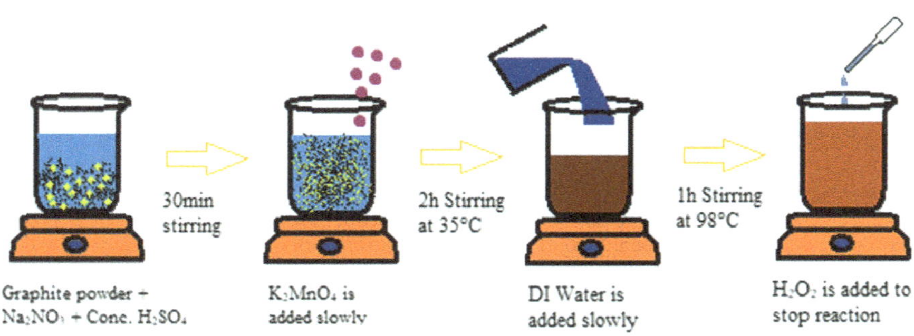


**Fig. 10.2** Graphical presentation of modified Hummer's method

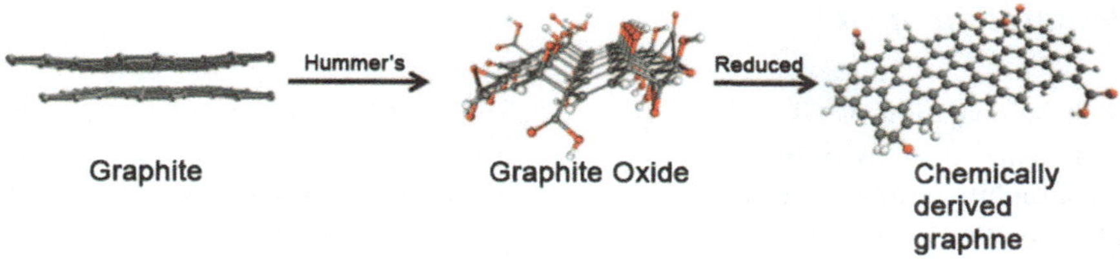


**Fig. 10.3** Model shows the chemical modification process of deriving rGO from graphene. Reprinted with permission from Ref. [5]. Copyright (2010) American Chemical Society

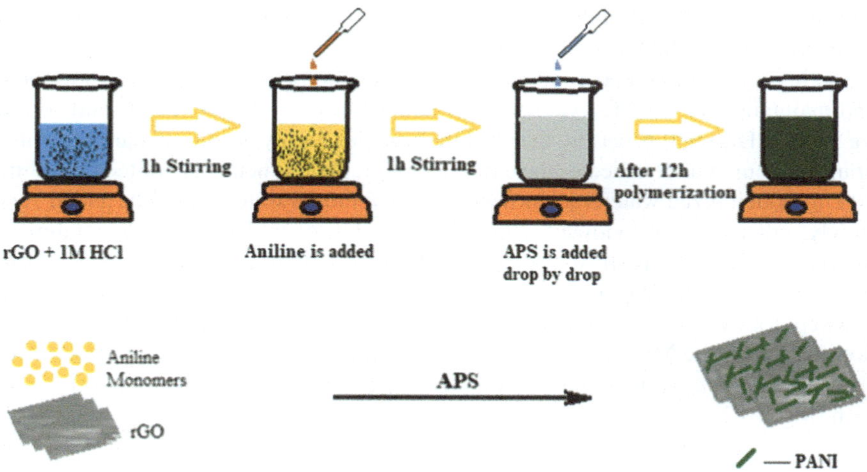


**Fig. 10.4** Schematic presentation of synthesis of PANI/rGO composite

GO can be dispersed easily in water because of the interaction of water with the oxygen-containing (epoxide and hydroxyl) functional groups introduced into the basal plane during oxidation. This hydrophilic nature of GO allows water molecules to intercalate between the layers and disperse them as separate sheets. But, directly adding the hydrazine hydrate into the aqueous solution of GO removes the oxygenated groups and reduces the hydrophilic nature which leads the individual graphene sheets to quickly agglomerate in solution. To avoid this problem of aggregation, the pH of the solution is raised before adding the reducing agent [5]. This makes the colloidal dispersion as charged-stabilized dispersion even after the reduction of GO.

## *Preparation of GO*

In the preparation of GO, nowadays, a slightly modified version of the original method described by Hummer [17, 7] is used. The chemicals required in this process are: Graphite powder or flakes, 98% concentrated $H_2SO_4$, $NaNO_3$, $KMnO_4$, $H_2O_2$ (30% solution), 10% HCl, and acetone and DI water for washing the product.

First, 5 g GO and 2.5 g $NaNO_3$ are taken in a 1000 ml beaker and 110 ml concentrated $H_2SO_4$ is added to it. The above mixture was stirred for 30 min in room temperature. After that everything is mixed properly and black slurry is formed. The reaction beaker is then kept in an ice bath to bring down the temperature of the reactants to below 20 °C. After the temperature goes well below 20 °C, 15 g of potassium permanganate ($KMnO_4$) is added slowly under vigorous stirring condition. During the addition of potassium permanganate, the temperature is to be carefully monitored and maintained at <20 °C. After that the temperature is raised to 35 °C and stirred for 15 min. During this period, the mixture becomes highly dense and dark brownish orange colour, which indicates the oxidation of graphite. Then, it is diluted by adding 300 ml of deionized water very slowly. Note that the DI water should be added very slowly; otherwise, un-oxidized graphite will be left over. Then, the temperature is raised to 98 °C and it is stirred for 30 min. Then 500 ml of deionized water is added. At the end, 30 ml hydrogen peroxide (30% $H_2O_2$) is added to terminate the reaction. It is kept for overnight and the precipitate is filtered and washed with 10% HCl solution to remove the $Mn^{2+}$ ions followed by washing five times with deionized water and acetone. The washed and filtered GO is dried keeping in vacuum at 60 °C for 12 h (Fig. 10.2).

## *Reduction of GO*

Chemical required are: Hydrazine hydrate (80% aqueous solution of $N_2H_4$), $NH_4OH$.

0.3 g as prepared GO is added to 300 ml DI water and ultrasonicated for 1 h to disperse properly. $NH_4OH$ is added to adjust the pH of the dispersion to 10. 600 μl hydrazine hydrate (2 μl/1 mg GO) is added and the temperature is increased up to 80 °C under stirring condition. It is kept stirring for 24 h under reflux condition. The black substance is then filtered and washed several times with DI water first and then with methanol. Then it is kept in oven at 80 °C for 6 h for drying [6] (Fig. 10.3).

## Preparation of PANI/Graphene Composite by In-Situ Polymerization

Required chemicals: Aniline ($C_6H_5NH_2$), Ammonium persulfate (APS) ($(NH_4)_2S_2O_8$), HCl.

Procedure: required amount of as prepared rGO is dispersed in 50 ml 1 M HCl by ultrasonicating for 2 h. In this dispersion, 2.3 ml aniline is added and stirred for 1 h (Fig. 10.4).

In another beaker, 5.7 g APS is dissolved in 50 ml 1 M HCl. The APS solution is added to the aniline rGO dispersion slowly under stirring condition. It is left overnight for complete polymerization of aniline. Next day, it is filtered and washed multiple times with DI water until clear water comes out from the filter paper. Finally, the filtrate is washed with ethanol and dried in air for 24 h [6, 8, 9].

## Different Characterizations of the PANI/Graphene Nanocomposite

### *Scanning Electron Microscopy (SEM) and Transmission Electron Microscopy (TEM)*

The morphology of PANI/graphene composite as seen in the SEM is shown in Fig. 10.5. The fiber-like thing as shown in Fig. 10.1b is actually the PANI nanofiber grown over the graphene sheets. The TEM image of the same material is shown in Fig. 10.6 which also shows the formation of PANI nanofibers with approximate dimension of 25–50 nm range on the graphene plane.

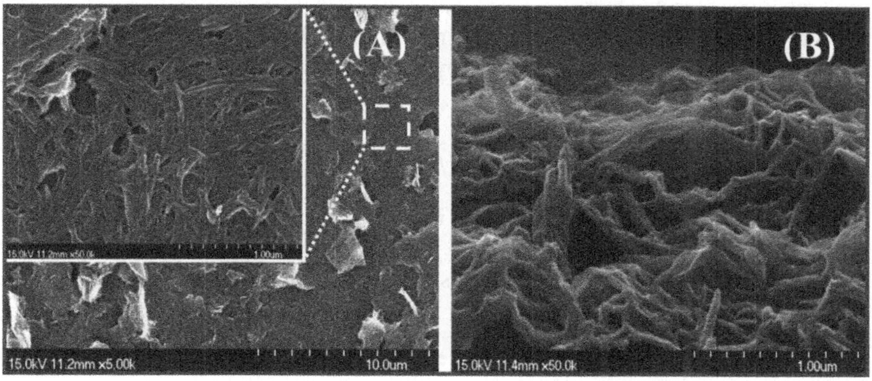


**Fig. 10.5** SEM image of PANI/graphene nanocomposite. **a** Top view **b** side view with a tilt angle of 45°. Reprinted with permission from Ref. [3]. Copyright (2010) American Chemical Society

**Fig. 10.6** TEM image of PANI/graphene nanocomposite. Reprinted with permission from Ref. [3]. Copyright (2010) American Chemical Society

## *X-Ray Diffraction (XRD)*

The XRD spectra of GO, rGO, pure PANI, and PANI/graphene are shown in Fig. 10.7. The graphite spectrum shows a sharp peak at $2\theta = 26.3°$ corresponding to the (002) plane. By calculating the full width at half maxima (FWHM) denoted by $\beta$ in Eq. 10.1, the crystalline size $D$ can be calculated by the Deby-Scherrer's formula:

$$D = \frac{k\lambda}{\beta \cos\theta} \tag{10.1}$$

Here, $k$ is an instrument constant and its value varies from 0.75 to 2 but, in most of the modern XRD machines, $k = 0.9$. $\lambda$ is the wavelength of the X-Ray. Interlayer distance $d$ is calculated using the Bragg's diffraction law given by:

$$d = \frac{\lambda}{2\sin\theta} \tag{10.2}$$

The interlayer distance for the graphite (002) plane is around 0.33 nm. Graphite shows another weak peak at 41° corresponding to (001) plane. GO peak is observed at around $2\theta = 11.6°$ associated with the (001) plane. The interlayer distance in GO is much larger than that of graphite presumably because of the insertion of the oxygen atoms between the graphitic layers. In rGO, a broad peak is found at $2\theta = 23°–26°$ attributed to (002) plane and the interlayer distance again becomes shorter ~0.34 nm due to the reduction process. Pure PANI shows three peaks in the XRD

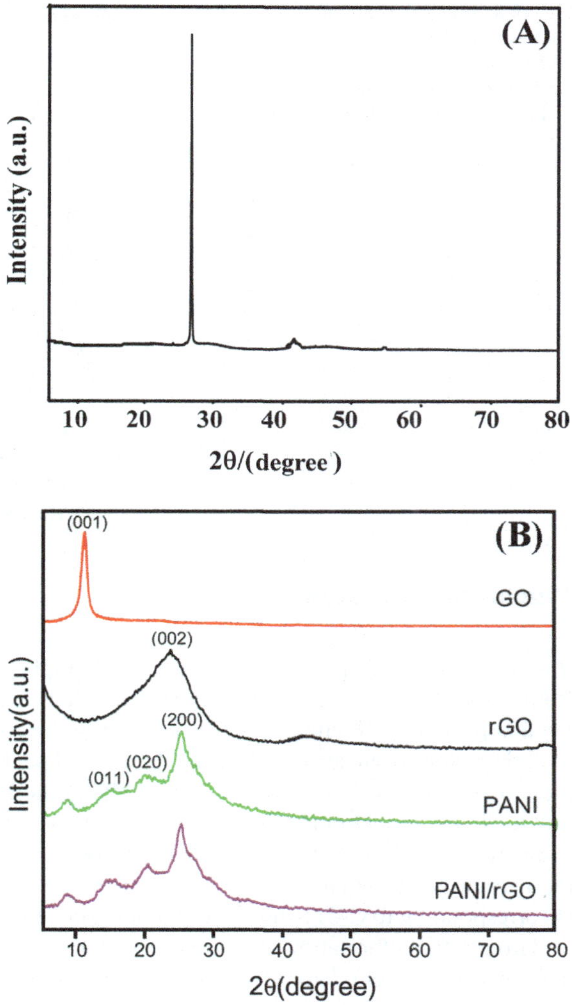


**Fig. 10.7** XRD spectra of **a** graphite, **b** GO, rGO, pure PANI, and PANI/graphene nanocomposite

spectrum, $2\theta = 14.7°, 20.2°, 25.4°$ which correspond to the (011), (020), and (200) planes. The PANI/graphene composite gives the XRD pattern almost similar to the pure PANI which indicates that during the formation of the nanocomposite PANI intercalates into the graphene layers [10–12].

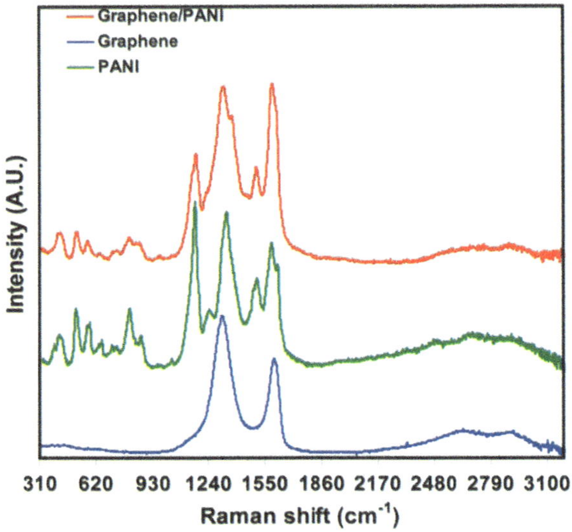


**Fig. 10.8** Raman spectra of rGO, pure PANI, and PANI/graphene nanocomposite. Reprinted with permission from Ref. [3]. Copyright (2010) American Chemical Society

## *Raman Spectroscopy*

Raman spectra of graphene, PANI, and PANI/graphene are shown in Fig. 10.8. The Raman spectra of graphene in PANI/graphene show D and G bands prominently. The D band peak at 1318 $cm^{-1}$ is corresponding to the conversion of the hybridization of carbon atom from $sp^2$ to $sp^3$, while the G band peak at 1592 $cm^{-1}$ is attributed to the vibration of $sp^2$-hybridized carbon. PANI in the PANI/graphene composite has polaronic and bipolaronic structures and that is why the vibrational peaks of PANI come from both the structures. C–N stretching vibration is found at 1372 $cm^{-1}$ and the bipolaronic N–H stretching vibration shows peak at 1504 $cm^{-1}$. The 1172 $cm^{-1}$ peak is corresponding to C–H bending for quinoid ring. Other minor peaks are, 856 $cm^{-1}$ (polaronic) for deformation of benzene ring, 806 $cm^{-1}$ (bipolaronic) for quinoid ring deformation, 712 $cm^{-1}$ for C–C ring deformation, 640 $cm^{-1}$ (bipolaronic) benzene ring deformation, 576 $cm^{-1}$ for (bipolaronic) amine ring deformation, 518 $cm^{-1}$ (polaronic) amine ring deformation, and 418 $cm^{-1}$ for (polaronic) C–N–C tortional vibration [13].

## *X-Ray Photoelectron Spectroscopy (XPS)*

XPS also called as electron spectroscopy for chemical analysis (ESCA) is used for the elemental analysis of the samples. The XPS spectrum for C(1 s) for PANI/graphene is shown in Fig. 10.9a. After convoluting the peak, four peaks are found. The main peak comes at around binding energy 284.6 eV which corresponds to C–C, C = C, and

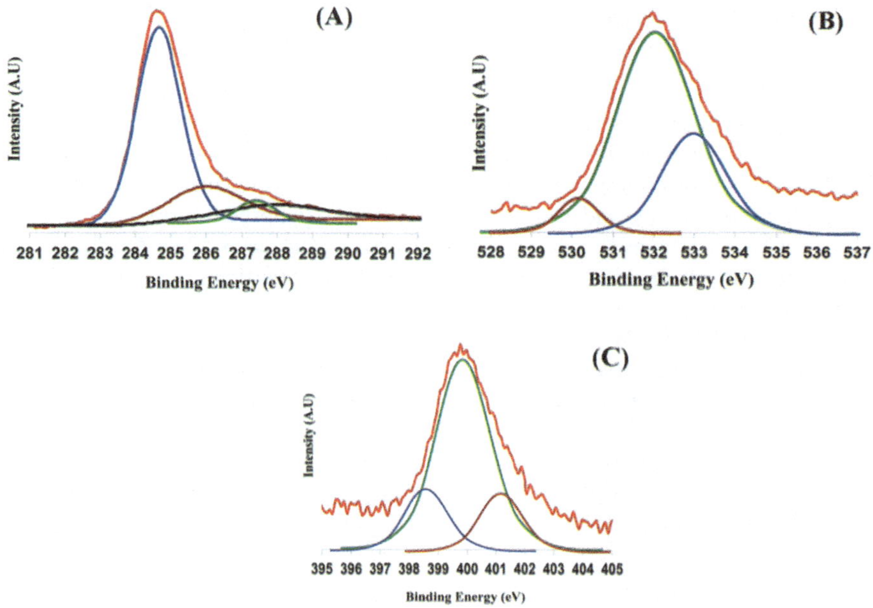


**Fig. 10.9** XPS spectra and the deconvolution graphs for **a** C(1 s), **b** O(1 s), and **c** N(1 s) of PANI/graphene nanocomposite. Adopted with permission from Ref. [3]. Copyright (2010) American Chemical Society

C–H. Another three peaks at binding energies 286.08 eV, 287.38 eV, and 288.18 eV are attributed to C–OH, C = O, and O = C–OH, respectively. The spectrum for O(1 s) is shown in Fig. 10.9b. The main peak at 532 eV binding energy comes from C–OH bonds. A lower intensity peak at around 530 eV corresponds to C = O and O = C–OH bonds. Figure 10.9c shows the spectrum for N(1 s) which shows three peaks; one is for the nitrogen coming from quinoid amine at the binding energy 398.3 eV, second is for the benenoid amine at 399.7 eV, third is for the cationic radicals of nitrogen ($N^{+}$) at 401.1 eV.

## PANI/graphene in Gas Sensing Application

In this following section, the application of PANI/graphene in different gas sensing is discussed.

## *Hydrogen Gas Sensor*

In 2010, Al-Mashat et al. [3] fabricated PANI/graphene based $H_2$ sensor. The construction of the sensor was as follows: At first, a conductometric transducer was prepared by depositing a thin (~50 nm) layer of chromium followed by another thin (~100 nm) layer of gold on a quartz surface. Then, the gold surface was patterned using photolithography and etching. The layer was patterned so that it had gold fingers of width 50 μm spaced at 50 μm distance from each other. On the top of this conductometric transducer, the PANI/graphene nanoparticles were sprayed with an airbrush with the help of $N_2$ gas. The layer thickness of the nanoparticles was approximately 500 nm. Now, for measuring the sensing property of the sensor, it was placed in an enclosed vessel connected with mass flow controller (MFC) to pass the hydrogenated air into the enclosure in an accurately controlled manner. The sensor was connected to a precise multimeter (Keithley 2001 multimeter) to measure its resistance during the flow of $H_2$ gas. $H_2$ was mixed with synthetic air containing 79% $N_2$ and 21% $O_2$ in different concentrations. The sensor was tested with 0.06, 0.12, 0.25, 0.50, and 1% $H_2$ concentration. The sensitivity S is calculated using the formula:

$$S = \left| \frac{R_{gas} - R_0}{R_0} \right| \times 100\% \tag{10.3}$$

where $R_{gas}$ is the resistance of sensor in presence of $H_2$ and $R_0$ is the resistance in synthetic air environment. The sensitivity at 1% $H_2$ was found to be 16.57%. Two separate sensors with graphene and pure PANI were also fabricated in this same work. The measure sensitivity at 1% $H_2$ of the graphene-based sensor was 0.83% and that was for the pure PANI-based sensor 9.38%.

The graphene produced by chemical exfoliation contains epoxide and carboxylic groups which attract the electrons and create holes in the valence band. Since $H_2$ is an electron donating (reducing) gas, when it comes in contact to the graphene, a depletion of holes is caused. Due to this deficiency of holes, the resistivity of graphene tends to increase in the presence of $H_2$.

In PANI, $H_2$ is chemisorped into the polymer chains and forms new N–H bonds by dissociating attaching of hydrogen with the amine nitrogens in the PANI chains. This formation of N–H bonds increases the conductivity of PANI. Hence resistance of PANI decreases in presence of $H_2$. So, from the discussion about the mechanisms of interaction of $H_2$ with graphene and PANI, it can be noticed that the resistive change in presence of $H_2$ is opposite to each other but still PANI/graphene shows better sensitivity to $H_2$ because PANI/graphene has a much higher porosity than that of the individual materials. In case of PANI/graphene composite, the resistance of the sensor decreases as reported in the article. Here the prominent mechanism becomes the interaction of PANI with $H_2$ [3].

## Ammonia Gas Sensor

In 2012, Huang et al. reported ammonia gas sensor using PANI/rGO [14]. In this work, the sensor was fabricated by spreading an ethanolic suspension of PANI/rGO (0.25 mg $ml^{-1}$ concentration) with the help of a microsyringe and evaporating the ethanol solvent by keeping in vacuum for 1 h at 80 °C. The sensor was tested in exposure of different concentration of $NH_3$ and the sensitivity or normalized change in resistance was calculated employing Eq. 10.3. In presence of 50 ppm $NH_3$, 59.2% change in resistance was observed. The response was compared to that of rGO and pure PANI nanofibers also. In rGO, the change in resistance was 5.2% while that was 13.4% in case of pure PANI. To analyze the recovery process of the sensors, i.e., how fast the resistance of the sensor achieves its initial value ($R_0$) after withdrawing the $NH_3$ gas, the sensor were illuminated with an IR lamp. In rGO, it is very easy to desorb $NH_3$ from its surface. It takes ~2.4 min to return to its normal normal resistance state. Due to the acid–base reaction in PANI, it takes a longer time for recovery after withdrawing $NH_3$ and not 100% recovery is possible. Around 3.2 min was needed for achieving 87.3% of the normal ambient resistance of PANI. In PANI/rGO, 81.5% resistance recovery was possible in ~4 min.

The process of interaction of $NH_3$ with rGO is similar as discussed in $H_2$ sensing. In normal ambient, rGO becomes slightly p-type because of the electron withdrawal behavior of epoxide and carboxylic groups. Thus, in exposure to the electron donating gas like $NH_3$, rGO becomes depleted of carriers and the resistance increases.

The acid-doped PANI, when comes in exposure to the $NH_3$ gas acid–base reaction takes place and PANI is dedoped. Due to this dedoping process, after exposure of $NH_3$, PANI is gradually transformed from conductive emeraldine salt to insulating emeraldine base and the resistance increases significantly. In PANI/rGO composites, PNAI nanoparticles are anchored in the surface of the rGO providing a higher surface area and thus better interaction with $NH_3$.

## Methane Sensor

In 2013, a methane gas sensor based on PANI/graphene operated at room temperature was fabricated and tested by Wu et al. [1]. The sensor was fabricated on an n-type silicon wafer. The electrical resistance of the sensor was measured using a precise digital multimeter. The sensitivity was estimated using formula given in Eq. 10.3. Here, $R_0$ denotes the resistance of the sensor at clean dry air. The sensor was tested with $CH_4$ concentration varying from 1–1600 ppm. The sensor made with the composite graphene 15 mg and PANI 186 mg showed a lower detection limit of 10 ppm and the response of the sensor saturates at 1600 ppm. When the sensor was alternatively exposed to 100 ppm $CH_4$ and air, it showed an approximate response time of 85 s and recovery time 45 s.

Graphene and PANI have conjugated π electrons. In the PANI/graphene composite, there is a possibility of formation of $\pi - \pi^*$ conjugated system as discussed in the article [1]. The area of conjugated $\pi$ electron cloud increases with the higher concentration of graphene in composite. The electron coming from the adsorbed $CH_4$ gas molecules interacts with the π electrons. Due to this charge transfer, the resistance of PANI/graphene decreases when it is exposed to $CH_4$ gas.

## *Toluene Gas Sensor*

A PANI/graphene-based toluene gas sensor was fabricated and tested by Parmar et al. in 2013 [15] To fabricate the sensor, PANI was first converted into emeraldine base by treating with $NH_4OH$. Then it is dissolved in N-methyl-2-pyrrolidone (NMP). With NMP-PANI solution graphene is mixed in graphene: PANI = 1:2 ratio. Graphene is properly dispersed into the solution by ultrasonication. After that, graphene-NMP-PANI was coated in layer-by-layer on a $SiO_2$ coated Si substrate. After depositing ten layers of PANI and graphene/PANI, the NMP was evaporated from the polymer layers by keeping it in 230 °C for 3 h. Then electrodes were deposited by DC sputtering of Al on two edges of the layers. The sensor showed a sensitivity of 8.4 for 100 ppm toluene gas. The response time of the prepared sample was measured as ~8 min in 100 ppm toluene. Recovery time was observed ~22 min.

Toluene is dehydrogenated when it comes to the contact of the PANI/graphene films and this process releases electrons. These electrons interact with PANI and change the conductivity. An increase in resistance is observed when the PANI/graphene film is exposed to toluene gas. In graphene/PANI, the presence of active sites and the swelling junctions in between the graphene flakes enhances the chemisorption and increases the sensitivity.

## Conclusion

The sensing property of PANI/graphene is discussed for four different gasses $H_2$, $NH_3$, $CH_4$, and toluene. Graphene and PANI show the sensing property individually but when they make composites, the sensitivity is significantly increased due to higher surface area, better chemisorption, etc. In Table 10.1, some PANI/graphene-based sensors are listed.

**Table 10.1** Gas sensor using PANI/graphene reported by various researcher groups

| Gas | Sensing material | Detection range | Operating temperature | Sensitivity | References |
|---|---|---|---|---|---|
| $H_2$ | PANI/rGO | 0.06–1% | RT | 16.57% @ 1% | [3] |
| $NH_3$ | PANI/rGO | 50 ppm | RT | 59.2% | [14] |
| | PANI:CNT | 1000 ppm | RT | 418% | [16] |
| $NH_3$ | PANI/CNT | 200 ppb–50 ppm | RT | 6 @ 10 ppm | [17] |
| $CH_4$ | PANI/rGO | 10–1600 ppm | RT | – | [1] |
| $H_2O$ | PANI/$SnO_2$/GO | 0–97% RH | RT | 29.1 Hz/%RH | [18] |
| $H_2O$ | PANI/GO | 11–97% RH | RT | 20 Hz/%RH | [19] |
| Toluene | PANI/rGO | 100 ppm | 50 °C | 8.4 | [15] |

**Acknowledgements** The authors acknowledge National Institute of Technology Manipur for providing various facilities such as library access, online access to different journals, etc.

## References

1. Wu, Z., Chen, X., Zhu, S., Zhou, Z., Yao, Y., Quan, W., Liu, B.: Room Temperature Methane Sensor Based on Graphene Nanosheets/Polyaniline Nanocomposite Thin Film. IEEE Sens. J. **13**, 777–782 (2013)
2. Schedin, F., Geim, A.K., Morozov, S.V., Hill, E.W., Blake, P., Katsnelson, M.I., Novoselov, K.S.: Detection of individual gas molecules adsorbed on graphene. Nat. Mater. **6**, 652–655 (2007)
3. Al-Mashat, L., Shin, K., Kalantar-zadeh, K., Plessis, J.D., Han, S.H., Kojima, R.W., Kaner, R.B., Li, D., Gou, X., Ippolito, S.J., Wlodarski, W.: Graphene/Polyaniline Nanocomposite for Hydrogen Sensing. J. Phys. Chem. C **114**, 16168–16173 (2010)
4. Huang, Z., Li, L., Wang, Y., Zhang, C., Liu, T.: Polyaniline/graphene nanocomposites towards high-performance supercapacitors: A review. Compos. Commun. **8**, 83–91 (2018)
5. Allen, M.J., Tung, V.C., Kaner, R.B.: Honeycomb Carbon: A Review of Graphene. Chem. Rev. **110**, 132–145 (2010)
6. Semwal, V., Gupta, B.D.: Highly sensitive surface plasmon resonance based fiber optic pH sensor utilizing rGO-Pani nanocomposite prepared by in situ method. Sensor Actuat. B-Chem. **283**, 632–642 (2019)
7. Hummers, W.S., Offeman, R.E.: Preparation of Graphitic Oxide. J. Chem. Soc. Chem. **80**(1958) 1339
8. Chen, X., Meng, F., Zhou, Z., Tian, X., Shan, L., Zhu, S., Xu, X., Jiang, M., Wang, L., Hui, D., Wang, Y., Lua, J., Gou, J.: One-step synthesis of graphene/polyaniline hybrids by in situ intercalation polymerization and their electromagnetic properties. Nanoscale **6**, 8140–8148 (2014)
9. Yan, J., Wei, T., Shao, B., Fan, Z., Qian, W., Zhang, M., Wei, F.: Preparation of a graphene nanosheet/polyaniline composite with high specific capacitance. Carbon **48**, 487–493 (2010)
10. Gui, D., Liu, C., Chen, F., Liu, J.: Preparation of polyaniline/graphene oxide nanocomposite for the application of supercapacitor. Appl. Surf. Sci. **307**, 172–177 (2014)
11. Moosa, A.A., Jaafar, J.N.: Green Reduction of Graphene Oxide Using Tea Leaves Extract with Applications to Lead Ions Removal from Water. J. Nanosci. Nanotechnol. **7**, 38–47 (2017)

12. Zhang, Y., Liu, J., Zhang, Y., Liua, J., Duan, Y.: Facile synthesis of hierarchical nanocomposites of aligned polyaniline nanorods on reduced graphene oxide nanosheets for microwave absorbing materials. RSC Adv. **7**, 54031–54038 (2017)
13. Luo, Y., Kong, D., Jia, Y., Luo, J., Lu, Y., Zhang, D., Qiu, K., Li, C.M., Yu, T.: Self-assembled graphene@PANI nanoworm composites with enhanced supercapacitor performance. RSC Adv. **3**, 5851–5859 (2013)
14. Huang, X., Hu, N., Gao, R., Yu, Y., Wang, Y., Yang, Z., Kong, E.S., Wei, H., Zhang, Y.: Reduced graphene oxide–polyaniline hybrid: Preparation, characterization and its applications for ammonia gas sensing. J. Mater. Chem. **22**, 22488–22495 (2012)
15. Parmar, M., Balamurugan, C., Lee, D.W.: PANI and graphene/PANI nanocomposite films–comparative toluene gas sensing behaviour. Sensors **13**, 16611–16624 (2013)
16. Eising, M., Cava, C.E., Salvatierra, R.V., Zarbin, A.J.G., Roman, L.S.: Doping effect on self-assembled films of polyaniline and carbon nanotube applied as ammonia gas sensor. Sensor Actuat. B-Chem. **245**, 25–33 (2017)
17. Xue, L., Wang, W., Guo, Y., Liu, G., Wan, P.: Flexible polyaniline/carbon nanotube nanocomposite film-based electronic gas sensors. Sensor Actuat. B-Chem. **244**, 47–53 (2017)
18. Zhang, D., Wang, D., Zong, X., Dong, G., Zhang, Y.: High-performance QCM humidity sensor based on graphene oxide/tin oxide/polyaniline ternary nanocomposite prepared by in-situ oxidative polymerization method. Sensor Actuat. B-Chem. **262**, 531–541 (2018)
19. Zhang, D., Wang, D., Li, P., Zhou, X., Zong, X., Dong, G.: Facile fabrication of high-performance QCM humidity sensor based on layer-by-layer self-assembled polyaniline/graphene oxide nanocomposite film. Sens. Actuators B **255**, 1869–1877 (2018)

# Chapter 11
# Investigation of Metal-Oxide/Reduced Graphene-Oxide Nanocomposites for Gas Sensor Applications

**Naorem Aruna Devi and Bibhu Prasad Swain**

## Introduction

Graphene, a two-dimensional (2D) carbon nanostructure having a crystal structure of single-atom-thick sheet of $sp^2$ hybridized carbon atoms that are packed in a hexagonal honeycomb, has become a new approach for the application of gas sensor due to its unique properties such as its electric, mechanical, and optical properties along with large surface area [1]. To design a gas sensor achieving high sensing performance like highly reproduction stability, sensitivity, and selectivity and fast sensing response and recovery time, graphene having large surface area combining with the sensitive mass device may be considered as the best option [2]. Graphene is considered as a promising candidate for gas sensors because of the absorption and desorption of the gas molecules which may lead to affect its electronic properties in a strong way regarding produce crystal lattice exhibiting high quality which creates quite low electronic noise along with good sensing area per unit volume forming the structure of 2D. There are various techniques to synthesized graphene. Mechanical exfoliation is the method in which single-layer graphene sheets are obtained but in this method, the production of layers cannot be controlled. Graphene synthesized by epitaxial growth and chemical vapor deposition (CVD) techniques are considered as the effective techniques for bottom-up synthesis of high-quality graphene sheets [3, 4] chemically synthesis of graphene oxide from graphite and the reduction of GO using chemical or thermal method are considered to be top-down approach which can produce a gas sensor to detect harmful or toxic gases [5, 6]. The graphene-based gas sensor was

N. A. Devi (✉) · B. P. Swain
Department of Physics, National Institute of Technology Manipur, Langol, Imphal, Manipur 795004, India
e-mail: aruna.nrm94@gmail.com

B. P. Swain
e-mail: bibhuprasad.swain@gmail.com; bpswain@nitmanipur.ac.in



B. P. Swain (ed.), *Nanostructured Materials and their Applications*, Materials Horizons: From Nature to Nanomaterials, https://doi.org/10.1007/978-981-15-8307-0_11

first reported by Schedin et al. [7] for the detection of gas molecules individually. It was observed that the sensor responses when the gas molecule attached or detached from the surface of the graphene. When there is adsorption of gas molecule, the local carrier concentration of graphene changes leading to a slight change in resistance. The author suggests that graphene as a low-noise material. Further, it was found that its resistance changes depending on the carrier concentration in graphene due to gas molecule's adsorption and further under vacuum, the sensor was annealed at 150 °C. They found the main drawback of having long recovery time which was further explained that it might have occurred due to the strong interaction of graphene and gas molecules at room temperature. To achieve fast response recovery time, external heat treatment should be given or UV illumination is needed. Yavari et al. [8] synthesized graphene films by using the CVD method for the utilization of the sensor to detect $NH_3$ and $NO_2$ gasses at room temperature. Excellent performance of sensing was observed against $NO_2$ and $NH_3$ at 500 ppb–100 ppb, respectively. Further, the joule heating treatment is given to desorb the gas molecules to achieve the whole recovery.

Nowadays, metal/metal oxide with graphene nanocomposites has developed a great interest in research. The composition of metal-oxide-based graphene was approved in producing gas sensors which show high sensing performance. Many researchers have reported a gas sensor based on metal-oxide in detecting toxic, explosive, and harmful gases [9]. Low-dimensional nanostructured with metal-oxide materials were considered to produce high-performance gas sensor to protect the environment and health of the human. Many factories and industries produce harmful and toxic chemicals or gases in the environment which lead to damage to our environment that causes issues of global warming. Hence, monitoring the quality of air is needed to make our environment safe. Gas sensor plays an important role in detecting toxic gases and to monitor the quality of the air to protect human health and the environment [10]. Recently, Kumar et al. [11] investigated the gas sensor based on graphene composite with different metal-oxide such as metal oxides like tin oxide, ferric oxide, zinc oxide, and indium oxide to detect toxic gases. The author observed that ZnO of 25 wt% exhibits a quick response against $NH_3$ with a short time to recover with comparison ZnO/rGO nanocomposite. The composite of $SnO_2$-rGO nanocomposite was also studied in which sensitivity was found to be 22 and 11 for the concentration $1.12 \times 10^{-6}$ toward ethanol and $H_2S$, respectively. Nevertheless, the gas sensor based on $SnO_2$ found to be 15 and 7. Hence, the study reveals that the gas sensor based on $SnO_2$-rGO nanocomposite shows better sensing performance than the bare $SnO_2$. Sensing response was observed improving with the addition of rGO. Moreover, the author investigated the sensing properties of $In_2O_3$-rGO nanocomposite synthesized by combining hard templet and ultrasonic mixing method. The study shows that the $In_2O_3$/rGO nanocomposite shows a fast sensing response toward ethanol in comparison with $In_2O_3$ but rGO did not. The author considered gas sensors based on $In_2O_3$-rGO nanocomposite as a good material to sense ethanol. The sensing performance of iron oxide with reduced graphene oxide was also investigated. The variation of sensing response was observed as the temperature is varying as of 350–450 °C. Against various concentrations of CO gas, the gas sensor was examined.

At 10–200 ppm, $Fe_3O_4$/rGO nanocomposite sensor responses and the response was found to be 1.8 and 1.9 against the CO gas concentration of 100–200 ppm, respectively, at 400 °C. Goutham et al. [12] design a gas sensor based on the composite of graphene/CdO nanocomposite prepared by a cost-effective ultrasonic-assisted process against the LPG gas for various situations. At room temperature, the sensing response was observed and was found able to detect changing LPG gas concentration as 100–800 ppm. The author observed the sensitivity at 600 ppm for 1 wt. % of the sensor toward the LPG at room temperature. Further, the author suggests that CdO/graphene nanocomposite sensor shows high selectivity, good sensitivity, and stability against the LPG of low concentration. Jyoti and her co-workers [13] investigated on rGO with Zinc oxide nanocomposites for the utilization to sense the different concentration of $NO_2$ gas in the range 4–100 ppm. The study on the effect of gas sensing with the increase of temperature was also done. It was found that at room temperature, the rGO-ZnO nanocomposite shows the highest percentage response. rGO/ZnO nanocomposite shows the limitation to detect at 4 ppm. At room temperature, the high response was obtained, i.e., 48.4% against the $NO_2$ of 40 ppm concentration. Further, the author suggested rGO-ZnO nanocomposite material as good material in sensing toward $NO_2$ gas sensor for room temperature. Mao et al. [14] found that the sensing performance of shows better sensing response than that of rGO toward $NO_2$. However, rGO/$SnO_2$ nanocomposite exhibits a weak response toward the exposure of $NH_3$. Song with his co-workers [15] reported $SnO_2$ quantum wire/rGO composites for the utilization in detecting $H_2S$ showed quick response and recovery time (2/292 s) ranging from 10–100 ppm. Gas sensors based on $SnO_2$ quantum wire/rGO nanocomposites showed a better response of about 33 than the rGO. Yang et al. [16] prepared the graphene/$CeO_2$ composites by using a facile solvothermal reaction for gas sensing applications. $CeO_2$ of 46.7 wt% composite with graphene nanocubes exhibits high sensing response at the limit of detection 5 ppm against the $NO_x$ gas. A higher sensitivity was observed of about 10.39% and quick response time was exhibited of 7.33 s toward $NO_x$ for concentration 100 ppm at room temperature in comparison with $CeO_2$. Zhang et al. [17] demonstrated the property of sensor toward hydrogen gas sensed using four different composites of CuO–rGO–CuO, $(CuO–rGO)_2$–CuO, pure rGO, and CuO. The study found that the CuO/rGO film sensor exhibits good response and recovery time and it was found to achieve the highest sensitivity among the other composites. For different conditions of exposure of hydrogen gas were examined such as for 50, 500, and 1500 ppm of hydrogen by the CuO–rGO–CuO sensor. The sensing response and recovery time were found to be less than 80 and 60 s which might be attributed as a good sensing property. Kumar et al. [18] reported the synthesis of ZnO/rGO nanocomposite for the detection of $NO_2$ gas. ZnO/rGO nanocomposite showed excellent sensing performance and high electrical conductivity than the pure rGO against $NO_2$ gas. The author explained that the strong interaction between ZnO and rGO composite might be responsible for exhibiting high electrical and sensing properties. At the exposure of 50 ppm concentration of $NO_2$, the high response was observed at 50 °C with the response value ca. 32% by ZnO/rGO nanocomposite sensor. Song and co-workers [15] design chemiresistive gas sensors using $SnO_2$ quantum wires/rGO nanocomposites by spin-coating on the substrates of ceramics materials. Sintering was not done further. The sensing performance was examined for $SnO_2$ quantum wire/rGO nanocomposites against $H_2S$ of concentration 50 ppm and was found fully reversible when the exposure of $H_2S$ begin with achieving sensor response as 33 in 2 s at 22°C. Moreover, the author explained that due to the transfer of charge in the interface of $SnO_2$/rGO and the excellent transport capability of rGO leads in enhancing the sensing properties of $SnO_2$ quantum wire/rGO nanocomposites. The author also suggests that a sensor made at room temperature with easy fabrication will exhibit a highly sensing response toward the ultrasensitive $H_2S$ gas with consuming less power. Many researchers have reported on graphene-based gas sensor showing excellent sensing performance (Fig. 11.1).

## *Different Methods for Preparing Graphene-Oxide and rGO/metal-Oxide Nanocomposites*

There are mainly three ways to form GO; they were Brodie method [19], Staudenmaier method [20], and Hummers method [21]. Hummer's method is a widely used method for synthesizing to GO. Different modified Hummer methods were reported for the preparation of GO [22]. For the reduction of GO, there are various methods such as micromechanical exfoliation, epitaxial growth, vapor deposition, and chemical reduction [23–26]. Novoselov et al. prepared single-layer graphene from graphite flakes by using scotch tapes which is deposited on a substrate of $SiO_2$/Si to obtain highly pure graphene [1]. The exfoliation of graphene from graphite using the micromechanical method produces fewer layers of graphene having a crystalline nature with large surface area [24]. The drawbacks of this method are the less production of graphene and give a small size. Nevertheless, this method is considered a simple method where there is no need for any other facilities. Good quality graphene with a large size is obtained by using the epitaxial growth method in comparison to that of the mechanical exfoliation process. However, there is the main drawback of this method was that it cannot produce a large amount of uniformly graphene thickness. For the production of graphene, the chemical vapor deposition (CVD) method is widely used as the size and structure of graphene produced by this method can be controlled.

Stankovich and his co-workers established a new approach to synthesized graphene which is considered in the bottom-up method, known as chemical reduction of graphene oxide which gives a new way of producing a large amount of graphene that can be used in various applications and modification [27]. The synthesis of GO using the chemical reduction method is hydrophilic and by sonication, it can be dispersed in various solvents [28]. The obtained GO was used for the preparation of different composites such as rGO powders, rGO film, and rGO with other composites. It is a cost-effective method. Many rGO-based gas sensor was fabricated using different methods in which nanolithography is included. However, it is an expensive method. And there found that the rGO sensor synthesized by this method shows poor sensing performance and lengthy recovery time of about 2 h. Further, to enhance the sensing properties, it should be given a heat treatment or illumination of UV is to be done [29]. Furthermore, the fabrication of the electrode using the nanolithography method led to contaminate the surface of the sample which strongly affects the sensing performance and the transport of charge in the specimen [30, 31]. Gu et al. [32] synthesized $In_2O_3$ with rGO nanocomposites by hydrothermal method followed by the reduction of GO using hydrazine hydrate. Ghule et al. [33] prepared $Fe_2O_3$ with graphene-oxide nanocomposite film as a sensing material by a simple solid-state method to detect liquefied petroleum gas (LPG) at 27 °C. The composite $Fe_2O_3$ with different concentrations of GO such as 0.1, 1.0, and 10 wt. % were synthesized and sintered for 2 h at 450 °C and utilize as a gas sensor material. It was found that at room temperature, the sensitivity of the composite 1.0 wt. % $Fe_2O_3$/rGO shows the response of approx. 35% for 100 ppm toward LPG as compared to GO (~15%) and $Fe_2O_3$ (~25%). Other composites exhibit a sensing response of about 20–30% and it was also explained that it can be attributed to its exhibiting large surface area and the reduction of charge transfer resistance which have surface leading to the adsorption of LPG molecules.

## Preparation of GO

GO is prepared by modified Hummer's method [34]. In this synthesis process, 5 g of graphite powder is mixed with an $H_2SO_4$ concentration of 98% and sodium nitrate of about 110 ml and 2.5 g, respectively. Stirring is to be done for about 30 min at room temperature. After that, a black slurry suspension was observed as the mixing is nicely done. The temperature of the beaker containing mixture should be controlled below 20 °C; to do this, the beaker is to be kept in a cold water bath. Potassium permanganate of 15 g is added very slowly into the dispersion when the temperature completely goes down below 20 °C under vigorously stirring condition. The temperature should be maintained below 20 °C to stay safe from the explosion and to not get overheated. Temperature is to be increased up to 35 °C after adding potassium permanganate and stirring is done for 2 h. Dark brownish color with highly dense suspension was observed for 2 h indicating the exfoliation of graphene oxide from graphite. Deionized water of 300 ml is added into the mixture. The temperature is to be raised to 98 °C and stirring is done for another 1 h. After the completion of 1 h, hydrogen peroxide having a concentration (30%) of about 15 ml is to be is added into the mixture to stop the reaction. The obtained mixture was filtered and the removal of metal ions from the resulting mixture filtration is to be done by washing the obtained product first time by HCl solution of 10% concentration and further is to be washed using deionized water for several times. After completing the washing and filtration of the resultants, it is to be kept at 60°C for 12 h to dry under vacuum. GO is obtained.

## Synthesis of rGO-Metal-Oxide Nanocomposite

Kumar et al. [18] reported the synthesis of rGO and zinc oxide–reduced graphene-oxide (ZrGO) nanocomposite powder samples by a hydrolysis method in which 85% hydrazine hydrate was utilized as the reducing agent. In the preparation of rGO from GO by chemical, different reductants have been used like hydroquinone, hydrazine, sodium borohydride, dimethylhydrazine, etc. rGO/metal-oxide was also prepared using the hydrothermal method. The rGO-CuO nanocomposite was prepared using the hydrothermal method which is depicted. Firstly, the obtained GO product was added to 40 ml of deionized water to make its suspension very well. After that $Cu(NO_3)_2 . 2.5H_2O$ of 0.02 g is put into the GO suspension under vigorously stirring condition and stir it for 15 min. Then, add NaOH (0.04 g) into the mixture and stir it for another 20 min. After that sealed the mixture suspension in a Teflon container of 100 ml having a steel enclosure and in the furnace, it is to be transferred. The action is to be done for 4 h at desired temperature viz., 100 °C. When the reaction was completed, the obtained product is to be kept to cool down at room temperature. Further centrifugation is done for the obtained substance and washing of the substance is required to removed metal ions using deionized water followed by ethanol several times. The drying of the obtained substance is to be done at 80 °C [35].

## Structural and Composition Characterization Techniques

### *Scanning Electron Microscope (SEM)*

SEM is a technique that gives information about the surface topography and the composition of the sample by scanning the surface with a focused beam of electrons. The image obtained of CuO/rGO nanocomposite from SEM is shown in Fig. 11.2, which reveals that confirms the CuO nanoparticles were well dispersed in the rGO sheet. The SEM image shows spherical shape morphology. The texture of the surface was observed as rough. Nevertheless, the size of the particle is found to be in the range of nano [36].

### *Transmission Electron Microscopy (TEM)*

TEM is a technique in which a beam of electrons is transmitted through a specimen to form an image. It gives us information about the size and morphology of the sample. Figure 11.3 shows the images obtained from TEM and HRTEM of CuO/rGO nanocomposite. The CuO nanoparticles were found to be agglomerated from the image. It obtained a particle size in the range of 60–65 nm. The image also confirms

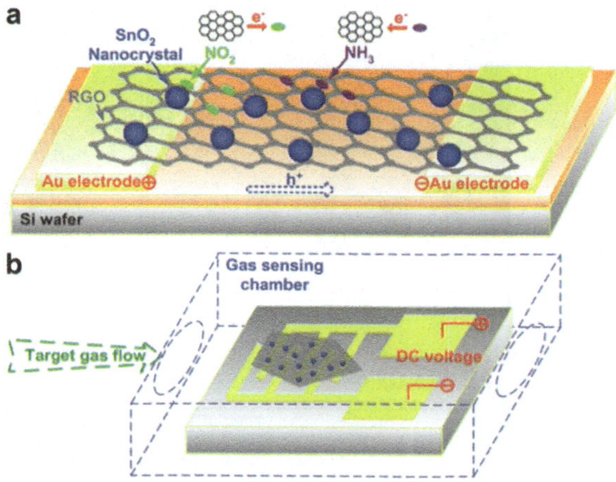


**Fig. 11.1** (i) Schematic of the novel gas-sensing platform of an rGO sheet decorated with metal-0oxide nanocomposite. (ii) Diagrammatic illustration for the testing system of the sensor. Reprinted with permission from Ref. [14]. Copyright (2012) Royal Society of Chemistry

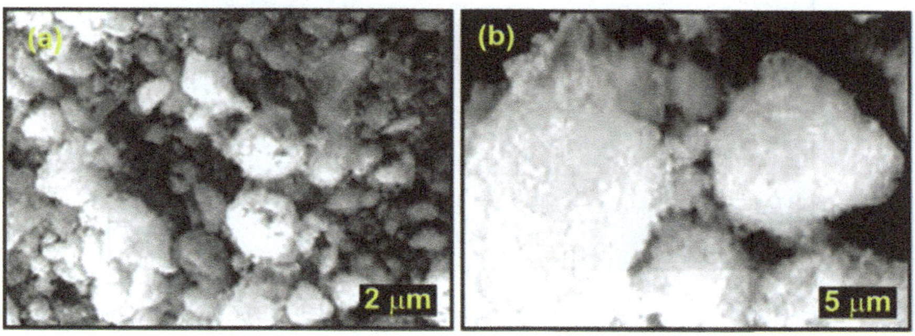


**Fig. 11.2** SEM image of CuO/rGO nanocomposite. Reprinted with permission from Ref. [36]. Copyright (2018) Taylor and Francis

that CuO/rGO nanocomposites are highly crystalline in nature. CuO is well in contact with the surface of rGO.

## *X-Ray Diffraction (XRD)*

X-ray diffraction is a non-destructive analytical technique mainly used for phase identification and structural characterization of crystalline materials. Bragg's Law is expressed as:

$$2dhkl\ \sin\theta = n\lambda$$

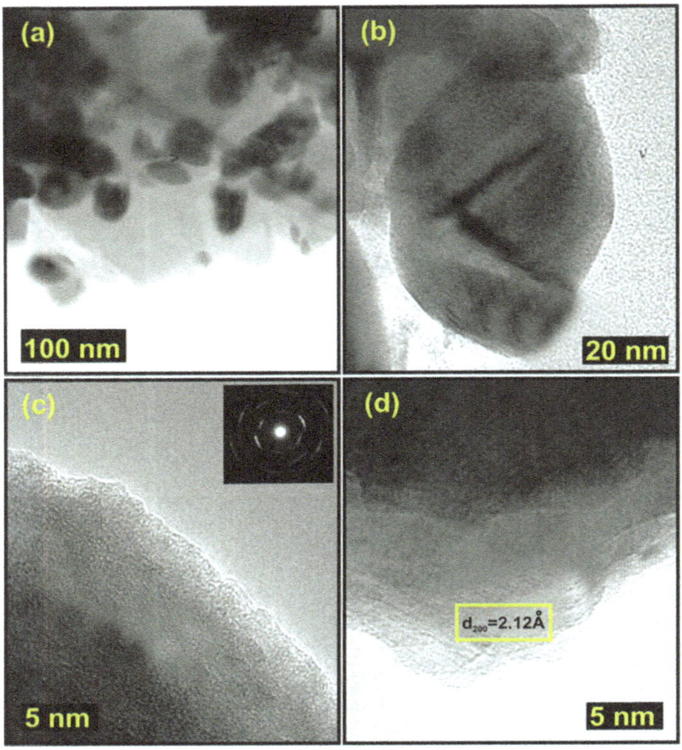


**Fig. 11.3** **a–b** TEM image of CuO/rGO nanocomposite **c–d** HRTEM image of CuO/rGO nanocomposite. Reprinted with permission from Ref. [36]. Copyright (2018) Taylor and Francis

where $\theta$ is the incident angle, $\lambda$ is the wavelengths of the x-rays, and $d$ $(hkl)$ is spacing between two crystal lattice planes.

The XRD pattern of the CuO/rGO composite is shown in Fig. 11.4. XRD pattern of the composition confirms the diffraction peak positions with the standard CuO and rGO specimens confirming the formation of the monoclinic phase for the CuO. The peaks were appeared corresponding to the planes (110), (−111), (−111), (−202), (020), (202), (−113), (−311), and (220). Other peaks were also observed at the planes (002) and (100). It is attributed due to the presence of rGO. The diffraction peak for GO was not observed as it was well reduced from GO. But the intensity of reflection peaks of rGO become weak in the composite while comparing with as-synthesized rGO. Less agglomeration and disordered stacking of the rGO sheets inside the composite might be responsible.

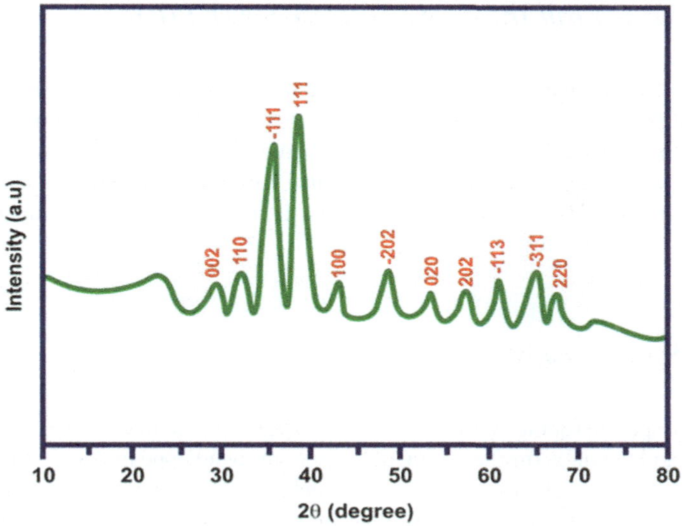


**Fig. 11.4** XRD patterns for CuO/rGO nanocomposites. Reprinted with permission from Ref. [36]. Copyright (2018) Taylor and Francis

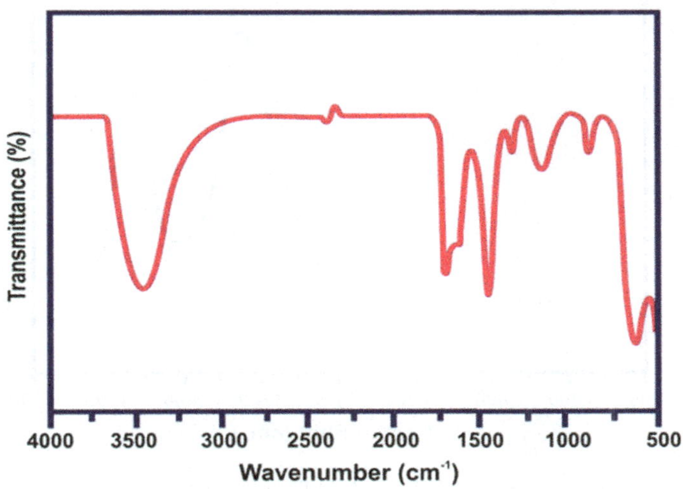


**Fig. 11.5** FTIR spectra of CuO/rGO nanocomposite. Reprinted with permission from Ref. [36]. Copyright (2018) Taylor and Francis

## *Fourier Transform Infrared Spectroscopy (FTIR)*

The FTIR spectra of CuO/rGO nanocomposite are depicted in Fig. 11.5. The spectrum for the CuO/rGO composite was found to be at 526 $cm^{-1}$ for Cu–O bond. It can be attributed that during the synthesis process, oxidation of cuprous ions was occurring due to GO. Another peak appears at 3433 $cm^{-1}$ which corresponds to the stretching vibration of OH. The results reveal that the composite was well reduced in the presence of cuprous ions.

## *Raman Spectroscopy*

The Raman spectra for the composite CuO/rGO are depicted in Fig. 11.6. It was found that the D and G bands for the rGO/CuO nanocomposite appeared at 1330 $cm^{-1}$ and 1594 $cm^{-1}$, respectively. In general, it is accepted that the Raman $I_D/I_G$ ratio (where $I_D$ and $I_G$ are the *D*- and *G*-peak Raman intensities, respectively) is related to the density of defects in graphene-based materials, and the graphitization degree of carbonaceous materials.

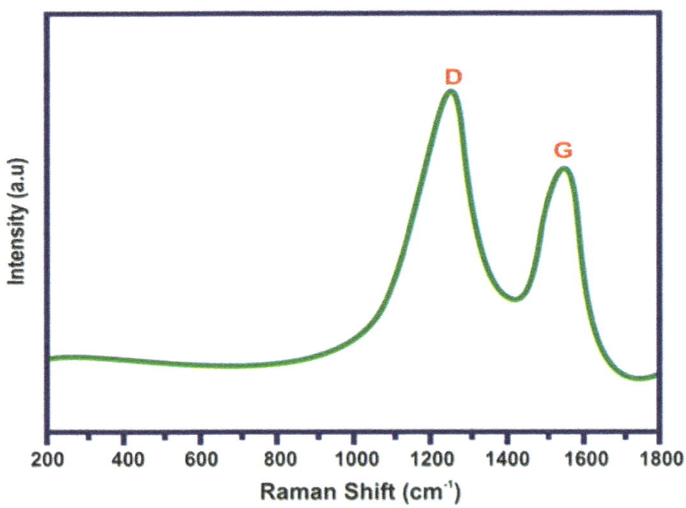


**Fig. 11.6** Raman spectrum of CuO/rGO nanocomposite. Reprinted with permission from Ref. [36]. Copyright (2018) Taylor and Francis

## *Cyclic Voltammetry*

It is an electrochemical technique that is used to measure the developing current in an electrochemical cell under conditions where voltage is more than that predicted by the Nernst equation. The CV is measured by varying different cycles of the potential of a working electrode and the obtained current is measured. The cyclic voltammetry measurement is done for CuO/rGO nanocomposite. The results for the CV measurement for the samples G-CuO composite, G, GO, rGO-CuO composites and CuO were depicted in Fig. 11.7. The range of the potential is set from −0.3 V to +0.8 V. Majumdar et al. achieved the values of specific capacitance by the CV measurement for all the samples were found to be 3.2 $Fg^{-1}$, 4.3 $Fg^{-1}$, 13.1 $Fg^{-1}$, and 263.6 $Fg^{-1}$ with respective to graphite, GO, graphene-CuO, and rGO-CuO composites samples at a scan rate of 5 $mVs^{-1}$. As compared to graphite, it was found that GO exhibits high specific capacitance. CuO/GO nanocomposite shows 4 times high specific capacitance than the pure graphite. The specific capacitance value for bare CuO was found

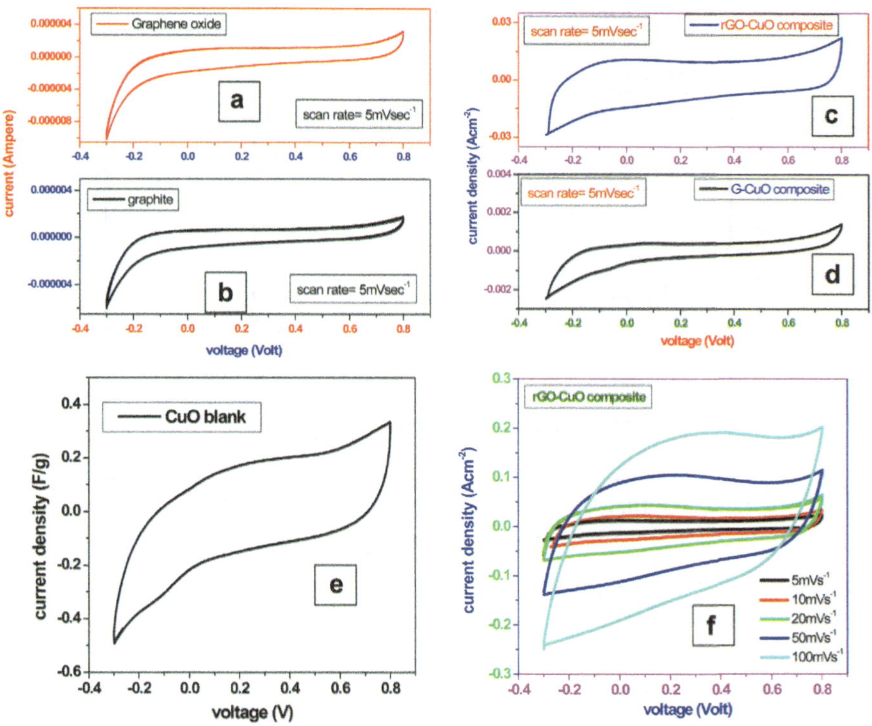


**Fig. 11.7** **a–e** Cyclic voltammetry measurement of GO, G, G–CuO composite and CuO in the potential range −0.3 V to +0.8 V at 5 $mVs^{-1}$. **f** Plots of current density versus voltage at different scan rates for the rGO/CuO composite. Reprinted with permission from Ref. [37]. Copyright (2017) Elsevier

to be 39.7 $Fg^{-1}$ which is 6 times lower as compared to rGO-CuO nanocomposite for the same scan rate. Hence, as compared to other composites, CuO/rGO nanocomposite found to achieve the highest specific capacitance. In Fig. 11.7, CV curves were plotted of voltage versus current density for different scan rates for all the composites [37].

## *X-Ray Photoelectron Spectroscopy (XPS)*

The surface composition and element analysis for the resulting products were characterized by an XPS technique. Figure 11.8 shows the XPS spectra of GO, rGO, and composites. It was observed that all reveal the two peaks. One peak appears at 285.5 and another at 531.0 eV that corresponds to $C_{1s}$ and $O_{1s}$, respectively. The peak for $Cu_{2p}$ and $Cu_{3p}$ was also observed which appears at 933.3 eV and 77.0 eV, respectively [38].

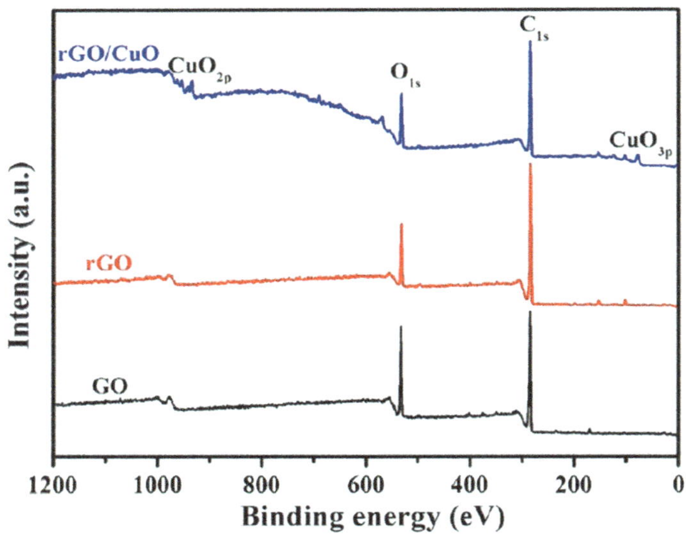


**Fig. 11.8** XPS spectra of CuO/rGO, rGO and GO. Reprinted with permission from Ref. [38]. Copyright (2014) American Chemical Society

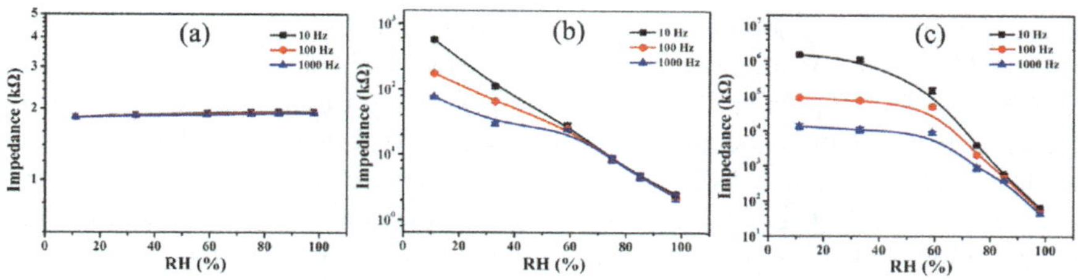


**Fig. 11.9** Impedances of the sensors based on **a** reduced graphene oxide, **b** metal-oxide and **c** rGO/metal-oxide composite under different RHs measured at different frequencies. Reprinted with permission from Ref. [38]. Copyright (2014) American Chemical Society

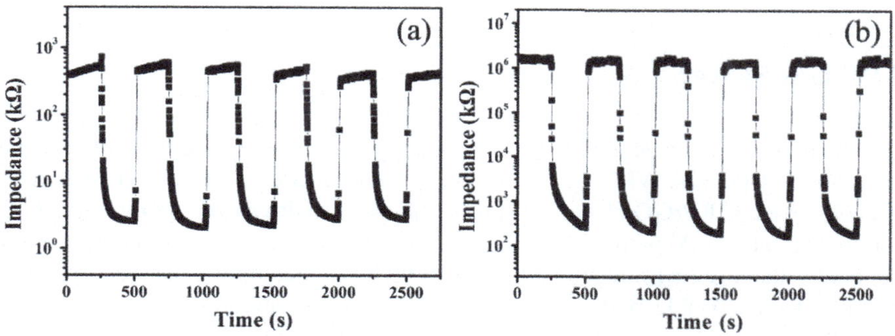


**Fig. 11.10** Response and recovery properties of the sensors fabricated with **a** metal-oxide and the composites **b** rGO/metal-oxide nanocomposite. Reprinted with permission from Ref. [38]. Copyright (2014) American Chemical Society

## Gas Sensing of Metal-Oxide/rGO Nanocomposite Toward Different Gases.

### *$NO_2$ Gas Sensor*

Recently, Jyoti et al. [13] investigated the composite of metal-oxide/rGO against the various concentration of $NO_2$ from 4–100 ppm. They also studied the sensing performance by varying the temperature and it was found that at room temperature, maximum percentage response of 48.4% at the concentration of $NO_2$ at 40 ppm was obtained. Mao et al. [14] reported that rGO-based $SnO_2$ gas sensor enhances the sensing properties toward $NO_2$ while for $NH_3$, the sensing becomes weak. $SnO_2$/rGO nanocomposite sensors exhibit a sensitivity of 1.29–2.87 for $NO_2$ (100 ppm) and sensitivity increases by 3.0–33.4% as compared to bare rGO. Whereas it shows 1.20–1.47 sensitivity for $NH_3$ and the sensitivity reduces by 13.4–23.8% with comparison to rGO. Kumar [18] and his co-workers investigated the chemically synthesized rGO/ZnO nanocomposite. The author suggested that the ZnO/rGO nanocomposite gas sensor exhibits a good sensing response toward NO2 gas. As compared to pure rGO, ZnO/rGO nanocomposite shows a high sensing response of ca. 32% against the $NO_2$ gas for 50 ppm at 50 °C. Gu et al. [32] demonstrated the sensing performance of $In_2O_3$/rGO nanocomposite. The results reveal that $In_2O_3$/rGO nanocomposite exhibits excellent selectivity, high response, and relatively short response and recovery time while detecting the $NO_2$ at room temperature. For the concentration 5–50 ppm of $NO_2$, there shows a rapid increase of response. While after concentration reached 50–100 ppm, it was found that response increases slowly which might be attributed to the fact that $NO_2$ reaches its saturation. The $In_2O_3$/rGO nanocomposite gas sensor shows a response at 8.25, response time in 4 min and recovery time in 24 min.

## *LPG Gas Sensor*

Goutham et al. [12] designed a gas sensor using CdO/rGO nanocomposite for the detection of LPG gas. When there was an exposure of LPG, a decrease in resistivity was observed. CdO/rGO (1 wt. %) sensor shows a sensitivity of LPG gas at 27 °C for concentration 600 ppm with the increasing response of 70%. Further, the author concluded that even at room temperature, CdO/rGO nanocomposite gas sensor shows highly selective, stable, and sensitive to the low concentration of LPG. Ghule et al. [33] investigated the sensing performance of $Fe_2O_3$/graphene-oxide nanocomposite toward the LPG gas. The sensitivity against LPG for 100 ppm concentration at room temperature shows that the 1.0 wt.% GO/$Fe_2O_3$composite sensor exhibits ~35% to that of other GO/$Fe_2O_3$ composite sensors, GO sensor and $Fe_2O_3$ sensor which shows 20–30%, ~15%, and ~25%, respectively. The author attributed to that of large specific surface area and reduced charge transfer resistance, which, in general, has more surface-active surface sites for adsorbing LPG molecules than those of individual and other low-surface area sensors.

## *Ammonia Gas Sensor*

Sakthivel et al. [35] synthesized CuO/rGO nanocomposite for the use in detecting ammonia. It was found that the rGO-CuO nanocomposite sensor showed a fast response in 12 s and a short recovery time of around 90 s for concentration 600 ppm toward $NH_3$ at 150 °C (Figs. 11.9 and 11.10).

## Conclusions

The previous work reported by the researchers on graphene-based metal-oxide nanocomposite was reviewed in this context. The sensing properties observed toward different gases like $NO_2$, LPG gas, $NH_3$, $H_2$, etc., by various rGO-based metal-oxide nanocomposite were discussed. The addition of rGO with metal-oxide improved the sensing properties and enhanced the response of the sensor. The results were obtained differently for different gas sensors prepared by different synthesis processes. The characterization techniques such as SEM, TEM, XRD, FTIR, Raman, and CV measurement give information about the structural, morphology, and behavior of the metal-oxide/rGO nanocomposite that were also depicted.

**Acknowledgements** The author would like to thank the Department of Physics, NIT Manipur for giving the Internet service which helps in accessing many online journals.

## References

1. Novoselov, K.S., Geim, A.K., Morozov, S.V., Jiang, D., Zhang, Y., Dubonos, S.V., Grigorieva, I.V., Firsov, A.A.: Electric Field Effect in Atomically Thin Carbon Films. Science **306**, 666–669 (2004)
2. Zhang, D., Chang, H., Li, P., Liu, R., Xue, Q.: Fabrication and characterization of an ultra-sensitive humidity sensor based on metal oxide/graphene hybrid nanocomposite. Sens. Actuat. B-Chem. **225**, 233–240 (2016)
3. Srivastava, A., Galande, C., Ci, L., Song, L., Rai, C., Jariwala, D., Kelly, K.F., Ajayan, P.M.: Novel Liquid Precursor-Based Facile Synthesis of Large-Area Continuous, Single, and Few-Layer Graphene Films. Chem. Mater. **22**, 3457–3461 (2010)
4. Gilje, S., Han, S., Wang, M., Wang, K.L., Kaner, R.B.: A Chemical Route to Graphene for Device Applications. Nano Lett. **7**, 3394–3398 (2007)
5. Mc Allister, M.J., Li, J.L., Adamson, D.H., Schniepp, H.C., Abdala, A.A., Liu, J., Herrera-Alonso, M., Milius, D.L., Car, R., Prud'homme, R.K.: Single sheet functionalized graphene by oxidation and thermal expansion of graphite. Chem. Mater. **19**, 4396–4404 (2007)
6. Park, S., Ruoff, R.S.: Chemical Methods for the Production of Graphenes. Nat. Nanotechnol. **4**, 217–224 (2009)
7. Schedin, F., Geim, A.K., Morozov, S.V., Hill, E.W., Blake, P., Katsnelson, M.I., Novoselov, K.S.: Detection of individual gas molecules adsorbed on graphene. Nat. Mater. **6**, 652–655 (2007)
8. Yavari, F., Koratkar, N.: Graphene-based chemical sensor. J. Phys. Chem. Lett. **3**, 1746–1753 (2012)
9. Goutham, S., Sadasivuni, K.K., Kumar, D.S., Rao, K.V.: Flexible ultra-sensitive and resistive $NO_2$ gas sensor based on nanostructured $Zn(x)\ Fe(1-x)_2O_4$. RSC Adv. **8**, 3243–3249 (2018)
10. Dieter, K.: Function, and applications of gas sensors. J. Phys. D Appl. Phys. **34** (2017)
11. Naveen Kumar, J.R., Shrinivasa Mayya, D., Savitha, M.B., Prasad P.: Graphene-based metal oxide nanocomposites for gas sensing application. Int. J. Appl. Eng. Manage. Lett. (IJAEML) **2** (2018)
12. Goutham, S., Jayarambabu, N., Sandeep, C., Sadasivuni, K.K., Kumar, D.S., Rao, K.V.: Resistive room temperature LPG sensor based on a graphene/CdO nanocomposite Microchim. Acta. **186**, 62 (2019)

13. Jyoti, Kanaujiya, N., Varma, G.D.: Highly selective room temperature $NO_2$ gas sensor based on rGO–ZnO Composite. In: AIP Conference Proceedings, vol. 1953, p. 030039 (2018)
14. Mao, S., Cui, S., Lu, G., Yu, K., Wen, Z., Chen, J.: Tuning gas-sensing properties of reduced graphene oxide using tin oxide nanocrystals. J. Mater. Chem. **22**, 11009 (2012)
15. Song, Z., Wei, Z., Wang, B., Luo, Z., Xu, S., Zhang, W., Yu, H., Li, M., Huang, Z., Zang, J., Yi, F., Liu, H.: Sensitive Room-Temperature $H_2S$ Gas Sensors Employing $SnO_2$ Quantum Wire/Reduced Graphene Oxide Nanocomposites. Chem. Mater. **28**, 1205–1212 (2016)
16. Yang, Y., Tian, C., Sun, L., Lü, R., Zhou, W., Shi, K., Kan, K., Wang, J., and Fu H.: Growth of small-sized $CeO_2$ particles in the interlayers of expanded graphite for high-performance room temperature $NO_x$ gas sensor. J. Mater. Chem. A. 1–8 (2013)
17. Zhang, D., Yin, N., Jiang, C., Xia, B.: Characterization of CuO–reduced graphene oxide sandwiched nanostructure and its hydrogen sensing characteristics. J. Mater. Sci. Mater. Electron. **28**, 2763–2768 (2017)
18. Kumar, N., Srivastava, A.K., Patel, H.S., Gupta, B.K., Varma, G.D.: Facile Synthesis of ZnO–Reduced Graphene Oxide Nanocomposites for $NO_2$ Gas Sensing Applications. Eur. J. Inorg. Chem. **11**, 1912–1923 (2015)
19. Brodie, B.C.: On the atomic weight of graphite. Philos. Trans. R. Soc. Lond. **149**, 249–259 (1859)
20. Staudenmaier, L.: Verfahren Zur darstellung der graphitsäure. Ber. Dtsch. Chem. Ges. **31**, 1481–1487 (1898)
21. Hummers, W.S., Offeman, R.E.: Preparation of graphitic oxide. JACS. **80**, 1339–1339 (1958)
22. Roy, I., Rana, D., Sarkar, G., Bhattacharyya, A., Saha, N.R., Mondal, S., Pattanayak, S., Chattopadhyay, S., Chattopadhyay, D.: Physical and electrochemical characterization of reduced graphene oxide/silver nanocomposites synthesized by adopting a green approach. RSC Adv. **5**, 25357–25364 (2015)
23. Subrahmanyam, K.S., Panchakarla, L.S., Govindaraj, A., Rao, C.N.R.: Simple method of preparing graphene flakes by an arc-discharge method. J. Phys. Chem. C. **113**, 4257–4259 (2009)
24. Subrahmanyam, K.S., Vivekchand, S.R.C., Govindaraj, A., Rao, C.N.R.: A study of graphenes prepared by different methods: characterization, properties, and solubilization. J. Mater. Chem. **18**, 1517–1523 (2008)
25. Paola, R., Anming, H., Giuseppe, C.: Synthesis, properties and potential applications of porous graphene: a review. Nano-Micro Lett. **5**, 260–273 (2013)
26. Yang, Z., Gao, R.G., Hu, N.T., Chai, J., Cheng, Y.W., Zhang, L.Y., Wei, H., Kong, E.S.W., Zhang, Y.F.: The prospective two-dimensional graphene nanosheets: preparation, functionalization, and applications. Nano-Micro Lett. **4**, 1–9 (2012)
27. Stankovich, S., Dikin, D.A., Dommett, G.H., Kohlhaas, K.M., Zimney, E.J., Stach, E.A., Piner, R.D., Nguyen, S.T., Ruoff, R.S.: Graphene-based composite materials. Nature **442**, 282–286 (2006)
28. Paredes, J.I., Villar-Rodil, S., Martínez-Alonso, A., Tascon, J.M.D.: Graphene oxide dispersions in organic solvents. Langmuir **24**, 10560–10564 (2008)
29. Yuan, W., Shi, G.: Graphene-based gas sensors. J. Mater. Chem. A **1**, 10078–10091 (2013)
30. Dan, Y.P., Lu, Y., Kybert, N.J., Luo, Z.T., Johnson, A.T.C.: Intrinsic response of graphene vapor sensors. Nano Lett. **9**, 1472–1475 (2009)
31. Lu, G.H., Park, S., Yu, K.H., Ruoff, R.S., Ocola, L.E., Rosenmann, D., Chen, J.H.: Toward practical gas sensing with highly reduced graphene oxide: a new signal processing method to circumvent run-to-run and device-to-device variations. ACS Nano **5**, 1154–1164 (2011)
32. Gu, F., Nie, R., Han, D., Wang, Z.: $In_2O_3$–graphene nanocomposite based gas sensor for selective detection of $NO_2$ at room temperature. Sensors Actuat. B. **219**, 94–99 (2015)
33. Ghule, B.G., Shaikh, F.S., Shinde, N.M., Sushil, S., Shinde, P.V., Mane, R.S.: Promoted room-temperature LPG gas sensor activity of graphene oxide@$Fe_2O_3$ composite sensor over individuals. Mater. Res. Express. **5**, 125001 (2018)
34. Marcano, D.C., Kosynkin, D.V., Berlin, J.M., Sinitskii, A., Sun, Z.Z., Slesarev, A., Alemany, L.B., Lu, W., Tour. J.M.: Improved synthesis of graphene oxide. ACS Nano **4**, 4806–4814 (2010)

35. Bhuvaneshwari, S., Gopalakrishnan, N.: Selective ammonia sensor based on copper oxide/reduced graphene oxide nanocomposite. J. Alloy Compd. **788**, 422–428 (2019)
36. Sagadevan, S., Zaman Chowdhury, Z., Johan, M.R.B., Aziz, F.A., Salleh, E.M., Hawa, A., Rafique, R.F.: A one-step facile route synthesis of copper oxide/reduced graphene oxide nanocomposite for supercapacitor applications. J. Exp. Nanosci. **13**, 302–313 (2018)
37. Majumdar, D., Baugh, N., Bhattacharya, S.K.: Ultrasound assisted the formation of reduced graphene oxide-copper (II) oxide nanocomposite for energy storage applications. Physicochem. Eng. Aspects. **512**, 158–170 (2017)
38. Wang, Z., Xiao, Y., Cui, X., Cheng, P., Wang, B., Gao, Y., Li, X., Yang, T., Zhang, T., Lu, G.: Humidity-sensing properties of urchinlike cuo nanostructures modified by reduced graphene oxide. ACS Appl. Mater. Interfaces. **6**, 3888–3895 (2014)

# Chapter 12
# Mechanical and Tribological Properties of Metal Incorporated DLC Thin Film

**Ranjan Kr Ghadai, Kanika Singh, Ashis Sharma, Manish Kr Roy, and Bibhu Prasad Swain**

## Introduction

Diamond-like carbon (DLC) films are defined as polycrystalline coatings consisting of a dense network of high-density carbon (Diamond) crystallites. The term generally refers to a family of thin films related more by their inherent properties (i.e., hardness and wear resistance). Diamond-like Carbon, or DLC, mainly consists of a mixture of crystalline and amorphous phases. Even though generally, it is called "diamond-like", it is in fact quite different from crystalline carbon: it is black, not as hard and not as brittle [1]. The work of Aisenberg and Chabot [2] in 1997 stimulated interest in the existence of hard carbon coatings. A carbon-ion beam was used to produce films exhibiting the properties of diamond. They called these films diamond-like. Preparation of such, and similar, films has been done ever since. Plasma activated chemical vapor deposition (CVD) is one such process explained in detail by T. Michler et al. and D. H. Lee et al. in their respective works [3, 4]. PIG plasma CVD is a process which exposes hydrogen gas to plasma patented by Ando et al. in 2015 [5]. Cathodic arc discharge has also been explored for producing DLC coatings [6, 7]. Magnetron Sputtering is another commonly used method [8, 9]. Some of such processes are listed in Table 12.1 complied by Moriguchi, Ohara, and Tsujioka given below.

A variety of coordination of carbon atoms are present in a DLC film, which itself can exist in seven distinct forms (listed below Fig. 12.1). All seven forms contain

R. K. Ghadai (✉) · K. Singh · A. Sharma · M. K. Roy
Department of Mechanical Engineering, Sikkim Manipal Institute of Technology, East Sikkim, Majitar, Rangpo, Sikkim 737136, India
e-mail: ranjankumarbls@gmail.com

B. P. Swain
Department of Physics, National Institute of Technology Manipur, Langol, Imphal, Manipur 795004, India


B. P. Swain (ed.), *Nanostructured Materials and their Applications*, Materials Horizons: From Nature to Nanomaterials, https://doi.org/10.1007/978-981-15-8307-0_12

**Table 12.1** Structures and characteristics of DLC production processes

| Processes | DLC structures | Characteristic features |
|---|---|---|
| Radio-frequency method | a-CH | Lower friction and smoother<br>Film formation on an insulating substrate<br>Adhesiveness is lower |
| PIG method | a-CH | Thick film formation<br>Stress and thickness is controllable<br>Lower distribution of film thickness |
| Self-discharge method | a-CH | Better coverage<br>Internal coating formation<br>Deposition rate and filling density is higher<br>Adhesiveness is lower |
| Arc method | ta-C | In oil, friction is lower<br>Diamond-like hardness<br>Surface sensitivity<br>Formation of thick films is difficult |
| Sputter method | a-C, ta-C | DLC conduction support<br>Hardness is lower |

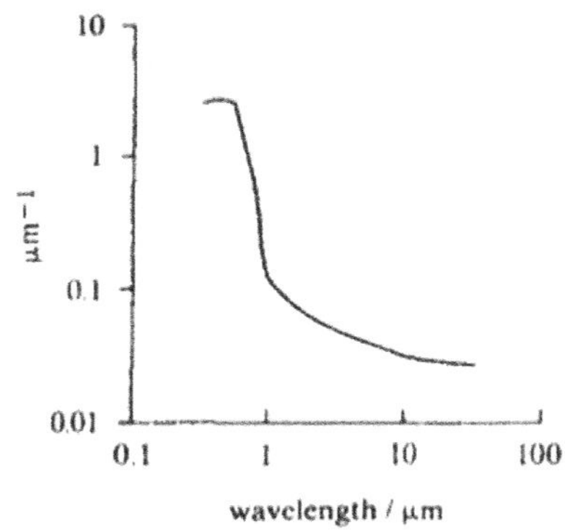


**Fig. 12.1** Absorption of DLC

$sp^3$ hybridized carbon atoms abundantly. Tetrahedral amorphous carbon (ta-C) is the "purest" form of DLC film, with 100% $sp^3$ hybridized carbon. But, hydrogen, $sp^2$ C graphite and other metals can be used as fillers to minimize expenses and impart unto the film some other properties [10].

The properties a coating of DLC may exhibit at any given point in time vary consistently with the change in the condition of deposition.

DLC films can be extensively hard, with values up to 3000 kg $mm^{-2}$ [11]. 0.05–0.20 is the general range for the friction coefficient against polished steel. It increases with the increase in relative humidity [12]. Internal stress varies with bias potentials and the thickness is limited to 2–3 μm. Heat treatment processes also change the coating properties, i.e., hardness, toughness, and rate of wear [13]. It has also been found to be hydrophobic in nature [14].

As shown in the figure below, diamond-like carbon films are absorbing in the visible and the IR range they provide good transmission of spectra.

The DLC coatings are chemically durable, electrically resistive, and resilient against abrasion.

Applications of DLC in the current day are multi-faceted.

A. From biomedical, i.e., for example, rigorous testing has given evidence of DLC's potential in joint replacements [15–17] and in heart valves which were traditionally made from LTI carbon [1],
B. to mechanical, for example, in magnetic discs [18, 19], work has even been done in applying DLC to optical fibres,
A. to electrical, for example in insulators in metal–insulator–semiconductor devices, [20]
B. and even to convert solar energy photo-thermally [21].

But there have been various issues when applying the DLC coatings. Some coatings were found to be too rough, so much so that excessive wear is produced on the surfaces, and the polishing processes may not be applicable to the given production requirements. One of the ways to deal with such issues is to incorporate different metals into the DLC films.

## *Structure of DLC Thin Film*

Diamond-like Carbon (DLC) thin films exist in two distinct forms, i.e., amorphous and crystalline. $Sp^3$, $sp^2$, and $sp^1$ chemical coordination exist for any carbon atom; enabling the formation of many structures, including ones with specks of other elements. The structure of DLC thin film has been discussed below with the help of necessary diagram.

Ion beam analysis and combustion analysis like Rutherford backscattering (RBS), along with elastic recoil detection (ERD), and nuclear reaction analysis (NRA) determine the amount of hydrogen in DLC films. Concentrations of H can go as high as 60% and as low as 10% but usually remains about 20–40% for regular conditions (Fig. 12.2).

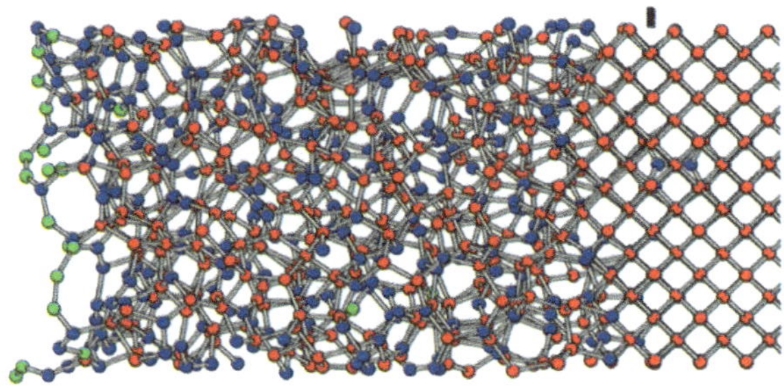

**Fig. 12.2** Structure of amorphous DLC thin film

## *Synthesis of DLC Thin Film*

### Physical Vapor Deposition Process

PVD is a vacuum deposition method to deposit thin films, the vaporized version of the materials needed are condensed on top of the substrate. Deposition of a luminescent thin film is commonly done by Pulsed laser deposition (PLD) and sputtering, two very common PVD methods. Working in vacuum, in the pulsed laser deposition, a focused pulsed laser is made to hit the unadulterated target material. This technique is quite exceptional for the oxide film deposition [22, 23]. In sputtering technique, atom is removed from the target material surface to deposit thin film. This method is applied to coat the films on numerous target materials [24, 25]. Many authors have synthesized metal incorporated Diamond-like carbon using PVD techniques. There are different types of PVD techniques that are extensively used.

#### Magnetron Sputtering

Magnetron sputtering is another plasma vapor deposition process wherein the plasma is made and the positively charged ions by accelerating an electrical field and superimposing it on the negatively charged electrode. For the primary process, a cathode (or target) plate is pounded by energetic ions produced in the discharge plasma. Bombardment process is the main cause of the sputtering, i.e., removal, of target atoms, which are then condensed on any substrate as a thin film [26]. In past decades, researchers have been doing their works using this technique and found out many enhanced properties of the DLC thin film (Fig. 12.3).

Some of the works are given below:

A. Zhang et al. prepared DLC multilayered films with various residual stress ratios and modulus ratios using Magnetron sputtering between adjacent sub-layers by alternating deposition bias [28].
B. Manninen et al. used Magnetron sputtering technique for the deposition of Ag content ranging from 1.3 to 13.1% into Ag-DLC coatings [29].

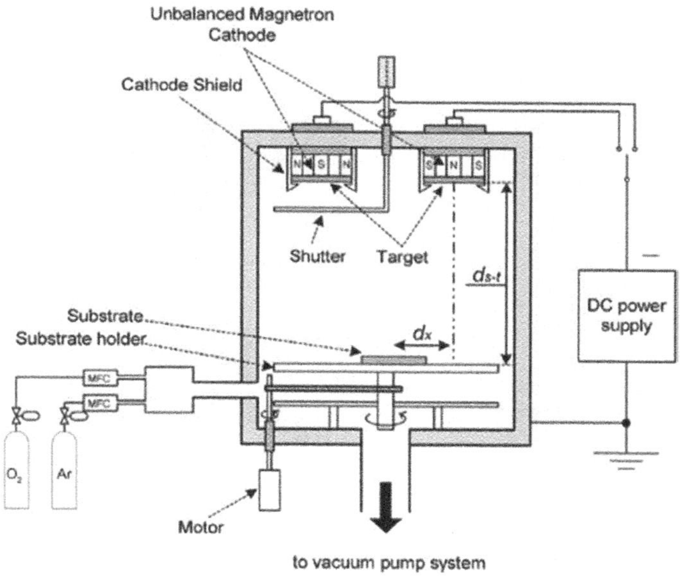


**Fig. 12.3** Magnetron sputtering diagram [27]

C. Dai et al. experimented on changing the microstructure and properties of films; using a hybrid ion beam system (consisting a magnetron sputtering unit and ion beam source), non-carbide and ductile Cu was assimilated into DLC thin films [30].

Arc Ion Plating

Ion plating is a deposition technique in which a substrate, i.e., a compound or a metal are deposited in target surface areas. At first, the material that is to be used is ionized and vaporized through the aid of an electric arc and then directed towards the target at high-speed bombardment during deposition is performed to obtain good adhesion modify residual stress, and to modify the structural and morphological properties of the deposition film. This process is typically performed within a vacuum chamber (Fig. 12.4).

Zou et al. deposited diamond-like carbon films using arc ion plating, as hard protective films on magnesium AZ91 alloy under the influence of negative pulse bias voltages fluctuating from 0-200 V [32]. Wong et al. performed a study where Zn-doped diamond-like carbon films KrF laser deposition method was used to fabricated it. Sources for the laser system were taken to be carbon targets containing 3.0, 5.0, and 10.0 at.% [33].

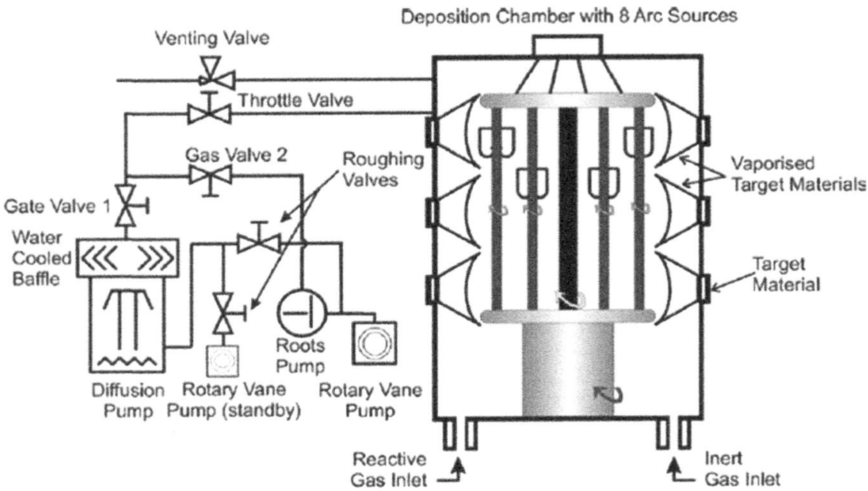


**Fig. 12.4** Schematic diagram of arc ion plating deposition technique [31]

### Pulsed Laser Deposition

Pulsed Laser Deposition (PLD) is also a type of PVD process which is carried out in a system of vacuum. Both molecular beam epitaxy and magnetron sputtering deposition have some characteristics common with PLD. In this technique, a pulsed laser is concentrated on the target to be placed. For high-density laser energy, each laser pulse creates plasma plume for vaporizing a small amount of material (Fig. 12.5).

## Chemical Vapor Deposition Process

The first works of van Arkel and de Boer can be seen as the origin of chemical vapor deposition [35]. Condensation of any compound (or multiple compounds) from gas phase, where reaction takes place, to produce a solid deposit on to a substrate is called as chemical vapor deposition (CVD) [36]. Chemicals are mixed in the vapor phase and heated in a chamber to very high temperatures causing chemical reactions [37].

### Plasma Enhanced Chemical Vapor Deposition Technique

PECVD or Plasma-enhanced chemical vapor deposition is one type of a Chemical Vapor Deposition Processes which is more suitable then the simple CVD process as all CVD processes needs their temperature to be high, limiting their applications for coating temperature-sensitive materials. Therefore, the thermal heat is reduced using

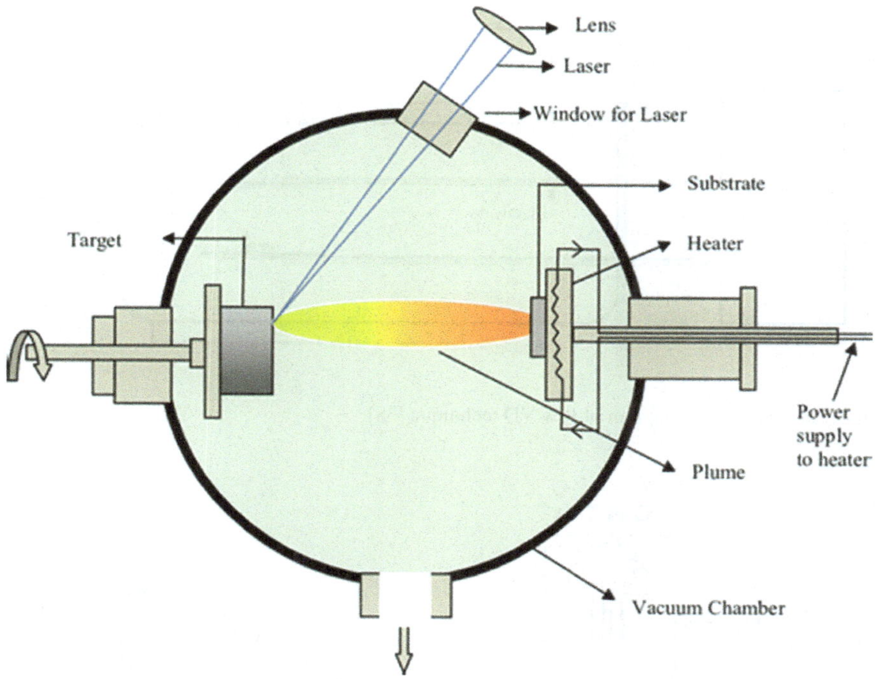


**Fig. 12.5** Schematic diagram of pulse laser deposition technique [34]

plasma source so that chemical reactions take place at equivalently high temperature. PECVD is essentially a gas-phase reaction in a lower temperature plasma that produces a thin solid film on a substrate [37]. Ahmed et al. produced various DLC and N-DLC samples on silicon substrate using a plasma-enhanced chemical vapor deposition technique (PECVD) [39]. Venkatesh et al. incorpotated Ag in a hydrogenated carbon thin film by a hybrid Plasma Vapor Deposition, i.e., Plasma-enhanced Chemical Vapor Deposition process [40] (Fig. 12.6).

Atmospheric Pressure Chemical Vapor Deposition

APCVD process occurs at atmospheric pressure (i.e., 1 atmosphere or 101,325 Pa or 760 Torr). The compatibility of Atmospheric Pressure Chemical Vapor Deposition with vaccum-free, continuous in-line processes makes it very attractive for cost-sensitive, high volume manufacturing applications [42] (Fig. 12.7).

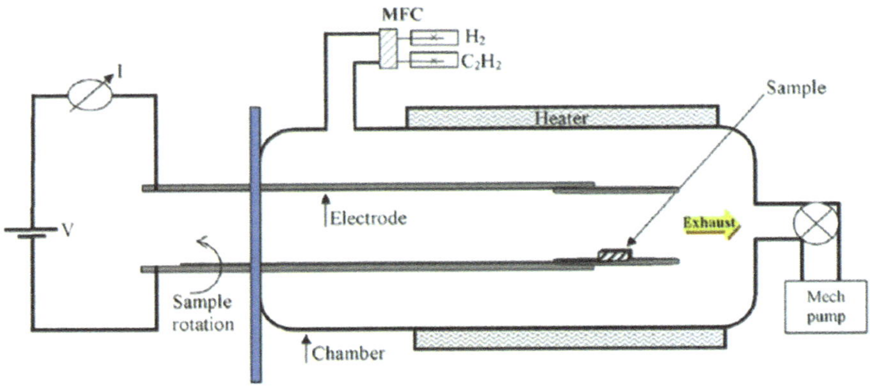


**Fig. 12.6** Schematic Diagram of PECVD technique [38]

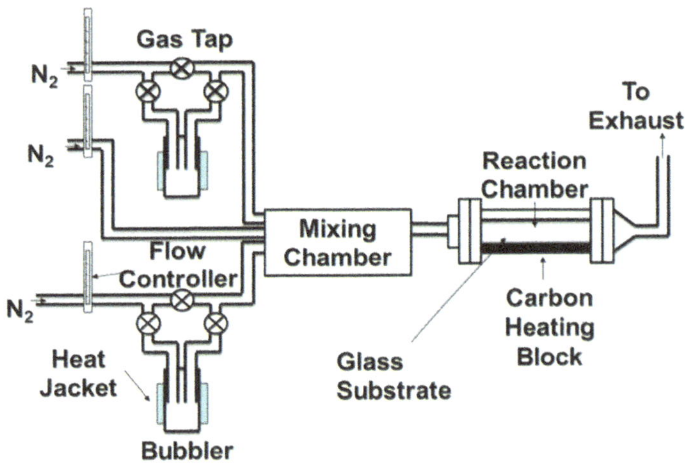


**Fig. 12.7** Schematic diagram of atmospheric pressure chemical vapor deposition [41]

## *Metal Incorporated DLC Thin Films*

So, to improve the properties of DLC thin films different alloying elements are incorporated [43]. Studies in the fields concerning DLC films have been majorly converged on metals (Cu, Mo, Al, Fe, Ag, Zn, Ni, Cr, etc.) and semiconductors (Si, Ge) incorporated DLC films which shows a significant decrease of residual stresses of the films [44].

## Mechanical and Tribological Properties

### Copper (Cu) Incorporated DLC

**Chaus et al.** deposited copper doped DLC coating on a silicon substrate using pulsed cathodic arc plasma consisting of a composite cathode of graphite inserted with copper rods. Good adhesion to the silicon substrate was observed without any micro-cracks, micro-voids and localized de-laminations at the substrate interface. The results of mechanical and optical properties were obtained and it showed that the Cu-doped DLC coatings had a roughness of 2–4 nm, resulting in a smooth surface; and it contained oxidized copper nano-inclusions. The micro-hardness was reduced by 15% and the internal stress was reduced by 2.5 times when compared with a single component carbon layer [45].

**Dwivedi et al.** explored the effect of metallic interlayers (namely Ti and Cu) and 5–20 mN indentation loads on mechanical properties of N-DLC, produced by RF-PECVD, and compared it to DLC thin films. Residual stress was found to be compressive with the maximum value being 1.8GPa; introduction of Ti and Cu reduces stress further. The value of hardness was found to be varying with penetration depth. Wear resistance was the minimum for N-DLC (0.088) and maximum for Ti-DLC (0.098) [46]. (Table 12.2).

**Dwivedi et al.** deposited Cu-DLC on various substrates with the help of a hybrid system combining DC-sputtering along with radio-frequency plasma-enhanced chemical vapor deposition to investigate the properties. The obtained residual stress value was between the limits 0.7 and 0.94 GPa, with the conductivity increasing drastically with incorporation. The elastic modulus and hardness values were found to be decreasing when the load increases from 4 to 10 mN. The hardness also increased depending on the penetration depth. Addition of Cu, degraded the mechanical properties, minimizing stress value and betters the electrical and optical properties for various applications [47].

**Gou et al.** incorporated copper onto DLC thin films with the help of a hybrid ion beam deposition system and deposited it on Si/glass substrate. Concentration of Cu was varied (0.1 to 39.7%) by changing the sputtering current. Residual stresses compared to DLC thin films (2.0 GPa) were significantly lower [48] (Fig. 12.8).

**Tsai et al.** synthesized a bioactive hydrogenated diamond-like carbon carrying distinct Cu fractions with the help of radio-frequency plasma-enhanced chemical

**Table 12.2** Obtained values for hardness and modulus of elasticity [46]

| | DLC | N-DLC | Cu-DLC | Ti-DLC |
|---|---|---|---|---|
| Hardness | 25.1 | 22 | 26.2 | 30.8 |
| Elasticity modulus | 257.2 | 250 | 280.9 | 314 |

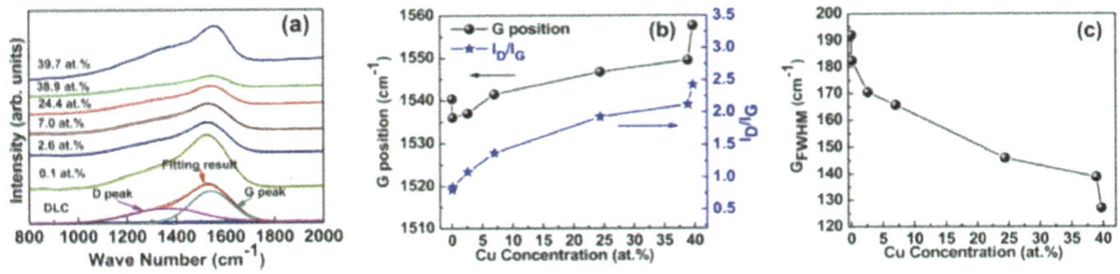


**Fig. 12.8** The Raman spectra, G peak position ID/IG, and GFWHM of the films with different Cu concentrations [48]

vapor deposition and a magnetron Cu electrode in various atmospheres. To improve surgical instruments, anti-bacterial performance, anti-sticking properties, mechanical strength, and corrosion resistance were tried to be enhanced. There was an irregularity found in the deposition attributed to the high working pressure, it increased with the argon flow ratio and plasma power in all atmospheres. Because of the Cu incorporation, all films were closer making them harder, and rougher. They were found to be hydrophobic in nature, due to a higher water contact angle [49].

**Gerhards et al.** made amorphous thin films containing Copper (0–50%) with the help of mass-selected ion beam deposition, synthesizing Cu into the diamond-like carbon. TEM images showed that Cu had been incorporated throughout the film depth. Assigning a modification of particle sizes was found to be difficult due to the impact of ion energy. There was also an increasing trend observed for cluster size with ascend in Cu concentration; mean cluster sizes increased (from 7 to 9 nm). The concentration curves indicated an inhomogeneous composition [50] (Fig. 12.9).

**Zhou et al.** used a pulsed cathode arc and direct current plasma technique to prepare diamond-like carbon bi-layers with Copper. At the annealing temperatures 100–400 degrees, the Cu/DLC layer possessed a greater hardness and lesser stress. Through thermal annealing, stress between the layers was decreased, resulting in an improved in the adhesion of the layers. The thickness of the DLC/Cu and Cu/DLC bilayers was found to be 108 and 90 nm. Pure C layer was soft, causing a decrease in annealed Cu/DLC hardness possibly due to graphite arrangement. The compressive stresses were seen to be 1.3 and 1.6 GPa, i.e., significantly lower than monolayer DLC film, the hard/soft and soft/hard layer concept was used to reduce the stress; annealing is another such known method [51].

**Dikshit et al.** prepared a 2–11% concentration Cu-DLC film by using the pulsed excimer laser deposition technique, taking the KrF radiation at laser energy density of 2–3 J/cm$^2$ and 50° substrate temperature on Corning (7059) glass and Si (100). Copper was seen in clusters at low concentration, and as it increases, the film acquires a disordered network. Layer size was 5 nm (at 2%) and 85 nm (at 11%). The increase in Cu content modifies the carbon network, inter-band transitions, and the reduction in bandgap [52] (Fig. 12.10).

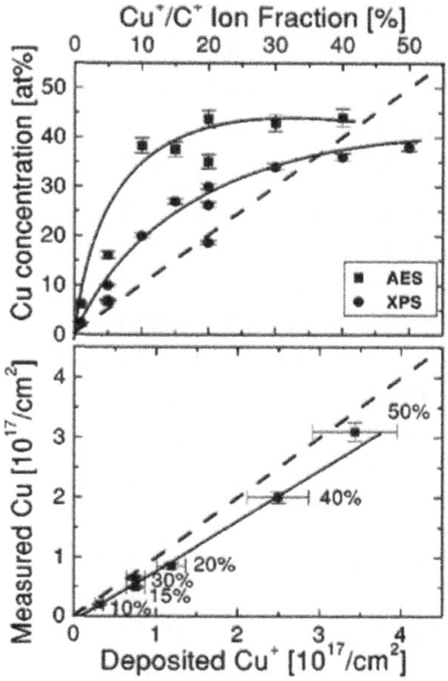


Fig. 12.9 Cu concentration curve: areal density of Cu as the constitution of bulk [50]

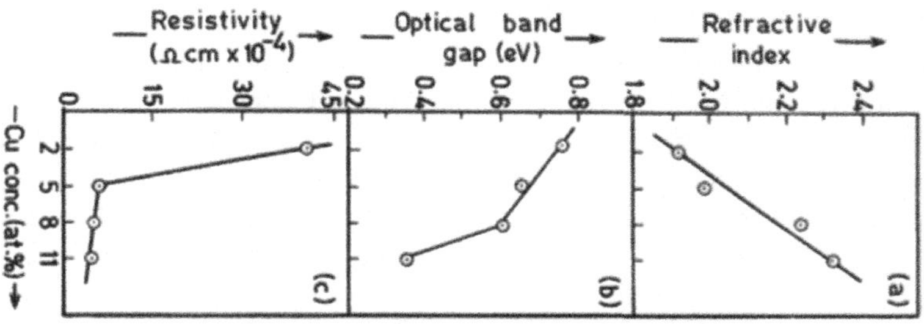


**Fig. 12.10** Variation of **a** refractive index, **b** optical band gap, and **c** electrical resistivity as a function of Cu concentration [52]

## *Silver (Ag) Incorporated DLC*

**Choi et al.** prepared Ag/DLC film on Si substrate using an end-Hall-typo hydrocarbon gun and DC magnetron sputter source for Ag. Interfacial surface tension between two planes was calculated by Young's equation. It was observed that under 89% Ar, Ag incorporation wasn't detected. In distilled water, with an increase in concentration, the contact angle hiked up from 89° to 99°. Ag/DLC was observed to be more hydrophobic than a DLC thin film. There was a 33–13 dynes/cm decrease in the total surface energy. Film is sensitive to hemo-compatibility, which refers is a measure of the thrombotic response induced by a material. Potential bio-material applications were explored [53].

**Marciano et al.** Obtained Ag-DLC films by plasma-enhanced chemical vapor deposition with a thin Si interlayer to find out the potential anti-bacterial properties. There3 was no substantial change observed due to incorporation. There was a slight decrease in hardness, i.e., 19.0–20.0 GPa for DLC thin film to 11.0–16.0 GPa for A-DLC, this is in turn compensated by a reduction in total compressive stress [54].

**Dwivedi et al.** proposed a simple approach for obtaining diamond-like carbon (DLC) film by sandwiching a silver layer between DLC and N-DLC films with low threshold electron field emission from a larger area (Fig. 12.11).

The root-mean-square roughness (Rq) was estimated at 0.12 nm, which seems like mirror-like smooth finish. For a different sample, measured Rq was 0.16 nm and in peak region was seen to increase up to 1.5 nm. The analysis made for roughness showed that the silver introduction increased the overall surface roughness. Ag-modified films were found to be potential candidates for field emission devices because of their exceptional field emission properties [55].

**Chen et al.** fabricated bio-ceramic diamond-like C/Ag films with the help of radio-frequency magnetron sputtering in an Ar/$CH_4$ atmosphere. Index od disorder degree initially decreases with the addition of Ag. The surface roughness was higher for more Ag and a larger nanocluster size. Above 200 W, excess Ag becomes a visible granular feature; such an increase in roughness changes other tribological, optical, and biomedical properties [56] (Fig. 12.12).

**Mazare et al.** presented a micro-structural, and corrosion-resistant film by Thermionic Vacuum Arc of Ag-diamond-like carbon. Contact angles were measured which depicted a rise in the wettability, which was because of the rise in the surface free energy. The resistance to corrosion was compared uncoated titanium which gave an increase in electrochemical stability. A wide and varying range of antibacterial activity as well as corrosion resistance was established. Plain Ti surface showed a relatively lower roughness: the average roughness, Ra was 1.66 nm and the root mean square roughness, Rq was 2.24 nm. But a nonhomogeneous, rough surface was observed. The area roughness increased by increasing the Ag ion concentration in the coating [57].

**Baba et al.** investigated the doping effect of silver on the various properties and the structure of DLC films. A combined process of acetylene plasma source ion

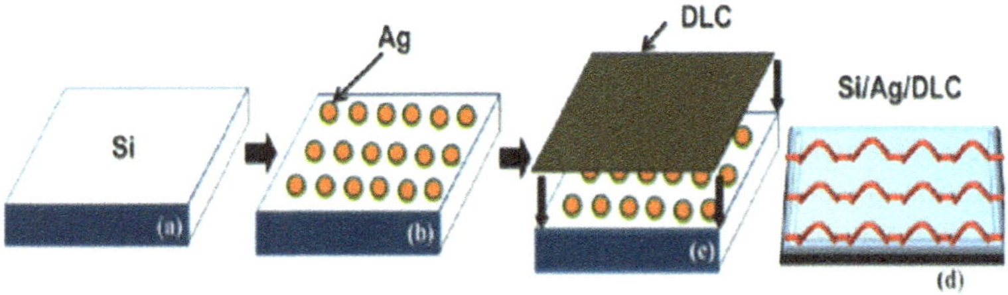


**Fig. 12.11** Schematic representation DLC films deposition containing Ag dots [55]

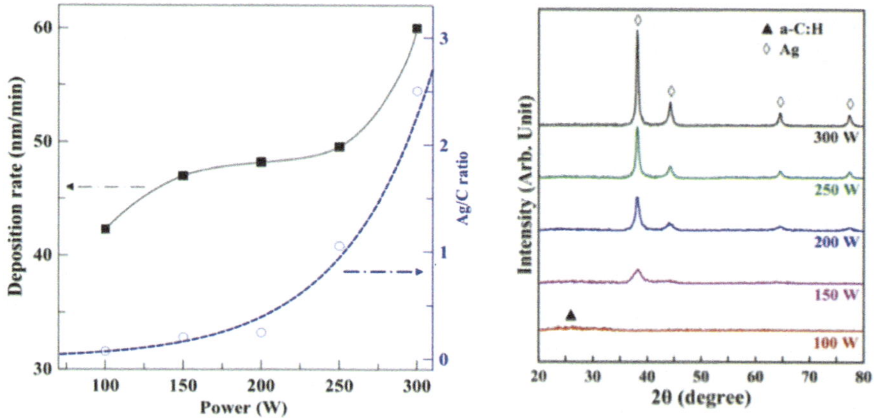


**Fig. 12.12** Dependence of deposition rate and Ag/C versus target plot, and the XRD patterns [56]

implantation and reactive magnetron sputtering of Agdisc was used to produce said films. The tribological properties were largely influenced by Ag concentration. Using scanning electron microscopy, surface roughness rose with silver concentration. With the increase in silver doping, the tribological properties of the films were seen to improve. Friction coefficient was around 0.06, and increased up to 0.28, for relatively low silver content. Wear property was found to be the best at 1.8% Ag-DLC films, a definite improvement from Ag-free films, even the surface roughness value rose with the addition of silver content in the films [58].

**Choi et al.** synthesized an Ag-implanted DLC film on Si wafer with the help of hybrid deposition: end-Hall-type hydrocarbon ion gun and a magnetron sputter source, this film was recommended for biomedical applications. Below 0.85 Ar concentration, the incorporation of Ag was undetectable. There was a noticeable reduction in residual stress in DLC films with the Ag incorporation without much degradation in the hardness; compressive residual stress reduces from 2.9 to 1.6 GPa with the Ag concentration changing from 0.1 to 1.7% [59]. (Fig. 12.13).

**Wu et al.** prepared Ag-DLC films with varying concentrations by medium frequency magnetron sputtering. 3.55% concentration contained a granular structure. Soften effect of Ag deteriorates mechanical properties, there was a noticeable decrease in hardness with the addition of Ag. At low concentrations, the addition of Ag improves the wear-resistance but for higher concentrations, a trend of worn surface increased with frictional losses [60] (Fig. 12.14).

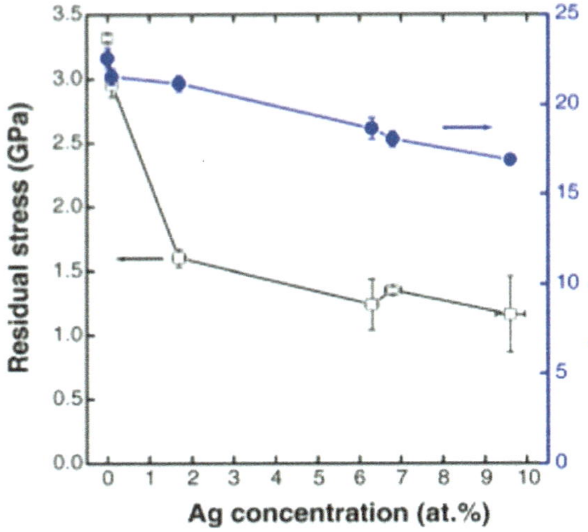


**Fig. 12.13** Change in the residual stress and hardness of Ag/DLC films with varying Ag concentrations [59]

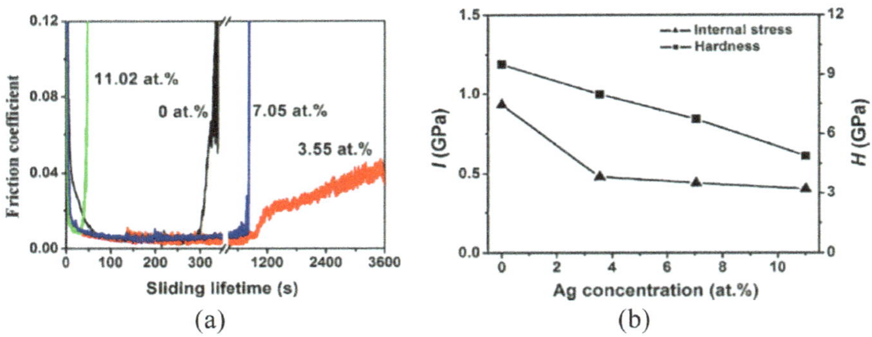


**Fig. 12.14** **a**Tribological performance in comparison to the sliding time, **b** the internal stress and hardness in comparison to Ag concentrations

## *Molybdenum (Mo) Incorporated DLC*

**Ji et al.** deposited a Mo/DLC film on stainless steel and Si wafer using unbalanced magnetron sputtering of Mo with radio frequency plasma vapor deposition (RFPVD) inductively coupled to analyze the effects of the sputtering currents and doping on the mechanical properties and microstructure. A minimal value of stress was found at 1 GPa at a 2 A sputtering current, exhibiting the tiniest sized grain and the maximum critical load. The hardness value was 10–11 GPa, and also the elastic modulus, which was found to be lower than that of DLC film (15 GPa) due to the formation of $sp^2$ sights with Mo atoms acting as the catalyst for the said reaction [61].

**Ji et al.** was set down a Molybdenum doped DLC film on Si substrates through unbalanced magnetron sputtering of Mo and plasma chemical vapor deposition (PCVD) of $CH_4$/Ar to compare surface various properties and structural composition with DLC films, combined with the effect of atomic oxygen (AO) irradiation. Internal compressive stresses in DLC film decreased drastically (5.5–0.7 GPa) compared to Mo/DLC film (1.4–1.2 GPa). The surface of AO radiated DLC film turns rougher, with a mean surface roughness of 0.977 after 30-min of irradiation. The wear scar grew deeper with the frictional coefficient and rate of wear increasing for the DLC film, while they stayed invariable for the Mo/DLC film; the experiment proved that the Mo deposited film demonstrated much better AO-resistance, which is a prerequisite for space applications; this is due to the formation of $MoO_3$ protective cladding on the surface [62].

**Huang et al.** investigated Mo-containing diamond-like carbon films settled by electron cyclotron resonance chemical vapor deposition (ECR-CVD). Permittivity of the film was seen to be increased by Mo introduction, with a peak decrease at lower frequency values. For a high frequency (1 MHz), the permittivity is slightly dependant on Mo below 145 K; also the relaxation time reduces with the rise in the Mo content. Relaxation times of the films and their distribution were also provided [63] (Fig. 12.15).

**Mandal et al.** made carbon-based coating contaminated with Mo and W using HIPIMS to be used in the automotive industry by overcoming the scant adhesion strength and meagre thermal stability that usually restricts its tribological applications. Coefficient of friction was found to be 0.335 for steel and wear resistance was, Kc = 3.87 × 10–16 $m^3\ N^{-1}\ m^{-1}$ after a 7.5 km distance. The wear mechanism of the Mo–W–C coating was observed to be oxidative, transferred layer consisted of two oxides. The Nano hardness and elastic modulus were estimated from the characteristic loading–unloading curves, we observed that the coating remains intact even after sliding. Tribological rendition of the Mo–W–C coating at ambient condition

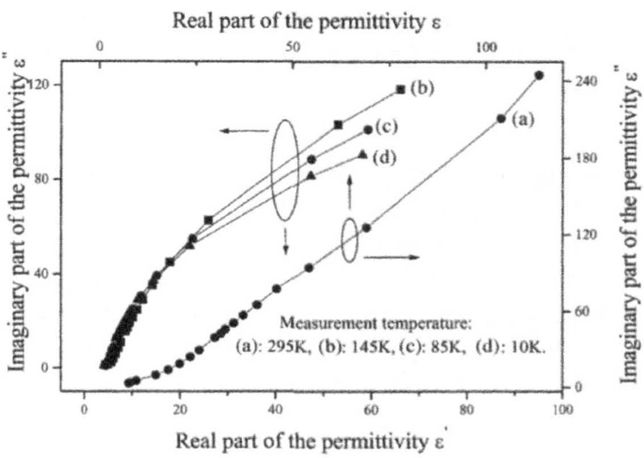


**Fig. 12.15** Plot of the imaginary part versus real part of the complex permittivity [63]

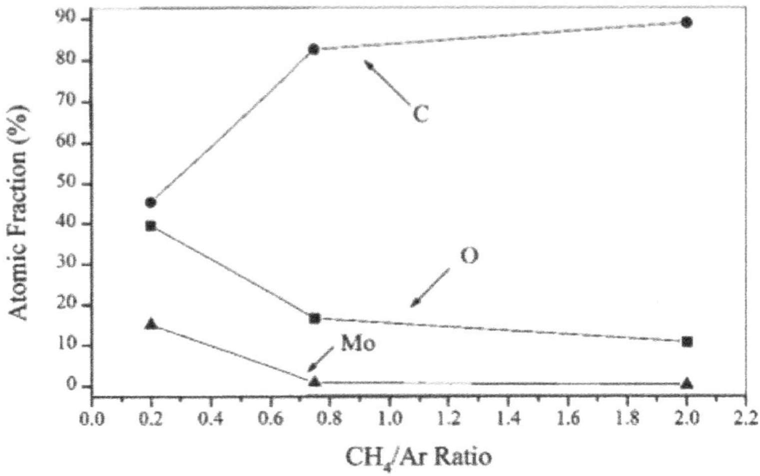


**Fig. 12.16** Change in atomic fraction of C, O, and Mo versus the air flow [65]

was found to be excellent. High critical loads (up to 80.8 N) were seen to be withstood before the substrate is rendered vulnerable because of adhesive failure [64].

**Yoon et al.** analyzed x-ray photoelectron spectroscopy (XPS) results for Mo-doped carbon films deposited using electron cyclotron resonance chemical vapor deposition (ECR-CVD). Variation of the Argon/methane flow ratio causes a change of the Mo concentration remains within the range of 0.32–15.11%. Deep threshold was observed, reactions involving cross-linking using ions and free radicals, which resulted in a tighter C-network. Highest Mo atomic fraction had 4 components whose binding energies were 283.05, 284.67, 286.22, and 288.17 eV, respectively [65] (Fig. 12.16).

**Butt et al.** made the use of off-axis pulsed laser deposition technique at room temperature to deposit a diamond-like carbon thin film and multilayer DLC:Mo films of various compositions on p-type Si substrate. Mo crystallite size and film thickness were seen to grow whereas micro-strain decreased with a rise in Molybdenum content [66] (Fig. 12.17).

**Constantinou et al.** used pulsed laser deposition to grow DLC and molybdenum-doped DLC up to 3.2% to test on Silicon substrates. The $sp^3$ hybridization inside the carbon was reduced upon doping. The DLC:Mo films display amplified nano-scratch resistance. The residual stresses were decreased with the increase in Mo concentration which can be related to the strain energy release. The compressive stress reduced close to 3 GPa for 3.2% of Mo, there was a +87% amplification in the desired load to delaminate the film [67] (Fig. 12.18).

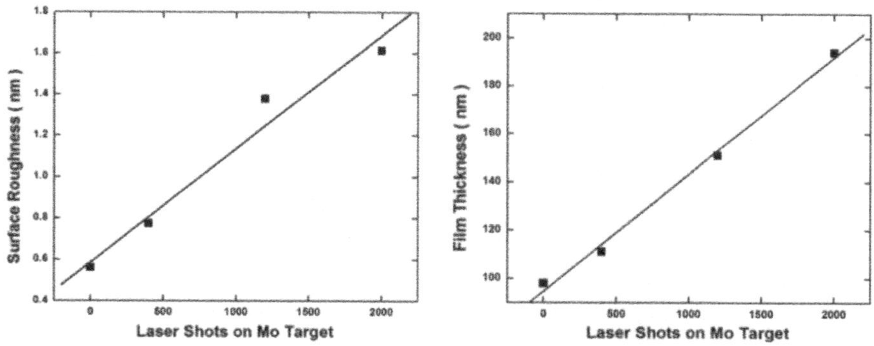


**Fig. 12.17** Surface roughness and film thickness plots with respect to Mo laser shots [66]

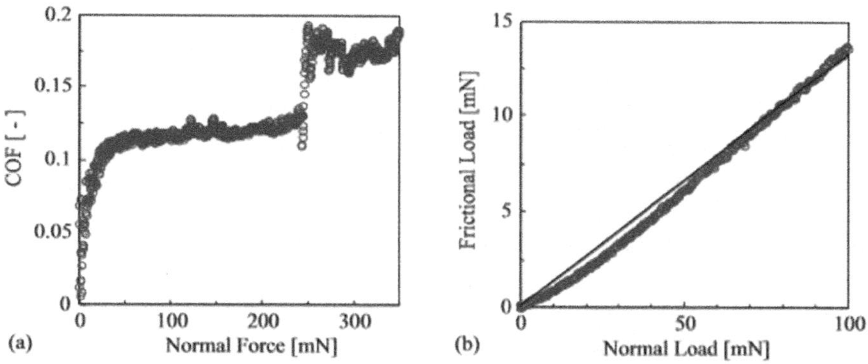


**Fig. 12.18** **a** At the time of the scratch test, coefficient of friction for the many normal loads employed, and **b** Frictional load instigated for the employed normal forces in the 0–100 mN range

## *Chromium (Cr) Incorporated DLC*

**Dai et al.** put on a small amount of Cr-DLC films on Si wafers by a hybrid deposition system. Concentration was controlled by air flow radius. Nano-indenter and ball-on-disc tribometer were used to measure up different mechanical and tribological properties. With Cr less than 0.3%, the films showed a low friction coefficient and a good wear resistance. Elastic modulus and hardness were seen to be decreasing with the increase of flux ratio. Even for a value of residual stress as low as 0.46 GPa, hardness value is still pretty high [68] (Fig. 12.19).

**Chiu et al.** synthesized Cr-doped DLC film using cathodic arc evaporation (CAE), with the film thickness controlled at 1.65 μm. In particular, the thermal oxidation behavior was investigated; due to thermal decomposition weight loss of film occurred at 290°–342°. TGA/DTA tests were mediated at a rate 10 °C/min for 25°–800°, the

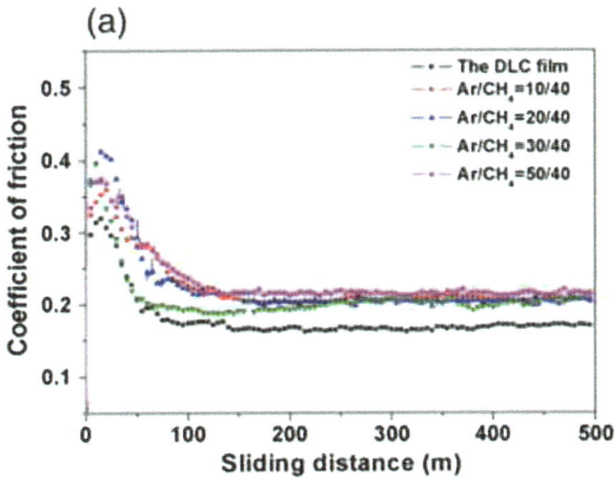


**Fig. 12.19** Coefficient of friction with respect to sliding friction [68]

Cr-DLC structures deteriorated up to 300° in ambient air. Binding energy shifted from 532.0 to 530.2 eV for annealed film Cr-doped was characterized at higher temperatures revealing a distinct decomposition [69] (Fig. 12.20).

**Singh et al.** made nanocomposite films containing Cr (varying up to 18%) with plasma-enhanced hybrid chemical and physical vapor deposition in an $Ar/CH_4$ mixture. Nanoindentation and pin-on-disc experiments were used to analyze the

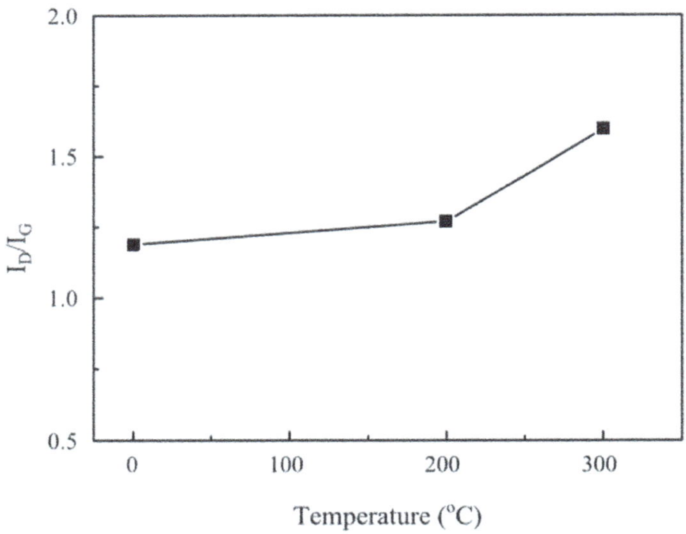


**Fig. 12.20** $I_D/I_G$ dependence on annealing temperatures [69]

mechanical and tribological properties of the films. There was a reduction in hardness with Cr content increase, for 1.5% Cr, the hardness was around 13.5 GPa. A relatively stable modulus of elasticity was observed. Variations in tribological behaviors were assessed with H/E variation. Low (<5%) concentration Cr-doped DLC films had similarities with tribological properties as pure DLC films, but films with >12%. For concentration less than 4.8%, coefficient of fraction was <0.14 and the order of wear rate were in the order of $10^{-7}$ $mm^3/N$ [70].

**Singh et al.** synthesized a hydrogenated Cr coated DLC film using hybrid PVD/CVD plasma-assisted deposition. For Cr content less than 12%, Cr/DLC was found to have excellent tribological properties. The local structure of around Cr content was studied with respect to Cr content. For higher concentrations, i.e., >1.5%, Cr nanoparticle size was less than 10 nm. The Cr solubility limit in Cr-DLC was found to be 0.4–1.5% of Cr [71].

**Fan et al.** produced Cr-doped diamond-like carbon films to use as an electrochemical nanoelectrode. Transmission electron microscopy was used to study the Z-contrast during the metal cluster formation within the Cr/DLC. For low doping <0.6% Cr, where the metal is seen to be evenly distributed in the matrix; and at the higher doping, i.e., 12%, for Cr, clusters are formed. It was found that Cr tends to be metal-like when it is distributed evenly C matrix. Low doping and more "C" at higher doping, which meant carbon is more like a diamond at lower doping [72].

**Wang et al.** deposited a diamond-like carbon film using a cathodic arc evaporation process. A 3824 Hv (which is 25 g) micro-hardness DLC films carbon was made intensive ion energy energetic Cr plasma. Cr-inclusion aids in the rise of film density and the stress release. An adhesion strength of 55 N was observed, with early chipping was seen at the start of scratch test; this damage is providing a low friction resistance. In contrast, a good wear performance was seen (3000 m of wear endurance <10 N at 0.3 m/s) [73] (Fig. 12.21).

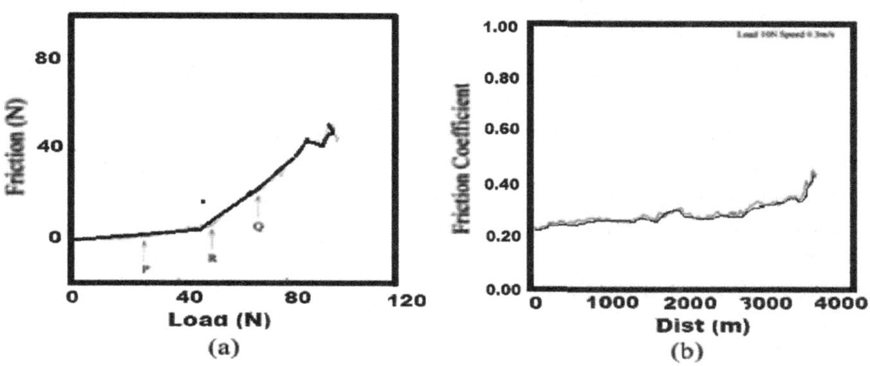


**Fig. 12.21** **a** Scratch performance before and after adhesive failure load, and **b** frictional coefficient versus the distance

**Table 12.3** Calculated internal stress for different concentrations using *G* peak shift [75]

| Samples | *G* peak position $\omega$ ($cm^{-1}$) | $I_D/I_G$ | FWHM ($cm^{-1}$) | Free standing film (reference) $\omega_0$ ($cm^{-1}$) | $\Delta\omega = \omega_0 - \omega$ ($cm^{-1}$) | Internal stress reduction $\sigma$ (GPa) |
|---|---|---|---|---|---|---|
| Pure DLC | 1555 | 0.53 | 218 | 1542 | | |
| DLC-Cr 5 at.% | 1527 | 0.37 | 201 | 1542 | 15 | 3 |
| DLC-Cr 10 at.% | 1529 | 0.43 | 193 | 1542 | 13 | 2 |
| DLC-Cr 15 at.% | 1534 | 0.48 | 185 | 1542 | 8 | 2 |
| DLC-Cr 20 at.% | 1539 | 0.49 | 179 | 1542 | 3 | 1 |

**Bujak et al.** prepared multilayer Cr/CrN/DLC coating by using filtered cathodic arc deposition (FCAD) to investigate pulsed high-voltage bias. Mechanical properties were influenced by substrate bias voltage by the changing chemical composition. For a 2 to 6 kV increase in bias voltage, Cr increases from 2 to 4% and the ratio of $sp^2/sp^3$ reduces from 1.06 to 0.44. A smooth surface was observed with a moderate hardness (8–11 GPa), and relatively low friction coefficient (0.13–0.15). Young's modulus was observed at 95–130 GPa. The wear resistance was because of intermediate adhesion strength. To improve the properties, parameters of the PVD process were changed [74].

**Gayathri et al.** produced amorphous DLC films with various concentrations for Cr at 5, 10, 15, and 20% using Ng-YAG pulsed laser deposition technique. Lower concentration Cr-DLC (i.e., 5%) reduced the internal stress up to 3 GPa. Frictional coefficient was as low as 0.04–0.09 for low Cr-content, which was increasingly higher for increased Cr coefficients. For adhesion and friction, the optimum level was found to be 10–15% Cr in DLC [75] (Table 12.3).

**Santana et al.** investigated structural and magnetic properties for Cr concentrations 0.1–20%. The carbide-hydrogenated and simple-hydrogenated Cr-DLC alloys were produced by plasma-assisted vapor deposition. At low temperatures, low Cr concentrations were seen to be ferromagnetic but the chromium carbide appeared to be antiferromagnetic accompanied by voluntary surface spinning (Fig. 12.22).

The negative magnetoresistance, saturating at 500 Oe, shown in the *I–V* curves, attributed to the spins at the Cr–C clusters surface. The material was seen to be suitable for spin-electronics applications [76].

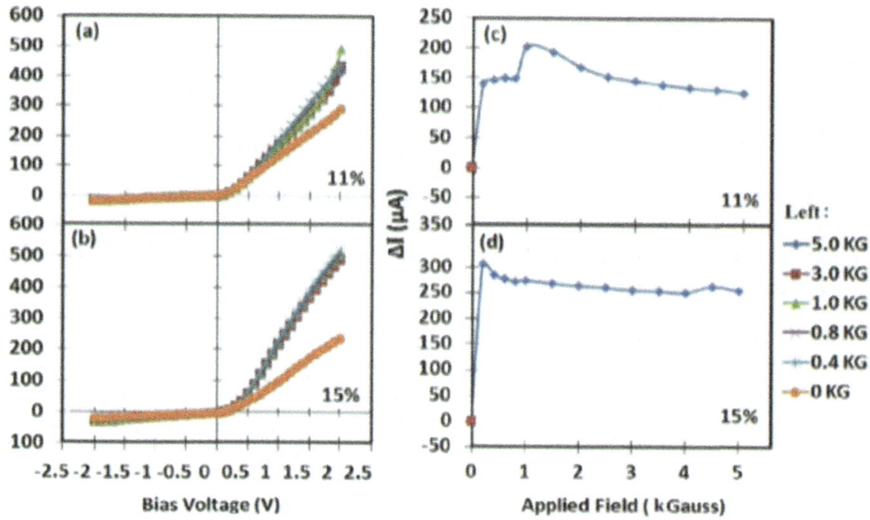


Fig. 12.22 The *I*–*V* curves for 11% Cr and 15% Cr [76]

## *Titanium (Ti) Incorporated DLC*

**Lin et al.** presented a simple model for synthesizing Ti-DLC films on Si substrates using filtered cathodic vacuum arc method (FCAV) to analyze the properties such as microstructure, mechanical strength, surface roughness. Ti-DLC multilayer exhibited better mechanical properties with very similar atomic concentrations as a Ti-implanted films, the hardness and electrical resistivity values of the multilayer show an increase, and in comparison the surface roughness exhibited a much lower value: Root mean square roughness of the DLC and Ti-DLC films were measured at 0.917 and 0.107 nm, Ti doing improves the smoothness; hardness of the multi-layered film increased to 57 from 36 GPa [77].

**Voevodin et al.** developed a hard and lubricious diamond-like carbon surface by a hybrid of magnetron sputtering and pulsed laser deposition to further analyze the wear properties and phase transitions between Ti–TiC–DLC. This was done with the intention of designing a wear-resistant coating. Maximum hardness 30 GPa was found in the TiC/aC region, i.e., 70% C; the trend for hardness is an increase from the self-lubricating steel substrate to 60–70 GPa and hard DLC layer at surface. Coating at the surface has a low friction value maintaining a friction coefficient below 0.1 without cracking related failures [78].

**Cui et al.** deposited a Ti-doped DLC film by the magnetron sputtering at atmospheric temperature Ti in Ar/$CH_4$ environment on Si substrates. For an air flow ratio of 60/190 and doping concentration 0.41%, a hard and low stress film was formed; an increase in air flow ratio caused decrease in hardness and an increase in stress. Wear rate was around 10–3 $mm^3/Nm$ and the friction coefficient was about 0.05 [79] (Figs. 12.23 and 12.24).

**Zhao et al.** prepared a Ti-doped DLC thin film by radio-frequency discharge and co-sputtering of Titanium targets with Oxygen. These films exhibited steady friction coefficients of about 0.008, exhibiting some sensitivity to relative humidity. Elastic modulus and nano hardness is 105.2 + −0.76 GPa and 12.9 + −0.07, respectively, which is higher than the film without doping. Shear strength in ambient air was around 10 MPa. Due to inherent physical properties, and transformations, the film showed excellent lubrication [80] (Fig. 12.25).

**Quiag et al.** produced Ti-doped DLC thin films on Si wafers by magnetron sputtering titanium twin-targets in a controlled $Ar/CH_4$ feed (0.2–0.42%). Low Ti content led to a significantly lower internal stress value with a slight decrease in hardness but it is still maintained at approximately 10 GPa, no further change was observed as the value of Ti was varied. Low content Ti was seen to improve wear resistance and

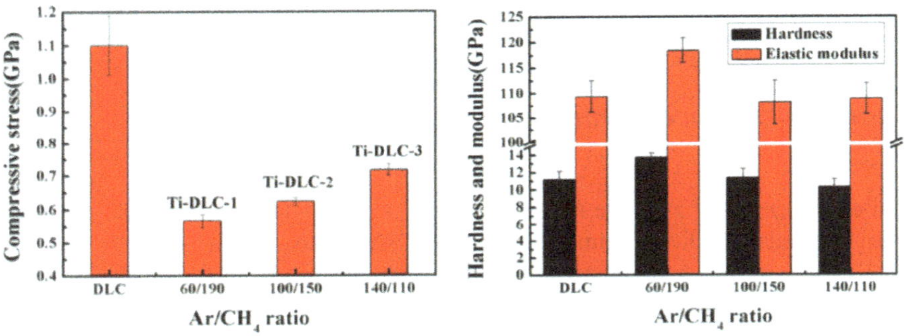


**Fig. 12.23** The compressive stress, hardness, and elastic modulus as a function of the air flow ratio [79]

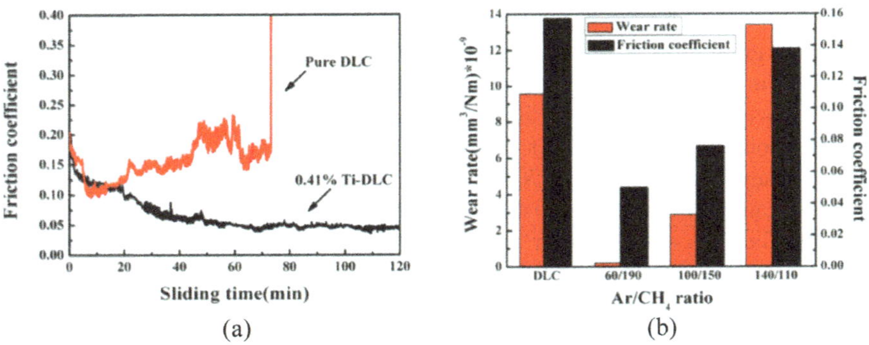


**Fig. 12.24** **a** Comparisons of frictional conduct for smaller concentration Ti-DLC and DLC films, **b** frictional coefficients and wear rates as related to Argon/Methane ratio [79]

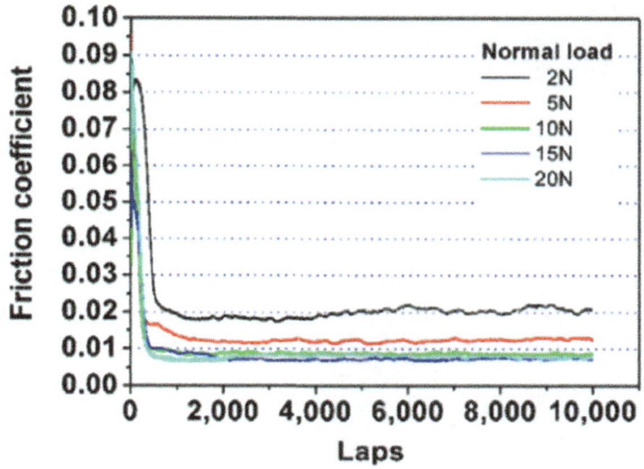


**Fig. 12.25** Frictional coefficient of Ti-DLC versus normal load [80]

coefficient of friction up to $10^{-8}$ $mm^3/N$ and 0.04, respectively. The amount of Ti doping was found to not be a major factor in the film performance so as the amount of doping is kept to a low value [81].

**Yang et al.** made a Ti-containing diamond-like carbon film using plasma decomposition with the introduction of tetrakis (dimethylamino) titanium (TDMAT, $Ti[(CH_3)2\ N]4$), which provided the titanium. Hardness value, along with other mechanical properties, was enhanced up to a substrate temperature of 350°. Increasing TiN volume fraction in the DLC matrix improved mechanical properties [82] (Fig. 12.26).

**Bharathy et al.** incorporated titanium into a DLC thin film and compared them to a DLC film using reactive-biased target ion beam deposition method. When the bias voltage was increase, there was an increase in the Ti content seen. At about 4.2%, Ti was seen to be almost uniformly distributed, with further exposure of the surface Ti was seen to be bonding with the atmospheric oxygen to get weak Ti–O bonding. Hardness value reduces with an rise in indentation depth [83] (Fig. 12.27).

## *Nickel (Ni) Incorporated DLC*

**Paul et al.** impregnated a diamond-like carbon film with nanocrystalline nickel and deposited the Ni-DLC film on glass substrate with the help of Plasma Chemical Vapor Deposition. Ni content was regulated by changing $Ar/CH_4$ content. The matrix became increasingly graphitic with an increase in Ni content, and with the increase in temperature, there was a decrease in magnetization. In the films containing 0.69% nickel, the coercive fields' difference increased significantly [84] (Fig. 12.28).

Fig. 12.26 The trends for the hardness and elastic modulus [82]

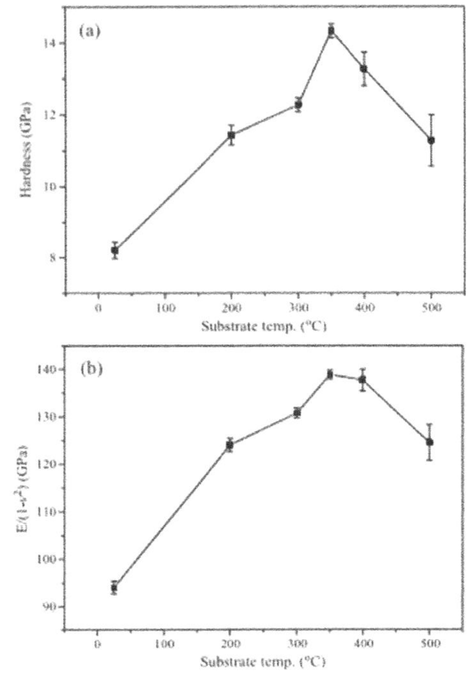


Fig. 12.27 Hardness trend with respect to indentation depth [84]

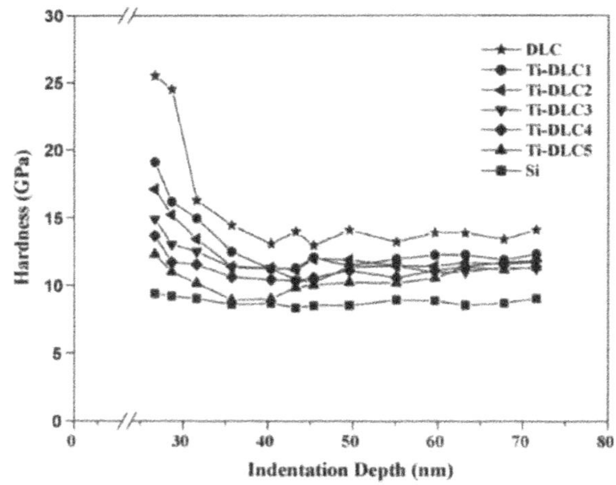


**Pandey et al.** synthesized DLC and Ni-DLC using electrodeposition, keeping the initial concentration was 4.76 × 10–4 M, the band gap energy was varied between 2.67 and 2.48 eV. Results showed that there was an increase in graphitization of DLC films with metal incorporation, which was evenly distributed and densely packed. Uniform and smooth films were produced, though the deposition directly deposits

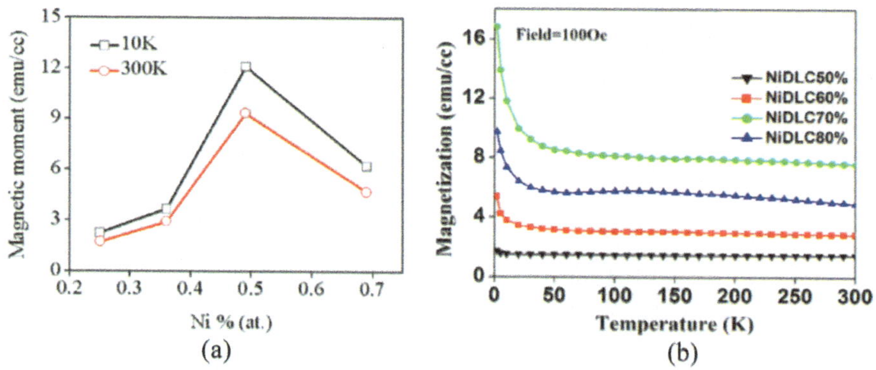


**Fig. 12.28** **a** Magnetic moment with respect to Ni%, and **b** magnetization with respect to temperature [84]

**Table 12.4** Results of the scaling analysis for roughness [85]

| Sample | $\omega$ (nm) | $L_\omega$ (nm) | $L_c$ (nm) | | $H$ | | $D$ | |
|---|---|---|---|---|---|---|---|---|
| | | | $L_{ca}$ | $L_{cp}$ | $H_\omega$ | $H_p$ | $D_\omega$ | $D_p$ |
| DLC 2 min | 4.5 | 555 | 70.8 | 70.9 | 0.67 | 0.69 | 1.33 | 1.31 |
| Ni-DLC 7 min | 8.4 | 500 | 126.9 | 127 | 0.79 | 0.86 | 1.21 | 1.14 |
| Ni-DLC 50 min | 3.4 | 425 | 62.1 | 62.2 | 0.62 | 0.72 | 1.38 | 1.28 |

where [31–34], $\omega$ = RMS roughness, $L_\omega$ = roughness saturation length, $L_{ca}$ = correlation length from ACF, $L_{cp}$ = correlation length from PCF, $H_\omega$ = roughness exponent from RMS roughness, $H_p$ = roughness exponent from PCF, $D_\omega$ = fractal dimension from RMS roughness, and $D_p$ = fractal dimension from PCF

the average grain size. Maximum transmittance was 78% but for Ni-DLC was 63%, this was correlated to spectrophotometry observations [85] (Table 12.4).

**Pandey et al.** deposited DLC films on ITO glass by incorporating Ni nanoparticles by low voltage electrodeposition method. With Ni addition, an increase in conductivity was observed, and continues so even for the dilution of Ni. The Ni is also responsible for the ferromagnetic contribution. Up to 300 K temperature, the ferromagnetic, superparamagnetic and paramagnetic features were exhibited [86].

**Mohammadinia et al.** incorporated nickel onto diamond-like carbon thin films using RF-Plasma enhanced chemical vapor deposition and RF Sputtering. Particle size and Ni content was controlled by adjusting applied power. Applied voltage was tested at different values, the Ni-proportion of the thin films was seen to increase for a higher voltage. The band gap energy was seen to be decreasing from 4.58 to 3.66 eV with a rise in applied power, and the Urbach energy decreases from 0.472 to 0.189 eV. Transmittance and reflection spectra were used to obtain the optical properties for the given Ni-DLC layers [87] (Fig. 12.29).

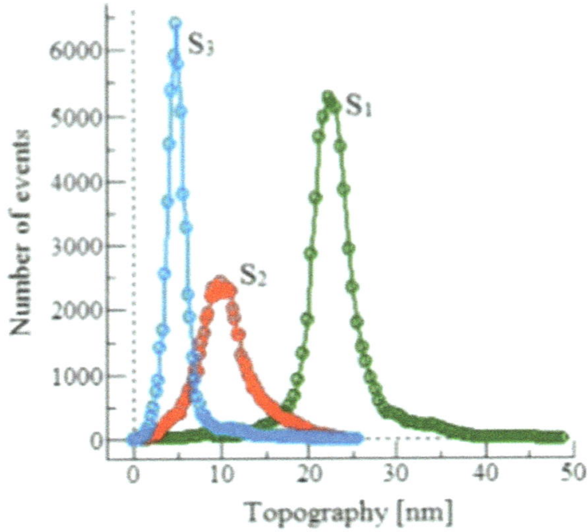


**Fig. 12.29** Distribution of the nanoparticles in the film [87]

**Lui et al.** produced a Ni-doped DLC thin film at ambient temperature using an unbalanced magnetron radio-frequency co-sputtering process for the detection of non-enzymatic glucose. Large electrochemical potential was exhibited around 2.12 V in 0.1 M $H_2SO_4$ electrolyte, which was noticed to have a drastically lower value for higher Ni doping. The Ni distribution was largely uniform with a fraction ranging from 3 to 11.5%. DLC thin film made with 7 W sputtering power with Ni-doping possessed up to a 10 mM sensitivity to glucose [88].

**Ma et al.** used an electro-deposition process to produce Ni-containing diamond-like carbon film. Friction and wear behavior was analyzed using a universal tribometer. Nickel incorporation reduced bonding strength and residual stress, which increased wear resistance [89] (Figs. 12.30 and 12.31).

**Benchikh et al.** fabricated metal embedded DLC matrix, alternatively ablating between graphite (99.997%) and nickel (purity 99.9%) using femtosecond laser deposition. Deposited Ni nanocrystal size was 100 nm. As an element, nickel is known to be "carbophobic". Even at a temperature as high as 400 degrees, Ni and C mixture consisted of metallic Ni and graphitic carbon, with a small amount of C [90] (Fig. 12.32).

## *Aluminium (Al) Incorporated DLC*

**Dai et al.** deposited Al-containing diamond-like carbon utilizing a hybrid ion beam system with various bias voltages, which had a crucial role on the microstructure of the films. A bias voltage of 100 V resulted in a relatively higher hardness, also the

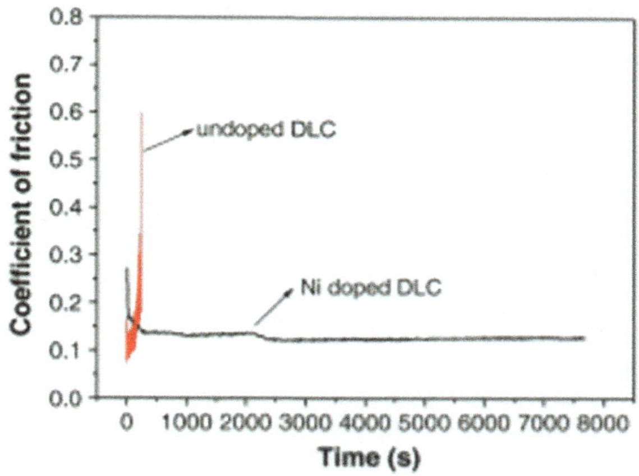


**Fig. 12.30** Variation of coefficient of friction with sliding time for doped and undoped DLC coating [89]

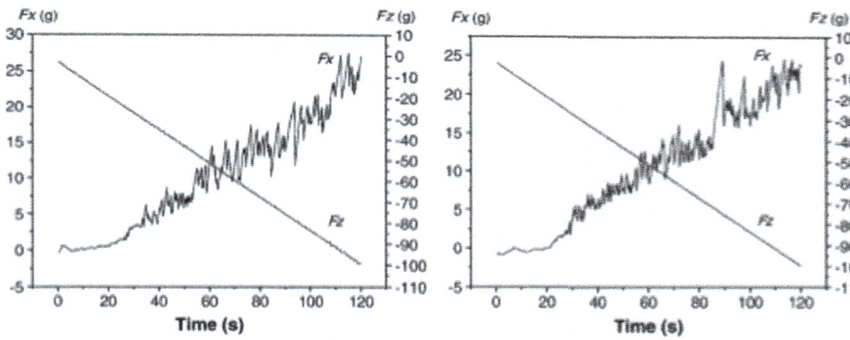


**Fig. 12.31** Variation in frictional forces for the DLC and the Ni-DLC films [89]

lowest wear rate. The wear rate was also corresponded to the mechanical properties. Friction coefficient was found to be around 0.02 (for values of sliding distance greater than 50 m), sporting a good frictional performance along with a good wear resistance, and excellent lubrication [91].

**Dai et al.** incorporated inert Al atoms into diamond-like carbon by a hybrid ion beams system, i.e., an anode-layer ion source and a magnetron sputtering unit. The produced film exhibited excellent friction performance. A lower coefficient of friction was exhibited in all Al-DLC compared to pure films. As concentrations rise from 0.68 to 17.6%, the friction coefficient changed from 0.089 to 0.024, wear rate of film increases monotonously. For 11.04%, a clear deep wear crack was visible. Internal stresses in the films lower with the rise in Al concentration; it falls more rapidly than the hardness in the given range of concentrations [92] (Figs. 12.33 and 12.34).

**Pu et al.** fabricated Al-diamond-like carbon films using the reactive magnetron sputtering >99.9% Al target in an $Ar/CH_4$ atmosphere in negative bias. Hardness level

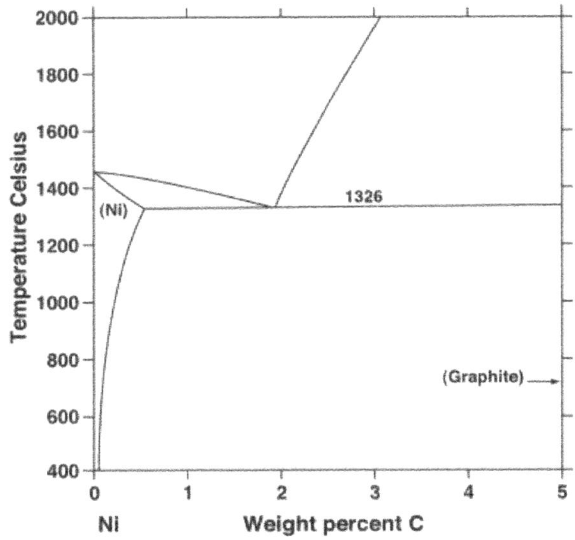


**Fig. 12.32** Thermodynamic Ni-C phase diagram [90]

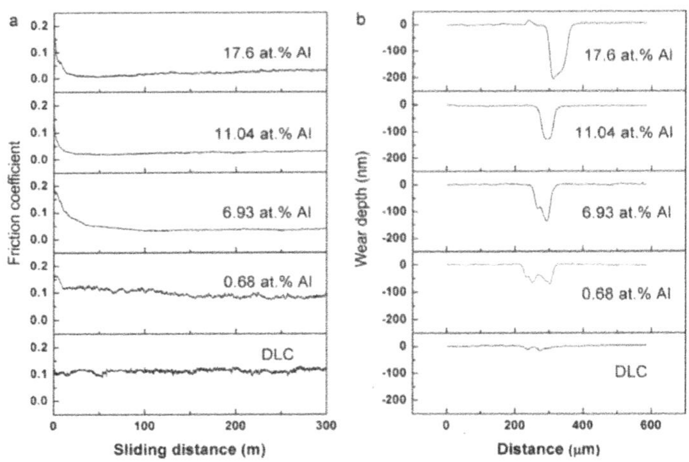


**Fig. 12.33** Frictional coefficient and wear depth as a function of distances in a range of concentrations [92]

was seen to be around 20 GPa, while internal stress decreases drastically with increase in the negative bias. Due to the mechanical properties and transfer layer formation, a prolonged wear life and lesser coefficient of friction was noted at 500 V bias for Al-DLC film; this makes it a good lubricating film [93] (Fig. 12.35).

**Chang et al.** investigated a cathodic arc deposited, TiAl-doped DLC coatings for their microstructural, mechanical properties. During steady-state sliding, a lower

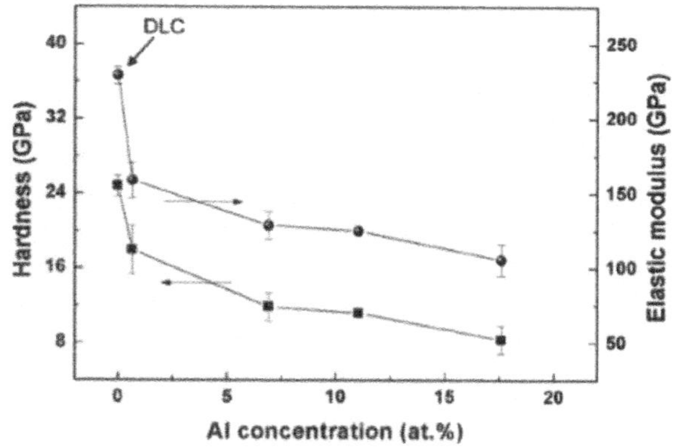


Fig. 12.34 Dependence of elastic modulus and hardness on Al concentrations [92]

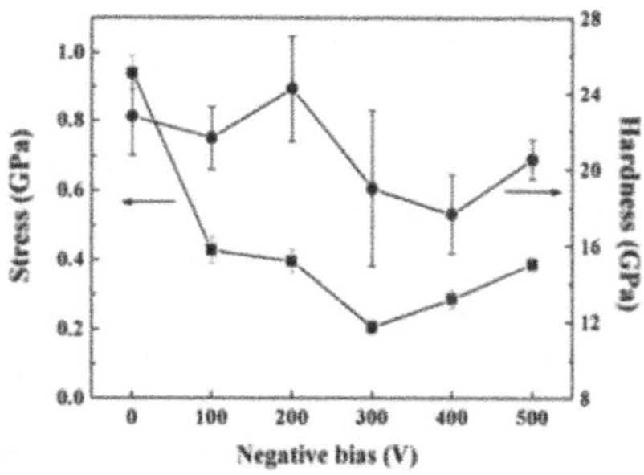


(a)

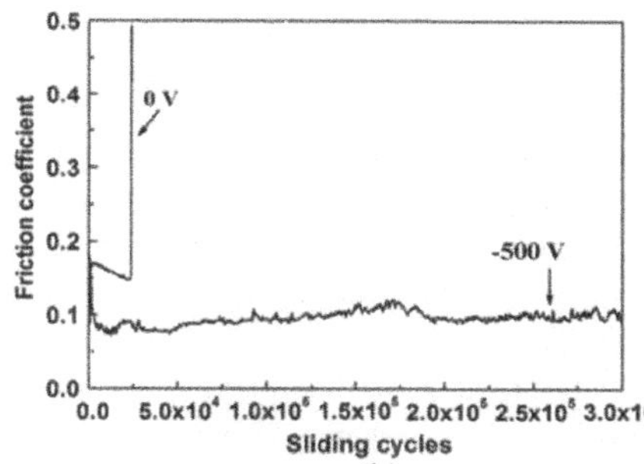


(b)

**Fig. 12.35** **a** Hardness and internal stress for Al-DLC films at varying bias voltages, and **b** the friction coefficient curves at 0 and 500 V bias voltage [93]

frictional coefficient was observed at 0.15. As bias voltage rose to 200 V, bonding energy was shifted to 284.2 eV. Lubricity and wear resistance decrease with the rise in bias voltage, with a consistently low friction coefficient ranging from 0.15 to 0.25 to the end a travel distance of 2 km. With the formation of a graphitized layer, low steady-state friction coefficient correlated to the optimum wear rate [94].

**Guo et al.** produced Ti/Al DLC films controlling the amount of metal and structure of C with the help of a hybrid ion beam system. Concentration and bond state were affected by the process parameters which accounted for the change of hardness, residual compressive stress, and toughness. The hardness and minimal residual stress were 12.1 ± 0.4 GPa and 0.41 ± 0.05 GPa respectively for a 1.5 A current. By augmenting the current further to 2.5 A, synthesized Al was found in both oxidized and uncontaminated clusters. At the same time, the Ti atoms reacted to form hard titanium carbide. At this point, the nominal residual stress was around 1.28 ± 0.1 GPa. The Ti and Al concentration in the films reduced a little then rose, causing constructional relaxation originated by C ions, of which the residual stress

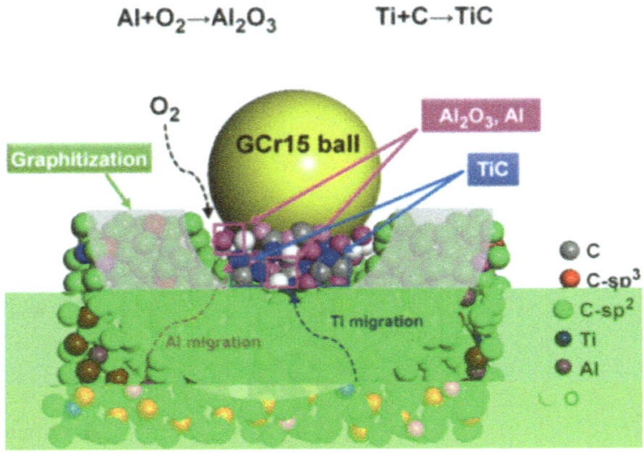


**Fig. 12.36** Diagrammatic depiction of tribological mechanism of Ti/Al-DLC films for the steel ball in ambient conditions [96]

was one of the major consequences; whereas the titanium carbide was responsible for the mechanical properties [95].

**Kong et al.** co-doped Al and Ti into DLC films, these exhibited a low residual stress and good hardness value. Hybrid ion beam system was used for incorporation, and further the tribological properties of films were investigated; doping concentrations were seen to have a vital role in the properties. At 2.5 A the film with th concentration: Ti 10.06% and Al 4.78% was deposited. The lowest coefficient of friction coefficient was about 0.05, with a corresponding wear rate of 1.56 × 10–16 $m^3$ $N^{-1}$ $m^{-1}$. Amorphous C and nano-crystalline formed a cross-linked transfer layer, directly relating to the outstanding tribological properties of the film at 2.5 A [96] (Fig. 12.36).

## Conclusion

Metal incorporated DLC films is now a day an attractive topic for the researcher due to its excellent properties. Many authors have synthesized different metal incorporated DLC film by using different deposition techniques like PECVD, APCVD, HWCVD, Sputtering, etc. It has been observed that by the incorporation of different metal into DLC film the mechanical and tribological properties of the films. It has been also observed that the incorporation of metal like Cu and Ag the residual stress of the film decreases significantly. Some authors have observed that by the incorporation of Nickel into the DLC thin significantly enhances the tribological properties.

## References

1. Dearnaley, G., Arps, J.: Biomedical applications of diamond-like carbon (DLC) coatings: a review. Surf. Coat. Techn. **200**(7), 2518–2524 (2005)

2. Aisenberg, S., Chabot, R.: Appl. Phys. (1971)
3. Michler, T., Grischke, M., Traus, I., Bewilogua, K.: HDLC Dimigen. Diam. Relat. Mater. **7**(2–5), 459–462 (1998)
4. Lee, D.H., He, X.M., Walter, K.C., Nastasi, M., Tesmer, J.R., Tuszewski, M., Tallant, D.R.: Appl. Phys. Lett. **73**(17), 2423–2425 (1998)
5. O'Neill, D.G., Storer, J.G.: US patent 5,401,543 (1995)
6. Takikawa, H., Izumi, K., Miyano, R., Sakakibara, T.: Surf. Coat. Technol. **163**, 368–373 (2003)
7. Ando, J., Suzuki, T., Ogawa, T., Iwai, A., Yoshimura, N., Hashitomi, H.: US Patent, 9,217,196 (2015)
8. Bewilogua, K., Wittorf, R., Thomsen, H., Weber, M.: Thin Solid Films **447**, 142–147 (2004)
9. Sanchez, N.A., Rincon, C., Zambrano, G., Galindo, H., Prieto, P.: Thin Solid Films **373**(1–2) (2000)
10. Krzan, B., et al.: Tribol. Int. **42**(2), 229–235 (2009)
11. Weissmantel, C., Schurer, C., Froehlich, F., Grau, P., Lehmann, H.: Thin Solid Films **61**, L5 (1979)
12. Enke, K., Dimigen, H., Hubsch, H.: Appl. Phys. Lett. **36**, 291 (1980)
13. Abdul Wasy Z.I.A., Zhifeng, Z., Shum, P., Li, L., Yan, K.: Surf. Coat. Technol. **320**, 118–125
14. Lettington, A.H.: Elsevier Science Ltd. **36**, 555–560 (1998)
15. Saikko, V., Altroos, T., Calonius, O.: Keranen. Biomaterials **22**, 1507 (2001)
16. Sheeja, D., Tay, B.K., Nung, L.N.: Surf. Coat. Technol. **190**, 231 (2005)
17. Xu, T., Pruitt, L.: J. Mater Sci. Mater. Med. **10**, 83 (1999)
18. Tsai, H.C., Bogy, D.B.: J. Vac. Sci. Technol. A. **5**(3287), 3312 (1987)
19. Marchon, B., Heiman, N., Khan, M.R., Lautie, A., Ager, J.W., Veirs, D.K.: J. Appl. Phys. **69**, 5748–5750 (1991)
20. Kapoor, V.J., Mirtich, M.J., Banks, B.A.: J. Vac. Sci. Technol. A **4**(3), 1013–1017 (1986)
21. Ball, G.J., Lettington, A.H.: DERA, Malvern Memorandum No. 3617 (1983)
22. Sohn, K., Park, D.H., Yoo, J.G., Kim, J.S.: Luminescence of pulsed laser deposited $Y_2SiO_5$: $Tb_3$ þ thin film phosphors on flat and corrugated quartz glass substrates. Jpn. J. Appl. Phys. **44**(4A), 1787–1791 (2005)
23. Lotin, A.A., Novodvorsky, O.A., Zuev, D.A., et al.: Influence of growth temperature on physical properties of ZnO films produced by pulsed laser deposition method. Opt. Mater. **35**(8), 1564–1570 (2013)
24. Luo, Q., Wang, L.S., Guo, H.Z., et al.: Blue luminescence from Ce-doped ZnO thin films prepared by magnetron sputtering. Appl. Phys. A **108**, 239–245 (2012)
25. Jong, M., Kesteloo, W., Kolk, E.V.D.: Deposition of luminescent NaCl: $Tm^{2+}$ thin films with a Tm concentration gradient using RF magnetron sputtering. Opt. Mater. **46**, 149–153 (2015)
26. Rossnagel, S.M.: Sputter deposition. In: Sproul, W.D., Legg, K.O. (eds.) Opportunities for innovation: advanced surface engineering. TechnomicPublishing Co., Switzerland (1995)
27. Pansila, P., Anun, N., Jamnongkan, T., Chaiyakun, S.: Preparation of pure anatase $TiO_2$ thin film by DC sputtering technique, study on the effect of oxygen partial pressure. Adv. Mater. Res. **463–463**, 1415–1419 (2012)
28. Zhang, et al.: Effect of microstructure and mechanical properties difference between sub- layers on the performance of alternative hard and soft diamond like carbon multilayers films. Surf. Coat. Technol. **232**, 575–581 (2013)
29. Manninen, et al.: Influence of Ag content on mechanical and tribological behaviour of DLC coatings. Surf. Coat. Technol. **232**, 440–446 (2013)
30. Dai, et al.: Microstructure and Mechanical property of diamond like carbon films with ductile copper incorporated. Surf. Coat. Technol. **272**, 33–38 (2015)
31. Huang, R.F., Chan, C.Y., Lee, C.H., Gong, J., Lai, K.H., Lee, C.S., Li, K.Y., Wen, L.S., Sun, C.: Wear-resistant multi-layered diamond-like carbon coating prepared by pulse biased arc ion plating. Diam. Relat. Mater. **10**, 1850–1854 (2001)
32. Zou, et al.: The microstructure and friction properties of protective diamond like carbon films on magnesium alloy. Appl. Surf. Sci. **258**, 1624–1629 (2011)

33. Wong, H., et al.: Improving the conductivity of diamond like carbon films with zinc doping and its material properties. Appl. Surf. Sci. **257**, 9616–9620 (2011)
34. Mittra, J., Abraham, G., Kesaria, M., Bahl, S., Gupta, A., Shivaprasad, S.M., Viswanadham, C.S., Kulkarni, U.D., Dey, G.K.: Role of substrate temperature in the pulsed laser deposition of Zirconium Oxide thin film. Mater. Sci. Forum **710**, 757–761 (2012)
35. Van Arkel, A.E., De Boer, J.H.: Z. Anorg, Altg. Chem. **148**, 345 (1925)
36. Bryant, W.A.: The fundamentals of chemical vapour deposition. J. Mater. Sci. **12**, 1285–1306 (1977)
37. Carey, P.: Plasma Enhanced Chemical Vapour Deposition of Thin Carbon Film. Dublin City University, School of Electronic Engineering (1989)
38. Baghgar, M., Abdi, Y., Arzi, E.: Fabrication of low pressure field ionization gas sensor using bent carbon nanotubes. J. Phys. D Appl. Phys. **42**, 135502 (2009)
39. Ahmed, et al.: Effect of surface structure and Wettability of DLC and N-DLC thin films on adsorption glycine. Appl. Surf. Sci. **258**, 5174 (2012)
40. Venkatesh, M., et al.: Synthesis of Ag-doped hydrogenated carbonth3in films by a hybrid PECVD process, Bull. Mater. Sci. **37**(7), 1669–1676 (2014)
41. Powell, M.J.: $VO_2$ Thin Films and Nanoparticles from Chemical Vapour Deposition and Hydrothermal Synthesis for Energy Efficient Applications. ResearchGate, Sept 2015
42. Davis, K.: Atmospheric Pressure Chemical Vapour Deposition of Functional Oxide Materials for Crystalline Silicon Solar Cells. University of Central Florida (2015)
43. Zhang, J., Gu, Y., Liu, J.: Tribol. Int. **90**, 263–269 (2015)
44. Singh, V., Jiang, J.C., Meletis, E.I.: Cr-diamond like carbon nanocomposite films: synthesis and characterization and properties. Thin Solid Films **489**, 150–158 (2005)
45. Chaus, A.S., Fedosenko, T.N., Rogachev, A.V.: Lubomin Cap Lovic, surface micro structure and optical properties of copper-doped diamond-like carbon coating deposited in pulsed cathodic arc plasma. Diam. Relat. Mater. **42**, 64–70 (2014)
46. Dwivedi, N., Kumar, S., Malik, H.K.: Nanoindentation measurements on modified diamond-like carbon thin films. Appl. Surf. Sci. **257**(23), 9953–9959 (2011)
47. Dwivedi, N., Kumar, S., Malik, H.K., Sreekumar, C., Dayal, S., Rauthan, C.M.S., Panwar, O.S.: Investigation of properties of Cu containing DLC films produced by PECVD process. J. Phys. Chem. Solids **73**(2), 308–316 (2012)
48. Guo, P., Li, X., Sun, L., Chen, R., Ke, P., Wang, A.: Stress reduction mechanism of diamond-like carbon films incorporated with different Cu contents. Thin Solid Films **640**, 45–51 (2017)
49. Tsai, M.Y., Huang, M.S., Chen, L.K., Shen, Y.D., Lin, M.H., Chiang, K.-L. Ou, Y.C., Ou, S.-F.: Surface properties of copper-incorporated diamond-like carbon films deposited by hybrid magnetron sputtering. Ceram. Int. **39**(7), 8335–8340 (2013)
50. Gerhards, I., Ronning, C., Hofsäss, H., Seibt, M., Gibhardt, H.: Ion beam synthesis of diamond-like carbon thin films containing copper nanocrystals. J. Appl. Phys. **93**(2), 1203–1207 (2003)
51. Zhou, B., Rogachev, A.V., Liu, Z., Jiang, X., Shen, R., Rudenkov, A.S.: Structure and mechanical properties of diamond-like carbon films with copper functional layer by cathode arc evaporation. Surf. Coat. Technol. **208**, 101–108 (2012)
52. Dikshit, S.J., Lele, P., Ogale, S.B., Kshirsagar, S.T.: Influence of in-process copper incorporation on the quality of diamond-like carbon films deposited by pulsed laser deposition technique. J. Mater. Res. **11**(9), 2236–2241 (1996)
53. Choi, H.W., Dauskardt, R.H., Lee, S.C., Lee, K.R., Oh, K.H.: Characteristic of silver doped DLC films on surface properties and protein adsorption. Diam. Relat. Mater. **17**(3), 252–257 (2008)
54. Marciano, F.R., Bonetti, L.F., Santos, L.V., Da-Silva, N.S., Corat, E.J., Trava-Airoldi, V.J.: Antibacterial activity of DLC and Ag–DLC films produced by PECVD technique. Diam. Relat. Mater. **18**(5–8), 1010–1014 (2009)
55. Dwivedi, N., Kumar, S., Carey, J.D., Tripathi, R.K., Malik, H.K., Dalai, M.K.: Influence of silver incorporation on the structural and electrical properties of diamond-like carbon thin films. ACS Appl. Mater. Interfaces. **5**(7), 2725–2732 (2013)

56. Chen, S.-Y., Ou, K.-L., Huang, W.-C., Chu, K.-T., Ou, S.-F.: Phase transformation of diamond-like carbon/silver composite films by sputtering deposition. Ceram. Int. **39**(3), 2575–2580 (2013)
57. Mazare, A., Anghel, A., Surdu-Bob, C., Totea, G., Demetrescu, I., Ionita, D.: Silver doped diamond-like carbon antibacterial and corrosion resistance coatings on titanium. Thin Solid Films **657**, 16–23 (2018)
58. Baba, K., Hatada, R., Flege, S., Ensinger, W.: Preparation and properties of Ag-containing diamond-like carbon films by magnetron plasma source ion implantation. Adv. Mater. Sci. Eng. **2012**, 1–5 (2012)
59. Choi, H.W., Choi, J.H., Lee, K.R., Ahn, J.P., Oh, K.H.: Structure and mechanical properties of Ag-incorporated DLC films prepared by a hybrid ion beam deposition system. Thin Solid Films **516**(2–4), 248–251 (2007)
60. Wu, Y., Chen, J., Li, H., Ji, L., Ye, Y., Zhou, H.: Preparation and properties of Ag/DLC nanocomposite films fabricated by unbalanced magnetron sputtering. Appl. Surf. Sci. **284**, 165–170 (2013)
61. Ji, L., Li, H., Zhao, F., Chen, J., Zhou, H.: Microstructure and mechanical properties of Mo/DLC nanocomposite films. Diam.Relat. Mater. **17**(11), 1949–1954 (2008)
62. Ji, L., Li, H., Zhao, F., Quan, W., Chen, J., Zhou, H.: Atomic oxygen resistant behaviors of Mo/diamond-like carbon nanocomposite lubricating films. Appl. Surf. Sci. **255**(7), 4180–4184 (2009)
63. Huang, Q.F., Yoon, S.F., Rusli, Z., Q., Ahn, J.: Dielectric properties of molybdenum-containing diamond-like carbon films deposited using electron cyclotron resonance chemical vapor deposition. Thin Solid Films **409**(2), 211–219 (2002)
64. Mandal, P., Ehiasarian, A.P., Hovsepian, P.E.: Tribological behaviour of Mo–W doped carbon-based coating at ambient condition. Tribol. Int. **90**, 135–147 (2015)
65. Yoon, S.F., Huang, Q.F., Rusli, Y., H., Ahn, J., Zhang, Q., Blomfield, C., Tan, L.Y.C.: X-ray photoelectron spectroscopy of molybdenum-containing carbon films. J. Appl. Phys. **86**(9), 4871–4875 (1999)
66. Butt, M.Z., Khaleeq-ur-Rahman, M., Ali, D., Akmal, A., Naseem, S.: Deposition and characterization of multilayer DLC: Mo thin films grown on silicon substrate by off-axis pulsed laser deposition technique. Appl. Surf. Sci. **331**, 407–414 (2015)
67. Constantinou, M., Pervolaraki, M., Koutsokeras, L., Prouskas, C., Patsalas, P., Kelires, P., Giapintzakis, J., Constantinides, G.: Enhancing the nanoscratch resistance of pulsed laser deposited DLC films through molybdenum-doping. Surf. Coat. Technol. **330**, 185–195 (2017)
68. Dai, W., Wang, A.: Synthesis, characterization and properties of the DLC films with low Cr concentration doping by a hybrid linear ion beam system. Surf. Coat. Technol. **205**(8–9), 2882–2886 (2011)
69. Chiu, M.C., Hsieh, W.P., Ho, W.Y., Wang, D.Y., Shieu, F.S.: Thermal stability of Cr-doped diamond-like carbon films synthesized by cathodic arc evaporation. Thin Solid Films **476**(2), 258–263 (2005)
70. Singh, V., Jiang, J.C., Meletis, E.I.: Cr-diamondlike carbon nanocomposite films: synthesis, characterization and properties. Thin Solid Films **489**(1–2), 150–158 (2005)
71. Singh, V., Palshin, V., Tittsworth, R.C., Meletis, E.I.: Local structure of composite Cr-containing diamond-like carbon thin films. Carbon **44**(7), 1280–1286 (2006)
72. Fan, X., Dickey, E.C., Pennycook, S.J., Sunkara, M.K.: Z-contrast imaging and electron energy-loss spectroscopy analysis of chromium-doped diamond-like carbon films. Appl. Phys. Lett. **75**(18), 2740–2742 (1999)
73. Wang, D.Y., Weng, K.W., Chang, C.L., Guo, X.J.: Tribological performance of metal doped diamond-like carbon films deposited by cathodic arc evaporation. Diam. Relat. Mater. **9**(3–6), 831–837 (2000)
74. Bujak, J., Michalczewski, R.: Characterization and properties of low-friction, multilayered Cr-doped diamond-like carbon coatings prepared by pulse biased filtered cathodic arc deposition. Proc. Inst. Mech. Eng. Part J J. Eng. Tribol. **225**(8), 875–882 (2011)

75. Gayathri, S., Kumar, N., Krishnan, R., Ravindran, T.R., Dash, S., Tyagi, A.K., Sridharan, M.: Influence of Cr content on the micro-structural and tribological properties of PLD grown nanocomposite DLC-Cr thin films. Mater. Chem. Phys. **167**, 194–200 (2015)
76. Santana, J.C., Skomski, R., Singh, V., Palshin, V., Petukhov, A., Losovyj, Y.B., Sokolov, A., Dowben, P.A., Ketsman, I.: Magnetism of Cr-doped diamond-like carbon. J. Appl. Phys. **105**(7), 07A930 (2009)
77. Lin, Y.H., Lin, H.D., Liu, C.K., Huang, M.W., Chen, J.R., Shih, H.C.: Structure and characterization of the multilayeredTi-DLC films by FCVA. Diam. Relat. Mater. **19**(7–9), 1034–1039 (2010)
78. Voevodin, A., Capano, M., Laube, S.J., Donley, M., Zabinski, J.: Design of a Ti/TiC/DLC functionally gradient coating based on studies of structural transitions in Ti–C thin films. Thin Solid Films **298**(1–2), 107–115 (1997)
79. Cui, J., Qiang, L., Zhang, B., Ling, X., Yang, T., Zhang, J.: Mechanical and tribological properties of Ti-DLC films with different Ti content by magnetron sputtering technique. Appl. Surf. Sci. **258**(12), 5025–5030 (2012)
80. Zhao, F., Li, H., Ji, L., Wang, Y., Zhou, H., Chen, J.: Ti-DLC films with superior friction performance. Diam. Relat. Mater. **19**(4), 342–349 (2010)
81. Qiang, L., Zhang, B., Zhou, Y., Zhang, J.: Improving the internal stress and wear resistance of DLC film by low content Ti doping. Solid State Sci. **20**, 17–22 (2013)
82. Yang, W.J., Sekino, T., Shim, K.B., Niihara, K., Auh, K.H.: Deposition and microstructure of Ti-containing diamond-like carbon nanocomposite films. Thin Solid Films **473**(2), 252–258 (2005)
83. Bharathy, P.V., Nataraj, D., Chu, P.K., Wang, H., Yang, Q., Kiran, M.S.R.N. Silvestre-Albero, J., Mangalaraj, D.: Effect of titanium incorporation on the structural, mechanical and biocompatible properties of DLC thin films prepared by reactive-biased target ion beam deposition method. Appl. Surf. Sci. **257**(1), 143–150 (2010)
84. Paul, R., Sharma, M.K., Chatterjee, R., Hussain, S., Bhar, R., Pal, A.K.: Ferromagnetism in nanocrystalline nickel incorporated diamond-like carbon thin films. Appl. Surf. Sci. **258**(15), 5850–5857 (2012)
85. Pandey, B., Pal, P.P., Bera, S., Ray, S.K., Kar, A.K.: Effect of nickel incorporation on microstructural and optical properties of electrodeposited diamond like carbon (DLC) thin films. Appl. Surf. Sci. **261**, 789–799 (2012)
86. Pandey, B., Das, D., Kar, A.K.: Electrical and magnetic properties of electrodeposited nickel incorporated diamond-like carbon thin films. Appl. Surf. Sci. **337**, 195–207 (2015)
87. Mohammadinia, E., Elahi, S.M., Shahidi, S.: Structural and optical properties of Ni-embedded hydrogenated diamond-like carbon (Ni-DLC) prepared by co-deposition of RF-Sputtering and RF-PECVD method. Mater. Sci. Semicond. Process. **74**, 7–12 (2018)
88. Liu, C.W., Chen, W.E., Sun, Y.T.A., Lin, C.R.: Fabrication and electrochemistry characteristics of nickel-doped diamond-like carbon film toward applications in non-enzymatic glucose detection. Appl. Surf. Sci. **436**, 967–973 (2018)
89. Ma, K., Yang, G., Yu, L., Zhang, P.: Synthesis and characterization of nickel-doped diamond-like carbon film electrodeposited at a low voltage. Surf. Coat. Technol. **204**(16–17), 2546–2550 (2010)
90. Benchikh, N., Garrelie, F., Donnet, C., Bouchet-Fabre, B., Wolski, K., Rogemond, F., Loir, A.S., Subtil, J.L.: Nickel-incorporated amorphous carbon film deposited by femtosecond pulsed laser ablation. Thin Solid Films **482**(1–2), 287–292 (2005)
91. Dai, W., Ke, P., Wang, A.: Influence of bias voltage on microstructure and properties of Al-containing diamond-like carbon films deposited by a hybrid ion beam system. Surf. Coat. Technol. **229**, 217–221 (2013)
92. Dai, W., Wang, A.: Deposition and properties of Al-containing diamond-like carbon films by a hybrid ion beam sources. J. Alloy. Compd. **509**(13), 4626–4631 (2011)
93. Pu, J., Zhang, G., Wan, S., Zhang, R.: Synthesis and characterization of low-friction Al-DLC films with high hardness and low stress. J. Compos. Mater. **49**(2), 199–207 (2013)

94. Chang, C.L., Jao, J.Y., Chang, T.C.: Influences of bias voltage on properties of TiAl-doped DLC coatings synthesized by cathodic arc evaporation. Diamond Relat. Mater. **14**(11–12), 2127–2132 (2005)
95. Guo, T., Kong, C., Li, X., Guo, P., Wang, Z., Wang, A.: Microstructure and mechanical properties of Ti/Al co-doped DLC films: Dependence on sputtering current, source gas, and substrate bias. Appl. Surf. Sci. **410**, 51–59 (2017)
96. Kong, C., Guo, P., Sun, L., Zhou, Y., Liang, Y., Li, P., Ke, K.-R. Lee, X., Wang, A.: Tribological mechanism of diamond-like carbon films induced by Ti/Al co-doping. Surf. Coat. Technol. **342**, 167–177 (2018)

# Chapter 13
# Low Temperature Sintering of Lithium Based Ferrites

**Mamata Maisnam**

## Introduction

Ferrites are compounds of iron oxide and some other metal oxide. They are ceramic in nature and are generally dark grey or black in appearance and very hard and brittle. The first ferrite material used by the ancient people is the naturally found magnetite ($Fe_3O_4$) known as "Loadstone" and it is a genuine ferrite. They exhibit ferrimagnetism and they are good dielectrics. Due to their unique electrical and magnetic properties, ferrites find ever-expanding advanced technological applications in the field of electronics, telecommunications, space research, biomedical research, water treatment, etc. The beautiful feature of this type of material is its high efficiency, low cost, great uniformity, and easy manufacture, which have made them replace other metals and garnets in many advanced technological applications. Research in ferrites is promising and much work is being carried out in the development of new ferrites, improvement in their manufacturing and enhancement of the ferrite characteristics.

Ferrites are broadly classified into three categories based on their crystal structure. They are soft or spinel ferrites, garnets or moderate ferrites and hard or hexaferrite. *Soft ferrites* crystallize into the spinel structure, which is named after the mineral spinel $MgAl_2O_4$ and hence they are generally known as spinel ferrites. They have the general formula $M^{2+}Fe^{3+}{}_2O^{2-}{}_4$, where $M^{2+}$ is a divalent metal ion like $Co^{2+}$, $Ni^{2+}$, $Fe^{2+}$, $Cu^{2+}$, $Zn^{2+}$, $Cd^{2+}$, $Mg^{2+}$, $Ca^{2+}$, $Mn^{2+}$ or a combination of metal ions whose average valency is two (2) viz. $Li^{1+}{}_{0.5}Fe^{3+}{}_{1.5}$. They find a wide range of uses in industrial and technological applications as radio and television receivers, magnetic antennas, carrier telephony, computer memories, etc. *Garnets* have the general formula $M^{3+}{}_3Fe^{3+}{}_5O^{2-}{}_{12}$ where $M^{3+}$ is a rare earth or yttrium ion. They

M. Maisnam (✉)
Department of Physics, National Institute of Technology Manipur, Langol, Imphal, Manipur 795004, India
e-mail: mmaisnam.nitm@gmail.com



B. P. Swain (ed.), *Nanostructured Materials and their Applications*, Materials Horizons: From Nature to Nanomaterials, https://doi.org/10.1007/978-981-15-8307-0_13

find use in microwave components such as bandpass and band-stop filters, harmonic generators, limiters and also as programmable sources/counters for high-resolution radar systems, etc. *Hard ferrites* have the general formula $M^{2+}Fe^{3+}{}_{12}O^{2-}{}_{19}$, where M is Ba, Sr, or Pb. They have a hexagonal structure. They are an important member of the permanent magnet family and are used as loudspeaker magnets, magnetic chucks, small electric motors, and focusing magnets. The spinel ferrites are of special interest because of their versatility in terms of compositions, characteristics, and applications. The $Fe^{3+}$ ions in spinel ferrites $M^{2+}Fe^{3+}{}_{2}O^{2-}{}_{4}$ maybe completely or partially replaced by trivalent ions like $Al^{3+}$, $Cr^{3+}$ $Mn^{3+}$ $Ga^{3+}$, $Rh^{3+}$, $Bi^{3+}$ etc.; tetravalent ions like $Ti^{4+}$, $Ge^{4+}$, $Si^{4+}$, $Sn^{4+}$ $Zr^{4+}$ etc., pentavalent ions like $Sb^{5+}$, $V^{5+}$, etc. thus resulting in a variety of what is called as substituted ferrites.

The spinel ferrite structure is determined by the rigid and closest possible packing of oxygen anions (face-centered cubic lattice). The oxygen anions leave two types of interstices, the tetrahedral or the A sites and the octahedral or the B sites, into which the metals ions or the cations get fitted. A metallic ion at the A sites is surrounded by four (4) oxygen ions while a metallic ion at the B sites has six (6) oxygen ions as nearest neighbors. Each oxygen anion on the other hand will have one A site and 3 B sites as nearest neighbors. The metallic ions at the A sites are said to be in tetrahedral coordination whereas the metallic ions in the B sites possess the octahedral coordination (Fig. 13.1c). In the spinel structure, the unit cell of the crystal lattice comprises of 8 formula units ($8 \times MFe_2O_4$). A unit cell comprising of 8 octants is seen in Fig. 13.1a. The 32 oxygen anions, therefore, constitute a unit cell of the fcc crystal lattice leaving 96 interstitial sites; 64 A sites and 32 B sites. Out of these only 8, A sites and 16 B sites are occupied by the metal ions. In order to understand the crystal structure more clearly, a unit cell with a cell edge ‘a’ is assumed to be divided into 8 octants with edge 1/2a. Each octant contains identically four oxygen ions at the corners of a tetrahedron. The occupied interstitial position in two adjacent octants is different but the same in the two octants sharing a common edge. The whole lattice can be constructed by the transition of any two such adjacent octants along the three

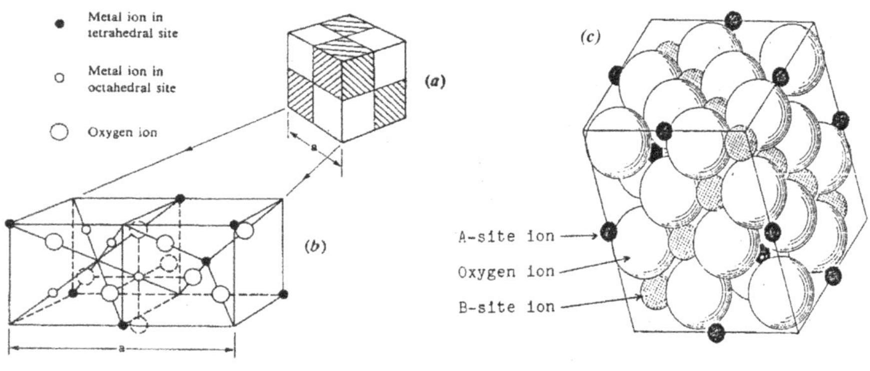


**Fig. 13.1** Schematic diagram of the unit cell of spinel lattice: **a** unit cell of spinel lattice, **b** two octants of the unit cell **c** closed FCC packing of Oxygen ions

(110) axes. A schematic diagram showing two octants of a spinel lattice can be seen in Fig. 13.1b. The statement that the oxygen ions form an fcc structure needs some refinement as there are slight deviations due to deformations caused by the metal ions. The A sites with radius $r_a \sim \leq 0.30$ Å and the B sites with radius $r_o \sim \leq 0.55$ Å are too small to accommodate the metallic ions which have ionic radii ranging from ~0.6 to 1.0 Å. The incorporation of such relatively larger metal ions in the A or B interstices tends to cause a slight deformation of the lattice of oxygen ions. Since the A sites are too small, a metal ion fitted in the A site causes the neighboring oxygen ions to expand along the body diagonal of the cube. The oxygen ions still occupy the corners of an enlarged tetrahedron so that the symmetry of each A site ion is still cubic. However, the four oxygen ions of the octahedral octants are shifted in such a way that this oxygen tetrahedron shrinks by the same amount as the first expands. Consequently, the symmetry of the oxygen ions surrounding the B site is no longer cubic and the oxygen ions are somewhat displaced. Quantitative measurement of the displacement is characterized by the oxygen parameter 'u', and is given by the distance between the oxygen ion and a face of the cube, which is put equal to 'au'. For an ideal fcc lattice, $u_{ideal} = 3/8$.

Spinel ferrites can be classified as normal, inverse, and mixed, according to the distribution of the divalent and trivalent cations in the A and B sites [1–5]. In *normal ferrites* the 8 A sites are occupied by divalent cations and the remaining 16 B sites by trivalent cations. They have the general formula $M^{2+}[Fe^{3+}]\ O^{2-}{}_4$, where $M^{2+}$ is the divalent cation. Examples are zinc ferrite and cadmium ferrite represented as $Zn^{2+}[Fe_2{}^{3+}]\ O^{2-}{}_4$ and $Cd^{2+}[Fe_2{}^{3+}]\ O^{2-}{}_4$ respectively. The *inverse spinel ferrites* have the general formula, written as $Fe^{3+}[M^{2+}Fe^{3+}]\ O^{2-}{}_4$. In this type of spinel ferrites, the eight A sites are occupied by the eight trivalent cations while the 16 B sites are occupied by eight trivalent and eight divalent cations. Magnetite, $Fe^{3+}\ [Fe^{2+}Fe^{3+}]\ O^{2-}{}_4$; nickel ferrites, $Fe^{3+}[Ni^{2+}Fe^{3+}]\ O^{2-}{}_4$; lithium ferrite, $Fe^{3+}\ [Li_{0.5}{}^{1+}Fe_{0.5}{}^{3+}Fe^{3+}]\ O^{2-}{}_4$ are some examples of inverse spinel ferrites. *Mixed spinel ferrites* are neither normal nor inverse with both the A and B sites randomly occupied by the divalent and trivalent metal ions and have the general formula $(M^{2+}{}_xFe^{3+}{}_{(1-x)})[M^{2+}{}_{(1-x)}Fe^{3+}{}_{(1+x)}]\ O^{2-}{}_4$. Example: Magnesium ferrite or $Mg^{2+}{}_xFe^{3+}{}_{1-x}\ [Mg^{2+}{}_{1-x}Fe^{3+}{}_{1+x}]\ O^{2-}{}_4$.

The preference of the metal ions on the A or B sites is influenced by several factors viz-the the ionic radii of the metal ions, electronic configuration, and electrostatic energy of the spinel structure. Information on the distribution of ions over the available sites can be acquired experimentally with the aid of neutron diffraction, Mossbauer studies, and magnitude of the saturation magnetization at the absolute zero temperature [1–3].

## Lithium Ferrites

Lithium ferrites which have a spinel structure are known to have remarkably attractive properties like high Curie temperature leading to unparalleled thermal stability, low dielectric and magnetic losses, rectangular hysteresis loop, and high saturation magnetization. They are also good dielectric materials. The properties of these materials are sensitively dependent on the method of preparation of the compound, type and amount of substituents, sintering temperature, and atmosphere. These materials are very versatile, being able to accommodate in its structure a variety of metal ions or substituents giving rise to a number of different types of ferrites with varied properties. Thus the properties of these materials can be tailored by controlling the above factors. Their Curie temperature ranges from 640 to 670 °C, saturation magnetization from 3000–4000 Gauss and d. c. resistivity ranging from a few ohm-cm to many orders of magnitude. Lithium ferrites are generally doped with various substituents to optimize the numerous properties of interest. The commonly used substituents are Zn, Mn, Ti, Co, Ni, Cr, Al, Cu, Mg, etc. Each ion plays a different role in controlling the properties of the material. Due attention needs to be addressed during processing in order to preserve the stoichiometry and important characteristics of lithium ferrites. The diverse properties of lithium ferrites originate from their ability to accommodate a variety of metal cations into their lattices; redistribute the cations among the tetrahedral and octahedral sublattices and subsequently alter the structural, optical, magnetic, and electrical properties; giving rise to ferrites of new and interesting properties or phenomena. Lithium ferrites find applications in both low and high-frequency electronic devices such as microwave devices, power transformers in electronics, rod antennas, read/write heads for high-speed digital tapes, memory cores, magnetic fluids, etc. Some characteristic problems faced by lithium ferrites, which prevent their broad utilization are low density, large loss characteristics, anisotropy and coercive force and difficulties related to their preparation technology [4–9].

## Ceramic Processing

Besides, the properties are very sensitive to various factors mainly the method of preparation, type and amount of substituents/additives, and processing parameters. This draws the attention of many researchers to develop ferrites of different kinds.

The preparation methodology of ferrites, in general, is the 'Conventional *Standard Double Sintering Ceramic Technique*', involving slow cooling to room temperature. This technique is the most widely used technique for the manufacture of ferrite material. It is also one of the easiest techniques for the production of polycrystalline ferrite materials on a large-scale basis. The properties of ferrites are very sensitive to the preparation methodology, which comprises of various stages. It is important that special care and attention be given to each and every stage of the preparation to produce ferrites with minimal defects. Therefore, a thorough understanding of the important typical sequential processing factors closely associated with this technique is essential.

## Conventional Double Sintering Ceramic Technique

The important processing factors are discussed as follows.

### *Raw Materials*

The raw materials are normally oxides or carbonates of the constituent metals. The purity, particle size, and cost factors of the raw materials are important characteristics. The purity of the raw materials is an important factor in ensuring high-quality ferrites. Another important factor is the particle size of the starting materials which has a profound effect on the behavior of the compound during preparation. The temperature at which good sintering takes place, the ease of mixing, the compressibility, the shrinkage, and the reactivity of the raw materials, all depend on the particle size. So it is important to check this parameter for uniformity of the products. Lastly, the cost of the prepared ferrite material should be reasonably priced for obtaining maximum and wide range uses.

### *Proportion*

The amount of the raw materials required for the composition to be prepared is to be weighed accurately in strict stoichiometric proportions. The relative proportions between the principal metal ions and the oxygen ions have a basic effect on the structural, electrical, and magnetic properties. The proportion of the constituent raw materials decides the properties of the finished products. However, allowances have to be made available for the anticipated loss or gain of the constituent metal during subsequent processing. In some ferrite compositions, trace elements are often added in order to modify the properties of the resulting ferrites. In such cases, since the elements are usually to be present in very small quantities as impurities in the main composition the purity of the raw materials is greatly encouraged.

### *Mixing*

The main purpose of this factor is to mix the starting materials thoroughly into a homogeneous mixture and obtain a uniform composition so that the constituents present in any microscopic volume of the bulk material are in correct proportions and subsequently the properties are uniform. The mixing should be done carefully with the greatest effort to obtain maximum homogeneity. The constituent raw materials are usually mixed in liquid suspension such as distilled water, acetone, alcohol, etc.

The purpose of the adoption of the liquid medium is for cooling and good mixing. Mixing or milling is done in a rotating chamber with steel balls or zirconia balls and distilled water is commonly used. After mixing/milling, the product obtained as slurry is poured off in a clean ceramic container or a glass beaker and is allowed to dry in an oven. Care should be taken so that the container/beaker used are clean and free from impurities.

## *Pre-sintering*

This factor is a calcining process in which the mixed and dried powder of the composition is subjected to a high temperature well above ~1000 °C to bring about the initial chemical reaction of the constituent materials. For lithium ferrite families, pre-sintering is usually performed at temperatures in the range of 700–900 °C [9, 10]. This process decomposes the carbonates and the higher oxides of the raw materials thereby reducing the evolution of gases during the main or the final sintering stage, which may cause the production of certain undesired properties in the final product. A large proportion of the solid-state reaction is completed in this stage. During this process, the constituents partially react to form ferrite-spinel and crystals begin to form and normally grow in size to about 1 μm, the extent however depends on the reactivity of the raw materials. The amount of shrinkage that occurs during the final sintering will be reduced and will, therefore, be more easily controlled. Pre-sintered material thus transforms into porous lumps of partially formed ferrite that can be crushed to be easily ready for the next process.

## *Processing of the Pre-sintered Powder*

Processing of the pre-sintered or calcined powder comprises the following steps.

(a) ***Milling***
This process involves milling of the pre-sintered powder in the presence of distilled water with steel balls or zirconia balls. At this stage, certain user agents may be added to the partially reacted powder and mixed homogeneously with it during milling. The agent will act as a sintering aid and enhance the structural and microstructural properties [10]. For instance, for the lithium ferrite families, a small amount of bismuth oxide ($Bi_2O_3$) as a fluxing agent or sintering aid is added to the pre-sintered powder before milling. This agent aids the sintering of lithium ferrites by lowering the sintering temperature and preventing material loss during sintering. The milling produces an initial rapid breakdown of the particle size. The extent of good milling affects the forming/shaping characteristics of the powder, the sintered density, and the magnetic properties such as permeability and losses.

(b) ***Granulation***

After wet ball milling, the slurry is drawn off, then oven-dried and powdered or granulated. At this stage, it is usual to add a binder to the dried compound and then granulate it to smooth powder. The choice of binders such as polyvinyl alcohol (PVA), dextrin, ammonium alginates, or waxes depends on the subsequent granulation process, the method of forming/shaping (by pressing or extrusion) of the compound, etc. The powder has to be granulated thoroughly, i.e., it has to be agglomerated uniformly and mostly into spherical particles to ensure a good flow of the spheres or the granules and a reproducible filling of the die for pressing into a definite shape.

(c) ***Forming***

The conventional die punch compacting is the most common procedural practice adopted for forming because it is very rapid and economical. The usual forming method is dry-pressing or extrusion and the main purpose of this step is compacting the powder into a definite shape which is as near as possible to the final shape required. The dimension of the forming tools needs to be large enough by a suitable factor allowing appreciable shrinkage that occurs during sintering. Another important purpose for forming is to force the crystallites into close proximity so that during sintering they are allowed to grow and form a dense material with low porosity. A suitably shaped die is filled with the correct quantity of granulated powder mixed with a binder and then compressed usually from above and below simultaneously with a certain amount of pressure, using appropriately shaped upper and lower die punches. The compressed sample is known as the *'green'* sample. While pressing, shapes having length/diameter ratios greater than about 5 are difficult to press successfully. It is also important that the density of the pressed sample should be as uniform as possible since inhomogeneously pressed density will cause the properties to vary from place to place within the sample volume. It also leads to non-uniform shrinkage during sintering and also causes shape distortion. The inhomogeneity in the density of the pressed sample leading to inhomogeneous properties within the sample volume will degrade the overall performance of the product. In addition, non-uniform shrinkage and distortion will make the sample difficult to hold acceptable dimensional tolerances. The simpler the shape the more successful is the pressing and the better it is to control the electrical and magnetic properties. Another method of forming is the extrusion process mainly used to form rods and tubes. Such shapes cannot be easily dry pressed, because if the pressing were axial the length/diameter ratio would be too great whereas if they were to be pressed transversely the die punches would have feather edges. Where very high density is more important than shaping, the isostatic-pressing technique is used. The powder is pressed into the simpler shape of high density from which the required final shape is cut after sintering. The pressed green samples are kept to be processed in the next step.

(d) ***Final sintering***

In this process of sample preparation, the green samples are loaded onto suitable trays and heated in a high-temperature furnace to a desired elevated temperature following a precise heating temperature/time cycle. The sintering atmosphere may be air, oxygen or nitrogen, etc. The generally adopted sintering temperature for lithium ferrite systems is in the range of 900–1050 °C. Sintering proceeds by growing together of the partially reacted powder particles to form crystallites or grains, the residual space between the particles being preferentially swept to the crystallite boundaries during the process, and the density rises. During the process, the linear dimensions of the green sample shrink to about 10–25% depending upon the powder and the pressing technology. The resulting grains in the polycrystalline mass typically range in size from 5 to 40 μm.

Again, the properties of the sintered ferrites are affected by changes occurring within the crystalline structure during sintering. To guarantee the reproducibility of the useful magnetic and electrical parameters of the final product, great care should be taken during the final sintering step especially on the temperature required and temperature homogeneity across the firing chamber. Also for obtaining the correct stoichiometry the right equilibrium of atmosphere has to be attained because sintering atmosphere too has a profound effect on the properties of ferrite. For example, all the high frequency and microwave ferrites are sintered in oxygen atmosphere during the full firing cycle to avoid as far as possible the loss of oxygen and formation of ferrous ions. Also, all divalent Mn containing ferrites is treated with nitrogen during the cooling period to avoid the re-oxidation of $Mn^{2+}$ to $Mn^{3+}$ and precipitation of $Mn_2O_3$ often leads to microcracks in the sintered samples. The majority of the high-quality ferrite is sintered in electrically heated continuous kilns, heated by elements consisting of either silicon carbide rods or wire such as kanthal placed in either side walls of the heating chamber. In this way the temperature maintained is uniform. The temperature is measured by means of a properly arranged thermos couple. The necessary firing temperature to obtain a dense final product depends on the chemical composition of ferrite.

The sintering of ferrite materials should ensure high-density ferrites free from micro-cracks and it should also control and preserve the correct stoichiometry of the samples. Sintering should complete the interdiffusion of the component metal ions into the desired crystal lattice, establish the appropriate valencies for the multivalent ions by proper oxygen control, and develop microstructure most appropriate for the application.

(e) ***Finishing***

The sample after sintering is hard and the surfaces are rough. For undergoing various measurements and to be ready for other applications the sintered samples need some finishing. The process consists of grinding or cutting to have smooth flat surfaces and also to remove the oxide layer formed on the surfaces during sintering.

(f) ***Use of additives***

Lithium ferrites faced difficulties from broad utilization due to various factors such as difficulties to prepare stoichiometric compositions, low densification, poor microstructural properties, large loss characteristics, high anisotropy, etc.

> Lithium ferrites with ideal properties and acceptable densities for applications require a processing temperature of the order of 1200 °C. Studies found that sintering lithium ferrites at a temperature above 1000 °C cause loss of constituents or evolution of secondary phase such as α-$Fe_2O_3$, LiO resulting in a low-quality product by affecting the structural, microstructural properties and hence the electromagnetic properties. Use of sintering aid like $Bi_2O_3$, $V_2O_5$, $Nb_2O_5$, $B_2O_3$, $P_2O_5$ etc. has been the general practice to lower sintering temperature <1100 °C. When the lithium-based ferrite systems are difficult to get sintered to the required density by solid-state sintering, a process known as liquid phase sintering may be adopted. In this process, additives are so chosen that they form a small amount of liquids between the grains which cause rearrangement of grains soon after the formation of the liquid phase and thus produce efficient packing of the grains and densification. The liquid phase act as a medium to reduce excessive uncontrolled grain growth hence the microstructure are more homogeneous. These liquid phase additive materials usually have low melting points and are often used to lower the sintering temperature. Care should be taken that the liquid phase additives which normally resides at the grain boundaries do not deteriorate the material properties or emerge as secondary phase making the material difficult to analyze.

The practice of using $Bi_2O_3$ has been done extensively in the preparation of ferrites. $Bi_2O_3$ not only lowers the sintering temperature to ~1000 °C but also enhances several technologically important properties like density, average grain size, remanence ratio, coercive force, and resonance line width and finally the mechanical properties. For maximum beneficial effects, the recommended amount of $Bi_2O_3$ is (0.25–1.0) wt%. The emergence of the secondary phase, abnormal grain growth leading to the deterioration of properties caused by the additives was a major concern for many researchers. Adoption of $V_2O_5$ with a low melting point of ~ 690 °C allow to sinter ferrites below 1000 °C in a shorter time. A combination of these aids with other metal oxides like Cao, $CaCO_3$, $Na_2O$, $SiO_2$, $Al_3O_3$, $ZrO_2$ etc. can also influence the microstructural properties and the electromagnetic properties. A combination of sintering aids with glass-like $Bi_2O_3$–$B_2O_3$, $V_2O_5$–$B_2O_3$ etc. or combinations of sintering aid, glass and other oxides such as $Bi_2O_3$–$B_2O_3$–ZnO–$SiO_2$ $Al_2O_3$–$SiO_2$–$B_2O_3$, $SiO_2$–CaO–$Na_2O$–$P_2O_5$ may also be done. A careful study of the crystallization conditions using the additives can adjust the electrical and magnetic properties. Another important factor that should be taken into account for studying the characteristics of lithium ferrites is the control of possible $Fe^{2+}$ ions formed during processing from the volatilization of lithia and oxygen. Besides improving the structural and microstructural properties and hence the domain structure, it is important to improve the electrical properties. $Fe^{2+}$ ions if found in the ferrite volume disturbs the stoichiometry. The microstructural properties play an important role in controlling many properties of spinel lithium ferrites and control of microstructural development is necessary during sintering. Excessive grain growth or uncontrolled grain growth should be avoided if controlled grain-sized ferrites are to be fabricated. The use of additives should benefit the densification, microstructure and stoichiometry [6–10].

## *AC Properties*

The use of the magnetic material in alternating current (AC) conditions requires addressing some important properties.

### Hysteresis Properties

The magnetic hysteresis or the delay in the response of the magnetization to the changes in the applied field is a key factor in determining the utility of the ferrimagnetic material. Ferrites exhibit hysteresis, which means that the materials retain memories of the past-induced magnetizations. This particular magnetic characteristic is the effect of an externally applied magnetic field on the magnetization Subjecting the material in a magnetic field, the material magnetized, reached saturation; reversing the field to zero and further reversing we get the hysteresis loop. For every one wave cycle of the AC current, the magnetization undergoes through one hysteresis loop cycle. The area of the hysteresis loop indicates the losses of the material during the cycle of magnetization in an AC field and the irreversible domain changes produce the magnetic hysteresis losses. The hysteresis properties of the ferrite material depend on various factors such as composition, stoichiometry, density, microstructural properties such as grain size, porosity, grain and grain boundary distribution, size, and shape of pores, etc.

A material in its virgin state, i.e., having no magnetic history is said to be magnetically neutral and its magnetic state is represented by the origin, O in Fig. 13.3. Initially, for small magnetizing field H, magnetization increases as OA, where, in this region, reversible domain wall motions predominate. At higher fields, i.e., from A to B, the domain walls are displaced beyond internal energy barriers and irreversible domain wall displacements occur as the domain breaks away from their pinning points and the curve rises steeply to B. At still higher fields, domain wall movements may be almost or entirely removed and further increase in the magnetization may be due to irreversible and reversible spin rotations and the magnetization attains saturation at C. The magnetization thus reaches the saturation value that is essentially equal to the spontaneous magnetization within the domain. The curve OABC is called the initial magnetization curve. When the applied field is decreased from its saturation value at C, the decrease in the magnetization lags behind the decrease in the field due to irreversible processes of magnetization and the domain magnetization rotates reversibly back to their nearest easy direction. When the field is zero, the flux density is left a certain magnetization at D of the curve. If H is then reversed, irreversible domain wall movements and domain rotations magnetized the material in the reversed direction and reach the saturation value at F. By performing the sequence of operations of increasing the field, reducing it to zero, increasing it in the reverse direction, decreasing it to zero and then increasing it to the initial value, a close curve CDEFGC may be obtained. The curve thus obtained is the hysteresis curve or loop [6, 7].

## Eddy Currents

When the ferrite material is magnetized by an AC field, a voltage is induced in the material in a direction opposite to the magnetizing current and the magnetizing field. The induced voltage then produces circular currents in the materials called internal current loops or eddy currents. These eddy currents produce eddy current losses and contribute to the total loss of the material. The eddy current losses play an important factor in determining the application of ferrite materials. The eddy current losses depend significantly on the resistivity of the ferrite material. In order to minimize the eddy current losses of the material, the resistivity of the ferrite material should be high. Ferrites play a very important role as an advantageous material for device applications and have immense industrial use because of their high resistivity. It helps in controlling the eddy current losses in high-frequency applications, being inversely proportional to resistivity. By nature, ferrites are semiconducting materials. Their resistivity varies from $10^{-3}$ Ω-cm to over $10^{11}$ Ω-cm and can cover appropriate applications over wide resistivity ranges. Many factors like composition, conditions of preparation, etc. control the resistivity of these materials. The electrical phenomenon occurring in ferrites is a complicated one. Their conduction mechanism is different and is much less fully understood than those of the elemental group IV semiconductors such as silicon or germanium. Unlike in semiconductors, wherein charge carriers occupy states in a wide energy band, the charge carriers in ferrites are localized at the magnetic atoms. In ferrites, the cations are surrounded by close-packed oxygen anions and as a first approximation can well be treated as isolated from each other. There will be a little overlap of the anion charge clouds or orbitals. Alternately, the electrons associated with the particular ion will largely remain isolated and hence localized electron model is more appropriate in the case of ferrites rather than the collective (band) model. The most widely accepted model to explain the electrical conduction occurring in these ceramic spinels the 'electron hopping model' of Verwey. This mechanism assumes the existence of ions of the same element in two different valency states situated randomly on equivalent lattice sites of the spinel crystal structure. The conduction is due to the exchange or hopping of electrons between such ions. In the case of ferrites, such electron-hopping takes place between the $Fe^{2+}$ and the $Fe^{3+}$ ions distributed randomly at the B sites. Thus the surplus or the deficiency of $Fe^{2+}$ ion content in the composition strongly influences the conduction mechanism and hence the resistivity. To improve the resistivity characteristics the ferrous ions in the ferrite should be kept at the minimum level, which can be done by controlling the type and amount of the doped impurity, processing conditions, and microstructures. The microstructural properties such as grain size, grain boundaries influence the resistivity of the material [6, 7, 10].

## Initial Permeability

Ferrites exhibit hysteresis. The ratio of the magnetic induction, B to that of the externally applied magnetic field, H at any point on the hysteresis loop is termed as the permeability, $\mu$ of the magnetic material and written as $\mu$ = B/H. It is the degree with which the magnetic lines of force can penetrate through the magnetic material and represents the state of magnetization of the material. It is dependent on factors such as static magnetization and frequency of the applied field. Ferrites are composed of regions known as domains where spontaneous magnetization occurs and these domains are separated by domain walls or Bloch walls. The existence of a strong force between the electrons of the constituent magnetic ions enforces a spatial parallel alignment of the electron spins within the domains. However, the domains are generally oriented in a random manner. When a static field H is applied, the domain magnetization tends to align parallel to the applied field producing an increase in magnetization equal to ($\mu - 1$) H.

The value of permeability, which is of much importance in ferrites, is the value of permeability for an initially unmagnetized material, the '*initial permeability,* $\mu_i$' observed in the presence of an infinitesimally small magnetic field. Referring to Fig. 13.3 if the material is in the symmetrical cyclic state about the origin, O, and H is vanishingly small, then the permeability is referred to as the initial permeability. So, $\mu_i$ is defined as $1/\mu_o$ times the slope at the origin. Initial permeability in ferrites is mainly controlled by the bulk properties like chemical composition, ionic distribution, post sintering density, impurity contents, microstructure, grain size, porosity, intra, and intergranular pores, and stoichiometry. Initial permeability is the simultaneous contributions from both the rotation of spins and the reversible displacement of the domain walls and is considered to decline when the coercive force is large. Permeability of ferrites is not loss-free and accompanied by some residual losses. Thus, this quantity is expressed by a complex quantity as,

$$\mu = \mu_i' - \mu_i''$$

where $\mu_i'$ and $\mu_i''$ are the real and imaginary part of the initial permeability. The first term is the inductive or dispersive part and it is what is known as initial permeability and the second part is the dissipative part contributing to the loss and both vary with the frequency. The ratio of the real part to the imaginary part of the complex initial permeability is known as the magnetic loss tangent of the initial permeability. This loss tangent is often expressed as the tangent of the loss angle $\delta_\mu$ i.e. tan $\delta\mu$ = $\mu_i'/\mu_i''$ and arises due to lag of domain wall motion with respect to the applied alternating magnetic field. The magnetization mechanism contributing to the initial permeability can be attributed to the rotation of spins in the Weiss domains and reversible displacement of the bulging walls. In the study of the magnetic spectrum, which is the frequency variation of initial permeability, two resonances may be observed. One resonance is normally observed in the radio frequencies; known as the *rf* resonance or low-frequency resonance and it is attributed to the domain wall oscillations. The other, which occurs at the Gega Hertz range is the microwave resonance and has been attributed to the Larmour precession of the electron spins or domain rotation.

The product of the resonance frequency and the initial permeability is approximately a constant known as the Snoek's limit. Thus when permeability decreases, resonance frequency increases due to their inverse proportionality. Resonance frequency represents the high-frequency upper limit up to which the material can be used in a device and this depends on the value of initial permeability as per Globus law: $(\mu_i)^{1/2}.f_r$ = constant. The frequency variation of initial permeability loss or the magnetic loss tangent, $tan\delta_\mu$ shows an identical behavior with that of the initial permeability [2, 3, 6].

The general practice of sintering ferrites using $Bi_2O_3$ evolves a secondary phase along the grain boundaries affecting properties like permeability, coercive field, anisotropy, microstructure, etc. Liquid phase sintering using $V_2O_5$ could sinter

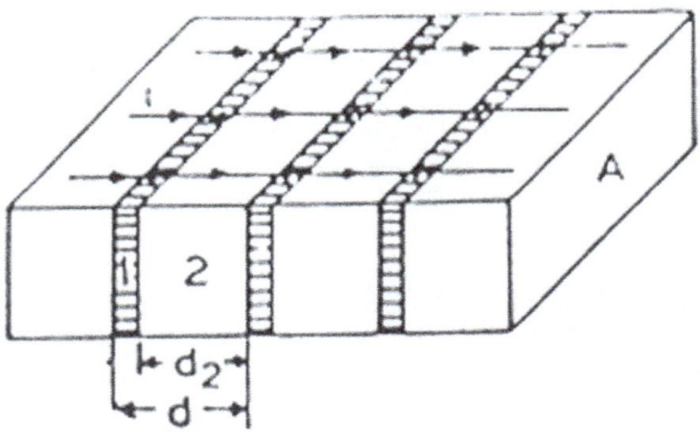


**Fig. 13.2** Inhomogeneous structure of polycrystalline ferrites comprising of poorly conducting grain boundaries and conducting grains

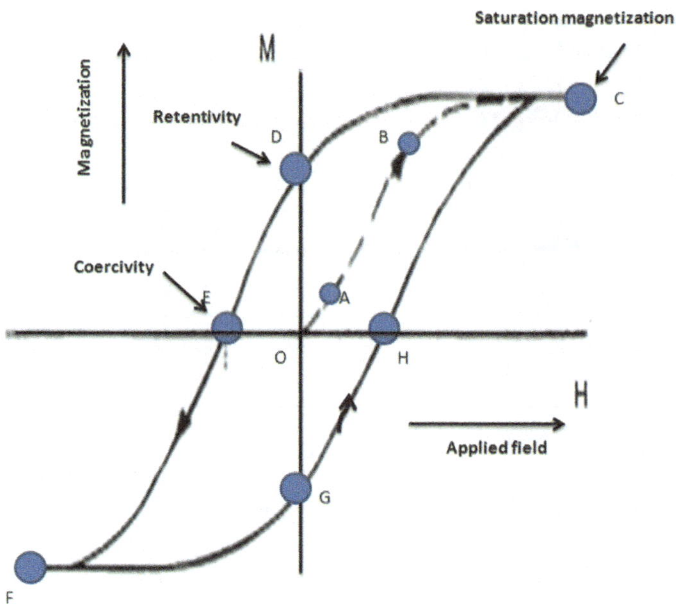


**Fig. 13.3** Schematic diagram of Hysteresis loop of the magnetic materials

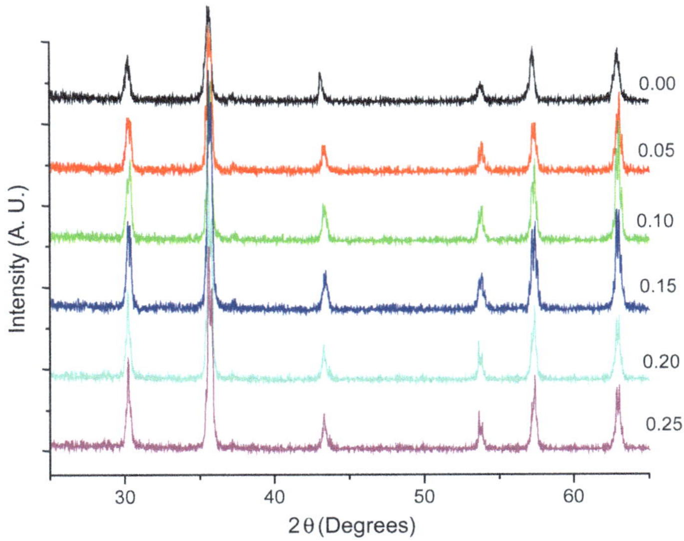


**Fig. 13.4** XRD patterns for $Li^{+}{}_{0.5+x}Zn^{2+}{}_{0.2}Ti^{4+}{}_{0.2}V^{5+}{}_{x}Fe^{3+}{}_{2.1-2x}O^{2-}{}_{4}$ for '$x$' ranging from 0.0 to 0.25 in steps of 0.05)

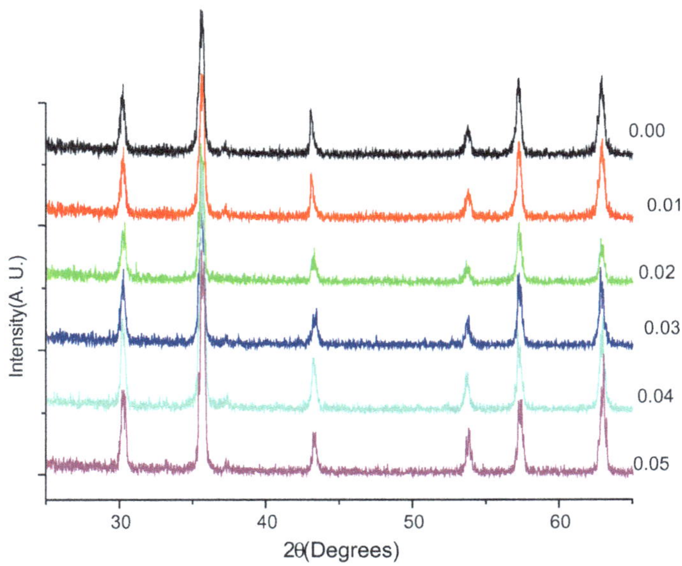


**Fig. 13.5** XRD patterns for $Li^{+}{}_{0.5+x}Zn^{2+}{}_{0.2}Ti^{4+}{}_{0.2}V^{5+}{}_{x}Fe^{3+}{}_{2.1-2x}O^{2-}{}_{4}$ for '$x$' ranging from 0.0 to 0.05 in steps of 0.01)

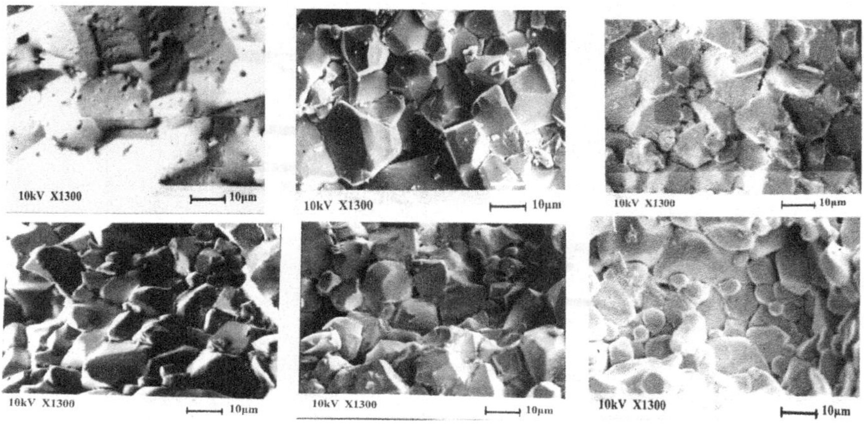


**Fig. 13.6** SEM micrographs for $Li^{+}{}_{0.5+x}Zn^{2+}{}_{0.2}Ti^{4+}{}_{0.2}V^{5+}{}_{x}Fe^{3+}{}_{2.1-2x}O^{2-}{}_{4}$ for '$x$' ranging from 0.0 to 0.25 in steps of 0.05

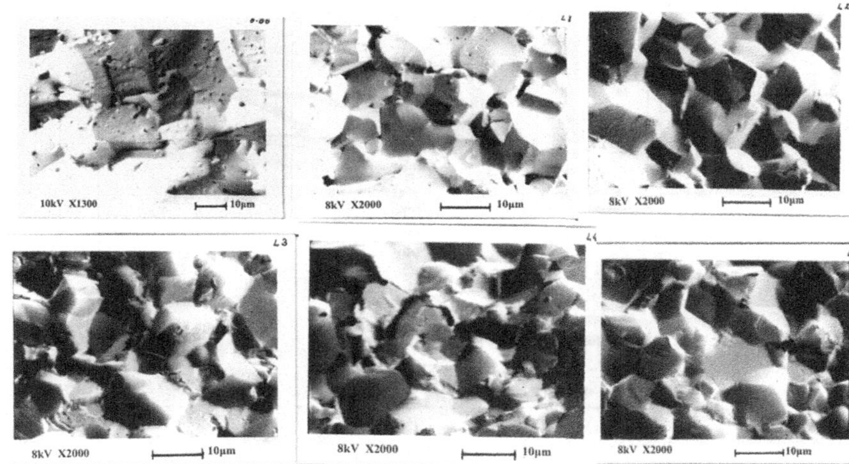


**Fig. 13.7** SEM micrographs for $Li^{+}{}_{0.5+x}Zn^{2+}{}_{0.2}Ti^{4+}{}_{0.2}V^{5+}{}_{x}Fe^{3+}{}_{2.1-2x}O^{2-}{}_{4}$ for '$x$' ranging from 0.0 to 0.05 in steps of 0.01)

lithium-based ferrites at reduced temperatures [11, 12]. It may be either used as substitution or addition. $V_2O_5$ has a low melting point ~ 690 °C and $V^{5+}$ with ionic radius 0.59 Å could enter the crystal lattice least affecting the crystal structure. $V^{5+}$ substitution in lithium zinc titanium ferrite ($Li^{+}{}_{0.5+x}Zn^{2+}{}_{0.2}Ti^{4+}{}_{0.2}V^{5+}{}_{x}Fe^{3+}{}_{2.1-2x}O^{2-}{}_{4}$ for '$x$' ranging from 0.0 to 0.25 in steps of 0.05) using $V_2O_5$ allowed sintering of the ferrite at reduced temperature 950 °C for reduced time 2 h only. The substitution does not seem to change the unit cell size within the experimental error for the range of substitution but produces a slight decrease in the Curie temperature

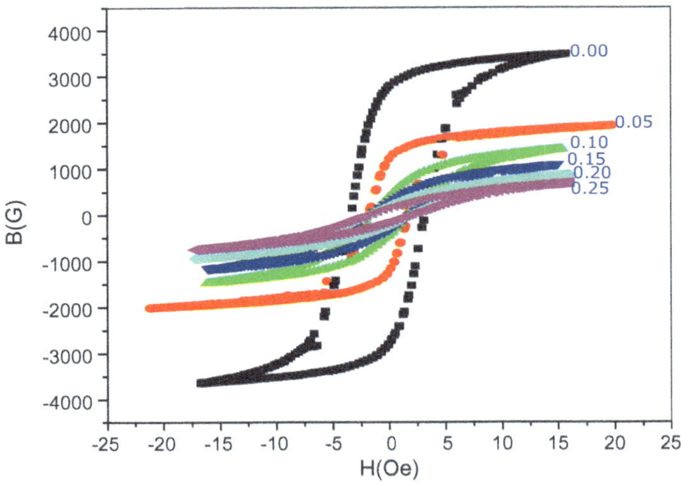


**Fig. 13.8** B-H loops traced for $Li^{+}{}_{0.5+x}Zn^{2+}{}_{0.2}Ti^{4+}{}_{0.2}V^{5+}{}_{x}Fe^{3+}{}_{2.1-2x}O^{2-}{}_{4}$ for '$x$' ranging from 0.0 to 0.25 in steps of 0.05

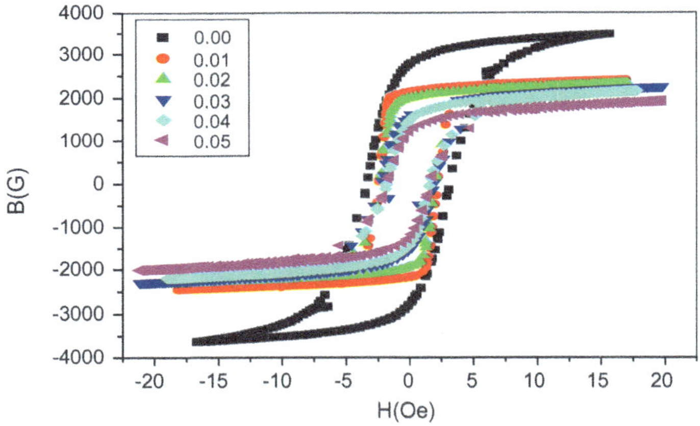


**Fig. 13.9** B-H loops traced for $Li^{+}{}_{0.5+x}Zn^{2+}{}_{0.2}Ti^{4+}{}_{0.2}V^{5+}{}_{x}Fe^{3+}{}_{2.1-2x}O^{2-}{}_{4}$ for '$x$' ranging from 0.0 to 0.05 in steps of 0.01

due to dilution of $Fe^{3+}$ linkages of the two sublattices of the spinel lattice. The SEM photomicrograph revealed homogeneous grains for smaller substitution and at higher substitution, the variation in grain size distribution increased where smaller grains are found to be embedded amongst the bigger grains with an increased number of larger pores. Hence the density decreased while the porosity increase. Studying the structural and microstructural properties of similar $V^{5+}$ substituted ferrites for substitution ranging from 0.01 to 0.05, the microstructural properties seemed to be nearly

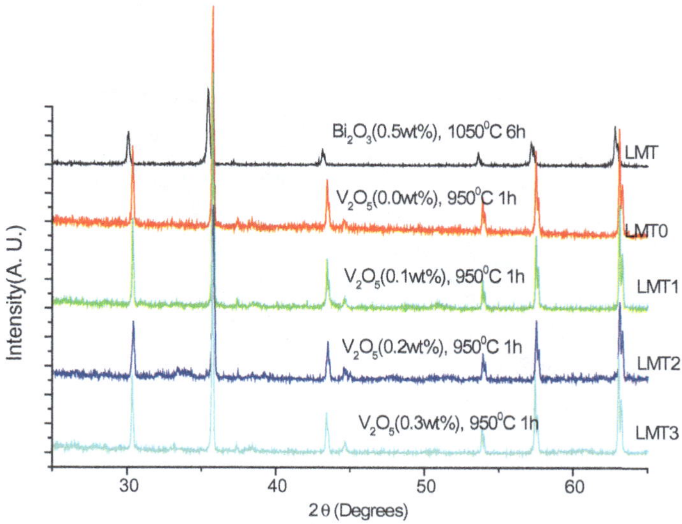


**Fig. 13.10** XRD patterns for $V_2O_5$ added $Li_{0.6}Mn_{0.1}Ti_{0.1}Fe_{2.20}O_4$ ferrites

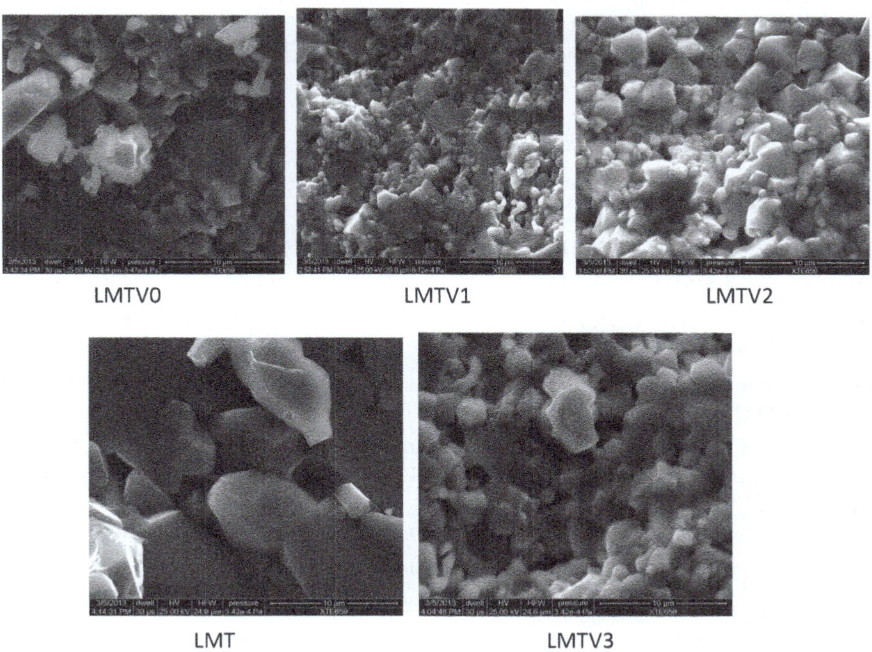


**Fig. 13.11** SEM micrographs for $V_2O_5$ added $Li_{0.6}Mn_{0.1}Ti_{0.1}Fe_{2.20}O_4$ ferrites

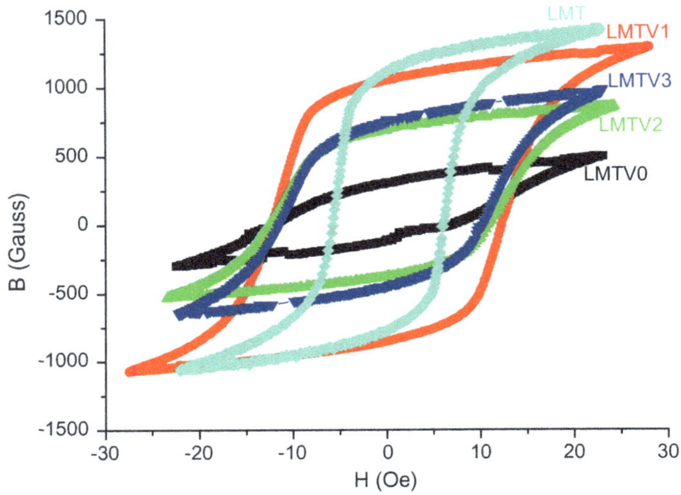


**Fig. 13.12** B-H loops traced for $V_2O_5$ added $Li_{0.6}Mn_{0.1}Ti_{0.1}Fe_{2.20}O_4$ ferrites

**Table 13.1** Structural properties and Curie temperature for $Li^{+}{}_{0.5+x}Zn^{2+}{}_{0.2}Ti^{4+}{}_{0.2}V^{5+}{}_{x}Fe^{3+}{}_{2.1-2x}O^{2-}{}_{4}$ for '*x*' ranging from 0.0 to 0.25 in steps of 0.05

| Sample '*x*' | The lattice constant, 'a' (Å) | Bulk density '$d_{exp}$' (g/cc) | *X*-ray density '$d_x$' (g/cc) | Porosity '*P*' (%) | Curie temperature (°C) |
|---|---|---|---|---|---|
| 0.00 | 8.3617 | 4.502 | 4.960 | 9.30 | 529 |
| 0.05 | 8.3523 | 4.404 | 4.942 | 10.93 | 462 |
| 0.10 | 8.3526 | 4.209 | 4.906 | 14.25 | 448 |
| 0.15 | 8.3534 | 4.191 | 4.868 | 13.96 | 440 |
| 0.20 | 8.3563 | 4.140 | 4.827 | 14.28 | 426 |
| 0.25 | 8.3566 | 4.135 | 4.790 | 13.57 | 421 |

**Table 13.2** Structural properties and Curie temperature for $Li^{+}{}_{0.5+x}Zn^{2+}{}_{0.2}Ti^{4+}{}_{0.2}V^{5+}{}_{x}Fe^{3+}{}_{2.1-2x}O^{2-}{}_{4}$ for '*x*' ranging from 0.0 to 0.05 in steps of 0.01

| Sample '*x*' | The lattice constant, 'a' (Å) | Bulk density '$d_{exp}$' (g/cc) | *X*-ray density '$d_x$' (g/cc) | Porosity '*P*' (%) | Curie temperature (°C) |
|---|---|---|---|---|---|
| 0.01 | 8.3559 | 4.5208 | 4.9650 | 8.94 | 498 |
| 0.02 | 8.3558 | 4.4955 | 4.9601 | 9.36 | 490 |
| 0.03 | 8.3530 | 4.4854 | 4.9557 | 9.49 | 472 |
| 0.04 | 8.3532 | 4.4358 | 4.9564 | 10.50 | 467 |
| 0.05 | 8.3522 | 4.3954 | 4.9462 | 11.12 | 463 |

**Table 13.3** Structural properties and Curie temperature for $V_2O_5$ added $Li_{0.6}Mn_{0.1}Ti_{0.1}Fe_{2.20}O_4$ ferrites

| Samples | Lattice constant (Å) | *X*-ray density, $d_x$ (g/cc) | Experimental density, d (g/cc) | Porosity (%) | Curie temperature (°C) |
|---|---|---|---|---|---|
| LMT | 8.334 | 4.925 | 4.389 | 10.98 | 580 |
| LMTV0 | 8.325 | 4.938 | 4.385 | 11.21 | 575 |
| LMTV1 | 8.326 | 4.937 | 4.492 | 9.01 | 570 |
| LMTV2 | 8.323 | 4.944 | 4.451 | 9.97 | 568 |
| LMTV3 | 8.335 | 4.922 | 4.429 | 10.01 | 567 |

the same. The B-H loop traced from these composition deviates from rectangularity for higher substitution. The substitution showed good structural, microstructural, and hysteresis properties preferably for $x < 0.05$ substitutions (Figs. 13.4, 13.5, 13.6, 13.7, 13.8 and 13.9; Tables 13.1 and 13.2). Again for $V_2O_5$ added lithium-based ferrites $Li_{0.6}Mn_{0.1}Ti_{0.1}Fe_{2.20}O_4 + (x)$, where '*x*' is the amount of $V_2O_5$ added and $x = (0.0, 0.1, 0.2$ and $0.3)$ wt% were prepared named as LMTV0, LMTV1, LMTV2, and LMTV3 respectively sintered at 950 °C for 1 h only. Another sample with chemical formula $Li_{0.6}Mn_{0.1}Ti_{0.1}Fe_{2.20}O_4 + 0.5$ wt% of $Bi_2O_3$ (LMT) was also prepared, calcined, and sintered at 850 °C and 1050 °C, respectively, for 4 h each. The lattice constant does not change significantly but density improved for $x < 0.3$ wt%. The hysteresis loop traced showed a large increase in the coercivity of the $V_2O_5$ containing ferrites [13, 14]. Thus $V_2O_5$ can either be used as a substitution or as an additive in the preparation of lithium-based ferrites, allowing to sinter these ferrites at a lower temperature for lesser time. The concentration however is preferably low for most cases (Figs. 13.10, 13.11 and 13.12; Table 13.3).

## References


1. Gorter, E.W.: Phillips Res. Rep. **9**, 295 (1954)
2. Snelling, E.C.: Soft Ferrites, p. 1. ILIFFE Books Ltd., London (1969)
3. Smit, J.: Magnetic Properties of Ferrites, p. 21. McGraw Hill Company (1971)
4. Sugimoto, M.: J. Am. Ceram. Soc. **82**, 269 (1999)
5. Baba, P.D., Argentina, G.M., Courtney, W.E., Dionne, G.F., Temme, D.H.: IEEE Trans. Magn. **MAG-8**, 283 (1972)
6. Smit, J., Wijn, H.P.J.: Ferrites, p. 136. Phillips Technical Library (1959)
7. Von Aulock, W.H.: Handbook of Microwave Ferrite Materials, p. 407. Academic Press (1965)
8. Blasse, G.: Phillips Res. Rep. **3** (1964); Soohoo, R.F.: Microwave Magnetics, p. 97. Harper & Row Publishers (1985)
9. Argentina, G.M., Baba, P.D.: IEEE Trans. Microw. Theory Tech. **MTT-22**, 652 (1974)
10. Kishan, P.: Microwave Materials, p. 141. Narosa Publishing House, New Delhi (1993)
11. Mamata, M., et al.: Phys. B **352**, 86 (2004)
12. Mamata, M., et al.: Mater. Lett. **58**, 2412 (2004)
13. Maisnam, M., et al.: Mod. Phys. Lett. B. **28**, 1450065 (2014)
14. Maisnam, M., et al.: Solid State Phys. **51**, 815 (2006)

# Chapter 14
# Optimization of the Precipitated Magnetite, Stoichiometry and Composites for Enhanced Stabilization

**N. Joseph Singh, L. Herojit Singh, and Vijayendra K. Garg**

## Introduction

The synthesis of magnetic nanoparticles has become an important area of research. Functional nanostructures such as superparamagnetic nanoparticles, quantum dots, noble metal nanorods, and nanoshells have developed in such a way to minimize the loss of energy, revolutionize in the field of technology, biomedical and agriculture, etc. Due to their unique properties such as high surface to volume ratio, easy separation methodology, and size-tunable optical properties these nanoparticles have attracted much consideration [1, 2]. As a result of these, there has been an increase in interest in various applications in color pigments, catalysis, bioapplications due to low toxicity, microwave devices, ferrofluids, advanced magnetic materials, and high-density magnetic record [3–5].

Iron oxide exists in different polymorphs. The earlier known polymorphs are FeO, $Fe_3O_4$, $\gamma$-$Fe_2O_3$, $\alpha$-$Fe_2O_3$, $\beta$-$Fe_2O_3$, $\varepsilon$-$Fe_2O_3$. Recently a polymorph of iron oxide was discovered and known as $\zeta$-$Fe_2O_3$. Iron oxides are known for its high applicability in modern technology. Among all the polymorphs of iron oxide, $Fe_3O_4$ and $\gamma$-$Fe_2O_3$ exhibits ferrimagnetism with saturation magnetization of approximately 80 emu/g. these two polymorphs is a family of cubic spinel ferrites ($AB_2O_4$, where A and B represent tetrahedral and octahedral cations). Cubic spinel ferrites (Fig. 14.1) have a wide range of properties depending upon the cations arrangement and nature of the

N. Joseph Singh · L. Herojit Singh (✉)
Department of Physics, National Institute of Technology Manipur, Manipur 795004, India
e-mail: loushambam@gmail.com

N. Joseph Singh
e-mail: ningthoujamjoseph44@gmail.com

V. K. Garg
Institute of Physics, University of Brasilia, Brasilia, DF 70919-970, Brazil
e-mail: vijgarg@gmail.com



B. P. Swain (ed.), *Nanostructured Materials and their Applications*, Materials Horizons: From Nature to Nanomaterials, https://doi.org/10.1007/978-981-15-8307-0_14

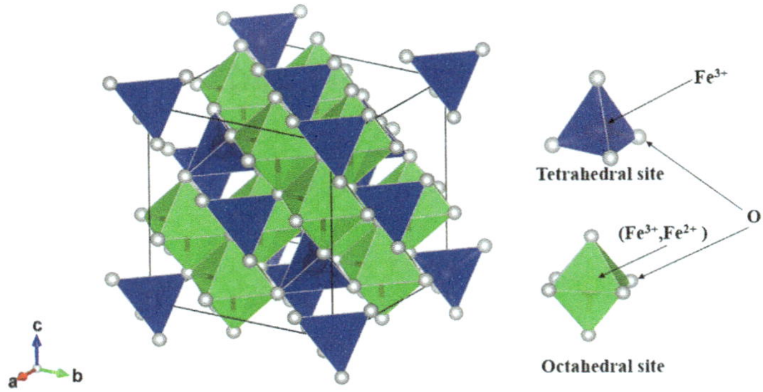


**Fig. 14.1** Crystal structure of magnetite nanoparticles.

element. In $ZnFe_2O_4$, the superexchange interaction of the tetrahedral Zn and octahedral Fe is negligible leading to antiferromagnetic below 10 K [6]. The occupancy of Zn in octahedral and Fe at tetrahedral increase the superexchange interaction and enhance the magnetic properties. $Fe_3O_4$ which is known as magnetite is the strongest candidate of inverse spinel ferrite. Interesting feature in magnetite is the existence in different stoichiometry. It is often represented by $(Fe^{3+})_A[Fe^{2+}_{1-3\delta}Fe^{3+}_{1+2\delta}\square_\delta]O_4$, where $\square$ is the $Fe^{2+}$ vacancies. The oldest form of iron oxide is the Hematite ($\alpha$-$Fe_2O_3$) which is widespread in nature in rocks and soils. Hematite is the most stable iron oxide under ambient condition [7, 8].The stability and semiconductor properties of $\alpha$-$Fe_2O_3$ allow it to be used as a photocatalyst [9]. Due to a small band gap (~2.1 eV), high resistivity, and relatively low cost the $\alpha$-$Fe_2O_3$ photoelectrode has received considerable attention as a solar energy conversion material [10].

Many chemical approaches have been used to prepare and synthesis magnetic nanoparticles. Synthesis technique such as co-precipitation technique, reduction of metal salt in aqueous phase, microemulsion approach, sol-gel technique, hydrothermal technique, solvothermal, etc. [11, 12] are used. Among all co-precipitation is the most favorable produce in huge amount which is favorable for industrial production. A disadvantage in co-precipitation is the coexistence of iron hydroxides along with magnetite while the synthesis parameter is slightly deviated. Some of the important issues face while synthesizing the nanoparticle are: particle size uniformity, size control, crystal structure, shape-control, and alignment for device applications [13]. These are directly influenced by the strategy employed to produce them.

# Coprecipitation of Magnetite/Composites

## *Co-Precipitation of Magnetite*

Figure 14.2 shows the synthesis of magnetite using co-precipitation technique. Co-precipitation technique is one of the most convenient ways to synthesize the magnetic nanoparticles from aqueous salt solutions by adding base as a precipitating agent. The main advantage of co-precipitation technique is that a large amount of nanoparticles can be synthesized for industrial production. Iron oxide nanoparticles were synthesized by the hydrolysis of $Fe^{2+}$ and $Fe^{3+}$ in a fixed molar ratio and generally precipitated in an alkaline solution such as ammonium hydroxide, potassium hydroxide, or sodium hydroxide [14].

Optimization of coprecipitation parameters has to be performed. A slight deviation from the optimized condition leads to coexistence of magnetite along with iron hydroxides. pH and temperature effect on the phase formation, which shows the high probability of coexistence [15]. The formation of iron hydroxide at optimized pH is due to oxidation of $Fe^{2+}$ and decrease the $Fe^{2+}/Fe^{3+}$ from 1:2 [16]. The phases nucleated at intermediate pH were mapped [17]. Aqueous co-precipitation of ferric and ferrous salts at alkaline conditions at high temperature can prepare magnetic nanoparticles [18]. A large amount of nanoparticles can be formed by the co-precipitation process but there is limitation in distribution of particle size as the growth of the crystal is controlled only by kinetic factors.

Superparamagnetic iron oxides particles were prepared by Massart [19] in a controlled way. Magnetite ($Fe_3O_4$) were spherical with diameter 8 mm (as measured by XRD). The influence of the base, PH value, added cations, and the ferrous to ferric ion ratio on the co-precipitation reaction yield and polydispersity and diameter of the nanoparticles can be demonstrated by studying the parameters of the process

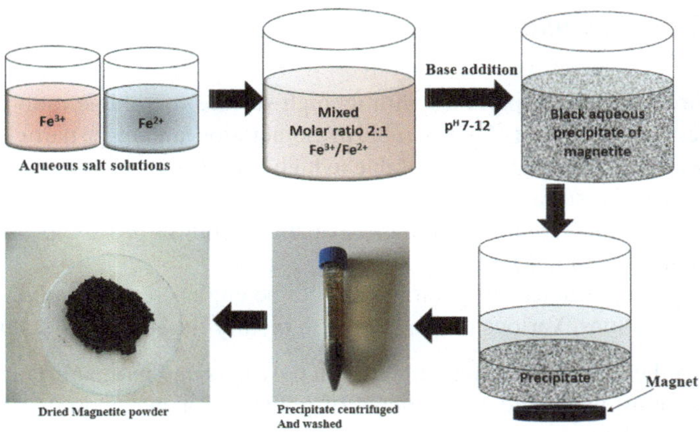


**Fig. 14.2** Co-precipitation of magnetite

[20]. Particles of size ranging from 16.6 to 4.2 mm can be obtained when all these parameters are changed [21].

Precipitate obtained from strong and weak base were studied by T. Ahn et al., Gnanaparkash et al. and Andujar et al. Precipitate at different molar ratio of ammonia to iron ions ($R = [NH_3]/[Fe^{2+} + Fe^{3+}]$) results in different polymorphs of iron hydroxides and oxide. When $R = 0$–$4.0$, the pH of the solution varies from 1.5 to 9.0. At the initial pH around 1.5, the Fe precipitates to akageneite (β-FeOOH). This precipitate starts dissolving at $R \approx 2$ reducing the particle size. The lepidocrocite (γ-FeOOH) was also formed at $R \approx 2$ and disappear totally from 2.6. Further increase in the $R$ to 2.5, goethite (α-FeOOH) was precipitated along with magnetite. A single-phase magnetite is found when $R > 3$. Microscopic studies show the magnetite forms at the tip of the rod shaped goethite and detached. Initial pH and the temperature of the solution also play an important role in the formation of oxides and hydroxides [22, 23]. Initial pH of the Fe precursor solution kept at 0.7–4.7 for 30 min and increase to 10 using $NH_3$ solution at room temperature precipitates to single-phase magnetite. Higher initial pH shows the coexistence of magnetite along with goethite. In the case when the solutions were heated at 333 K results in the evolution of goethite peaks. At high initial pH and temperature the $Fe^{2+}$ gets oxidized which diverts from the required ratio of $[Fe^{2+}]$: $[Fe^{3+}]$, i.e., 1:2. Further increase in the pH to 12.5 nucleates a trace amount of goethite.

Weak base (sodium carbonate) coprecipitation of $Fe^{2+}$ and $Fe^{3+}$ precursor solutions lead to goethite around 6.5 pH. The mechanism for formation of iron hydroxides is known due to olation or condensation of hydroxo and aquohydroxo complexes in solution [24]. Whereas formation cubic spinel particles are through oxolation, which is a two-step reaction. Here, initial condensation of hydroxo complexes is followed by β-elimination of a proton which leads to the formation of the oxo-bridge. The second step is favored at high temperatures.

## *Magnetite in Templates/Composites*

Magnetic iron oxide nanoparticles with its high surface to volume ratio have high surface energies. As a result of which rapid agglomeration among the particles occurs. The nanoparticle oxidizes easily in air due to high chemical reactivity. So it is necessary to modify the surface of iron-oxide nanoparticles to enhance their stability by preventing certain agglomeration and oxidation and also provide possibility for further functionalization [25].

Magnetic nanoparticles are often required to be engineered depending upon the mode of applications. Various methods exist to modify the surface of magnetite.

1. Capping the magnetite nanoparticles with polymers and inorganic compounds.
2. Incorporating magnetite in the pores of polymers or mesoporous/nanoporous carbon template.
3. Incorporating magnetite in the pores of inorganic compounds.

## Capping the Magnetite Nanoparticles with Polymers and Inorganic Compounds

Surface of nanoparticles exist huge point defects. Surface defects have very interesting properties such as magnetism and catalytic properties etc. [26]. The antiferromagnetic materials possess ferromagnetism when surface defect were introduced [27, 28]. This enhances the interactions among neighbor nanoparticles. The surface stabilization is achieved by chemical reaction (i.e., oxidations) and Ostwald ripening (spontaneous process of forming larger particles from smaller particles) [29]. The interactions of the external agents which oxidize and agglomerate are huge disadvantages for bio applications. For utilizing magnetite nanoparticles for bio medications such as target drug delivery and magnetic hyperthermia, etc., a stabilized ferrofluids is necessary. For prevention of nanoparticles from oxidation and agglomerations, the surfaces of the nanoparticles were modified by capping with suitable materials.

Capping of magnetite with polymers involves two steps. In the first, the magnetite nanoparticles were nucleated using various synthesis methods (e.g., Coprecipitation, sol–gel, combustion, solvothermal, citrate method, etc.). Later after nucleation the nanoparticles were dispersed in the polymeric solution where the binding of the nanoparticles takes place [30]. Capping the nanoparticles reduces the magnetic interaction among the nanoparticles. The binding of the polymers depends on the surface defects and oxidation state of the Fe. TGA-DTA of Oleic acid capped magnetite nanoparticles studied by Zhang et al. observed two exothermic peaks which is due to the existence of two different binding energies [31]. Two different binding energies are due to coexistence of two different oxidation state of iron. This is again confirmed by thermal studies of oleic acid capped magnetite and goethite composite [32]. The goethite components get reduced around 573 K whereas the magnetite nanoparticles were remained unmodified. Only after increment of the temperature above 773 K, the magnetite nanoparticles started reduction to wustite (FeO). The changes in the structure and the stoichiometry depend on the reaction of the dissociated components of oleic acid.

The binding energy of the capping agent and the nanoparticles can be categorized into low, medium, and high. The anchors of the capping agent can both lead to too weak reversible binding and to too strong binding depending on their affinity to the cations of the investigated oxide [33]. Binding strength of the anchors affects the colloidal nature of the nanoparticles. Too low binding strength leads to nanoparticle aggregation and results in nanoparticle dissolution when the binding strength is too high. Therefore an optimization in the binding energy of the capping agent and the iron oxides nanoparticles is of high priority. To understand the binding mechanisms, Esther Amstad et al. work on different derivatives of catechol with magnetite nanoparticles [34]. The colloidal stability of the nanoparticles capped with different derivatives of catechol is listed below.

Magnetite nanoparticles stabilized with the mentioned capping agent shows various degree of stability. The magnetite with low-affinity agent started to irreversibly agglomerate below the body temperature. Nanoparticles with intermediate stability agglomerate between 60 and 80 °C as indicated by the increase in the hydrodynamic radius measure by dynamic light scattering (DLS).The colloidal particle stability was also observed to have high correlation with the high packing density with the capping agent.

Apart from polymeric capping various inorganic compounds were used for capping the nanoparticles. $SiO_2$ and Au coated magnetic nanoparticles are frequently reported. Capping of $SiO_2$ was performed mainly in two different approaches [35]. The first method is based on Stöber process. This method was discovered by Werner Stöber and his team in 1968. In this method uniform and tuned size silica was formed in-situ through the hydrolysis and condensation of a sol–gel precursor [36]. This method was employed to cap rod-shaped ferromagnetic nanoparticles, then to micrometer-sized hematite [37]. It is then extended to synthesize core@shell of iron oxide@$SiO_2$ nanoparticles [38]. The second method is based on microemulsion synthesis [39]. In this method, micelles or inverse micelles were used to confine and control the coating of silica on core nanoparticles.

The dipole interaction among the magnetite nanoparticles was systematically investigated by capping with silica of different thickness. Mossbauer investigations on these nanoparticles show the evolution of a superparamagnetic doublet as the thickness increases. Au capping on the nanoparticles also reduces the magnetic interactions. Capping the nanoparticles prevents the oxidation of the nanoparticles and therefore the stoichiometry is maintained.

## Incorporating Magnetite in the Pores of Polymers or Mesoporous/Nanoporous Carbon Template

Depending upon the purpose of the applications, magnetite nanoparticles were often required to embed in a matrix. Incorporation of magnetite was carried out with single divalent or trivalent Fe precursor as well as $Fe^{2+}/Fe^{3+}$ precursors [40, 41]. Micron-sized spheres of mesoporous Sty-DVB copolymer produced by suspension polymerization in the presence of inert diluents [42] were dispersed in $FeSO_4$ aqueous solution for 1 h at room temperature. The polymers were separated by filtration and washed with water till the iron was removed from the surface using o-phenanthroline test. The separated and washed iron-containing resin was immersed in the oxidizing aqueous solution containing KOH and $NaNO_3$ [41]. The obtained nanocomposites are the Sty-DVB and $Fe_3O_4$.

$Fe^{3+}$ precursor's solution was used to synthesize magnetite. The synthesis of magnetite composites was carried out in two methods. The first is to synthesize the magnetite nanoparticles [43]. In the second method the $Fe^{3+}$ precursors were dispersed along with the materials with which the composite is to be prepared [44].

As from the literature, coprecipitation of $Fe^{2+}$ and $Fe^{3+}$ precursors at high pH results to magnetite. Synthesis of magnetite from $Fe^{2+}$ precursor requires partial oxidation to have $[Fe^{2+}]/[Fe^{3+}] = ½$, which is the stoichiometric ratio to obtain $Fe_3O_4$. The precipitate from $Fe^{3+}$ precursor results in iron hydroxides as a major product. To obtain magnetite from these hydroxides, reduction of the precipitate is needed.

### Incorporating Magnetite in the Pores of Inorganic Compounds

Nanocomposites are the multiphase materials that possess properties emerging from the combination of each component. The nanocomposites of magnetite with polymers, organic and inorganic compounds can lead to improvements in several areas, such as optical, mechanical, electrical, and magnetic properties [45]. The incorporation of inorganic particles with magnetite not only provide mechanical and thermal stability, but also provides new functionalities that depend on the chemical nature, the structure, the size, and crystallinity of the inorganic nanoparticles [46].

Zeolite is one of the potential inorganic materials for making composites with magnetite. Zeolite belongs to a group of crystalline alumino-silicate with the basic tetrahedral structural units of $SiO_4$ and $AlO_4$ with Si or Al in the center of the tetrahedron [47]. The $SiO_4$ and $AlO_4$ are joined to form a larger unit called sodalite unit (SU). The formation of sodalite unit favors with [Si]/[Al] ratio lying close to 1.23. The cavity form inside sodalite unit is known as β-cage. SU arrangement forms a cage with the entrance windows having a diameter close to 8 Å while the maximum diameter of the empty cage is around 13 Å. In the aluminosilicate structure, each tetrahedral group of $AlO_4$ has an uncompensated charge of $-1$ and hence remain attractive to positive ions resulting in the occupation of $Na^+$, $Ca^{+2}$ at sites nearby $AlO_4$ groups. This property of zeolite has been extensively used for binding heavy cations in pores and cages. Zeolite nanoparticles localized the magnetite during synthesis providing a rigid template that stabilized the nanoparticles. When the nanoparticles are subjected to charging and discharging agglomeration is likely to take place. To improvise this promising material in terms of thermal stability and capacity, the nanoparticles can be embedded in a robust, charge imbalance and porous structured template [48]. The imbalanced charge can accommodate extra Li to neutralize the environment. Zeolite nanoparticle which is a porous material which can accommodate and easily transport the charge. The structure of the zeolite plays an important role in the ion exchange and the rate of exchange. It is biocompatible and won't have any environmental issues. The high thermally stable zeolite will preserve the desired phase and particle size at elevated temperature. The introduction of zeolite along with iron-oxide nanoparticles enhances the properties of supercapacitor since zeolite nanocomposites have high specific capacitance [49].

## Reduction of Iron Hydroxides to Magnetite

Iron hydroxides exist in different polymorphs. The list of iron hydroxides with structural details are listed in Table 14.2.

**Table 14.1** Magnetite stability with different capping agent [34]

| Capping agent | Colloidal stability |
|---|---|
| PEG(5)-nitroDOPA | High |
| PEG(5)-nitrodopamine | High |
| PEG(5)-mimosine | Intermediate |
| PEG(5)-DOPA | Intermediate |
| PEG(5)-dopamine | Low |
| PEG(5)- | Low |
| hydroxydopamine | Low |
| PEG(5)-hydroxypyridine | Low |
| PEG(5)-hydroxypyrone | Low |
| PEG(5)- | |

**Table 14.2** Polymorphs of iron hydroxides

| Sl. No | Composition | Name |
|---|---|---|
| 1 | $Fe(OH)_2$ | Green Rust I |
| | $Fe(OH)_3$ (Variables) | Green Rust II |
| 2 | α-FeOOH | Goethite |
| 3 | β-FeOOH | Akageneite |
| 4 | γ-FeOOH | Lepidocrocite |
| 5 | δ-FeOOH | Feroxyhyte |

The $Fe(OH)_2$ is the initial product while precipitating $Fe^{2+}$ precursor. The $Fe(OH)_2$ gets oxidized to $Fe^{3+}$ compounds and results in different polymorphs of iron hydroxides and oxides [50]. The major product from precipitation method using single iron precursor (i.e., $Fe^{2+}$(aq) and $Fe^{3+}$(aq)) is goethite. The changes in the local structure of the goethite at elevated temperature were studied by Herojit et al. [15]. The dissociated $H^+$ and $OH^-$ from $H_2O$, attached on the nanoparticles played an important role in the transformation of α-FeOOH to $\gamma$-$Fe_2O_3$ in the temperature regime 400–460 K. The transformed $\gamma$-$Fe_2O_3$ got reduced to off-stoichiometric magnetite in the temperature range 500–625 K, respectively. $H^+$ desorption in the process of thermal dehydration of α-FeOOH is understood to create oxygen vacancies. This increases the concentration of Fe ions locally and diffuses to result in a partial conversion of α-FeOOH to $\gamma$-$Fe_2O_3$. Topotactical transformation of α-FeOOH to $\alpha$-$Fe_2O_3$ is observed after annealing treatments above 500 K and this effect becomes predominant for annealing above 550 K.

The reduction of goethite and other hydroxides will result into magnetite. The reduction of hydroxides is often carried out to obtain composites of magnetite. Reduction of iron hydroxides is performed in various ways. The first is to anneal in the presence of reducing gas (CO and $H_2$, etc.). Heat treatment of hydrocarbon or derived hydrocarbon-coated iron hydroxide in reduced atmosphere creates oxygen vacancies and transforms into magnetite. L. Herojit et al. and S. Ayyappan et al. have investigated the reduction/oxidation of the nanoparticles capped with hydrocarbon during heat treatment [51]. Oleic acid capped α-FeOOH coexisting with $Fe_3O_4$ was annealed in air and reduced atmosphere of $10^{-6}$ mbar. When an iron hydroxides coexisting with $Fe_3O_4$ is annealed in reduced atmosphere ($10^{-6}$ mbar), topotactical transformation of α-FeOOH to γ-$Fe_2O_3$ is observed from 523 K. and obtained its maximum, i.e., 75% of α-$Fe_2O_3$ at 773 K. The same nanoparticles after capping with oleic acid, the heat treatments show a huge divergent as that from the uncoated nanoparticles. At 523 K, 10% of α-$Fe_2O_3$ was obtained, but no α-$Fe_2O_3$ was observed after 523 K. This is due to the reduction of α-FeOOH during the heat treatment. The capped hydrocarbon gets dissociated into $H_2$, CO, $O_2$, $H_2$ and $CO_2$ [51]. If the annealing were carried out in air, oxidation of particles occur leading to α-$Fe_2O_3$. The mechanism of reduction of the capped particles in reduced atmosphere is understood from the gas effluents released during thermal decomposition of the oleic acid as TGA-coupled mass spectrometry. The oleic acid coated $Fe_3O_4$ nanoparticles exhibited a four-step weight loss in the TGA measurements. The first step in the temperature range of 373–473 K was due to desorption of loosely bound $H_2O$. The second step at 501 K was due to the dissociation of weakly bound functional groups (COOH) from surfactant. A fraction of $CO_2$ initiates evolving before 473 K and maximum is achieved at around 513 and 603 K. $H_2$ was among the gas product evolve from dissociation of hydrocarbon chains along with CO above 593 K accompanied by a trace amount of $H_2O$, which may be due to the reaction between $H_2$ and $CO_2$. The third step occurs at 887 K, and it is understood due to release of $CO_2$ gas. And the fourth weight loss observed at 953 K was mainly due to release of CO and $CO_2$. The released gas reacts with the iron hydroxides and drives away the oxygen. The reduced particles are required to prevent oxidation. To prevent oxidation, nanoparticles are needed to keep at reduced atmosphere or inert gas.

## Vacancies and Cationic Distribution

One-third of total cations in $Fe_3O_4$ are contributed by $Fe^{2+}$ which is metastable. It often oxidized, resulting in cation vacant system. In magnetite which is represented by $(Fe^{3+})_A[Fe^{2+}_{1-3\delta}Fe^{3+}_{1+2\delta}\square_\delta]O_4$, becomes stoichiometric magnetite ($Fe_3O_4$) when $\delta = 0$ and transformed into maghemite (γ-$Fe_2O_3$) as $\delta = 0.33$. Maghemite exhibits similar structural and magnetic properties as that of magnetite at room temperature. Low-temperature magnetization/specific heat/resistivity studies of stoichiometric magnetite show a transition at 120 K [52]. This transition is known as Verwey transition. The transition temperature is affected by $Fe^{2+}$ vacancies. It shifts to lower temperature and found no transition in magnetite with $\delta < 0.15$. The cationic vacancies due to oxidations affect the electrical, magnetic, and thermal properties. These require characterizing the stoichiometry of the magnetite to get our desired properties. Various techniques exist to determine the stoichiometry. The techniques that can be mentioned are electron paramagnetic resonance (EPR), UV–vis, photoacoustic and Mossbauer spectroscopy, etc. [53, 54]. As the magnetite get oxidized, the g-value determined from EPR decreases [54]. The absorption intensity at higher wavelength, i.e., 500–1000 nm decreases as the $Fe^{2+}$ vacancies increases.

Nature of cationic defects existing in the bulk and the nano magnetite has different form. In the bulk magnetite, the cationic defect is understood well due to the oxidation of $Fe^{2+}$ to $Fe^{3+}$ creating $Fe^{2+}$ vacant sites. A quantitative analysis of the cationic distribution can be precisely determined using Mossbauer spectroscopy. Mossbauer spectrum of bulk magnetite consists of two sextets [55, 56]. The tetrahedral Fe consists of 33% of the total Fe having isomer shift of 0.2 mm/s and hyperfine field of 48.8 T. The remaining 2/3 of the Fe recites in octahedral with Mossbauer parameter 0.65 mm/s Isomer shift and hyperfine field 45.6 T. Intensity of sextet corresponding to the octahedral Fe atoms decreases as the oxidation takes place in the particles. The magnetite with different stoichiometry measured by Mossbauer was correlated with photoacoustic spectra. Two broad absorption peaks in Photoacoustic spectra were observed in the temperature range 400–750 nm and 750–1000 nm. The peak intensity in the wavelength 750–1000 nm decreases as the stoichiometric magnetite transform to off-stoichiometric as a result of oxidation [*q*]. The photoacoustic spectra of the $Fe_3O_4$, $\gamma$-$Fe_2O_3$ and $\alpha$- $Fe_2O_3$ shows similar pattern with the UV spectra reported by He et al. [57]. The electronic transition of $\gamma$-$Fe_2O_3$ and $\alpha$-$Fe_2O_3$ is listed in Table 14.3. The photoacoustic spectra correlating to the Mossbauer spectra, the range of defect ($\delta$) in magnetite is determined.

In the case of nanoparticles, random arrangement of $Fe^{2+}$ and $Fe^{3+}$ cations are present along with the cationic vacancies. As the particle size decreases the magnetic spin of the atoms on the surface becomes disordered and the relaxation time decrease. These collapses the sextet corresponding to Fe atoms from the octahedral and tetrahedral sites resulting in broad low hyperfine field sextet [58]. Further decrease in the particle size, the sextet transform to a doublet. These limit the employment of Room temperature Mossbauer spectra for determining the cationic distribution. However, the trend of cationic rearrangement is investigated in the magnetite nanoparticles as a

**Table 14.3** Electronic transitions of $\gamma$-$Fe_2O_3$ and $\alpha$-$Fe_2O_3$

| Sample | Band position (nm) | Transition/origin |
|---|---|---|
| Maghemite ($\gamma$-$Fe_2O_3$) | 434 | $^6A_1$ to $^4E_1$, $^4A_1$ |
| | 510 | $2(^6A_1)$ to $2(^4T_1)$ |
| | 666 | $^6A_1$ to $^4T_2$ |
| | 934 | $^6A_1$ to $^4T_1$ |
| Hematite ($\alpha$-$Fe_2O_3$) | 404 | $^6A_1$ to $^4T_2$ |
| | 444 | $^6A_1$ to $^4E^4A_1$ |
| | 529 | $2(^6A_1)$ to $2(^4T_1)$ |
| | 649 | $^6A_1$ to $^4T_2$ |
| | 884 | $^6A_1$ to $^4T_1$ |

result of annealing. Mossbauer spectrum of magnetite nanoparticles of nearly 12 nm was resolved into two equally populated tetrahedral and octahedral Fe sites.

As the particle coarsen in the reduced atmosphere of $10^{-6}$ mbar, the population of the octahedral components increases with concomitant decrease in tetrahedral Fe [59]. Apart from the population, the isomer shift of both the sextet in the as-prepared nanoparticles has similar isomer shift (~0.44 mm/s). Subjection of the particles in high temperature coarsened the particles and the isomer shift of the tetrahedral sextet decreases. On the other hand, the isomer shift of the octahedral sextet increase. The decrease in the isomer shift of the tetrahedral sextet is due to the increase in the oxidation state from <3 to $3^+$. The increase in the isomer shift is understood as due to the decrease in the oxidation state from $3^+$ to $2.5^+$. The changes in the isomer shift were interpreted as the migration of $Fe^{2+}$ from the tetrahedral sites to octahedral sites. These conclude the magnetite with particle size near to superparamagnetic regime deviates from inverse spinel structure to mixed state.

In addition to the random occupation of $Fe^{2+}$ and $Fe^{3+}$ cations in tetrahedral and octahedral sites, nanoparticles often oxidized in normal atmosphere. As discussed earlier, Mossbauer spectroscopy fails to identify the transformation of stoichiometric magnetite to maghemite for the nanoparticles close to superparamagnetic regime. The determination of the $Fe^{2+}$ vacancies for the nanoparticles can be carried out with reference to the bulk magnetite photoacoustic spectra of different stoichiometry determined by Mossbauer spectroscopy. The rate of oxidation of the nanoparticles depends on the size of the particles. Photoacoustic spectra of the magnetite nanoparticles with size near to 9 nm show a similar spectrum of stoichiometric bulk magnetite. Decrease in the particle size, i.e., below 8.6 nm, the spectra resembles that of magnetite having $\delta < 0.30$. Further decrease in the size (~3 nm) the photoacoustic spectrum have equivalent features as that of maghemite having $\delta \sim 0.33$.

## Stability of the Magnetite Nanoparticles

Magnetite is a metastable polymorph of iron oxide. Moderate external thermal energy (>423 K) often oxidized to maghemite [60]. Heat treatment at elevated temperature (>673 K) transforms the structure from cubic spinel to orthorhombic hematite [61–63]. Apart from oxidation, the nanoparticles get agglomerated as time elapse and at the process of heat treatments [64]. The chemical stability of the magnetite can be enhanced by doping a trace amount of transition elements [65]. The maghemite to hematite transition temperature is observed to enhance with few decades of Kelvin after doping trace amount of transition elements. Trapping of alkali ions in the nanoparticles is also reported to enhance the stability of the magnetite nanoparticles [66].

The trapped nanoparticles in porous materials prevent agglomeration. These result to maintain its morphology at high temperature. The magnetite nanoparticles nucleated in the presence of zeolite maintains its particle size till 823 K [67]. The mean size of the particles derived from the XRD analysis shows an increase of about 1.8 nm in the annealing interval 300–823 K. A sharp increase is observed for annealing treatments beyond 850 K. Zeolite belongs to a group of crystalline aluminosilicate with the basic tetrahedral structural units of $SiO_4$ and $AlO_4$ with Si or Al in the center of the tetrahedron. Wherever the Si is replaced by Al, an electron has to be balanced. The aluminum sites act as cationic vacant sites. These provide an affinity to the cations and enhance stability. S. S. Pati et al. and L. Herojit Singh et al. investigated the composites of magnetite nucleated in the presence of zeolite [68]. The structural transformation from $\gamma$-$Fe_2O_3$ to $\alpha$-$Fe_2O_3$ in pure magnetite was obtained at 863 K. Introduction of trace amount of zeolite increase the transition temperature. Addition of 25 mg zeolite shifts the transition temperature by 72 K. Increase in the zeolite further enhances the transition temperature to 988 K by 75 mg of zeolite.

. Iron oxides nanoparticle due to strong magnetic attraction among the particles, the van der wall forces, and its high surface energy tend to agglomerate [69]. Iron oxide NPs have been shown to change their solubility, stability, and aggregation behavior in response to changes in pH, temperature, and organic matter concentration. The impact of pH on particle size distribution provides valuable information that could be used to distinguish the colloidal stability from the aggregation state of the various NPs. It is reported that the magnetite nanoparticles compared to the size distribution showed colloidal stability at pH = 6 and 7.5[70]. Depending upon the pH of the solution the surface of the magnetite will be positive or negative. Electrophoretic mobility measurement suggests that iron oxide NPs are highly positively charged at low pH values (i.e., pH 2–6) and that electrophoretic mobility (EPM) decreases with the increase in pH from 6 to 9 and becomes negative at pH 10, with a point of zero charge at approximately pH 9.1[71]. In iron oxide nanoparticles, the surface iron atoms act as Lewis acids. This coordinates with molecules that donate lone-pair electrons. Therefore, the Fe atoms coordinate with water in aqueous solutions, which dissociates readily to make the iron oxide surface hydroxyl functionalized. These hydroxyl groups may react with acids or bases as they are amphoteric in nature.

Sae-Yeol-Rim Paik et.al reported that the value of zeta potential at neutral pH of SMNPs was lower than that of IONPs. The zeta potential value far from zero indicates that the particles disperse well in the media since the electrostatic repulsive force among the particles is large, and thus, the particles have high aqueous stability. The dispersion of nanoparticles with a value of zeta potential far from zero was stable or relatively Monodisperse, while it is reported that a value close to zero showed that the monodispersity is poor [72].

## Applications of Magnetite and Its Composites

Engineering or designing different shape and size of nanoparticles further expand the scope of applications. S. Mathur et al. have synthesized vertical and horizontal nanorods of magnetite using catalysis assisted chemical vapor deposition process

[73]. Nano needle shape magnetite was synthesized by T. Togashi et al. by one-pot hydrothermal synthesis [74]. Many more techniques were discovered to synthesize one-dimensional magnetite [75, 76]. Solvothermal synthesis method is also adopted to obtain different shape of magnetite nanoparticles [77]. Spherical nanoparticles of various sizes are tuned using the organic solvent [78]. Circular nanoclusters of magnetite nanoparticles were prepared by solvothermal process [79]. The sodium acetate (NaOAc) plays an important role in the reduction of $Fe^{3+}$ and size of the nanoclusters. Coprecipitation method generally leads to spherical or edges shaped nanoparticles [80, 81]. However, double step synthesis, where precipitated iron hydroxides were subjected to high temperature in pressure with desired ionic solution results to different aspect ratio nanoparticles. Different shape nanoparticles have advantages over one another.

## *Bioapplications*

Distribution of drug-loaded nanoparticles to different parts of the body requires a path. Iron oxide nanoparticles can be used in drug delivery [82]. For these applications, the properties of iron oxide nanoparticles depend on their size and shape. Laurent S, Forge D, Port M et al. reported that the required size of the nanoparticles (inorganic core) for biomedical applications is below 100 nm [18]. Also, these applications required appropriate surface coating of the magnetic nanoparticles, which is non-toxic and biocompatible and must allow targetable delivery of drugs with particle in a specific area. The major disadvantage of most chemotherapies is that they are relatively non-specific which damage both sick and healthy cells [83]. MNPs can be used to solve this problem by delivering the drugs to the chosen target area, keeping the particles at the specific site during the drug release with the application of an external magnetic field is applied. MNPs possess a large surface-to-volume ratio that adsorbs proteins or load drugs. In order to prevent the magnetic nanoparticles from agglomeration and oxidation the nanoparticles are coated with a biocompatible polymer such as PVA. Also, inorganic coatings such as silica have been developed recently. The coating acts to shield the magnetic particle from the surrounding environment and can also be functionalized by attaching carboxyl groups, biotin, avidin, carbodi-imide, and other molecules [84]. Magnetic nanoparticles will target cancer cells and when exposed to an alternating magnetic field. As a result of which the nanoparticles incline to align with the applied field which absorbs energy. With the fluctuation of an external magnetic field, the magnetic nanoparticles convert this electromagnetic energy into heat and thus increasing the temperature at this region, destroying the cancer cells by the increased temperature (hyperthermia) or releasing of the anticancer healing agents at the specific targeted sites by the drug delivery base on hyperthermia technique [85].

## *Photocatalysis*

Iron-oxide nanoparticles have attracted much interest in many field including environmental purification [86]. Photocatalyst can be prepared from semiconductor materials such as magnetic nanoparticles with favorable optical band gap. Iron oxides nanoparticles are one of the promising material due to its photocatalytic properties, chemical stability, nontoxicity, and natural availability for separations in waste-water treatment [87, 88].Among the various oxides of iron, such as hematite ($\alpha$-$Fe_2O_3$), maghemite ($\gamma$-Fe2O3) and wustite ($Fe_{1-x}$ O), a-$Fe_2O_3$ is of special interest due to its unique catalytic properties [89]. $Fe_2O_3$ with suitable band-gap of ~2.2 eV is an interesting n-type semiconducting material which is suitable for photodegradation under visible light condition [90]. The better photocatalytic activity of iron oxide nanoparticles can be attributed to considerable generation of electron-hole pairs through the narrow band-gap illumination. These nanoparticles possessing catalytic properties are used in photocatalysis for safe and wastewater treatment. An example is the photodegradation of Congo red (CR) dye using iron oxide nanoparticles which were synthesized by thermal evaporation and co-precipitation approach [91].

Photocatalyst is defined as a material usually metal oxide that is capable of absorbing light, producing electron-hole pairs that enable chemical transformations of the reaction participants and regenerate its chemical composition. The photocatalytic process is focused to convert the highly toxic chemicals to either less toxic chemicals or $CO_2$ and $H_2O$. Two sources mainly constituted the photocatalytic activity of metal oxide. They are, (1) generation of OH radicals by oxidation of $OH^-$ anions. (2) Generation of $O_2{}^-$ radicals by reduction of $O_2$. In the last few decades, heterogeneous binary metal oxides photocatalyst such as $TiO_2$, $V_2O_5$, ZnO, $Fe_2O_3$, CdO, CdS, and $Al_2O_3$ [92] have been studied extensively for the removal of organic colored pollutants such as azo dye, acid orange 7 [93], methylene blue, alizarin S, Congo red, orange G, etc. $TiO_2$ was also used for the full oxidation, detoxification, and complete mineralization of dyes like alizarin S, methyl red, Congo red, methylene blue, crocein orange G dyes in aqueous solutions under irradiation of UV light [94].

## *Magnetic Nanoparticles Heavy Metal Removal*

Fresh water consists of only 2.5% of the earth's water and a meager amount is available for human consumption. Due to the increasing population and rising cost of portable water, the remediation of waste water for utilization is very important. At present, water pollution has alerted the researchers and forced to search for various activities to purify the contaminated water. The polluted waters are mainly from the waste products from the rapidly increasing industries, factories, laboratories, and chemicals such as insecticides and pesticides from agriculture. This polluted water mainly contains heavy metals which are highly toxic [95] and carcinogenic [96]. As a result, removal of heavy metal from waste water is a much for domestic use and consumption.

The removal of heavy metals from waste water is not an easy task. Various technologies have been developed for the removal of contaminants in waste water such as heavy metals such as arsenic, chromium, lead, copper, etc.[97]. Carbon-based nanomaterial, zerovalent metals, metal oxide base materials, and nanocomposites have proven efficient material for the removal of heavy metal from wastewater. Iron oxide based nanomaterials have received much attention in the recent years in removing heavy metals from waste water.

Iron oxide-based nanoparticles such as magnetite, maghemite, hematite, goethite, etc. have been reported in removing heavy metals from waste water in the recent years [98]. Maghemite and hematite ($\gamma$-$Fe_2O_3$) nanoparticles which possess large surface area which made it possible for high adsorption capacity has been reported extensively to treat heavy metals in wastewater [99]. Magnetite nanomaterial with its modified surfaces are also widely used adsorbent due to their physiochemical property, small size low cost, simplicity of use, and easily available [100]. The advantages of using magnetite and maghemite over hematite is that the magnetite and maghemite nanoparticles can be easily separated from wastewater after treatment with the addition of external magnetic field [101]. Moreover, the synthesis of magnetite and maghemite nanoparticles is simple and can be produced in large quantity using co-precipitation method or by Sol–Gel approach. They are eco-friendly without producing secondary pollution [102].

## Summary

Magnetic nanoparticles with its unique properties such as superparamagnetism, high surface-to-volume ratio, easy separation methodology, etc. have attracted much attention for various applications in bio application and water treatment, etc. For the applications, one needs to characterize the stoichiometry, stability, and morphology. These factors can be controlled by the synthesis process by optimizing the pH and temperature, capping with suitable polymers or inorganic compounds, and incorporating the magnetite in the pores of inorganic compounds. Zeolite nanoparticles localized the magnetite during synthesis providing a rigid template that stabilized the nanoparticles.

## References

1. Ali A., Zafar H., Zia, M., Haq2, I., Pulls, A.R., Ali, J.S., Hossain, A.: Synthesis, characterization, applications, and challenges of iron oxide nanoparticles. Nanotechnol. Sci. Appl. **9**, 49–67

2. Thapa, D., Palkar, V.R., Kurup, M.B., Malik, S.K.: Properties of magnetite nanoparticles synthesized through a novel chemical route. Mat. Lett. **58**, 2692–2694 (2004)
3. Wu, W., Wu, Z., Yu, T., Jiang C., Kim, W.-S.: Recent progress on magnetic iron oxide nanoparticles: synthesis, surface functional strategies and biomedical applications. Sci. Technol. Adv. Mater. **16**, 023501 (43 pp) (2015)
4. Ali, A., Zafar, H., Zia, M., Haq, I, Phull, A.R., Ali, J.S., Hussain, A.: Synthesis, characterization, applications, and challenges of iron oxide nanoparticles. Nanotechn. Sci. Appl. **9**, 49–67 (2016)
5. Kim, J.S., Yoon, T.J., Yu, K.N., Kim, B.G., Park, S.J., Kim, H.W., Lee, K.H., Park, S.B., Lee, J.-K., Cho, M.H.: Toxicity and tissue distribution of magnetic nanoparticles in mice. Toxicol. Sci. **89**, 1 338–347 (2006)
6. Nakashima, S., Fujita, K., Tanaka, K., Hirao, K., Yamamoto, T., Tanaka, I.: First-principles XANES simulations of spinel zinc ferrite with a disordered cation distribution. Phys. Rev. B. **75**, 174443 (2007)
7. Islama, M.S., Kurawakia, J., Kusumotoa, Y., Abdulla-Al-Mamuna, M., Bin Mukhli, M.Z.: Hydrothermal novel synthesis of neck-structured hyperthermia-suitable magnetic ($Fe_3O_4$, $\gamma$-$Fe_2O_3$ and $\alpha$-$Fe_2O_3$) nanoparticles. J. Sci. Res. **4**(1), 99–107 (2012)
8. Jie-feng, L., Tsai, C.J.: Hydrothermal phase transformation of hematite to magnetite. Nano Scale Res. Lett. **9**, 230 (2014)
9. Lassoued, A., Lassoued, M.S., Dkhil, B., Ammar, S., Gadri, A.: Photocatalytic degradation of methylene blue dye by iron oxide ($\alpha$-$Fe_2O_3$) nanoparticles under visible irradiation. J. Mater. Sci.: Mater. Electron. **29**, 8142–8152
10. Ohmori, T., Takahashi, H., Mametsuka, H., Suzuki, E.: Photocatalytic oxygen evolution on films using $Fe_3{}^+$ ion as a-$Fe_2O_3$ a sacrificial oxidizing agent. Phys. Chem. Chem. Phys. **2**, 3519–3522 (2000)
11. Nikumbh, A.K., Pawar, R.A., Nighot, D.V., Gugale, G.S., Sangale, M.D., Khanvilkar, M.B., Nagawade, A.V.: Structural, electrical, magnetic and dielectric properties of rare-earth substituted cobalt, ferrites nanoparticles synthesized by the co-precipitation method. J. Magu. Magn. Mater (2014). https://doi.org/10.1016/j.jmmm.2016.08.027
12. Koseoglu, Y., Alan, F., Tan, M., Yilgin, R., Ozturk, M.: Low temperature hydrothermal synthesis and characterization of Mn doped cobalt ferrite nanoparticles. Ceram. Int. (2012). https://doi.org/10.1016/j.ceramint.2012.01.001
13. Lane, M.K.M., Zimmerman, J.B.: Controlling metal oxide nanoparticle size and shape with supercritical fluid synthesis. Green Chem. **21**, 3769 (2019)
14. Daoush, W.M.: Co-precipitation and magnetic properties of magnetite nanoparticles for potential biomedical applications. J. Nanomed. Res. **5**(3), 00118 (2017)
15. Herojit Singh, L., Govindaraj, L., Mythili, A.R., Amarendra, G., Sondra, C.S.: Atomic scale study of thermal reduction of nano goethite coexisting with magnetite. AIP Adv. **3**, 022101 (2013)
16. Gnanaprakash, G., Mahadevan, S., Jayakumar, T., Kalyanasundaram, P., Philip, J., Raj, B.: Effect of initial pH and temperature of iron salt solutions on formation of magnetite nanoparticles. Mater. Chem. Phys. **103**, 168–175 (2007)
17. Jolivet, J.-P., Chanéac, C., Tronc, E.: Iron oxide chemistry. From molecular clusters to extend solid networks Chem. Commun. 481–487 (2004)
18. Laurent, S., Forge, D., Port, M., Roch, A., Robic, C., Vander Elst, L., Muller, R.N.: Chem. Rev. **108**, 2064–2110 (2008)
19. Cannas, C., Gatteschi, D., Musinu, A., Piccaluga, G., Sangregorio, C.: Structural and magnetic properties of $Fe_2$ nanoparticles disposed over a silica matrix, C. J. Phys. Chem. **102**, 7721 (1998).
20. Răileanu, M., Crian1, M., Petrache, C., Crian1, D., Jitianu1, A., Zaharescu1, M., Predoi D., Kuncser2 V., Filoti, G.: Sol-Gel $Fe_lO_n$–$Sio_2$ nanocomposites. Physics **50**(5–6), 595 (2005)
21. Gupta, A.K., Gupta, M.: Synthesis and surface engineering of iron oxide nanoparticles for biomedical applications. Biomaterials **26**, 3995–4021 (2005)

22. Cudennec, Y., Lecerf, A.: The transformation of ferrihydrite into goethite or hematite, revisited". J. Solid State Chem. **179**, 716–722 (2006)
23. Hamzaoui, A., Mgaidi, A., Megriche, El Maaoui, M.: Kinetic study of goethite formation from ferrihydrite in alkaline medium. Ind. Eng. Chem. Res. **41**, 5226–5231 (2002)
24. Blanco-Andujar, C., Ortega, D., Pankhurstab, Q.A., Thanh, N.T.K.: Elucidating the morphological and structural evolution of iron oxide nanoparticles formed by sodium carbonate in aqueous medium. J. Mater. Chem. **22**, 12498 (2012)
25. Ling, W., Wang, M., Xiong, C., Xie, D., Chen, Q., Chu, X., Qiu, X., Li, Y., Xiao, X.: Synthesis, surface modification, and applications of magnetic iron oxide nanoparticles. J. Mater. Res. **11**, 34 (2019)
26. Peng, W., Huang, X., Yu, J., Jiang, P., Liu, W.: Electrical and thermophysical properties of epoxy/aluminum nitride nanocomposite: effects of nanoparticle surface modification. Composites: Part A **41**, 1201–1209 (2010)
27. Wu, J., Mao, S., Ye, Z.G., Xie, Z., Zheng, L.: Room-temperature weak ferromagnetism induced by point defects in $\alpha$-$Fe_2O_3$. ACS Appl. Mater. Interfaces **2**(6), 1561–1564 (2010)
28. Eid, C., Luneau, D., Salles, V., Asmar, R., Monteil, Y., Khoury, A., Brioude, A.: Magnetic properties of hematite nanotubes elaborated by electrospinning process. J. Phys. Chem. C **115**, 17643–17646 (2011)
29. Verma, S., Kumar, S., Gokhale, R., Burgess, D.J.: Physical stability of nanosuspensions: Investigation of the role of stabilizers on Ostwald ripening. Int. J. Pharm. **406**, 145–152 (2011)
30. Phan, C.M., Nguyen, H.M.: Role of capping agent in wet synthesis of nanoparticles. J. Phys. Chem. A **121**, 3213–3321 (2017)
31. Zhang, L., He, P., Hong-Chen, G.: Oleic acid coating on the monodisperse magnetite nanoparticles. Appl. Surf. Sci. **253**, 2611–2617 (2006)
32. Herojit Singh, L., Govindaraj, R., Amarendra, G., Sundar, C.S.: Atomic scale study on the thermal evolution of local structure and magnetic properties in oleic acid coated iron oxide nanoparticles. J. Phys. Chem. C **117**, 25042–25051 (2013)
33. Amstad, E., Gillich, T., Bedeck, I., Tester, M., Reimhult, E.: Ultra stable iron oxide nanoparticle, colloidal suspensions using dispersants with catechol-derived anchor groups. Nano Lett. **9**(12), 4042–4048 (2009)
34. Amstad, E., Gehring, A.U., Fischer, H., Nagaiyanallur, V.V., Hähner, G., Tutor, M., Reimhult, E.: Influence of electronegative substituents on the binding affinity of catechol-derived anchors to $Fe_3O_4$ nanoparticles. J. Phys. Chem. C **115**, 683–691 (2011).
35. Lu, Y., Yin, Y., Mayers, B.T., Xia, Y.: Modifying the surface properties of superparamagnetic iron oxide nanoparticles through a sol-gel approach. Nano Lett. **2**, 183–186 (2002)
36. Stöber, W., Fink, A., Bohn, E.: Controlled growth of monodisperse silica spheres in the micron size range. J. Colloids Interface Sci. **26**, 62–69 (1968)
37. Ohmori, M., Matijević, E.: Preparation and properties of uniform coated colloidal particles. VII. Silica on hematite. J. Colloids Interface Sci. **150**, 594–598 (1992)
38. Gan, Q., Lu, X., Yuan, Y., Qian, J., Zhou, H., Lu, X., Shi, J., Liu, C.: A magnetic, reversible pH-responsive nanogated ensemble based on $Fe_3O_4$ nanoparticles-capped mesoporous silica. Biomaterials **32**, 1932–1942 (2011)
39. Santra, S., Zhang, P., Wang, K., Tapec, R., Tan, W.: Conjugation of biomolecules with luminophore-doped silica nanoparticles for photostable biomarkers. Anal. Chem. **73**(20), 4988–4993 (2001)
40. Luiz, C.A., Oliveira, R., Rios, V.R.A., Fabris, J.D., Garg, V., Sapag, K., Lag, R.M.: Activated carbon/iron oxide magnetic composites for the adsorption of contaminants in water. Carbon **40**, 2177–2183 (2002).
41. Rabelo, D., Lima, E.C.D., Reis, A.C., Nunes, W.C., Novak, M.A., Garg, V.K., Oliveira, A.C., Morais, P.C.: Preparation of magnetite nanoparticles in mesoporous copolymer template. Nano Lett. **1**, 105–108 (2001)
42. Coutinho, F.M.B., Rabelo, D.: Scanning electron microscopy study of styrene-divinylbenzene copolymers. Eur. Polym. J. **28**, 1553–1557 (1992)

43. Do, B.P.H., Nguyen, B.D., Nguyen, H.D., Nguyen, P.T.: Synthesis of magnetic composite nanoparticles enveloped in copolymers specified for scale inhibition application. Adv. Nat. Sci.: Nanosci. Nanotechnol. **4**, 045016 (7 pp) (2013)
44. Chen, M.L., He, Y.J., Chen, X.W., Wang, J.H.: Quantum dots conjugated with $Fe_3O_4$-filled carbon nanotubes for cancer-targeted imaging and magnetically guided drug delivery. Langmuir **28**, 16469–16476 (2012)
45. Breiner, J.M., Mark, J.E.: Preparation, structure, growth mechanisms and properties of siloxane composites containing silica, titania or mixed silica-titania phases. Polymer **39**, 5486–5493 (1998)
46. Sanchez, C., Julián, B., Belleville, P., Popall, M.: Applications of hybrid organic-inorganic nanocomposites. J. Mater. Chem. **15**, 3559–3592 (2005)
47. Cvetkov1, A.V., Gango1, S.E., Ivanova, M.S., Khanin, S.D., Lukin1, A.E., Solovyev1, V.G., Trifonov, S.V., Veisman, V.L.: Electrical and optical properties of iodide nanoparticles embedded into zeolite porous matrices. J. Phys.: Conf. Ser. **929**, 012042 (2017)
48. Herojit Singh, L., Govindaraj, R., Amarendra, G., Sundar, C.S.: Local structure and magnetic properties of cubic iron oxide nanoparticles formed in zeolite as deduced using Mossbauer spectroscopy. Appl. Phys. Lett. **103**, 193104 (2013)
49. Cao, L., Xu, F., Liang, Y.-Y., Li, H.-L.: Preparation of the novel nanocomposite $Co(OH)_2$/ultrasatble T-zeolite and its application as a supercapacitor with high energy density. Adv. Mater. **16**(20 Oct 18) (2004)
50. Ahn, T., Kim, J.H., Yang, H.M., Lee, J.W., Kim, J.D.: Formation pathways of magnetite nanoparticles by coprecipitation method. J. Phys. Chem. C **116**, 6069–6076 (2012)
51. Ayyappan, S., Gnanaprakash, G., Panneerselvam, G., Antony, M.P., Philip, J.: Effect of surfactant monolayer on reduction of $Fe_3O_4$ nanoparticles under vacuum. J. Phys. Chem. C **112**, 18376–18383 (2008)
52. Klencsár, Z., Ábrahám, A., Szabó, L., Szabó, E.G., Stichleutner, S., Kuzmannd, E., Homonnay, Z., Tolnai, G.: The effect of preparation conditions on magnetite nanoparticles obtained via chemical co-precipitation Mater. Chem. Phys. **223**, 122–132 (2019)
53. Gorski, C.A., Cherer, M.M.: Determination of nanoparticulate magnetite stoichiometry by Mossbauer spectroscopy, acidic dissolution, and powder X-ray diffraction: a critical review. Am. Mineral. **95**, 1017–1026 (2010)
54. Herojit Singh, L., Pati, S.S., Guimarães, E.M., Oliveira, A.C., Garg, V.K.: Synthesis, structure, morphology and stoichiometry characterization of cluster and nano magnetite. Mater. Chem. Phys. **178**, 182–189 (2016)
55. Shipilina, M.A., Zakharovab, I.N., Shipilinc, A.M., Bachurin, V.I.: Mössbauer studies of magnetite nanoparticles. J. Surf. Invest. X-ray Synchr. Neutron Tech. **8**(3), 557–561 (2014)
56. Gee, S.H., Hong, Y.K., Erickson, D.W., Park, M.H.: Synthesis and aging effect of spherical magnetite, $Fe_3O_4$... nanoparticles for biosensor applications. J. Appl. Phys. **93**(10), Parts 2 & 3 (2003)
57. He, Y.P., Miao, Y.M., Li, C.R., Wang, S.Q., Cao, L., Xie, S.S., Yang, G.Z., Zou, B.S.: Size and structure effect on optical transitions of iron oxide nanocrystals. Phys. Rev. B **71**, 125411 (2005)
58. Joosa, A., R¨umenappa, C., Wagnerb, F.E., Gleicha, B.: Characterization of iron oxide nanoparticles by Mossbauer spectroscopy at ambient temperature. J. Magn. Magn. Mater. **399**, 123–129 (2015)
59. Herojit Singh, L., Panda, A., Govindaraj, R., Mythili, R., Amarendra, G.: Implications of linear correlation between hyperfine parameters in iron oxide nanoparticles. Mater. Chem. Phys. **214**, 440–448 (2018).
60. de Faria, D.L.A., Venancio Silva, S., de Oliveira, M.T.: Microspectroscopy of some iron oxides Raman and oxyhydroxides. J. Raman Spectrosc. **28**, 873–878 (1997)
61. Chamritski, I., Burns, G.: Infrared- and Raman-active phonons of magnetite, maghemite, and hematite: a computer simulation and spectroscopic study. J. Phys. Chem. B **109**, 4965–4968 (2005)

62. Campos, E.A., Pinto, D,V.P.S., de Oliveira, J.I.S., Mattos, E.C., Dutra, R.C.L.: Synthesis, characterization and applications of iron oxide nanoparticles—A short review. J. Aerosp. Technol. Manag. **7**(3), 267–276
63. Yen, F.S., Chen, W.C., Yang, J.M., Hong, C.T.: Crystallite size variations of nanosized $Fe_2O_3$ powders during $\gamma$- tor-phase transformation. Nano Lett. **2**(3) (2002).
64. Lu, A.H., Salabas, E.L., Schuth, F.: Magnetic nanoparticles: synthesis, protection, functionalization, and application. Angew. Chem. Int. Ed. **46**, 1222–1244 (2007)
65. Bliem, R., Pavelec, J., Jamba, O., McDermott, E., Wang, Z., Gerhold, S., Wagner, M., Osiecki, J., Schulte, K., Schmid, M., Blaha, P., Diebold, U., Parkinson, G.S.: Adsorption and incorporation of transition metals at the magnetite $Fe_3O_4$(001) surface. Phys. Rev. B **92**, 075440 (2015)
66. Tajabadi M., Khosroshahi, M.E.: Effect of alkaline media concentration and modification of temperature on magnetite synthesis method using $FeSO_4/NH_4OH$. Int. J. Chem. Eng. Appl. **3**(3), (2012)
67. Singh, L.H., Pati, S.S., Sales, M.J.A., Guimarães, E.D.M., Oliveira, A.C., Garg, V.K.: Facile method to tune the particle size and thermal stability of magnetite nanoparticles. J. Braz. Chem. Soc. **26**(11) (2015)
68. Herojit Singh, L., Pati, S.S., Coaquira, J.A.H., Matilla,J., Guimarães, E.M., Oliveira, A.C., Kuzmann, E., Garg, V.K.: Magnetic interactions in cubic iron oxide magnetic nanoparticle bound to zeolite. J. Magn. Magn. Mater. **416**, 98–102 (2016)
69. Xia, T., Wang, J., Wu, C., et al.: Novel complex-coprecipitation route to form high quality triethanolamine-coated $Fe_3O_4$ nanocrystals: their high saturation magnetizations and excellent water treatment properties. Cryst. Eng. Comm. **14**(18), 5741–5744 (2012)
70. Demangeat, E., Pédrot, M., Dia, A., Bouhnik-le-Coz, M., Grasset, F., Hanna, K., Kamagate, M., Cabello-Hurtado, F.: Colloidal and chemical stabilities of iron oxide nanoparticles in aqueous solutions: the interplay of structural, chemical and environmental drivers. Environ. Sci.: Nano **5**, 992–1001 (2018)
71. Baalousha, M., Manciulea, A., Cumberland, S., Kendall, K., Lead, J.R.: Aggregation and surface properties of iron oxide nanoparticles: influence of pH and natural organic matter. Environ. Toxicol. Chem. **27**(9), 1875–1882 (2008)
72. Paik, S.-Y.-R., Kim, J.S., Shin2, S.J., Ko, S.: Characterization, quantification, and determination of the toxicity of iron oxide nanoparticles to the bone marrow cells. Int. J. Mol. Sci. **16,** 22243–22257 (2015)
73. Mathur, S., Barth, S., Werner, U., Hermandez-Ramirez, F., Romano-Rodriguez, A.: Adv. Matter **20**, 1550 (2008)
74. Togashi, T., Umetsu, M., Naka, T., Ohara, S., Hatakeyama, Y., Adschiri, T.: One-pot hydrothermal synthesis of an assembly of magnetite nanoneedles on a scaffold of cyclic-diphenylalanine nanorods. J. Nanopart. Res. **13**, 3991 (2011)
75. Dawei Su, H.J.A., Wang, G.: One-dimensional magnetite $Fe_3O_4$ nanowires as electrode material for Li–ion batteries with improved electrochemical performance. J. Power Sources **244,** 1–52 (2012)
76. Jianchao Ma, J.C., Hongzhu Ma, D.Z., Qingliang Ma, S.W.: Lengthy one-dimensional magnetite (Fe3O4) sub-microfibers with excellent electrochemical performance. J. Colloid Interface Sci. **498**, 282–291 (2017)
77. Tan, Y., Binbin, Yu., Li, X., Li, K.: Facile solvothermal synthesis of monodisperse $Fe_3O_4$ nanocrystals with precise size control of one nanometer as potential MRI contrast agents. J. Mater. Chem. **21**, 2476 (2011)
78. Chen, F., Xie, S., Zhang, N., Liu, R.: Synthesis of spherical $Fe_3O_4$ magnetic nanoparticles by co-precipitation in choline chloride/urea deep eutectic solvent. Mat. Lett. **112**, 177–179 (2013)
79. Sergio, I., Madrid, U., Pal, U., Sánchez-De Jesús, F.: Controlling size and magnetic properties of $Fe_3O_4$ clusters in solvothermal process. Adv. Nano Res. **2**(4), 187–198 (2014)
80. Hua, C.C., Zakaria, S., Farahiyan, R., Khong, L.T., Nguyen, K.L., Abdullah, M., Ahmad, S.: Size-controlled synthesis and characterization of $Fe_3O_4$ nanoparticles by chemical coprecipitation method. Sins Malaysiana **37**(4), 389–394 (2008)

81. Rashida, H., Mansoor, M.A., Haiderd, B., Nassir, R., Hamid, S.B.A., Abdulrahman, A.: Synthesis and characterization of magnetite nano particles with high selectivity using in-situ precipitation method. Separ. Sci. Technol. J. Homepage (2019). ISSN 0149–6395 (Print) 1520–5754 (Online)
82. Mahdavi, M., Ahmad, M.B., Haron, M.J., Namvar, F., Nadi, B., Rahman, M.Z., Amin, J.: Synthesis, surface modification and characterisation of biocompatible magnetic iron oxide nanoparticles for biomedical applications. Molecules **18**, 7533–7548 (2013)
83. Khiabania, S.S., Farshbafb, M., Akbarzadeha, A., Deveron, S.: Magnetic nanoparticles: preparation methods, applications in cancer diagnosis and cancer therapy. Artif. Cells Nanomed. Biotechnol. (2016). https://doi.org/10.3109/21691401.2016.1167704
84. Kopčanský, P., Tomašovičová, N., Koneracká, M., Timko, M., Závišová, V., Tomčo, L.: Magnetic nanoparticles in magnetic fluids. Acta Electrotechnica Et Informatica **10**(3), 10–13 (2010)
85. Abenojar, E.C., Wickramasinghe, S., Bas-Concepciona, J., Samia, A.C.: Structural effects on the magnetic hyperthermia properties of iron oxide nanoparticles. Progr. Nat. Sci.: Mater. Int. **26,** 440–448 (2016)
86. Santhosh, C., Malathi, A., Dhaneshvar, E., Bhatnagar, A., Grace, A.M., Madhavan, J.: Iron oxide nanomaterials for water purification. Nanoscale Mater. Water Purif. 431–446 (2019). https://doi.org/10.1016/B978-0-12-813926-4.00022-7
87. Thangave, S., Raghavan, N., Venugopal, G.: Magnetically separable iron oxide-based nanocomposite photocatalytic materials for environmental remediation. In: Photocatalytic Functional Materials for Environmental Remediation, 1st edn. Wiley (2019)
88. Piao, X., Zeng, G.M., Huang, D.L., Feng, C.L., Hu, S., Zhao, M.H., Lai, C., Wei, Z., Huang, C., Xie, G.X., Liu, Z.F.: Use of iron oxide nanomaterials in wastewater treatment: a review. Sci. Total Environ. **424**, 1–10 (2012)
89. Show, B., Mukherjee, N., Mondal, A.: $\alpha$-$Fe_2O_3$ nanospheres: facile synthesis and highly efficient photo-degradation of organic dyes and surface activation by nano-Pt for enhanced methanol sensing (2012)
90. Mishra, M., Chun, D.M.: $\alpha$-$Fe_2O_3$ as a photocataltic material—A review. Appl. Catal. A **498**, 126–141 (2015)
91. Khedr, M.H., Abdel Halim, K.S., Soliman, N.K.: Synthesis and photocatalytic activity of nano-sized iron oxides. Mater. Lett. **63**, 598–560 (2009).
92. Karunakaran, C., Senthilvelan, S.: $Fe_2O_3$-photocatalysis with sunlight and UV light: oxidation of aniline. Electrochem. Commun. **8**, 95–101 (2006)
93. Vinodgopal, K., Wynkoop, D.E., Kamat, P.V.: Environmental photochemistry on semiconductor surfaces: Photosensitized degradation of a textile azo dye, Acid Orange 7, on $TiO_2$ particles using visible light. Environ. Sci. Technol. **30**, 1660–1666 (1996).
94. Lachheb, H., Puzenat, E., Houas, A., Ksibi, M., Elaloui, E., Guillard, C., et al.: Photocatalytic degradation of various types of dyes (Alizarin S, Crocein Orange G, Methyl Red, Congo Red, Methylene Blue) in water by UV-irradiated titania. Appl. Catal. B Environ. **39**, 75–90 (2006)
95. Afroze, S., Sen, T.K.: A review on heavy metal ions and dye adsorption from water by agricultural solid waste adsorbents. Water Air Soil Pollut. **229**, 225 (2018)
96. Cocârţă, D., Neamţu, S., Deac, A.R.: Carcinogenic risk evaluation for human health risk assessment from soils contaminated with heavy metals. Int. J. Environ. Sci. Technol., **13**, 2025–2036 (2016)
97. Yang, J., How, B., Wang, J., Tian, B., Bi, J., Wang, N., Li, X., Huang, X.: Nanomaterials for the removal of heavy metals from wastewater. Nanomaterials **9**, 424 (2019). https://doi.org/10.3390/nano9030424.
98. Chowdhury, S.R., Yanful, E.K.: Arsenic and chromium removal by mixed magnetite maghemite nanoparticles and the effect of phosphate on removal. J. Environ. Manage. **91**, 2238–2247 (2010)
99. Dave1, P.N., Chordae, L.V.: Application of iron oxide nanomaterials for the removal of heavy metals. J. Nanotechnol. Article ID 398569, 1–14 (2014). https://doi.org/10.1155/2014/398569

100. Giraldo, L., Erto A., Moreno-Pirajań, J.C.: Magnetite nanoparticles for removal of heavy metals from aqueous solutions: synthesis and characterization. Adsorption **19**, 465–474 (2013)
101. Hu, J., Chen, G., Lo, I.M.C.: Removal and recovery of Cr(VI) from wastewater by maghemite nanoparticles. Water Res. **39**, 4528–4536 (2005)
102. Tuutijärvi, T., Lu, J., Sillanpää, M., Chen, G.: As (V), adsorption on maghemite nanoparticles. J. Hazard. Mater. **166**, 1415–1420 (2009)

# Chapter 15
# New Insights on MXenes: Synthesis and Their Uses in Energy Storage and Environmental Applications

**Konthoujam Priyananda Singh and L. Herojit Singh**

## Introduction

2D materials are believed to give a tremendous impact in the future. Though the unique physio-chemical properties of 2D materials are widely studied and known at the research level, the mass scale production of 2D materials are at infant stage. There are always several technological and practical challenges lying in this venture of exploiting full potential of 2D materials known so far. For instance, graphene is one such material which is studied most in the last few decades but, their application in large scale is quite limited [1].

### *Why Are 2D Materials Promising?*

2D materials have been emerged as one of the most promising candidates for electronic and energy applications, mainly due to their unique and special properties. For example, recent studies in 2D carbon materials like carbon nanotubes (CNT), fullerene, and graphene have reported enhanced electrochemical performance owing to their high electrical conductivity, large specific surface areas, and outstanding mechanical strength. In addition to this, 2D materials, such as TMDCs and black phosphorene, are also known for exhibiting comparatively higher energy storage potential.

K. Priyananda Singh · L. Herojit Singh (✉)
Department of Physics, National Institute of Technology Manipur, Manipur, India
e-mail: loushambam@gmail.com

K. Priyananda Singh
e-mail: pknandasingh@gmail.com



B. P. Swain (ed.), *Nanostructured Materials and their Applications*, Materials Horizons: From Nature to Nanomaterials, https://doi.org/10.1007/978-981-15-8307-0_15

Here, it's pertinent to discuss why the 2D materials are so special and different from their bulk counterparts [2]. For instance, how do fullerenes (0D), nanotubes (1D), graphene (2D), and graphite (3D) are so different from each other? The answer lies in the fact that it's not only the size but also dimensionality that determines material properties. Restrictions of size in certain dimension or dimensions effectively alter material properties- optical, electronic, chemical, mechanical, etc. This is mainly due to confinement of electrons only in limited dimensions, absence of interlayer interactions, high surface to bulk ratio, etc. This unique property of tunability of their material properties opens up a new avenue for 2D materials for various interesting applications.

## *Mxenes-A New Class of 2D*

Some of the most studied 2D materials are graphene, phosphorene, COFs (covalent Organic Frameworks), black phosphorous, hexagonal boron nitride, TMDs (transition metal dichalcogenides) like Molybdenum disulfide, tungsten disulfide [3], etc. The top-down exfoliation technique using suitable etchants is one of the most common synthesis methods. In this backdrop, some MAX family has been reported to be exfoliated into a new class of 2D materials of metal carbides or nitrides called MXenes. The combination of hydrophilicity and high electrical conductivities has rendered MXenes promising candidates for energy storage application, electromagnetic interference shielding (EIS), transparent conductive electrodes, environmental applications, etc.

## *2D Materials and Applications*

High energy storage capacity batteries are of great demand in this energy-driven modern world. In this context, the poor rate of charge and low energy density are two major challenges. They are mainly due to the use of graphite as negative electrode. Because graphite electrode forms Solid Liquid interface (SLI) with the electrolyte during charging. To resolve this shortfall, massive research is going on to find the alternatives to this material. 2D MXenes one of the pioneering competitors in this front [4].

Also, MXenes based 2D nanomaterials can play a vital role in combating the menace of environmental degradations. Many studies have reported the exceptionally high adsorption capacity of MXenes in the removal of various organic and inorganic materials, mainly due to large surface areas endowed with various functional groups. They also exhibit higher water permeability capacity than other commercial membranes. MXenes can be used in several other potential applications, which will be discussed broadly in this chapter.

## MXenes

Historically, the study of 2D metal carbides, nitrides, and carbonitrides was started in 2011 for the first time, when a new class of family called MXenes was synthesized. The first MXene $Ti_3C_2$ was discovered by a group of researchers in Drexel university through the chemical exfoliation of $Ti_3AlC_2$. Since then, there had been mounting interest in these materials significantly. These materials were used for the first time in lithium-ion batteries for energy storage purposes. Since then, there has been rapid expansion on the number of new MXenes and exploration of various interesting applications.

In order to understand this relatively new but quick growing 2D family, let us first study the parent compound from which they are being produced, i.e., MAX phases.

### *What Are MAX Phases?*

MAX is a big family of ternary (composed of three) carbides. They possess the qualities of both metal as well as ceramics [5]. MAX phases are composed of hexagonally layered structure of transition metal carbides, nitrides, or carbonitrides. MAX phases have general formula $M_{n+1}AX_n$, where $n = 1, 2, 3$; M is transition metals namely, Sc, Ti, Ta, Nb, W, Mo, V, Mn, Hf, Mn, Y, and Zr. The potential A elements are Al, Si, P, S, Ga, As Pb, Tl, etc.

Depending upon the possible value of '*n*', MAX phases are broadly classified into three types viz, 211 ($M_2AX$) type, 312 ($M_3AX_2$) type and 413 ($M_4AX_3$) type structures. For example, considering $Ti_2AlC$, the small difference between these three phases namely, MAX 211, 312, and 413 phases lies in the number of Titanium (Ti) layers between each two Aluminum (Al) layers, i.e., two, three, and four Titanium (Ti) layers, respectively. Moreover, the number of C layers is always one layer less than the number of Ti layers. It is from these parent phases that MXenes can be exfoliated.

MAX phases are of broadly classified into two groups- oMAX (the phases with out of plane ordered double transition metals) and iMAX (in plane ordered double transition metals). Out of the two MAX phases viz, oMAX and iMAX, each atom in oMAX forms hexagonal lattice whereas iMAX atoms exhibit deviation from that of hexagonal lattice positions. This can be explained as an effect of different sizes of transition metal atoms occupying the same layers.

### *MXenes Versus MAX Comparison*

Compared to MXenes, the applications of MAX phases are majorly confined to their temperature stabilities, good metallic, and better mechanical properties. Whereas, the 2D MXenes exfoliated from such bulk phases exhibit entirely vast, different yet interesting properties. Many such unique electronic, optical, magnetic, chemical, and electronic properties have been predicted theoretically but are yet to be realized experimentally. In fact, the qualities and end-properties of MXenes so exfoliated have a close connection to that of parent MAX phases.

## *What Are MXenes?*

MXenes are a new class of solid 2D transition metal carbides, nitrides or carbonitrides, discovered in 2011 [6]. They have the formula $M_{n+1}X_nT_x$, where $n = 1, 2$ or 3, M is an early transition metal from group III B to VII B, such as Ti, V, Nb, Ta, Cr, and Mo, X is carbon and/or nitrogen, and $T_x$ refers to surface functional groups such as OH, O, and/or F. So far, nearly 30 MXenes have been reported. Some of them are $Ti_2CT_x$, $NbC_3T_x$, $V_2C$, $Hf_3C_2T_z$ [7–9] etc. The term "MXene" was particularly given to distinguish these new 2D materials from graphene and considering its connection to parent MAX phase. The name applies to both the original MAX phases and MXenes fabricated from them [10].

## *Crystal Structure of MXenes*

MXenes are hexagonal 2D materials, derived from MAX phases. They have layered hexagonal, similar to that of graphene and a space group symmetry of P63/mmc (No. 194). Each layer of X (carbon or nitrogen) atoms is sandwiched between two layers of transition metals and every two $M_{n+1}X_n$ layers are interleaved with a layer of A atoms. During an acid treatment, the "A" atoms are washed from the MAX phase structure, resulting in multiple layers of $M_{n+1}X_n$, which can be dispersed into monolayers or a few layers by using sonication.

In similar to their MAX precursors, M atoms in these 2D materials are close-packed and X atoms occupy the octahedral interstitial sites. Although the overall crystal structure of MXenes is hexagonal close-packed, the ordering of M atoms varies from $M_2X$ to $M_3C_2$ and $M_4C_3$. For example, M atoms in $M_2X$ follows ABABAB ordering (hexagonal close-packed stacking), while M atoms in $M_3C_2$ and $M_4C_3$, have ABCABC ordering (fcc stacking) [11].

## *Rational Behind Treating MXenes as New Class of Solids*

Unlike other 2D materials like graphene and phosphorene, which respectively have graphite and black phosphorus natural 3D bulk precursors, MXenes do not have a straightforward 3D precursor in nature. Rather they are produced by selective etching of A atoms from the MAX phases. In fact, MXenes have gained nickname "conductive clays" as they are the only polycrystalline solids in which single grains can deform by a combination of slip, kink band formation, and de-laminations. The extent end members that were first synthesized in powder form by which grains can delaminate and deform at room temperature is unique.

Basal plane dislocations are mobile and multiply at room temperature, which is a unique quality. They also exhibit excellent conduction of heat and elasticity, high monolayer elastic properties and bending rigidity, thereby making them potential nanofillers for reinforced composites.

## Synthesis

The two distinct synthesis techniques of MXenes can be categorized as, i.e., top-down and bottom-up. They are usually utilized for the synthesis of different MXenes as single, few, or multilayer sheets top-down methods especially liquid-based processes have been extensively employed relative to bottom-up methods that have seen rare utilization.

Though chemical exfoliation is more widely used, mechanical exfoliation also can be applied for synthesizing MXenes [12]. $Mo_2CT_x$ and $Zr_3C_2T_x$ are examples of MXenes synthesized from the non-MAX phases $Mo_2Ga_2C$ and $Zr_3Al_3C_5$.[13, 14] (Table 15.1).

**Table 15.1** Timeline: Emergence of MXenes

| Year | Development | Synthesis method/technique | Example | Refs. |
|---|---|---|---|---|
| 2011 | Discovery of $Ti_3C_2$ (First MXene) | Wet etching Method using HF | $Ti_3C_2T_x$ | [6] |
| 2012 | Different variety of new MXenes | Wet etching method | $(Ti,Nb)_2CT_x$, $(V,Cr)_3C2T_x$, $Ta_4C_3T_x$,etc | [38] |
| 2013 | Isolated single layer MXene | Intercalation and delamination | | [15] |
| 2014 | Clay-like MXenes | Clay method using in-situ LiF/HCl etchants | Clay-like $Ti_3C_2T_x$ | [42] |
| 2015<br>2016<br>2017 | Large scale MXene synthesis<br>Large flake MXenes<br>Development of ordered divacancies | Amine-assisted TBAOH method<br>MILD (without sonication) method<br>Synthesized through in-plane double transition metal MAX phase | $Mo_2TiC_2T_x$, $Cr_2TiC_2T_x$<br>$Mo_2Ti_2C_3T_x$<br>Single flake $Ti_3C_2T_x$ having size greater than 2 μm<br>($Mo_{1.3}CT_x$) | [44]<br>[41]<br>[3] |

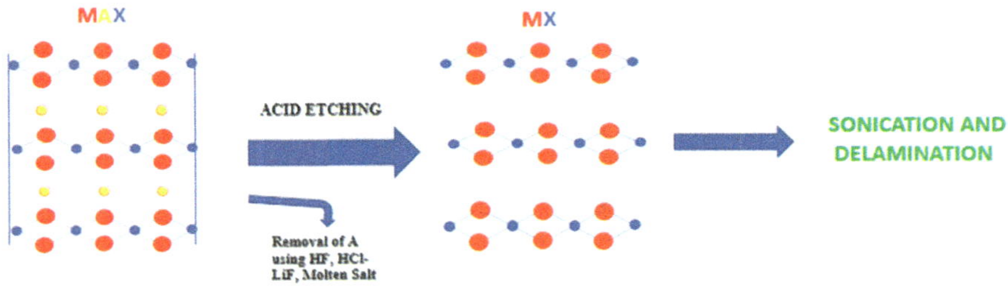


**Fig. 15.1** Synthesis process of MXenes: Starting from MAX precursor, then selective removal of A atom through acid etching and finally followed by sonication and delamination

## *Top-Down Wet Etching Method*

Like many other 2D material synthesis, MXenes are also synthesized by selectively etching out a single or multiple layer from a big class of ternary metal carbide, nitride, and carbonitride family, generally known as MAX family. But, unlike the other 2D material which are found by etching out weak covalent bond from the layered bulk materials, the MXenes are synthesized by etching out relatively strong interlayer bonds.

This top-down synthesis method comprises three major steps. First, the synthesis starts with etching the A-element from the MAX phases, thereby producing multi-layered MXene powder. These layers are held together by hydrogen and weak van der Waals bonds [15]. Second, washing through repeated centrifugation is carried out and a pH value of around 6 is being maintained. Third, a suitable intercalation and delamination method is being chosen to isolate the delaminated material from the multilayered structure [16].

In this process, different etchants are being used. Depending on different etchant used, the quality of MXenes will be different. Some of the most commonly used etchants are HF, LiF-HCl mixture, $NH_4HF_2$ etc.

Generally, $M_4AX_3$ (413 phase) demands more aggressive etching conditions than $M_3AX_2$ (312 phase) and $M_2AX$ (211 phase). Out of 12 possible A (transition metal) elements present in MAX phases, only Aluminum (Al) and Silicon (Si) have been successfully etched extracted from the parent MAX phases to produce MXenes. Recently, Etching Si from $Ti_3SiC_2$ to form $Ti_3C_2T_x$ MXene was reported (Fig. 15.1).

## *Bottom-Up Method*

The bottom-up method of MXenes synthesis mainly involves two broad steps viz; deposition of MAX thin film along with compositional atoms, followed by the removal of A atom resulting in subsequent formation of MXene thin film. One such example of epitaxial growth of MXene thin film is formation of $Ti_3C_2T_x$ in which MAX $Ti_3AlC_2$ thin films were first deposited on sapphire substrate using sputtering technique, followed by treatment of the MAX deposited substrate with suitable etching agents. Lai et al. [17] have synthesized $Mo_2CT_x$, with a starting MAX thin film comprising $Mo_2GaC$ phase as a hexagonal ternary laminated carbide [18]. L. Wang et al. also reported the synthesis of Molybdenum carbide MXenes film over a Cu/Mo alloyed substrate through Chemical Vapor Deposition (CVD) method.

### *Recent Attempts for Large-Scale Preparation of Mxenes*

The use of greater efficient intercalants like TBAOH (tetra-butylammonium hydroxide), iso-propylamine, etc. in addition to earlier intercalants like HF, LiF, HCl, $NH_4HF_2$, etc. led to the emergence of two new families of MXenes containing two transition metals. This amine and TBAOH assisted delamination techniques have not only expanded this 2D family but also increased their yields.

In order to break the strong bond between M and A elements in MAX phases, etching is necessary for synthesizing MXenes. This can be done either by adding chemical etchants like HF, LiF + HCl, $NH_4HF_2$, etc. or mechanically. But, mechanical exfoliation is comparatively difficult and used.

Many computational studies have theoretically predicted the possibility of breaking M–A–M bonds to etch out A-atoms by applying suitable mechanical stress in such a way the M–X–M bonds are kept untouched. To succeed this, the ideal choice of material will be a MAX precursor having weak M–A bond as well as strong M–X bond at the same time.[15].

### *Synthesis of MXenes from Non-MAX Phases*

$Mo_2Ga_2C$ and $Zr_3Al_3C_5$ are examples of two non-MAX phases, from where MXenes have been successfully synthesized. $Mo_2CT_x$ is the first MXene of this kind and it was synthesized by extracting Ga layers from $Mo_2Ga_2C$. This phase, despite its similarity to MAX phases, has two A-element layers (Ga) separating the carbide layers [19]. Similarly, $Zr_3C_2T_x$ was prepared from $Zr_3Al_3C_5$, a yet another non-MAX-phase precursor, by etching aluminum carbide ($Al_3C_3$) layers rather than removing just Al layers [20].

## Properties of MXenes

General Properties of MXenes are (1) Structure (2) Electronic structures (3) Magnetic states (4) Photonic/Optical properties (5) Surface Properties.

## *Structure*

As transition metal ions have co-ordination numbers of six, it is assumed MXenes has hexagonal lattice structure with six-fold symmetry. The transition metals in MXenes form six chemical bonds to X atoms and surface chemical groups, causing production of $M_2XO_2$, $M_2XF_2$, and $M_2X(OH)_2$.

## *Electronic Structures*

In similar to parent MAX phases, pristine MXenes are mostly metallic. Electronically, MXenes can be metallic, semi-metallic, or semi-conducting. Because the p bands in X of most MXenes occur beneath the d-bands and a new band starts forming on functionalization due to the hybridization of the p orbitals of functionalized group and d orbital of transition metal M present. Thus, the d band is raised above the Fermi level, resulting in the transformation of metallic MAX phases into semi-conductors.

Though spin-orbit coupling does affect their electronic structure, these effects are not very significant. However, addition of functional groups $T_x$ (–F, –OH, –O) can transform some of metallic MXenes into semiconductors due to the modification crystal's local field and Fermi level shift. Upon functionalization, a new band is generated below the Fermi energy, hybridized with d orbital of M. for example, most 2D transition metal carbides or nitrides are metallic with the Fermi energy located at the d-bands of the transition metals whereas $Sc_2CT_2$ (T=F, OH, O) and $M_2CO_2$ (M=Ti, Zr, Hf) are semiconductors.[24].

## *Magnetic*

MXenes are expected to be non-magnetic due to its strong covalent bond between metal and corresponding carbides or nitrides attached to it [25]. However, external strain results in the discharge of D-orbitals and triggers the magnetic properties. Thus, MXenes, mostly made up of Cr and Mn, exhibit magnetic properties owing to the different number of available electrons for the transition metals and different electron configuration of the d-orbitals [26]. Only those electrons that occupy non-bonding D-orbitals, formed near the fermi level, can majorly contribute to their magnetic properties. It has been reported that majority of the MXenes are antiferromagnetic at the ground state whereas Manganese based MXenes exhibits ferromagnetism [27, 28].

### *Optical Properties*

Optical properties of MXenes are evaluated using thin films of MXenes and their composites. These films can be produced from the layer-by-layer deposition of MXene dispersions. Using $Ti_3C_2T_x$ thin film, the optical properties are studied showing that MXenes thin films are mostly transparent. Titanium carbides exhibit transmittance greater than 97% of visible light per nanometer [29]. These unique optical properties paved the way for the application of MXenes to several photonic based applications and opto-electronic applications like optical switch, photothermal evaporation system, conductive transport electrodes, polarization driven photovoltaics, conductive coating, etc.[30–32].

## General Applications of MXenes

In the recent years, MXenes have gained lots of popularity owing to its unique properties like safety, big interlayer spacing, high surface reactivity, good biocompatibility and environment flexibility, hydrophilicity, outstanding absorption capacity, etc. These properties pave the way for the possibility of applications in vast areas like Ferroelectricity; Piezoelectricity, Thermoelectricity; EMI shielding (electromagnetic interference shielding) [33], Superconductivity; Catalyst; Photocatalysis; Ion batteries; Gas Sensors; Hydrogen storage; optoelectronic devices, triboelectric nanogenerators, supercapacitors, lithium ion batteries, lithium-sulfur batteries, etc.

### *Energy Conversion*

Energy conversion happens through electrochemical reactions, in which oxygen evolution and reduction are critical. So, the absence of effective oxygen reducing catalyst is a bottleneck in this regard. Q. Xue et al. reported the outstanding oxygen reduction of evenly distributed $Mn_3O_4$ nanoparticles on $Ti_3C_2$ MXenes layer, giving a greater stable and positive potential composite. This $Mn_3O_4$–$Ti_3C_2$ nanocomposite was perhaps the first non-noble metal facilitating the $O_2$ reduction mechanism [34].

Hydrogen is yet another key renewable energy source. Most electrocatalysts for hydrogen evolution is Pt based; but, recently Safizadeh F. et al. have reported better effective catalysts like metal carbides, sulfides, nitrides, borides, phosphides [35] etc.

Water splitting (generation of hydrogen gas) through various types of photocatalysis is another potential renewable energy source. Recently, Su, T. M and his group have reported that a new photocatalyst $Nb_2O_5$–C–$Nb_2C$ Mxene exhibits higher water splitting capacity (almost fourfold hydrogen generation capacity as that of $Nb_20_5$) [36]. $Ti_3C_2$ MXenes and Pt photocatalyst combined with g-$C_3N_4$ also shows greater hydrogen evolution compared to man other parent semiconductor photocatalysts.[37]

This enhanced photocatalysis (hydrogen evolution) may be due to increased separation of charge carriers and electron transfer among conduction band of $C_3N_4$ and the photocatalysts used [38, 39].

## *Energy Storage Applications*

The first application that was investigated for MXenes application was the energy storage, which was used in lithium-ion batteries. Energy storage mainly takes place through various means like batteries, supercapacitors, etc. The versatile physio-chemical properties of MXenes and their tunability enable them to use for various applications including energy storage.

### Batteries

The culmination of various properties like high electrical conductivity, environment friendliness, high inter-layer ion diffusion capacity, large exposure of surface to the electrolyte, high inter-layer ion diffusion capacity, etc. have resulted in making MXenes as a good electrode. Though supercapacitors have edges over the traditional batteries because of their higher power densities, relatively short recharging duration, and long-lasting usability, they are expensive and remain less efficient. Li-ion batteries mainly use graphite as its anode, but it has got many shortfalls due to their low conductivity and longer diffusion length. So, in the quest of searching new materials, MXenes emerge as forefront.

### Super Capacity

D Wang et.al has recently reported the great possibility of using $Ti_2N$ as anode materials. N atom in $Ti_2N$ requires one electron less than that of Titanium (Ti) to form an ionic bond, thereby developing a non-bonding electron-rich surface [40]. This negative electron cloud facilitates the diffusion of cations. Despite all these potential benefits, these 2D films are densely packed, resulting in near-zero capacities. Because this electron-rich property, many researches have been carried out to tap these transition metal MXenes' properties for battery purposes, e.g., $V_{n+1}C_n$, $Nb_2C$, $Cr_2CO_2$, and $Mo_2C$ [43].

### Superconductivity

$Mo_2GaC$, a parent MAX precursor of $Mo_2C$ MXene is a superconductor with a critical temperature ($T_c$) of 4 K, whereas this 2D MXene was reported to be superconducting with $T_c < 4$ K. This MXene has been exfoliated experimentally from this non-MAX-phase structure, has already been fabricated with various thicknesses through chemical vapor deposition. The theoretical DFT calculation and BCS theory have predicted the corresponding $T_c$ for pristine $Mo_2C$ and $Mo_2C$ functionalized with –H, –OH, –O, –S, –Se, and –Br [41].

#### Thermoelectricity

Generation of electricity due to application of heat or temperature gradient and vice versa is known as thermoelectricity. The thermoelectric properties of MXenes are studied theoretically using First principle calculation and Boltzman theory. Despite most metallic MXenes has very good electrical conductivities, their Seebeck coefficients are very small, resulting in a poor power factor ($S_2$), whereas semiconducting MXenes exhibit very high Seebeck coefficients and comparatively greater electrical conductivity and hence, possess a high-power factor. Therefore, semiconducting MXenes like $Sc_2CT_2$ (where T is F, OH or O) and $M_2CO_2$ (where M is Ti, Zr or Hf) can be considered as potential materials for thermo-electric applications [46] Moreover, A.N. Gandi et.al reported that the thermal conductivities of MXenes are in the range of 10–60 $Wm^{-1}$ $K^{-1}$, which are comparable to that of transition metal dichalcogenides [47].

#### Piezoelectricity

Piezoelectricity is the materials that generate electricity due to mechanical strain developed by shifting the centers of negative and positive charges. Owing to the presence of non-centrosymmetric, the semiconducting MXenes have the potential applications in piezoelectricity. Mohammad Khazaei et.al revealed that the semiconducting iMXenes possessed giant piezoelectric coefficients as high as $45 \times 10^{-1}$ c/m [48] Among the MXene members, $(Mo_{2/3}Sc_{1/3})_2CO_2$, $Sc_2CO_2$, $(W_{2/3}Y_{1/3})_2CO_2$ etc. are some of the materials without in-plane centro-symmetry whose piezoelectric stress and strain coefficient ($e_{11}$,$d_{11}$) are (3.333, 4.137), (44.55, 29.24), (35.53, 24.98), respectively.[49].

## *MXenes for Environment Application*

#### Photocatalysis

Semiconducting photocatalyst in the presence of MXenes can find applications in enhancing photocatalysis, environmental remediation, water cleaning, and splitting, etc. Various computational studies like DFT, etc. have predicted some MXenes to have various narrow band gap, that can be tuned by changing the surface group in respect to transition metal M atom [50, 51].

These transition metal carbides in combination with some semiconductor (like $Ti_3C_2$ with $TiO_2$) can be used as excellent photocatalysts, showing a sharp improvement of nearly 400% [52] and 2.8 times greater than that of pure $TiO_2$ for reducing carbon dioxide.[53] This is mainly due to transferring of electron to $Ti_3C_2$ MXenes with lesser barrier potential. Recently, a study has successfully prepared Ti-based MXenes viz. $CdS/Ti_3C_2$ heterostructure and found an enhanced photocatalytic activity. The photocatalytic activity of CdS is as low as 105 $\mu molh^{-1}$ $g^{-1}$, which varies linearly with the amount of metal carbide present, showing a very high activity of 14,322 $\mu molh^{-1}$ $g^{-1}$.

## Anti-Microbial Applications

In general, the control of microbial action is difficult, owing to complexities arising out of surface exposure of compounds for long periods to complex media rich in bacteria and nutrients. It is through complex interaction between nanoparticle and biological microorganisms that inhibit bacterial growth. Ag (Silver) based MXene nanoparticles are often used for anti-microbial applications such as inhibiting *E-coli* [54], selective removal of species like Rhodamine B methyl green, bovine serum, albumin [55] etc. $Ti_3C_2T_x$ membrane on a PVDF (Polyvinylidene Difluoride) substrate inhibited Bacillus subtilis and E. coli growth (>99%) much more effectively than did the control polyvinylidene difluoride membrane (73% for Bacillus subtilis and 67% for E. coli [56]. Rasool et al., has successfully studied the antibacterial activity of a $Ti_3C_2T_x$ MXene against Gram-negative *E. coli* and Gram-positive Bacillus subtilis was explored [57].

## Adsorption

The most studied MXene, $Ti_3C_2T_x$ adsorbed methylene blue (a cationic dye used in photocatalysis experiment) from water to the extent of nearly 40 mg/g. This quantity was lesser than that of commercial activated carbons (around 1,000 mg/g) [22], but comparable to that of graphenes-$Fe_3O_4$ (approximately 45 mg/g). Thus, different structures of MXenes having diverse surface chemistries may adsorb with different adsorption capacity depending on functional group attached [58].

Various studies have reported the effective removal of heavy metals (like U, Cu, Cr, Hg, etc.) from water. A. K. Fard et al. used $Ti_3C_2T_x$ nanosheets for the successful removal of $Ba^{2+}$ ions from water. It was evident from the study that adsorption capacity was dependent on pH value (maximum at pH 6 to 7), owing to its effect on charge of adsorbent surface [59].

### Other Potential Applications of MXenes

Ceaseless exploration of new interesting properties of MXenes has opened up a new avenue for the MXenes' applications in many other diverged areas. In this context, some emerging potential topics include SERS (Surface-Enhanced Raman Spectroscopy), study of correlation between MXene structure and their properties, anisotropic, electronic conduction, bipolar magnetic properties, magnetic properties, multipolar surface plasmons, elastic properties, thermoelectric properties, transport properties for understanding free carrier dynamics and thermal conductivity.

Other areas are formation of heterostructures and nano-composites of MXenes, innovative structure analysis techniques like structural engineering, nanopore engineering, phase transition by oxidation, surface alloying, liquid exfoliation, etc.; MXene-based sensors like glucose biosensors, piezoresistive sensors, and force sensors;

## Summary

MXenes, with their rich Physical and chemical properties, provide a fertile ground for further academic research. In many applications, it has been reported that MXenes outperform many other existing 2D materials. In this chapter, we have described some common practices of synthesizing MXenes, their basic properties, and overall applications for energy storage and environmental applications. However, as of now, synthesizing whole new and mass-scale production of MXenes is an experimental challenge. Exploring new MAX phase for synthesizing their 2D counterparts is a hurdle in realizing their full potential.

Despite the huge endeavor and attention among scientific communities on these 2D materials, many areas remained unexplored. This chapter highlights some of them. For instance, production of theoretically predicted new ternary carbide experimentally can be a potential research area. Characterization of mechanical and surface chemistry of MXenes need more advanced and new standard methods. MXenes without any surface functional groups can be produced by CVD or other thin film epitaxial techniques in vacuum, but the study in this area is still very limited. Moreover, the deeper understanding and controlling of ions dynamics in MXenes can help us in improvising the present energy storage and conversion devices. In the nutshell, the potential of MXenes is truly enormous.

**Acknowledgements** The author Konthoujam Priyananda Singh is thankful to UGC for providing financial support under UGC JRF scheme.

## References

1. Natalie B., Subramanian, S., Lin, Z., et al.: A roadmap for electronic grade 2D materials. 2D Mater. **6**, 022001 (2019)
2. Rub M.B., Gomez-N., C., Gomez-H, J., Felix Z.: 2D materials: to graphene and beyond. Nanoscale **3**, 20–30 (2011)
3. Tao, Q., et. al.: Two-dimensional Mo1.33C MXene with divacancy ordering prepared from parent 3D laminate with in-plane chemical ordering. Nat. Commun. **8**, 14949 (2017)
4. Novoselov, K.S., Mishchenko, A., Carvalho, A., Castro, A.H.: 2D materials and van der Waals heterostructures. Science 353–9439 (2016)
5. Barsoum, M.W.: The $M_{n+1}AX_n$ phases: a new class of solids: thermodynamically stable nanolaminates. Prog. Solid State Chem. **28**, 201–228 (2000)
6. Naguib, M., Kurtoglu, M., Presser, V., Lu, J., Niu, J., Heon, M., Hultman, L., Gogotsi, Y., Barsoum, M.W.: Two dimensional nanocrystals produced by exfoliation of $Ti_3AlC_2$. Adv. Mater. **23**, 4248–4253 (2011)
7. Tang, Q., Zhou, Z., Shen, P.W.: Are MXenes promising anode materials for Li ion batteries? Computational studies on electronic properties and Li storage capability of $Ti_3C_2$ and $Ti_3C_2X_2$ (X = F, OH) monolayer. J. Am. Chem. Soc. **134**, 16909–16916 (2012)
8. Liu, F.F., Zhou, J., Wang, S.W., Wang, B.X., Shen, C., Wang, L.B., Hu, Q.K., Huang, Q., Zhou, A.G.: Preparation of high-purity V2C MXene and electrochemical properties as Li–ion batteries. J. Electrochem. Soc. **164**, A709–A713 (2017)
9. Mashtalir, O., Lukatskaya, M.R., Zhao, M.Q., Barsoum, M.W., Gogotsi, Y.: Amine-assisted delamination of $Nb_2C$ MXene for Li–ion energy storage devices. Adv. Mater. **27**, 3501–3506 (2015a)
10. Alhabeb, M., Maleski, K., Anasori, B., Lelyukh, P., Clark, L., Sin, S., Gogotsi, Y.: Guidelines for synthesis and processing of two-dimensional titanium carbide ($Ti_3C_2T_X$ MXene). Chem. Mater. **29**, 7633–7644 (2017)
11. Babak, A., et.al.: 2D Metal Carbides and Nitrides (MXenes) for Energy Storage, vol 2, Art. No. 16098. Macmillan Publishers Limited, Springer (2017)
12. Khazaei, M., Ranjbar, A., Arai, M., Sasaki, T., Yunoki, S.: Electronic properties and applications of MXenes: a theoretical review. J. Mater. Chem. C **5**, 2488–2503 (2017)
13. Halim, J., Kota, S., Lukatskaya, M.R., Naguib, M., Zhao, M.Q., Moon, E.J., Pitock, J., Nanda, J., May, S.J., Gogotsi, Y., Barsoum, M.W.: Synthesis and characterization of 2D molybdenum carbide (MXene). Adv. Funct. Mater. **26**, 3118–3127 (2016)
14. Zhou, J., Zha, X., Chen, F.Y., Ye, Q., Eklund, P., Du, S., Huang, Q.: A two-dimensional zirconium carbide by selective etching of $Al_3C_3$ from nano laminated Zr3Al3C5. Angew. Chem. **128**, 5092–5097 (2016)
15. Naguib, M., Unocic, R.R., Armstrong, B.L., Nanda, J.: Large scale delamination of multi-layers transition metal carbides and carbonitrides "MXenes." Dalton Trans. **44**, 9353–9358 (2015)
16. Mashtalir, O., Naguib, M., Mochalin, V.N., Dall'Agnese, Y., Heon, M., Barsoum, M.W., Gogotsi, Y.: Intercalation and delamination of layered carbides and carbonitrides. Nat. Commun. **4**, 1716 (2013)
17. Lai, C.C., Fashandi, H., Lu, J., Palisaitis, J., Persson, P.O.A., Hultman, L.P., Eklund, R.: Phase formation of nano laminated $Mo_2GaC$ and $Mo_2(Au_{1-x}Ga_x)_2C$ by a substitutional reaction within Au-capped $Mo_2GaC$ and $Mo_2Ga_2C$ thin films. Nanoscale **9**, 17681–17687 (2017)
18. Wang, L., Xu, C., Liu, Z., Chen, L., Ma, X., Cheng, H.M., Ren, W., Kang, N.: Magneto transport properties in high quality ultra thin two-dimensional superconducting $Mo_2C$ crystals. ACS Nano **10**, 4504–4510 (2016)
19. Meshkian, R., et al.: Synthesis of two-dimensional molybdenum carbide, $Mo_2C$, from the gallium based atomic laminate $Mo_2Ga_2C$. Scripta Mater. **108**, 147–150 (2015)
20. Zhou, J., et al.: A two-dimensional zirconium carbide by selective etching of $Al_3C_3$ from nano laminated $Zr_3Al_3C_5$. Angew. Chem. Int. Ed. **128**, 5092–5097 (2016)
21. Naguib, M., Mochalin, V.N., Barsoum, M.W., Gogotsi, Y.: Two dimensional MXenes: A new family of two dimensional materials; Adv. Mater. **26**, 992 (2014)

22. Mashtalir, O., Cook, K.M., Mochalin, V.N., Crowe, M., Barsoum, M.W., Gogotsi, Y.: Dye adsorption and decomposition on two-dimensional titanium carbide in aqueous media. J. Mater. Chem. A **2**, 14334–14338 (2014)
23. Gusev, A., Rempel, A.: Atomic ordering and phase equilibria in strongly nonstoichiometric carbides and nitrides. In: Gogotsi, Y., Andrievski, R.A. (eds.) Materials Science of Carbides, Nitrides and Borides, pp. 47–64. Springer, 1999
24. Khazaei, M., Arai, M., Sasaki, T., Chung, C.Y., Venkataramanan, N.S., Estili, M., Sakka, Y., Kawazoe, Y.: Novel electronic and magnetic properties of two-dimensional transition metal carbides and nitrides. Adv. Funct. Mater. **23**, 2185–2192 (2013)
25. Hu, J.P., Xu, B., Ouyang, C.Y., Yang, S.A., Yao, Y.G.: Investigations on V2C and $V_2CX_2$ (X = F, OH) monolayer as a promising anode material for Li ion batteries from first-principles calculations. J. Phys. Chem. C **118**, 24274–24281 (2014)
26. Zhao, S.J., Kang, W., Xue, J.M.: Manipulation of electronic and magnetic properties of $M_2C$ (M = Hf, Nb, Sc, Ta, Ti, V, Zr) monolayer by applying mechanical strains. Appl. Phys. Lett. **104**, 133106 (2014)
27. Lu, J., Thore, A., Meshkian, R., Tao, Q., Hultman, L., Rosén, J.: Theoretical and experimental exploration of a novel in-plane chemically-ordered, (Cr2/3M1/3)2AlCi-MAX Phase with M = Sc and Y, Cryst. Growth Des. **17**, 5704–5711 (2017)
28. Sun, W., Xie, Y., Kent, P.R.C.: Double transition metal MXenes with wide band gaps and novel magnetic properties. Nanoscale **10**, 11962–11968 (2018)
29. Hantanasirisakul, K., et al.: Fabrication of $Ti_3C_2T_x$ MXene transparent thin films with tunable optoelectronic properties. Adv. Electron. Mater. **2**, 1600050 (2016)
30. J. Guo, Sun, Y., Liu, B., et al.: Two-dimensional scandium-based carbides (MXene): Band gap modulation and optical properties. J. Alloys Compd. **712**, 752–759 (2017)
31. Chandrasekaran, A., Mishra, A., Singh, A.K.: Ferroelectricity, antiferroelectricity, and ultrathin 2D electron/hole gas in multifunctional monolayer MXene. Nano Lett. **17**, 3290–3296 (2017)
32. Ying, G., Kota, S., Dillon, A.D., Fafarman, A.T., Barsoum, M.W.: Conductive transparent $V_2CT_x$ (MXene) films. FlatChem **8**, 25 (2018)
33. Li, R., Zhang, L., Shi, L., Wang, P.: MXene $Ti_3C_2$: an effective 2D light-to-heat conversion material. ACS Nano **11**, 3752–3759 (2017)
34. Shahzad, F., Alhabeb, M., Hatter, C.B., Anasori, B., Man Hong, S., Koo, C.M., Gogotsi, Y.: Electromagnetic interference shielding with 2D transition metal carbides (MXenes). Science **353**, 1137–1140 (2016)
35. Xue, Q., Pei, Z.X., Huang, Y., Zhu, M.S., Tang, Z.J., Li, H.F., Huang, Y., Li, N., Zhang, H.Y., Zhi, C.: Y :$Mn_3O_4$ nanoparticles on layer-structured $Ti_3C_2$ MXene towards the oxygen reduction reaction and zinc-air batteries. J. Mater. Chem. A **5**, 20818–20823 (2017)
36. Safizadeh, F.; Ghali, E.; Houlachi: G. Electrocatalysis developments for hydrogen evolution reaction in alkaline solutions - A review. Int. J. Hydrog. Energy (2015), 40, 256–274
37. Su, T.M., Peng, R., et al.: One-step synthesis of $Nb_2O_5$/C/$Nb_2C$ (MXene) composites and their use as photocatalysts for hydrogen evolution. Chem Sus Chem. **11**, 688–699 (2018)
38. Halim, J., Lukatskaya, M.R., Cook, K.M., Lu, J., Smith, C.R., Naslund, L.A., May, S.J., Hultman, L., Gogotsi, Y., Barsoum, E.P., M. W. : Transparent conductive two-dimensional titanium carbide epitaxial thin films. Chem. Mater. **26**, 2374–2381 (2014)
39. Wang, A.X.Q., Wang, W., Duan, H, Shi, J., Yu, X.: The synergetic effects of $Ti_3C_2$ MXene and Pt as co-catalysts for highly efficient photocatalytic hydrogen evolution over g-$C_3N_4$.; Phys. Chem. Chem. Phys. **20**, 11405–11411 (2018)
40. Wang, D., Gao, Y., Liu, Y., Jin, D., Gogotsi, Y., Meng, X., Du, F., Chen, G., Wei Y.: First principles calculations of $Ti_2N$ and $Ti_2NT_2$ (T = O, F, OH) monolayers as potential anode materials for Lithium–ion batteries and beyond. J. Phys. Chem. C **121**, 13025–13034 (2017)
41. Anasori, B., Xie, Y., Beidaghi, M., Lu, J., Hosler, B.C., Hultman, L., Kent, P.R., Gogotsi, Y., Barsoum, M.W.: Two-dimensional, ordered, double transition metals carbides (MXenes). ACS Nano **9**, 9507–9516 (2015)
42. Ghidiu, M., et al.: Conductive two-dimensional titanium carbide clay with high volumetric capacitance. Nature **516**, 78–81 (2014)

43. Lukatskaya, M.R., Mashtalir, O., Ren, C.E., Dall'Agnese, Y., Rozier, P., Taberna, P.L., Naguib, M., Simon, P., Barsoum, M.W., Gogotsi, Y.: Cation intercalation and high volumetric capacitance of two-dimensional titanium carbide. Science **341**, 1502–1505 (2013)
44. Mashtalir, O., Lukatskaya, M.R., Zhao, M.Q., Barsoum, M.W., Gogotsi, Y.: Amine-assisted delamination of $Nb_2C$ MXene for Li–Ion energy storage devices. Adv. Mater. **27**, 3501–3506 (2015b)
45. Zhang, J.J., Lin, L., Zhang, Y., Wu, M., Yakobson, B.I., Dong, S.: Type-II Multiferroic $Hf_2VC_2F_2$ MXene monolayer with high transition temperature. J. Am. Chem. Soc. **140**, 9768–9773 (2018)
46. Khazaei, M., Arai, M., Sasaki, T., Estili, M., Sakka, Y.: Two-dimensional molybdenum carbides: potential thermoelectric materials of the MXene family. Phys. Chem. Chem. Phys. **16**, 7841–7849 (2014)
47. Gandi, A.N., et al.: Thermoelectric performace of the MXenes $M_2CO_2$ (M = Ti, Zr, or Hf). Chem. Mater. **28**, 1647–1652 (2016)
48. Mohammad, K., Vei W., Cem, S., Ahmad, R., Masao, A., Seiji, Y.: Electronic structures of iMAX phases and their two-dimensional derivatives: a family of piezoelectric materials. Phys. Rev. Mater. **2**, 074002 (2018)
49. Khazaei, M., Wang, V., Sevik, C., Ranjbar, A., Arai, M., Yunoki, S.: Electronic structures of iMAX phases and their two-dimensional derivatives: a family of piezoelectric materials. Phys. Rev. Mater. **2**, 074002 (2018)
50. Enyashin, A.N., Ivanovskii, A.L.: Structural, electronic properties and stability of MXenes $Ti_2C$ and $Ti_3C_2$ functionalized by methoxy groups. Comput. Theor. Chem **989**, 27–32 (2012)
51. Khazaei, M., et al.: OH terminated two-dimensional transition metal carbides and nitrides (MXenes) as ultralow work function materials. Adv. Funct. Mater. **23**, 2185–2192 (2013)
52. Low, J., Zhang, L., et al.: $TiO_2$/MXene $Ti_3C_2$ composite with excellent photocatalytic $CO_2$ reduction activity. J. Catal. **361**, 255–266 (2018)
53. Ran, J., Gao, G., Li, F.T., Ma, T.Y., Du, A., Qiao, S.Z.: Ti3C2 MXene co-catalyst on metal sulfide photo-absorbers for enhanced visible-light photocatalytic hydrogen production. Nat. Commu. **8**, 13907 (2017)
54. Pandey, R.P., et al.: Ultrahigh-flux and fouling-resistant membranes based on layered silver/MXene ($Ti_3C_2T_x$) nanosheets. J. Mater. Chem. A **6**, 3522–3533 (2018)
55. Cui, Y.Q., Zhang, Z.X., et al.: Ultrasound assisted fabrication of AgBr/$TiO_2$ nano-tube arrays photoelectrode and its enhanced visible photocatalytic performance and mechanism for detoxification of 4-chlorphenol. Sep. Purif. Technol. **197**, 189–196 (2018)
56. Rasool, K., Mahmoud, K.A., Johnson, D.J., Helal, M., Berdiyorov, G.R., Gogotsi, Y.: Efficient antibacterial membrane based on two-dimensional $Ti_3C_2T_x$ (MXene) nanosheets. Sci. Rep. **7**, 1598 (2017)
57. Rasool, K., Helal, M., Ali, A., Ren, C.E., Gogotsi, Y., Mahmoud, K.A.: Antibacterial activity of $Ti_3C_2T_x$ MXene. ACS Nano **10**, 3674–3684 (2016)
58. Zheng, W., Zhang, P.G., Tian, W.B., Qin, X., Zhang, Y.M., Sun, Z.M.: Alkali treated $Ti_3C_2T_x$ MXenes and their dye adsorption performance. Mater. Chem. Phys. **206**, 270–276 (2018)
59. Fard, A.K., McKay, G., Chamoun, R., Rhadfi, T.P., Homme, H., Atieh, M.A.: Barium removal from synthetic natural and produced water using MXene as two dimensional (2-D) nanosheet adsorbent. Chem. Eng. J. **317**, 331–342 (2017)

# Chapter 16
# ZIFs Recent Development and Its Role in Photocatalysis

**Boris Wareppam and L. Herojit Singh**

## Introduction

Metal-organic framework (MOFs) are playing a huge role in today's world global economy as large amount of crystalline mesoporous nanomaterials are being employed in separation of gas, water treatment, petrochemical cracking, etc [1]. Continuous research and development have been done recently to achieve a unique porous material with various methods and techniques. ZIFs is one such porous material for which research has been going on for decades showing promising properties giving simultaneous properties of both MOFs and Zeolite [2]. They have well organized mesoporous structure consisting of transition metal ions and Imidazole organic linkers which is a 3D tetrahedral network. Its characteristics features such as high thermal and chemical stability, hydrophobicity, tunable pore size, and crystallinity nature is what drives the researcher to engineer this material and bring forward the best of it in many applications [3, 4]. An in-depth study by many researchers found that some key factors, like the type of metal atoms and organic linkers, as well as the cage geometry and flexibility, play a major role in the thermodynamics and dynamics of guest molecules [5]. Its flexibility that is the opening of gate and swinging effect is what makes this material attracted to a wide field of application [6]. The synergetic effect caused by the incorporation or addition of other functional materials into the ZIFs building blocks has paved a way to established new innovative research of multifunctional material in this area.

Making world, a better place to live is what science seeks but the rise of organic and inorganic pollutants day by day makes that impossible. Although many solutions

B. Wareppam · L. Herojit Singh (✉)
Department of Physics, NIT Manipur, Langol 795004, India
e-mail: loushambam@gmail.com

B. Wareppam
e-mail: brswareppam@gmail.com



B. P. Swain (ed.), *Nanostructured Materials and their Applications*, Materials Horizons: From Nature to Nanomaterials, https://doi.org/10.1007/978-981-15-8307-0_16

have been developed in the past years to control the problems of pollution. It seems the problem of pollutants far exceeds the solutions and this is the thing to be highly concern about in present. One such solution leads to photocatalysis which is a technique that involves sunlight irradiation to accelerate the reaction mechanism with the use of a catalyst. In this regard, ZIFs research carries forward to the photocatalysis process. They can act as a major catalyst in the degradation of organic dye and other pollutants whose particle size is less than the ZIFs pore size. Main research is going on for water treatment and separation application in ZIFs using this technique [7, 8]. Herein, recent ZIFs development and notable research area are being discussed which will help in building perfect multifunctional MOFs in the near future. This chapter gives a picture about the ZIFs development, its properties, and its role and mechanism in photocatalysis.

## Structure of ZIFs

ZIFs framework is built with transition metal centers and imidazole linkers which acts as a bridge between the metals center. They have an analogous structure with that of a Zeolites having M-IM-M structure with bond angle of 145° as shown in Fig. 16.1 [1, 9].

The Imidazole ring consists of a five membered and the bridge between imidazole and the metal is solely connected by Nitrogen in 1,3 position of the ring. Some ZIF structures are generated with the use of Methyl, Benzyl, or Phenyl group of the Imidazole. This can be related to the pore size variation of the ZIFs pore window [1]. Park et.al established the structure of ZIF-1 to ZIF-12 in their work thus showing different structural variations of the compounds with the change of metal center or the imidazole linkers. ZIFs structure is not restricted to tetrahedral network, the ring network may consist of four membered, six membered and eight membered or simultaneously two groups of the said membered ring at one framework. Table 16.1 shows the list of ZIFs (ZIF-1 to ZIF-10) topologies, crystal system, and its pore metrics. Earlier report shows the synthesize of variety of ZIFs structure which possesses the Zeolite like topology. Figure 16.2 show the typical structure of ZIFs (ZIF-1 to ZIF-10) which resembles zeolite topologies. Most of the ZIFs structure are

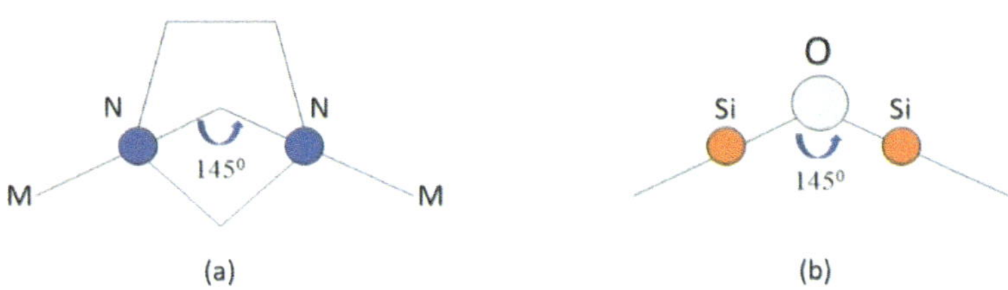


**Fig. 16.1** Schematic diagram of ZIF and aluminosilicate structure (a) M-IM-M and (b) Si–O–Si respectively

**Table 16.1** Composition, Crystal space group, network and pore diameter of ZIFs (ZIF-1 to ZIF-10)

| Sl. No | ZIF-n | Composition | Crystal space group | Network topology[a] | Crystal system | Pore diameter in (Å) |
|---|---|---|---|---|---|---|
| 1 | ZIF-1 | $Zn(IM)_2$ | P2(1)/n | crb | Monoclinic | 6.94 |
| 2 | ZIF-2 | $Zn(IM)_2$ | P b c a | crb | orthorhombic | 6.00 |
| 3 | ZIF-3 | $Zn(IM)_2$ | P4(2)/mnm | dft | tetragonal | 8.02 |
| 4 | ZIF-4 | $Zn(IM)_2$ | Pbca | Cag | orthorhombic | 2.04 |
| 5 | ZIF-5 | $In_2Zn_3(IM)1_2$ | Ia $\overline{3}$ d | Gar | Cubic | 3.03 |
| 6 | ZIF-6 | $Zn(IM)_2$ | I4(1)/amd | Gis | tetragonal | 8.80 |
| 7 | ZIF-7 | $Zn(PhIM)_2$ | R $\overline{3}$ | Sod | hexagonal | 4.31 |
| 8 | ZIF-8 | $Zn(MeIM)_2$ | I $\overline{4}$ 3 m | Sod | cubic | 11.6 |
| 9 | ZIF-9 | $Co(PhIm)_2$ | R $\overline{3}$ | Sod | hexagonal | 4.31 |
| 10 | ZIF-10 | $Zn(IM)_2$ | I4/mmm | Mer | tetragonal | 12.12 |

Data in the table is adopted from Refs. [1, 10]
[a]RCSR symbols crb-B net of the $CrB_4$; dft-DAF-2; cag- named after $CaGa_2O_4$ (Variscite); gar-garnet; gis- gismondine; sod-Sodalite; mer-merlinoite

formed by linkage of the transition metal element with the imidazole ring through tetrahedral network topology. Basically, complexity of the heterolink structure arises due to the introduction of other organic moieties into its framework thereby modifying the network [10]. ZIF-5 is a special case where there is both tetrahedral and octahedral arrangement formed by the Zn (II) and In (III), respectively.

Looking to the generalize network topology and its resemblance with zeolite, it is being reported that zeolite net found in the ZIFs structure is vertex-transitive, i.e., unimodal. Of the four coordinated ZIF framework structure which are unimodal, only four types (**frl, lcs, neb, zni**) possess entirely new structure with no resemblance to zeolite or any aluminosilicate compound [10]. So many ZIFs structures have been discovered in such a short span of time that an enormous field of synthetic organic material chemistry is waiting to be explored.

## Synthesis Methods

Many synthesis methods have been developed in recent years for the production of ZIFs in powder or thin film/membrane form. Most widely used technique includes hydrothermal or solvothermal process for the synthesis of ZIFs material whether it be solvent-free or solvent-based formation [2]. And establishment of different ongoing research towards this material leads to the development of different methods per the requirement. It is found from earlier research that the synthesis criteria play a major role in the formation of ZIFs. So, a clear understanding of the synthesis mechanism becomes an essential task.

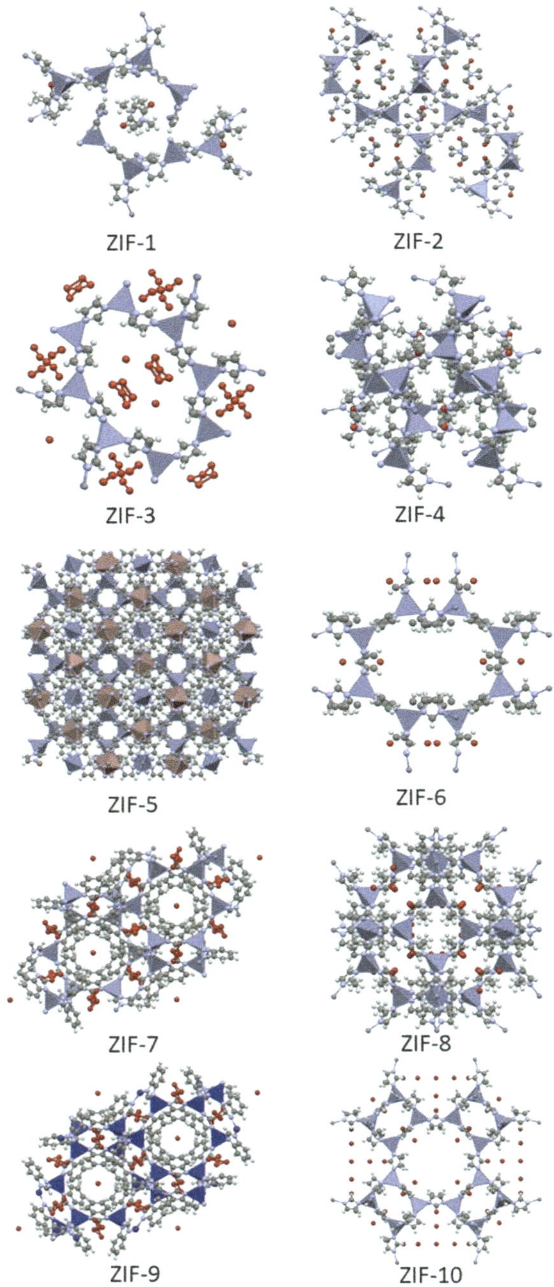


**Fig. 16.2** Structure of some ZIFs (ZIF-1 to ZIF-12) showing its tetrahedral arrangement and framework

### *Solvothermal Synthesis*

MOFs have been reported to synthesized using solvothermal technique via conventional electric heating in small scale in vials or sealed NMR tubes [11]. Conventional method of ZIFs synthesis use solvothermal method with organic solvent as the reaction medium. Many researchers have used different solvent medium to examine the structural and morphological changes using this method. In 2006, Yaghi et al. pioneered the synthesis of twelve ZIF crystals, termed as ZIF-1 to ZIF-12, in organic solvent systems such as N,N-dimethylformamide (DMF), N,N-diethylformamide (DEF), N-methylpyrrolidine (NMP), and as a continuation of their work they have also established ZIF materials including ZIF-60 to ZIF-77, ZIF-78 to ZIF-82, ZIF-90, ZIF-95 and ZIF-100 using the same procedure[1, 12–15]. Different modified version of the synthesis was emerging later on and the use of methanol as solvent was one of the most successful addition in the synthesis of ZIFs using this method.

### *Hydrothermal Synthesis*

Solvothermal technique is the most widely used synthesized technique at the beginning of the ZIF research, indubitably, organic solvents are expensive, flammable, and not environmentally friendly. Recently, much effort has been devoted to the fabrication of ZIFs in green and facile ways via using fewer organic solvents or eventually avoiding the use of organic solvents. Pan et al. for the first time reported the generation of ZIF-8 in an aqueous system at room temperature via a simple procedure [16], the zinc nitrate solution mixed with the 2-methylimidazole (MeIm) solution was stirred for a very short time of about 5 min and the product were centrifuged and collected. But the stoichiometric molar ratio of zinc ions and MeIm in ZIF-8 is 1:2, respectively, it is understandable that an excess amount of MeIm (molar ratio of Zn: MeIm = 1: 70) in this method was wasted. Therefore, to overcome this, people are searching for more effective way of synthesizing ZIFs in aqueous systems. In particular, Miyake et al. [17] had synthesized ZIF-8 with the molar ratio of Zn: MeIm = 1: 20. [18] which is more effective than the previous method. In addition, Qian and Co. have prepared ZIF-67 nanocrystals with the molar ratio of Co: MeIm: $H_2O$ = 1: 58: 1100 in aqueous solutions at room temperature [19]. Interestingly, Chen et al. [20] had reported to synthesized a new type of ZIF which is known as ZIF-L which is 2-D structure having very similar properties with ZIF-8. In the traditional preparation of ZIF-8, they have slightly modified the recipe i.e. the solvent is taken as water [20] (Fig. 16.3).

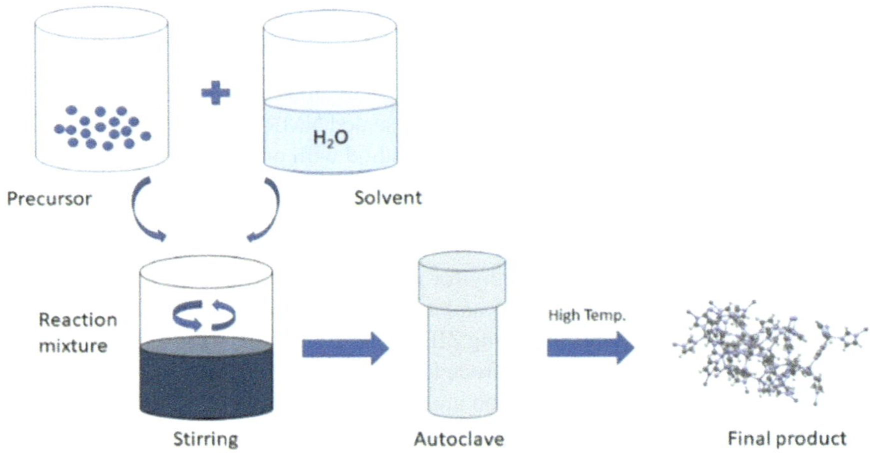


**Fig. 16.3** Schematic diagram showing the synthesis procedure of Hydrothermal process

## *Ionothermal Synthesis*

Although solvothermal and hydrothermal techniques are widely used there are certain prospects in which Ionothermal synthesis work better than the said methods. The ionothermal process relies on the green solvents viz ionic solvents and eutectic mixture in the productions of ZIFs [2]. Ionic solvents act as both solvents and templates simultaneously during the reaction and they can be recycled for further use. Synthesis of ZIFs using ionothermal method was first implemented by Morris and his Co-workers with the use of ionic liquid 1-ethyl-3-methylimidazolium bis[(trifluoromethyl) sulfonylimide [16]. Now, the use of various Ionic liquid has been implemented to develop ZIF materials in an environmentally friendly way.

## *Sonochemical Synthesis*

MOFs are also prepared through sonochemical synthesis where the chemical reaction takes place under the influence of ultrasound radiation at a certain time interval. Sonochemical method has some advantage over the conventional synthesis method for the preparation of ZIFs as they support the formation of nucleation uniformly [2]. The process involves generation of high temperature and pressure in the local environment due to a phenomenon known as acoustic cavitation which arises due to collapse of bubbles [21, 22]. Sonocrystallization of ZIF-7, ZIF-8, ZIF-11, and ZIF-20 were reported by Seoane's et al. and they found that pure crystals of ZIFs can be accomplished at lower temperature and shorter duration in comparison to the conventional methods [23]. They also highlighted the formation of crystal having small and narrower particle size distribution. Recently people have tried this technique to achieve high yield which will be helpful for industrial application [2].

## *Mechanochemical Synthesis*

Recently solvent minimization synthesis of ZIFs have been studied and one such technique is mechanochemical methods. The reaction mechanism in this method is induced by mechanical energy like compression, shear or friction [24]. Partial ZIFs structure formation was first reported back in 2006 when ZnO was manually grinded with excess amount of Imidazole [25]. Other modified mechanochemical synthesis was carried out in which the use of liquid or ionic liquid was introduced thereby increasing the mobility and enhance the formation of ZIFs [26, 27]. But completely solvent-free synthesis was achieved by Tanaka and his co-workers using ball milling of nano-sized ZnO and MeIM [28]. It is being noticed that the final product contains large unaltered ZnO and small nanoparticles which does not resemble ZnO (Fig. 16.4). The smaller nanoparticles most likely formed ZIF-8 due to the reaction between the ZnO and MeIM. Further, Ball milling synthesis have shown the amorphous nature of ZIFs instead of crystalline ZIFs which is a good achievement in the exploration of ZIFs framework [29].

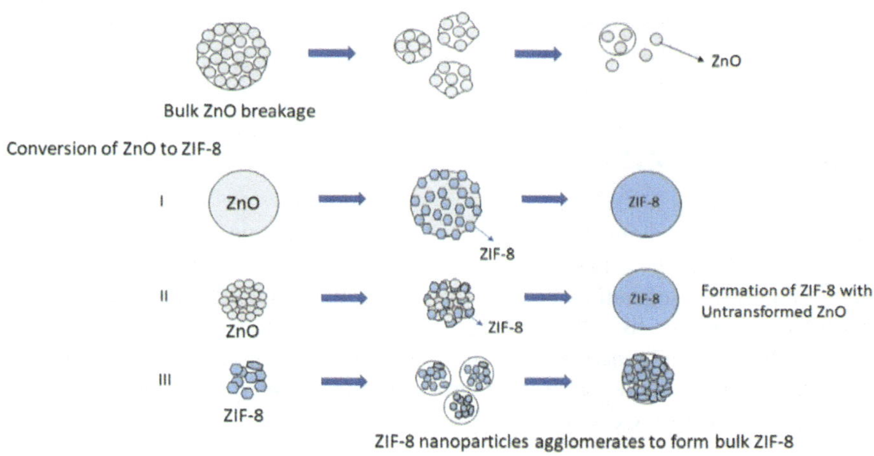


**Fig. 16.4** Model mechanism of mechanochemical synthesis dry conversion of ZnO to ZIF-8

## *Membranes and Thin Film Synthesis*

Film and membranes based synthesis is another area where the scope of ZIFs material expand to great height. Basically, there are two methods for the production of ZIFs membranes viz secondary growth crystallization and in situ crystallization [2]. Secondary growth crystallization is a two steps synthesis process where the crystal seeds are deposited by rubbing, thermal seeding, dip coating, or repeated growth and then pre-treating the support by using reactive seeding or organic functional groups [2]. Shu et.al have reported the growth of ZIF-8 membrane on porous $\alpha$-alumina tubes [30]. For this method, the interaction between the support and the seeding is the key factor in establishing a good membrane. Besides this, new techniques like reactive seeding and contra-diffusion synthesis have also emerged. Reactive seeding of ZIF-8 membrane was reported on a ZnO support which is assisted by hydrothermal synthesis [31]. The in-situ crystallization method on the other hand uses chemically modified support assisted by one step or one-pot hydrothermal or solvothermal synthesis process without the need of any seeding process [2]. Hyuk and his co-workers have synthesized a one-step in situ method based on counter diffusion concept to prepared ZIF-8 membrane with enhances microstructure [32]. Another area which highly emerging at present is the mixed matrix membranes (MMM) where ZIFs based composites act as fillers due to compatibility with polymers, and high sieving effect [2]. Roberto et.al [33] demonstrated the fabrication of MMM using Matrimid-PEG 200 and ZIF-8 and his study shows more effective CO2 permeability in the separation. Thus, membrane-based ZIFs applications are also rapidly advancing.

# Applications of ZIFs

The whole idea of studying these materials is to apply its unique properties in potential applications. Although it has a wide range of applications, recent ZIFs applications mainly focuses on separation, catalyst, and adsorption capability. Its tunable nature and high porosity act like a catalyst for many researchers and thus both pure ZIFs and ZIFs-based membrane/films or composites have turn into multifunctional material even extending its research in biomedical drug delivery. Here the most recent applications of ZIFs are discussed and highlighted.

## *Separation*

If a topic is brought up for global issues, natural gas purification, and $CO_2$ capture is what everyone wants to deal it now. Although this has been said there are a lot of techniques already develop for these issues but the main problem is green and efficient way of solving these issues. ZIFs are non-toxic material possessing quite a special feature like chemical stability, hydrophobic nature, tunable pore size, variable structures, and wide functionalities [2]. Thus, they are one of the best candidates for separation applications. Most of separation for gas are focus on the $H_2$ and $CO_2$ separation due to great application in ZIFs membrane/thin films. However, separation of a gas is dependent on the properties of the target gas and the environment conditions like temperature and pressure and has to be checked carefully. Technically, ZIF-8 has pore size of ~0.34 nm [34] and ZIF-7 has ~0.33 nm [1] so based on this the target gas cannot have a size greater than this otherwise it will be difficult to separate. For example, $H_2$ gas molecules size is ~0.29 nm and that of $CO_2$ is ~0.33 nm so both ZIF-7 and ZIF-8 can be used to separate $H_2$ gas whereas only ZIF-8 is possible for $CO_2$ gas separation from other larger molecules. Likewise, other ZIFs species can be used for separation application based on the target size and its own tunable pore size. Fernando et al. [35] have reported the insertion of ZIF-9 and ZIF-7 in Ni hollow fiber, they have found increase in selectivity of $H_2/CO_2$ gas uptake at low temperatures due to the contraction of the framework. Further mixed matrix membranes incorporating with polyethyleneimine (PEI) grafted ZIF-8 showing 79.9 of selectivity for $CO_2/N_2$ (15/85 vol%) separation, and 40.7 of selectivity for $CO_2/CH_4$ (10/90 vol%) separation, under 0.30 MPa gas pressure was reported by yongqiang and his group [36]. Thus, gas separation is widely used with ZIFs materials and still further research is going on. Separation of unwanted liquids from mixed liquid is another interesting area people are working. Selective oil separation from water is possible through the use of highly hydrophobic and superoleophilic ZIF-8 in a tea bag which can further be reused for oil–water separation up to 20 cycles [37]. The flexibility of ZIFs material such as pore window and gate opening phenomenon is the major factor that will further enhance these applications forward.

## *Sensors and Electronics*

The intrinsic properties of ZIFs materials such as high surface to volume ratio, tunable pore size, high adsorbent, hydrophobic nature, and easy functionalization makes it a good candidate for application sensors and electronic devices. Chemical and bio-sensors are widely explored field for ZIFs applications. G.Lu et.al first reported the used of ZIF-8 as selective sensors for chemical vapors and gas [38]. Fisher and his co-workers fabricated ZIFs thin film on SiO2 coated quartz crystal microbalance (QCM) which can detect vapor phase volatile organic compounds like alcohol/water, benzene, toluene, ethylbenzene, and xylene isomers [39]. This can lead to its potential application in biofuel detection and the separation of the said organic compounds which will be quite useful. Recently, an innovative portable eNose was developed by Fabio A. Bahos and Co. which has surface acoustic wave (SAW) sensors fabricated using ZIF-8 and ZIF-67 nanomaterials [40]. This can be used as non-invasive device that can detect several diseases marker such as acetone, ethanol, and ammonia related to the diagnosis and control of diabetes drug mellitus through exhale breath with high sensitivity, selectivity, and reproducibility [40]. This device can be used as a prototype for future detection of the mellitus drug. ZIF-8 had also been reported to use in luminescence probe sensors to detect metal ions and small molecules such as $Cu^{2+}$ and $Cd^{2+}$ ions since its luminescence intensity is very sensitive to the ions [41]. ZIF materials high porosity has led them to the application in electronic devices due to its low-k dielectrics [42, 43]. Frequency-dependent dielectric response of ZIF compounds were studied by Mathew and Co. using synchrotron-based infra-red reflectivity combined with DFT calculations [43]. This can pave the way for infrared sensing and high-speed wireless communication in the future.

## *Drug Delivery*

ZIF materials transition to biomedical applications is because of its special properties such as high chemical and thermal stability, tunable pore size, multifunctionality, and its pH-sensitive capabilities [44]. Many researches are going on for anti-cancer drug delivery agent and ZIFs are treated as capable agents [2]. Sun et.al for the first time found ZIF-8 can be used as a drug delivery agent because of its pH-responsive release properties (Fig. 16.5). ZIF-8 as known from the literature is stable in solvent like water and sodium hydroxide solution but highly unstable in acid medium, this is where it is found as a promising drug delivery agent [1]. ZIF also has high capacity for loading drugs as it is highly porous. Zheng et al. used one pot synthesized of ZIF-8 encapsulated with anti-cancer drugs Doxorubicin (DOX) released under low pH response [45]. They have found high efficiency and some synergetic effect which will be helpful in future development of biomedical drug delivery applications. Another

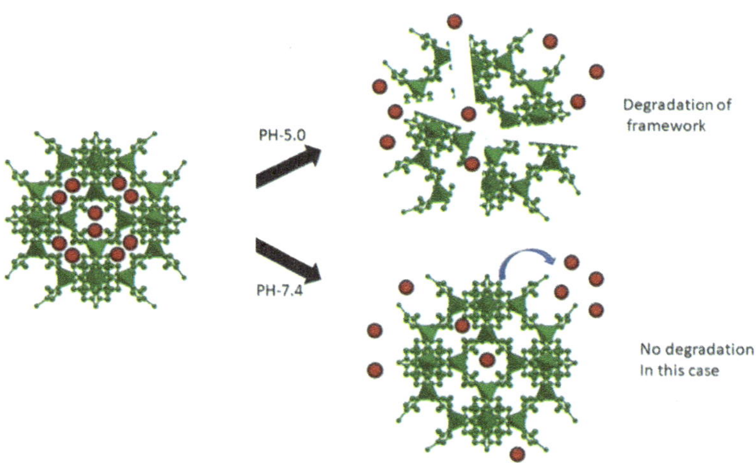


**Fig. 16.5** Schematic diagram showing release of encapsulated drugs in two different ways

group have developed folic acid functionalized DOX@ZIF-8 nanoparticles and they had found that it has shown higher anticancer efficiency in HepG2 cells than pure DOX and DOX@ZIF-8 nanoparticles [46]. This is interesting because with proper functional material further efficiency can be brought. Recently, anti-cancer drugs 5-fluorouracil embedded in carbon nanodots@ZIF-8 shows slow released of the drug at early stage while a fast release at later stage [47]. This hints that drug released time can be controlled. It is not only beneficial to anti-cancer drugs but also other drugs like caffeine which is an amphiphilic drug. Caffeine can be encapsulated in ZIF-8 thus controlling the release of drug and also at the same time provide thermal protection in high-temperature environment as ZIF-8 is highly thermal stable compound [48].

## ZIFs Role in Photocatalytic Application

Removal of pollutants is a worldwide issue as it does not bound to certain areas or land. Due to overpopulation, the basic needs of people are increasing and day by day disposal of waste and hazard products is increasing. Even chemical by-products from industrial waste and agricultural activities have become a major issue in this era and this is a major concern for the future environment. Photocatalysis is one such technique that will eradicate some portion of this pollution problem. It is the process which change the rate of chemical reaction under the influence of light with the introduction of a catalyst which in turn affects the reactants [49]. It has been reported that photocatalytic activity depends on the shape, size, and surface area of the catalyst as well as reaction temperature and pH of the medium [49]. Photocatalysis is not only used in water treatment application but also in hydrogen generation and $CO_2$ adsorption due to redox effect [7]. The mechanism of photocatalytic process can be explained by the theory of HOMO and LUMO state shown in Fig. 16.6. Hydroxyl radical is the active species that is playing a major role in photocatalytic process [8]. It is the concentration of these (OH) radical produced that can be detected using fluorescence intensity. If we consider the photocatalyst to be ZIF-8, the HOMO is occupied by N 2p bonding orbitals and the LUMO is occupied by the empty Zn orbital. During light irradiation, electron transfer is taking place from the HOMO to the LUMO state of the ZIFs. The electron excited at the LUMO can be easily knock out and an electron-deficient at the HOMO needs to be filled to attain its stable configuration. Thus, one electron is taken from the water molecules that is oxygenated into OH radical. The degradation of pollutants like dye (methylene blue, Rhodamine 6G), As, etc. can be provided with the release of these OH radical efficiently during the photocatalytic activity.

Recently, ZIFs material had been widely used as a photocatalytic reagent. Jing et.al have reported the used of ZIF-8 as catalyst in the degradation of methylene blue (MB) under UV irradiation [8]. ZIF-8 can worked in a wide range of pH and show high absorption capacity and efficient degradation of MB [8]. This shows that the ZIF-8 is a promising candidate for the degradation of organic pollutants. Reports show that ZIF-8 do not exhibit any photocatalytic activity towards MB under visible light

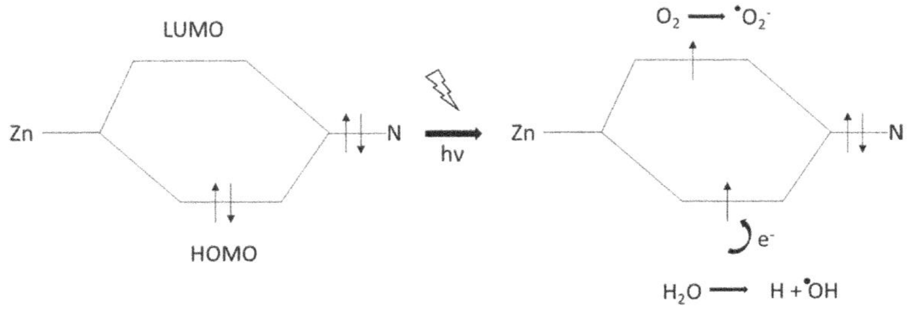


**Fig. 16.6** Schematic diagram depicting the mechanism of photocatalytic effect in ZIF-8

irradiation [50]. So, fabrication of hybrid materials of ZIF-8 is essential for a visible light driven photocatalytic applications. Mai Thi Thanh and Co. have fabricated transition metal doped ZIF-8 i.e. Fe doped ZIF-8 for sunlight driven photocatalytic application [51]. Doping of Fe(II) in high band gap ZIF-8 material leads to decrease in band gap energy thereby shifting the absorption edge to higher visible wavelength. Anindita Chakraborty and Co. have reported CuO deposited ZIF-8 nanocomposites for the degradation of Rhodamine 6G (Rh6G) dye upto 96%. The nanocomposite catalyst worked under wide range of pH ranging from (5–12) indicating effective degradation environment in alkaline environment. Thus, hybrid multifunctional ZIF based compound shows promising result in eradicating the organic pollutants in our environment.

## Summary

The development of ZIFs material in recent times is highlighted and discussed. Its wide area of applications and the possibility of forming more unique materials with varieties of organic linkers are very attractive for researchers. More initiatives can be taken up to look deeper into this material in biomedical areas as it shows promising aspects up to now. Interestingly, a more fascinating development of these materials can be hope to come up in the near future.

## References

1. Park, K.S., Ni, Z., Côté, A.P., Choi, J.Y., Huang, R., Uribe-Romo, F.J., Chae, H.K., O'Keeffe, M., Yaghi, O.M.: PNAS **103**, 10186–10191 (2006)
2. Chen, B., Yang, Z., Zhu, Y., Xia, Y.: J. Mater. Chem. A. **2**, 16811–16831 (2014)
3. Hu, M., Lou, H., Yan, X., Xiaoyan, Hu., Feng, R., Zhou, M.: Microporous Mesoporous Mater. **271**, 68–72 (2018)
4. Fenga, Y., Haiqiang, L., Gua, X., Qiu, J., Jia, M., Huang, C., Yao, J.: J. Phys. Chem. Solids **102**, 110–114 (2017)
5. Zheng, B., Sant, M., Demontis, P., Suffritti, G.B.: J. Phys. Chem. C. **116**, 933–938 (2012)

6. Maul, J., Ryder, M.R., Ruggiero, M.T., Erba, A.: Physical Review B. **99**, 014102 (2019)
7. Wang, X., Liu, J., Leong, S., Lin, X., Wei, J., Kong, B., Yongfei, Xu., Low, Z.X., Yao, J., Wang, H.: ACS Appl. Mater. Interfaces. **8**, 9080–9087 (2016)
8. Jing, H.P., Wang, C.C., Zhang, Y.W., Wang, P., Li, R.: RSC Adv. **4**, 54454–54462 (2014)
9. Tan, J.C., Bennett, T.D., Cheetham, A.K.: PNAS **107**, 9938–9943 (2010)
10. Phan A., Doonan, C.J.,Uribe-Romo, F.J., Knobler, C.B., O'Keeffe, M., Yaghi, O.M.: Acc. Chem. Res. **43**, 58–67 (2010)
11. Lee, Y.R., Kim, J., Ahn, W.S.: Korean J. Chem. Eng. **30**, 1667–1680 (2013)
12. Banerjee, R., Phan, A., Wang, B., Knobler, C., Furukawa, H., O'Keeffe, M., Yaghi, O.M.: Science **319**, 939–943 (2008)
13. Banerjee, R., Furukawa, H., Britt, D., Knobler, C., O'Keeffe, M., Yaghi, O.M.: J. Am. Chem. Soc. **131**, 3875–3877 (2009)
14. Morris, W., Doonan, C.J., Furukawa, H., Banerjee, R., Yaghi, O.M.: J. Am. Chem. Soc. **130**, 12626–12627 (2008)
15. Hayashi, H., Cote, A.P., Furukawa, H., O'Keeffe, M., Yaghi, O.M.: Nat. Mater. **6**, 501–506 (2007)
16. Martins, G.A.V., Byrne, P.J., Allan, P., Teat, S.J., Slawin, A.M.Z., Li, Y., Morris, R.E.: Dalton Trans. **39**, 1758–1762 (2010)
17. Tanaka, S., Kida, K., Okita, M., Ito, Y., Miyake, Y.: Chem. Lett. **41**, 1337–1339 (2012)
18. Kida, K., Okita, M., Fujita, K., Tanaka, S., Miyake, Y.: Cryst. Eng. Commun. **15**, 1794–1801 (2013)
19. Qian, J., Sun, F., Qin, L.: Mater. Lett. **82**, 220–223 (2012)
20. Chen, R., Yao, J., Gu, Q., Smeets, S., Barlocher, C., Gu, H., Zhu, D., Morris, W., Yaghi, O., Wang, H.: Chem. Commun. **49**, 9500–9502 (2013)
21. Suslick, K.S., Hammerton, D.A., Cline, R.E.: J. Am. Chem. Soc. **108**, 5641–5642 (1986)
22. Suslick, K.S.: Science **247**, 1439–1445 (1990)
23. Seoane, B., Zamaro, J.M., Tellez, C., Coronas, J.: Cryst. Eng. Commun. **14**, 3103–3107 (2012)
24. Achar, T.K., Bose, A., Mal, P.: Beilstein J. Org. Chem. **13**, 1907–1931 (2017)
25. Fernández-Bertŕan, J.F., Hernández, M.P., Reguera, E., Yee-Madeira, H., Rodriguez, J., Paneque, A., Llopiz, J.C.: J. Phys. Chem. Solids. **67**, 1612–1617 (2006)
26. Beldon, P.J., Fábзián, L., Stein, R.S., Thirumurugan, A., Cheetham, A.K., Friscic, T.: Angew. Chem. Int. Ed. **49**, 9640–9643 (2010)
27. Friscic, T., Childs, S.L., Rizvi, S.A.A., Jones, W.: Cryst. Eng. Commun. **11**, 418–426 (2009)
28. Tanaka, S., Kida, K., Nagaoka, T., Ota, T., Miyake, Y.: Chem. Comm. **49**, 7884–7886 (2013)
29. Bennett, T.D., Saines, P.J., Keen, D.A., Tan, J.-C., Cheetham, A.K.: Chem.–Eur. J. **19**, 7049–7055 (2013)
30. Shu-Hui, Du., Liu, Y.G., Kong, L.Y., Zhang, J.: J. Inorg. Mater. **27**, 1105–1111 (2013)
31. Dong, X., Huang, K., Liu, S., Ren, R., Jin, W., Lin, Y.S.: J. Mater. Chem. **22**, 19222–19227 (2012)
32. Kwon, H.T., Jeong, H.-K.: J. Am. Chem. Soc. **135**, 10763–10768 (2013)
33. Castro-Muñoz, R., Fıĺa, V.: Chem. Eng. Technol. **42**, 1–10 (2019)
34. Yao, J., Wang, H.: Chem. Soc. Rev. **43**, 4470–4493 (2014)
35. Cacho-Bailo, F., Etxeberría-Benavides, M., David, O., Téllez, C., Coronas, J.: ACS Appl. Mater. Interfaces. **924**, 20787–20796 (2017)
36. Gao, Y., Qiao, Z., Zhao, S., Wang, Z., Wang, J.: J. Mater. Chem. A. **6**, 3151–3161 (2018)
37. Pan, Y., Gao, Z., Zhan, S., Xia, F.: Sep. Purif. Technol. **206**, 186–191 (2018)
38. Lu, G., Hupp, J.T.: J. Am. Chem. Soc. **132**, 7832–7833 (2010)
39. Min, Tu., Wannapaiboon, S., Khaletskaya, K., Fischer, R.A.: Adv. Funct. Mater. **25**, 4470–4479 (2015)
40. Bahos, F.A., Sainz-Vidal, A., Sánchez-Pérez, C., Saniger, J.M., Gràcia, I., Saniger-Alba, M.M., Matatagui, D.: Biosensors **9**, 4 (2019)
41. Liu, S., Xiang, Z., Hu, Z., Zheng, X., Cao, D.: J. Mater. Chem. **21**, 6649–6653 (2011)
42. Eslava, S., Zhang, L., Esconjauregui, S., Yang, J., Vanstreels, K., Baklanov, M.R., Saiz, E.: Chem. Mater. **25**, 27–33 (2012)

43. Ryder, M.R., et.al.: J. Phys. Chem. Lett. **9**, 2678–2684 (2018)
44. Sun, C.Y., Qin, C., Wang, X.L., Su, Z.M.: Expert Opin. Drug Delivery **10**, 89–101 (2013)
45. Zheng, H., Zhang, Y., Liu, L., Wan, W., Guo, P., Nyström, A.M., Zou, X.: J. Am. Chem. Soc. **138**, 962–968 (2016)
46. Bi, J., Yishan, L., Dong, Y., Gao, P.: J. Nanomater. **1357812**, 1–5 (2018)
47. He, L., Wang, T., An, J., Li, X., Zhang, L., Li, L., Li, G., Wu, X., Su, Z., Wang, C.: Cryst. Eng. Comm **16**, 3259–3263 (2014)
48. Liédana, N., Galve, A., Rubio, C., Téllez, C.: J. Coronas ACS Appl. Mater. Interfaces. **4**, 5016–5021 (2012)
49. Saravanan, R., et al: Basic principles, mechanism, and challenges of photocatalysis. In: Nanocomposites for Visible Light-induced Photocatalysis. Springer Series on Polymer and Composite Materials, pp. 19–40 (2017)
50. Yang, H., He, X.W., Wang, F., Kang, Y., Zhang, J.: J. Mater. Chem. **22**(2012), 21849–21852 (1851)
51. Thanh, M.T., Thien, T.V., Du, P.D., Hung, N.P., Khieu, D.Q.: J Porous Matter. **25**, 857–869 (2018)

# Chapter 17
# Biopolymer Based Nano-Structured Materials and Their Applications

**Thoudam Vilip Singh and Lenin S. Shagolsem**

## Introduction

A world without polymers, both synthetic and natural, seems unimaginable today. Both our biotic and abiotic environments are packed with them. The DNA in our body, the myosin in our connective tissues, and the haemoglobin in our blood cells are all biopolymers that are essential to our living. It is hard to imagine a house without plastics. For instance, the household items we use, the books we read, the pens we write with, the electronic devices that seem inseparable in our day-to-day life, the foods we consume and many more have polymers in them.

The term *polymer* comes from the Greek word "polymers" meaning *having many parts*. Polymers are made up of repeating units called monomers. Man-made polymers are called artificial polymers and the ones found naturally are known as natural polymers. Although fossil fuel based polymers like plastics have their usefulness, they pollute the land, the ponds, the rivers, the oceans, and even the atmosphere when burnt. Their degradation rate is quite slow. Plastic pollution also has a great impact on the marine fauna—from ingestion to entanglement [1–3]. Because of the growing environmental concerns all over the world, there has been a hike in the research interest towards biopolymers and the nanomaterials derived from them due to their environment-friendly nature, their versatile properties like non-toxicity and biocompatibility. Thus, the biopolymer-based materials find applications in diverse fields.

*Biopolymers* are bio-macromolecular chains of covalently bonded monomers. In simple terms, they are polymers which are derived from living matter and are found

T. V. Singh · L. S. Shagolsem (✉)
National Institute of Technology Manipur, Langol, Imphal, Manipur 795004, India
e-mail: slenin2001@gmail.com

T. V. Singh
e-mail: thoudamsinghsingh@gmail.com



B. P. Swain (ed.), *Nanostructured Materials and their Applications*, Materials Horizons: From Nature to Nanomaterials, https://doi.org/10.1007/978-981-15-8307-0_17

naturally in microorganisms, plants and animals. There are also synthetic biopolymers which are created because we cannot always depend on the availability of petroleum feedstocks. There are certain characteristics that biopolymers possess. They spontaneously fold into a well-defined three-dimensional shapes and structures. These structures are vital to their proper functioning. They also degrade spontaneously in aqueous environment typical of biological systems, although they can persist due to kinetic trapping. Biopolymers such as DNA and proteins have often been used as templates for the synthesis of nanostructures [4].

By incorporating a wide range of techniques and methods available in the realm of nanotechnology, nano-structured materials can be derived from biopolymers. These are called *biopolymer-based nano-structured materials.* Having synthesized these novel biomaterials, many things that seemed impossible a decade ago are today a reality. It has opened a world of various possibilities. Furthermore, they can be considered "green" in the sense that they are renewable and biodegradable. And this makes us move towards a more positive and greener future.

In this chapter, we cover different types of biopolymers and nano-structured formed by them, and their applications ranging from drug delivery, tissue engineering to sensors which are very relevant for today's ever-increasing demand for alternative biodegradable substitute.

## Classification of Biopolymers

Biopolymers can be classified in different ways based on their biodegradability, their structures, and monomeric units, their applications, their polymer backbones, and even on their origins. For instance, depending on their biodegradability, they are classified as biodegradable and non-biodegradable biopolymers. As far as their backbones are concerned, they can be polysaccharides, polyamides, polyesters, polycarbonates, and vinyl polymers. Here we shall discuss briefly the biopolymers classified on the basis of their origins and focus mainly on those that have applications towards nanotechnology.

So, based on the origin, biopolymers can be classified as *natural* and *synthetic* biopolymers. The natural ones are carbohydrates, proteins, and nucleic acids, while synthetic ones are constituted by biopolymers synthesized with the help of microbes and with the aid of chemicals (Fig. 17.1).

### *Natural Biopolymers: Carbohydrates, Proteins, and Nucleic Acids*

In the context of biopolymers, carbohydrates are biopolymers of monosaccharides called polysaccharides. The term *carbohydrate* originates from two words

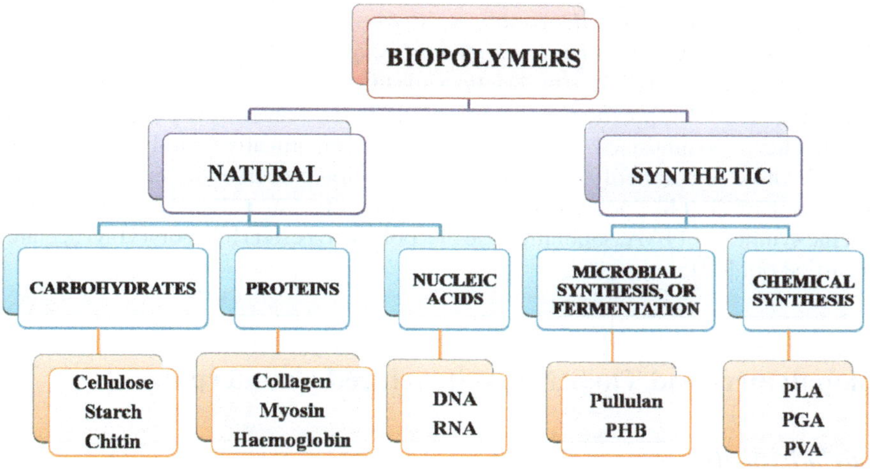


**Fig. 17.1** Classification of biopolymers on the basis of their origin

"carbo-" meaning *carbon* and "hydrate" indicating its formation when combined with water. Polysaccharides are further classified into homopolysaccharides and heteropolysaccharides. Cellulose, starch, glycogen, chitin, and agar are some examples of homopolysaccharides. They have only one type of monosaccharide. On the other hand, heteropolysaccharides consists of more than one monosaccharide. They are also known as heteroglycans. Hyaluronic acid, which is also called hyaluronate, is one of them.

Proteins are biopolymers made up of amino acids. The term *protein* comes from the word "proteios" meaning *the first quality*. There are 20 standard amino acids, although there are as many as 300 amino acids found naturally like collagen and histone.

Nucleic acids are biopolymers made up of a number of monomers called nucleotides. They are one of the most important biopolymers in the sense that they store, transmit, and express genetic information. Primarily, they are of two types-deoxyribonucleic acid (DNA) which acts as a warehouse for genetic information, and ribonucleic acid (RNA) which plays diverse roles in the expression of those information.

## *Synthetic Biopolymers: Via Microbes and Chemicals*

Some of the biopolymers are synthesized by microbes or produced through fermentation. Pullulan is produced by fungus *Aureobasidium pullulans* [5], curdlan by bacterium *Alcaligenes faecalis* var. *myxogenes* [6], polyhydroxybutyrate (PHB) by bacteria like *Pseudomonas oleovorans*, *Ralstonia eutropha* and *Bacillus megaterium* [7], and xanthan by bacterium *Xanthomonas campestris* pv. *campestris* [8].

Poly-glutamic acid (PGA), polylactic acid (PLA), polyvinyl alcohol (PVA), polycaprolactone (PCL), polyhydroxybutyrate (PHB) are some of the biopolymers that are chemically synthesized. PGA can be produced chemically by polycondensation of $\gamma$-glutamic acid dimer [9]. Various polymerization processes are employed in the production of PLA like polycondensation and ring-opening polymerization [10–12]. For the synthesis of PVA, saponification of a poly(vinyl ester) like poly(vinyl acetate) is carried out [13].

## Biopolymers and Their Nano-Structured Counterparts

### *Carbohydrates*

Carbohydrates are primarily made up of carbon, hydrogen, and oxygen, and they function as a source of energy. Their general formula is $C_m(H_2O)_n$ where $m$ and $n$ maybe equal or not, with two functional groups called carbonyl group and hydroxyl group. They can be divided into monosaccharides, oligosaccharides, and polysaccharides. Monosaccharides do not decompose on hydrolysis, while oligosaccharides do so by giving a definite number. Polysaccharides also undergo hydrolysis yielding a lot of monosaccharide molecules.

Carbohydrate-based nanostructures are of great interest and a great deal of research is being carried out as these nano-carbohydrates can be used in the applications of cancer treatment [14], drug delivery [15], biosensors [16], tissue engineering [17] and can even be used as functional biomaterials [18].

Our focus will be on natural polysaccharides in which a number of monosaccharides are linked by glycosidic bonds [19, 20]. These biomacromolecules can be branched or unbranched.

*Cellulose*: Being made by plants, cellulose is one of the most copious carbohydrates. They are present in the cell wall and so they offer vigor for the upward growth of trees against the gravitational pull. They also give resistance to environmental hassles. Cellulose is a straight polysaccharide chain of $\beta$-D-glucose units linked by glycosidic bonds between carbon-1 of one glucose unit and carbon-4 of the next unit. It has a general formula $(C_6H_{10}O_5)_n$ [19] (Fig. 17.2).

Owing to the versatile nature of cellulose-based nanomaterials, they are used in the fields of medical science and material science. Nanocellulose is but a minute fiber. Now research in nanocellulose-protein composites is going on to bring out the superior qualities from both the materials [21]. Korhonen and coworkers have synthesized highly porous nanocellulose aerogels which are both recyclable and reusable as oil absorbents [22]. Nanofibrilated cellulose and cellulose nanocrystals also exist [19] (Fig. 17.3).

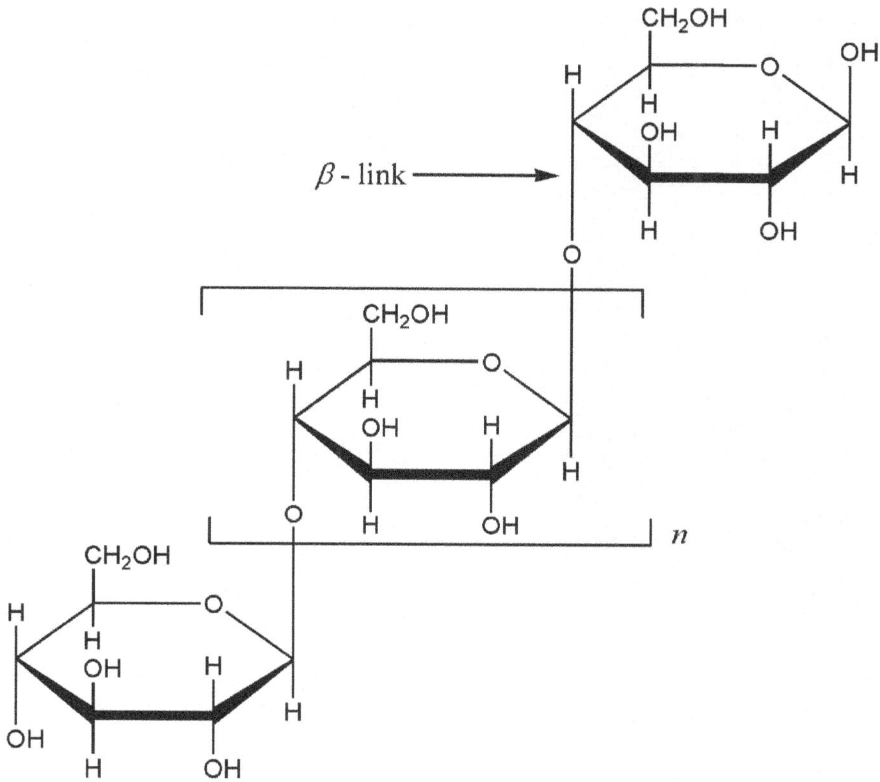


**Fig. 17.2** Structure of cellulose

*Starch*: Starch has a granular structure in its native form. The general formula for starch is given by $(C_6H_{10}O_5)_n$. Comprising of two polysaccharides-amylopectin and amylose, they are usually found in plants, particularly in plant seeds which serve as a food reserve. We find them in our daily life in the form of rice, potatoes, bread, cakes, and so on. Amylopectin is a branched polysaccharide made up of chains of 24–30 D-glucose units connected by $\alpha$-glycosidic linkage between carbon-1 of one glucose unit and carbon-4 of the next. Furthermore, these chains are linked by 1,6 linkages [20]. Amylose, a linear polysaccharide, consists solely of D-glucose units. They are connected by $\alpha$-glycosidic linkages between carbon-1 of one glucose unit and carbon-4 of the next glucose unit (Fig. 17.4).

Starch nanoparticles with diverse properties, shape, and even crystallinity can be synthesized by precipitation of amorphous starch [24, 25], combining complex formation and enzymatic hydrolysis [26], and microfluidization [27]. There are various starch-based nanocomposites reinforced by cellulose nanowhiskers, chitin/chitosan nanoparticles, carbonaceous nanofillers, etc. such that their shapes, their sizes as well as their surface chemistry are not the same [28]. These

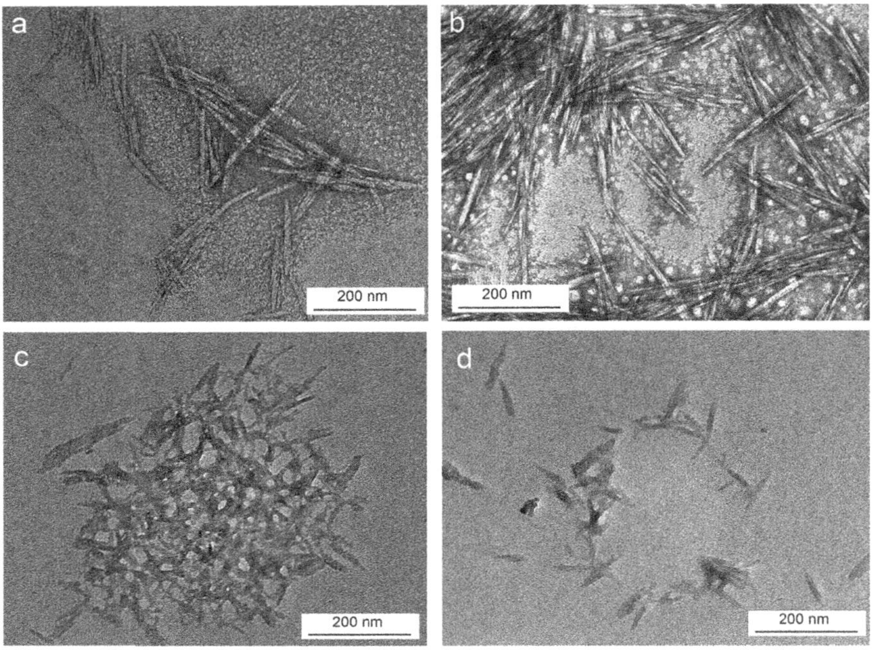


**Fig. 17.3** Transmission electron microscopy images of **a** cellulose nanocrystal, **b** silane-treated cellulose nanocrystal, **c** 6% cellulose nanocrystal-unsaturated polyester resin nanocomposites, and **d** 6%) silane-treated cellulose nanocrystal nanocomposites (with permission from Ref. [23])

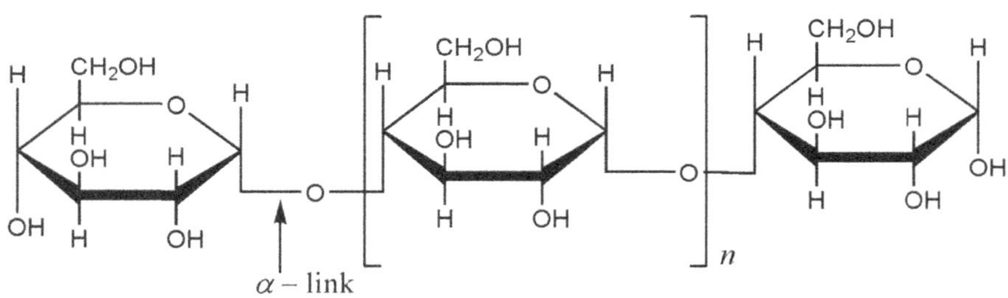


**Fig. 17.4** Structure of amylose (a starch)

nano-biocomposites can be used in many promising applications of biomedicine (Fig. 17.5).

*Chitin*: Chitin is a biopolymer of *N*-acetyl D-glucosamine units connected by glycosidic bonds between carbon-1 of a unit and carbon-4 on the next unit [30]. This biopolymer is an important component of the tough exoskeleton we see in many animals like crabs and lobsters. Its general formula is $(C_8H_{13}O_5N)_n$ (Fig. 17.6).

Nanofibrils and nanowhiskers derived from chitin have been used as drug delivery scaffolds [31]. Its positively charged derivate, chitosan, also has immense potential

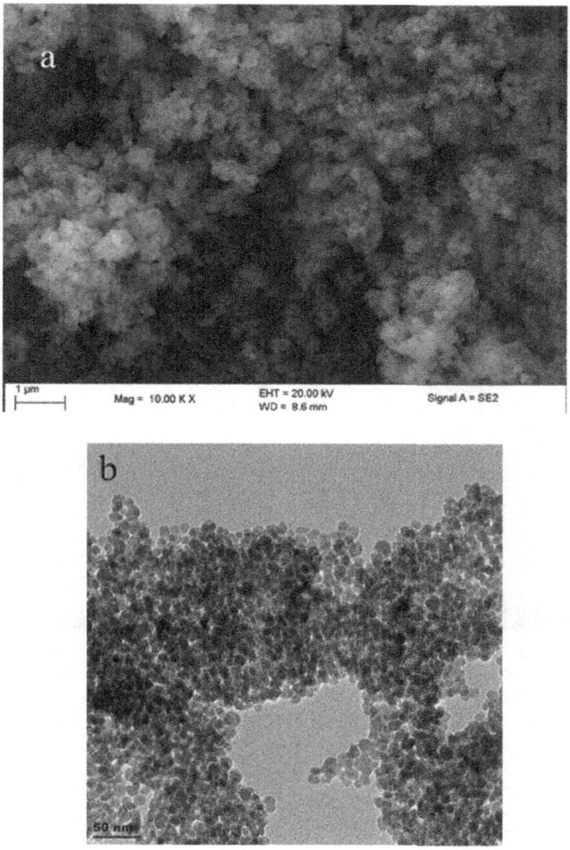


**Fig. 17.5** **a** Scanning electron micrograph, and **b** transmission electron micrograph of a magnetic starch nanocomposites. Such biomaterials have emerging industrial, analytical, environmental, and biomedical applications (with permission from Ref. [29])

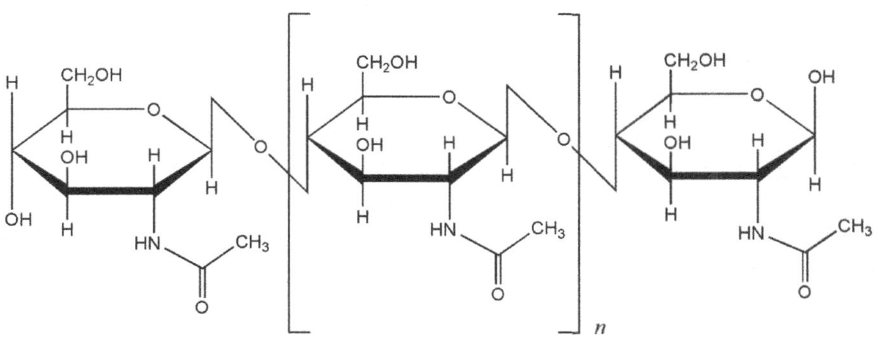

**Fig. 17.6** Structure of chitin

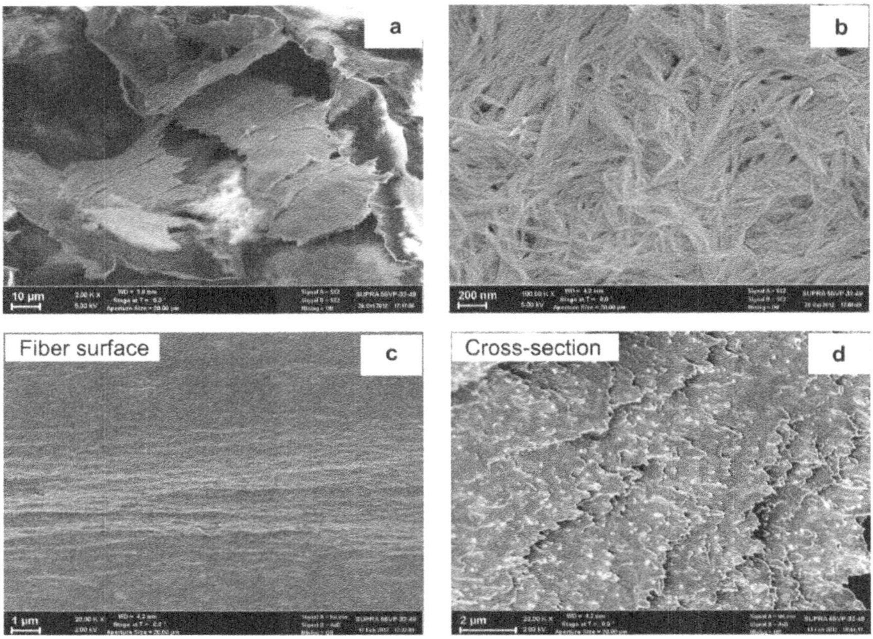


**Fig. 17.7** Scanning electron microscope images of the film acquired from chitin nanofibril water dispersion through freeze-drying (**a** and **b**). The cross-section of the composite monofilament chipped in liquid nitrogen, fiber has 1 wt% of chitin nanofibrils (**c** and **d**) (with permission from Ref. [32]

in the nanotechnology of biomedicine as that positive charge not only accelerates wound healing, but it also provides antibacterial, anticoagulation, and antimicrobial properties [17] (Figs. 17.7 and 17.8).

## *Proteins*

The basic structure of protein is known as the *primary structure*. It consists of amino acids held together by peptide bonds [33]. The sequence of each amino acid is determined genetically through transcription and translation.

In a cell, proteins are produced in ribosomes. When they are released in the cytoplasm, they undergo a spontaneous transition called "protein folding" resulting in their native states. These states are important for proper functioning of the organism, and their two most common types of structures are $\alpha$-helices or $\beta$-sheets [33]. They are known as the *secondary structures* of proteins. In $\alpha$-helix secondary structure, the chain is twisted spirally, usually in the right-hand manner. This structure is established by formation of large number of hydrogen bonds and it is found in proteins like keratin

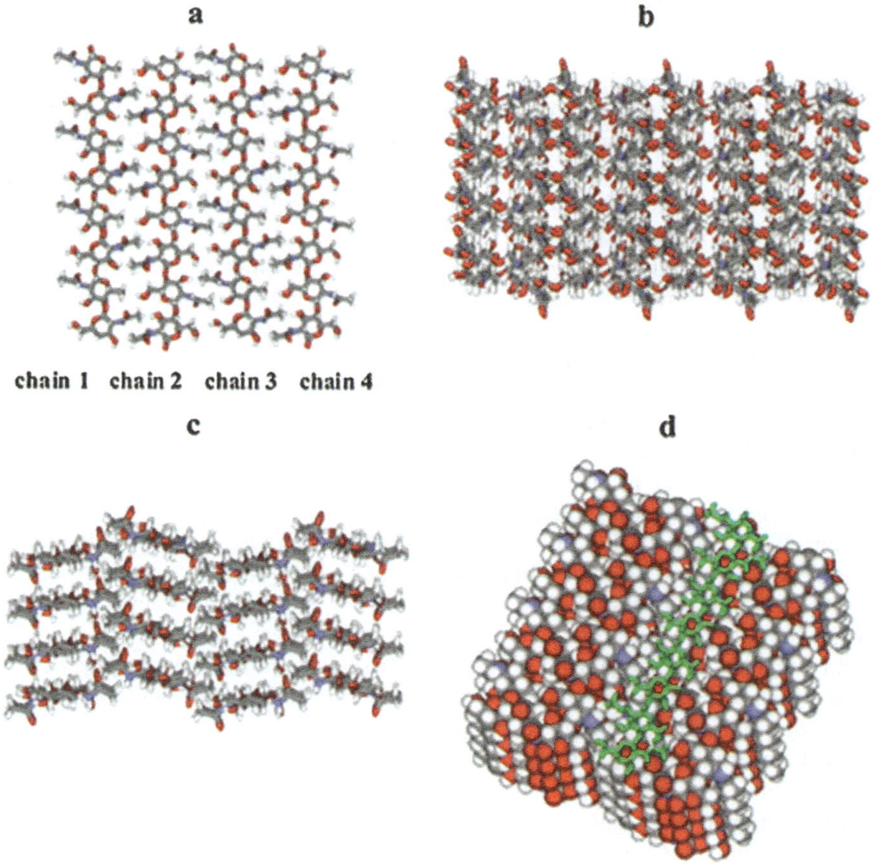


**Fig. 17.8** Here, the chitin nanocrystallite consists of sixteen chains with eight monomer each: **a** top view; **b, c** views from the side (surface layer of the crystallite consists of four chitin chains (chain 1-chain 4); **d** one chitosan chain (green) on surface of the chitin crystalline. Carbon atoms are grey, oxygen atoms are red, nitrogen atoms are blue and hydrogen atoms are white. (With permission from Ref. [32])

which is present in hairs, tropomyosin present in muscles, epidermin in skins, and so on. The $\beta$-pleated secondary structures of proteins result when polypeptide chains are linked by hydrogen bonds in parallel or antiparallel fashion.

The three-dimensional arrangement of the helices and the sheets results in the *tertiary structures* of proteins [34]. A lot of bond come to play while stabilizing this structure. Peptide bonds and disulphide bonds are the covalent bonds involved. Electrostatic bonds, hydrogen bonds, hydrophobic interactions, and Van der Waals interactions also help in establishing this structure.

Stabilized by electrostatic bonds, hydrogen bonds, and hydrophobic interactions, the *quaternary structures* typically comprise of more than one polypeptide chains linked into a single protein having multiple sub-units [35].

Similar to nano-carbohydrates, there are nanomaterials derived from proteins. The research towards protein-based biopolymers has increased considerably in the past few years. Many of these materials are seen in biomedical applications like drug delivery [36], antitumor photodynamic and photothermal therapy [37]. They are also often seen in food applications [38].

*Collagen*: The native structure of this biopolymer is stabilized by its triple-helical domain [39]-the three polypeptides winding in an $\alpha$-helix. It repeats a pattern of glycine-proline-*R* where *R* is an amino acid. They are present in large extent in connective tissues and so it is the most abundant proteins in animals.

Using biomimetic synthesis, Lio and co-workers developed a nano-hydroxy apatite/collagen/PLA composite that can be used as a bone scaffold material [40]. This bone-resembling substitute material, nano-hydroxyapatite/collagen, is biodegradable as well as bioactive [41] (Fig. 17.9).

*Haemoglobin*: Haemoglobin is a biopolymer which is responsible for the red color of our blood. Not only does it help in the transportation of oxygen from the lungs to the body tissues, it also helps in bringing back the $CO_2$ molecules and $H^+$ ions to the lungs. It has four sub-units that have cofactors containing iron atom at the centre [43] (Fig. 17.10).

In humanity's fight against deadly viruses like HIV and Ebola virus, researchers have developed a nano-biomaterial formed by crosslinking haemoglobin, superoxide dismutase, catalase, and carbonic anhydrase called polyHb-SOD-CAT-CA bio-nanomaterial complex, which is a soluble complex with improved red blood cell functions and it does not have any blood group antigens [44–46]. There are also biopolymer-based nanocapules that have haemoglobin as well as red blood cell enzymes in them [46] (Fig. 17.11).

## Nucleic Acids

The fundamental unit of nucleic acids is called a nucleotide, which is a product of a pentose sugar, a nitrogenous base, and phosphoric acid. Ribose ($C_5H_{10}O_5$) and deoxyribose ($C_5H_{10}O_4$) are two types of pentose sugar that occur in the nucleotides. RNA with D-ribose and DNA with D-deoxyribose are the two types of nucleic acids. The nitrogenous bases are also of two types-pyrimidines and purines. Pyrimidines are six-membered cyclic compounds with nitrogen at 1 and 3 positions, numbered in clockwise direction. Their bases are cystosine (C), thymine (T), and uracil (U). On the other hand, purines are nine-membered double-rings with nitrogen at 1, 3, 7, and 9 positions, numbered in anti-clockwise directions. Adenine (A) and guanine (G) are its two types.

Based on the X-ray diffraction pattern of deoxyribose nucleic acid produced by Rosalind Franklin, which suggested that DNA possessed helical structure, Watson

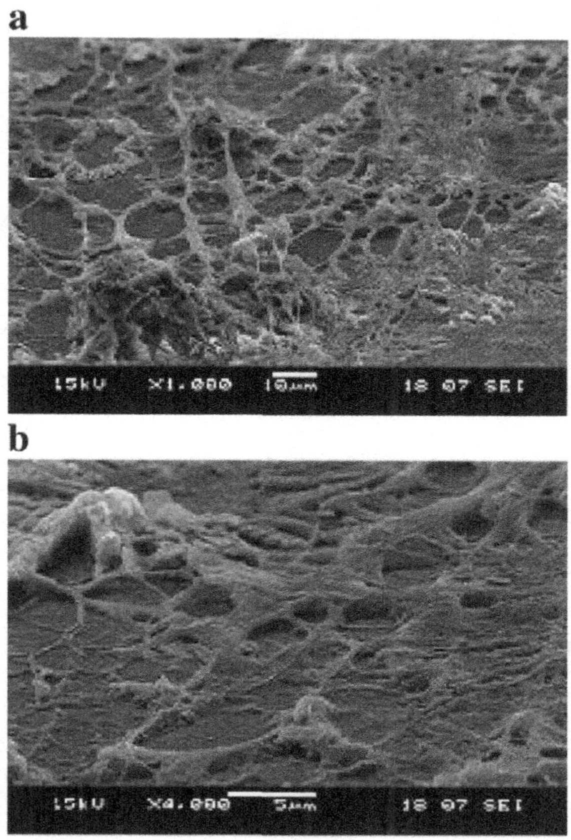


**Fig. 17.9** Scanning electron micrographs of cell-scaffold constructs, after seven days in human osteoblastic cell culture, for two different magnifications. The bars correspond to 10 μm in **a** while 5 μm in **b**. The fibrous porous scaffold was prepared by self-assembled nano-hydroxy apatite-Si(0.2%)-Mg(0.6%)-Zn(0.2%), chitin, and collagen type I layers (with permission from Ref. [42])

and Crick derived at their *DNA double helix model* [48, 49]. While the bases face toward one another inside the double helix, the sugar back-bonds are on the outside. With a diameter of 2 nm, the helix is right-handed containing 10 nucleotide pairs per turn. The phosphate group is linked to carbon 5′ of the sugar residue of its own nucleotide and carbon 3′ to that of the next nucleotide by phosphodiester bonds. The two DNA chains also run antiparallel to each other. The number of pyrimidines present is equal to the number of purines present in the DNA. This is the *Chargaff's rule* [50–52] and its proof lies in the complementary base pairings. This is an ideal structure of DNA known as B-DNA, but the naturally occurring ones diverge from this structure time and again (Figs. 17.12 and 17.13).

RNA is formed by ribonucleotides joined together by 3′–5′ phosphodiester bonds. Not only is the sugar ribose in RNA, it also has uracil instead of thymine, and can exist as a single-stranded structure. Some of the types of RNA are messenger RNA (mRNA), transfer RNA (tRNA), and ribosomal RNA (rRNA) [54, 55] (Fig. 17.14).

In DNA-based nanostructures, nuclear bases are inserted and deleted in targeted areas of the DNA, and as a result, the stress and the flexibility of these nanostructured

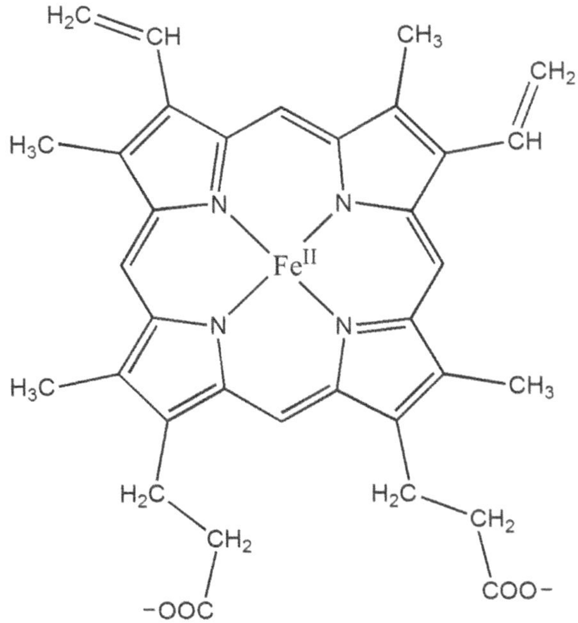


**Fig. 17.10** Structure of deoxyhaemoglobin

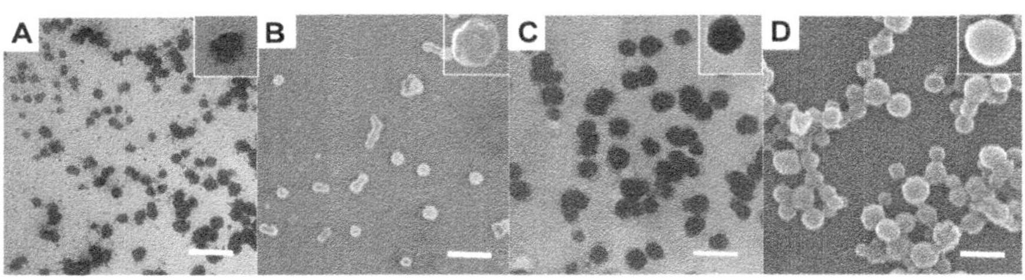


**Fig. 17.11** Transmission electron microscope images (**a**, **c**), and scanning electron microscope images (**b**, **d**) of functionalized nanogels (**a**, **b**), and haemoglobin-loaded nanogels. The haemoglobin loaded nanogels showed high affinity for oxygen. They were stable as well as haemo-compatible (with permission from Ref. [47])

biomaterials can be controlled in order to create bio-complex structures [56]. They have also been used as biosensors [57, 58]. There are also DNA origami nanostructures with which intricate nanostructures in one, two, and three dimensions can be engineered [57]. For over a decade, design methods for DNA and RNA nanostructures have been developed [59–63]. DNA nanostructures can be used as templates for bottom-up fabrication of nanodevices [64].

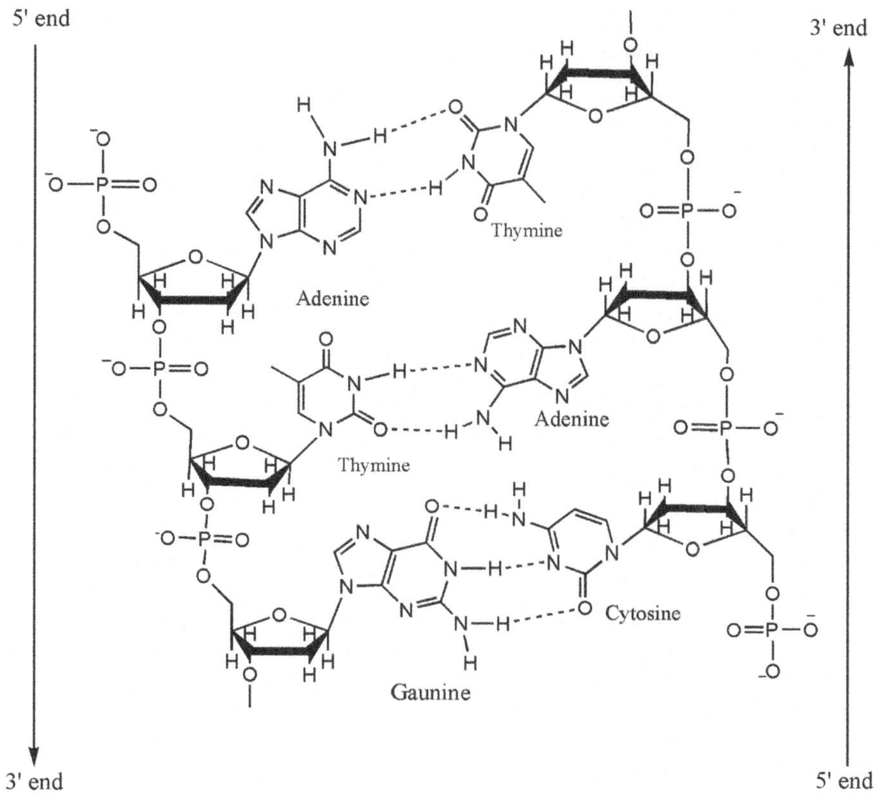


**Fig. 17.12** The bonding in nucleic acid structure. The hydrogen bonds (dashes) between adenine and thymine, and between guanine and cytosine account for the AT and CG base pairing of DNA. While the AT pair is held by two hydrogen bonds, there are three hydrogen bonds in the CG pair. If this structure were to be RNA, then there will be uracil (U) as the pairing partner for adenine

## *Synthetic Biopolymers*

When biopolymers are produced through the use of chemicals or with the help of microbes, they are called synthetic biopolymers.

*Pullulan*: Pullulan is a linear polysaccharide of maltotriose components. When three saccharide molecules are joined by $\alpha$-1,4 glycosidic linkages, we call it maltotriose. Furthermore, in pullulan, $\alpha$-1,6 glycosidic bonds links successive maltotriose units. It has a chemical formula of $(C_6H_{10}O_5)_n$. Because of its unique structure, it has huge potential for applications in biomedicine [65] (Fig. 17.15).

Upon investigation of pullalan-based nanoparticles, it was seen that they hold huge potential in the application of protein delivery [66]. Lu and coworkers have made a pH-sensitive drug delivery system based on pullulan/doxorubicin conjugate [67].

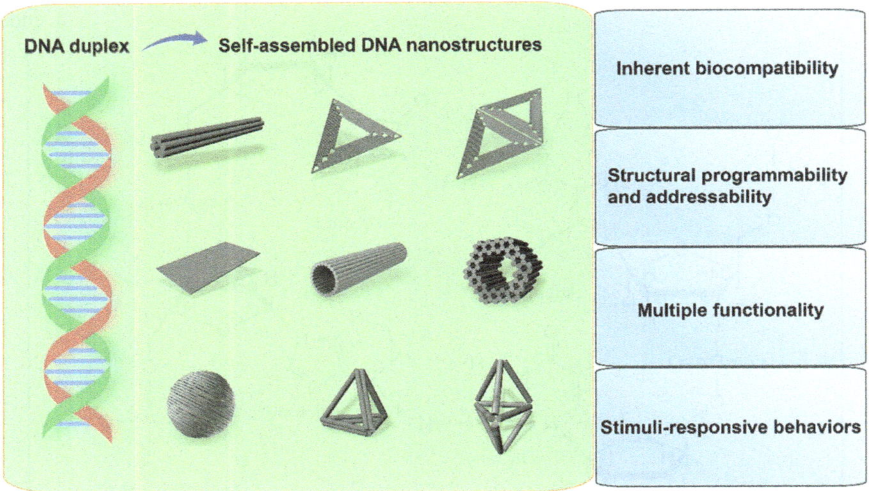


**Fig. 17.13** DNA nanostructures (with permission from Ref. [53])

*Polyhydroxyalkanoate (PHA)*: Polyhydroxyalkanoates constitute a family of polyesters [68] and it can be divided according to its length as short PHAs and medium PHAs [69]. Short PHAs have C-monomers ranging from three to five while that of medium PHAs ranges from six to fourteen C-monomers in the 3-hydroxyalkanoate units. They are considered as green plastics with positive influence on the environment [70], and as such, they also have potential applications in packaging [71] (Fig. 17.16).

PHA-based nanostructured materials such as nanofibre scaffolds are suitable for repairing the damages in the spinal cord [72]. Nanofibres-based peptide modified PHA scaffold also have a huge impact in tissue engineering. There are also PHA-based nanocomposites that have high-quality mechanical properties [73].

*Polyhydroxybutyrate (PHB)*: Polyhydroxybutyrate is a rigid crystalline biopolymer. It is hydrophobic in nature. Its chemical structure is given below (Fig. 17.17).

Nanoparticles based on PHB with an average diameter of 55 nm can be created through dialysis method of nanofabrication. When therapeutic drugs are loaded, these sizes are not affected [74]. The PHB nano-drug carriers are efficient in distributing the drugs uniformly and it also reduces local irritation [75].

*Polylactic acid (PLA)*: Polylactic acids are linear aliphatic thermoplastic polyesters with hydroxyl group next to the carboxyl group [76]. It is said to be the most common synthetic polymers studied till today, due to its biodegradable and biocompatible nature (Fig. 17.18).

Even though PLA are regarded as a good substitute to petroleum-based plastic materials, many applications are restricted by properties like high crystallinity, time-consuming degradation, and stiffness [77]. PLA-based nanomaterials provide a wide

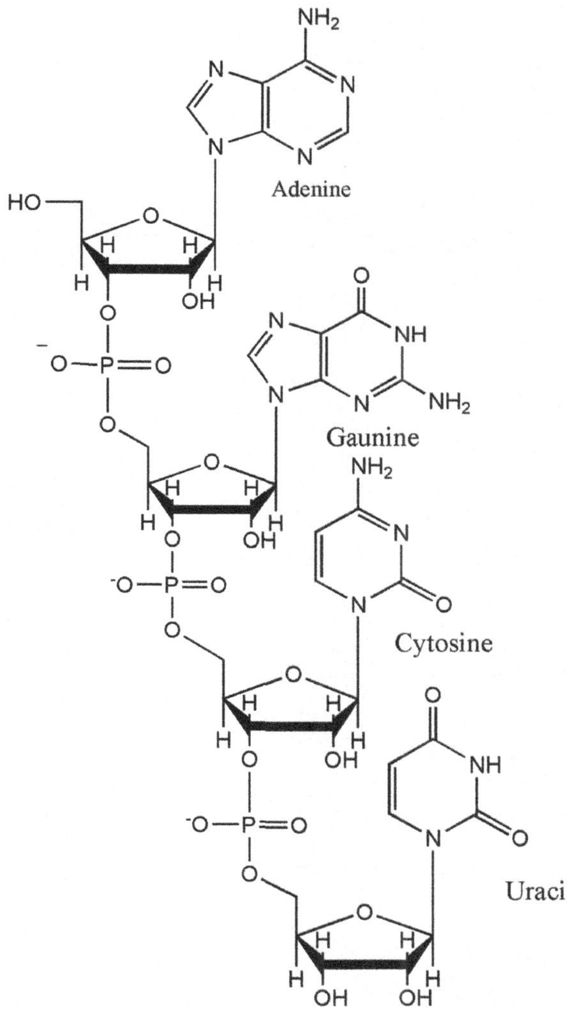


**Fig. 17.14** Structure of single-stranded RNA

range of better qualities offering applications to bioplastics that can be used in packaging food and agricultural products. There are plenty of PLA-based nano-structured biomaterials such as PLA/clay nanocomposites, PLA/caly nanotube nanocomposites, and PLA/hydroxyapaite nanocomposites [78]. Bhardwaj and coworkers developed a PLA-based novel hyperbranched nano-structure that had significant improvement in its toughness (~570%) and elongation (~847%) in comparison to ordinary PLA [79].

*Polyglycolide* (*PGA*): Polyglycolide, a thermoplastic biopolymer, has the chemical structure of $(C_2H_2O_2)_n$. It has high crystallinity of about 45–55%. They are

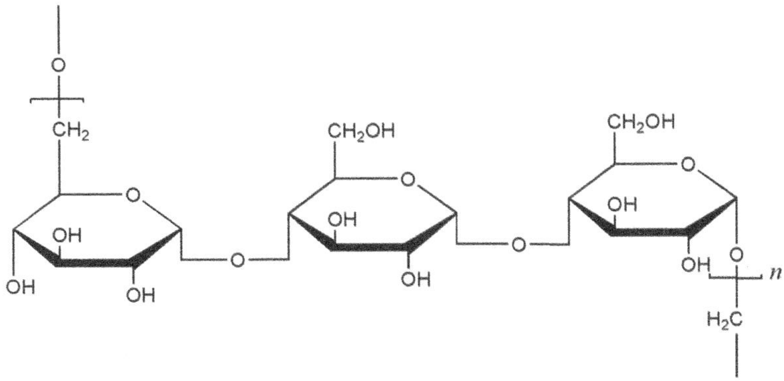


**Fig. 17.15** Structure of pullulan

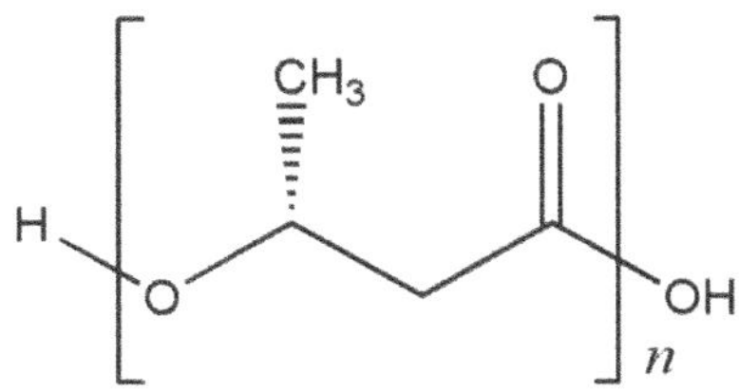


**Fig. 17.16** Structure of PHA

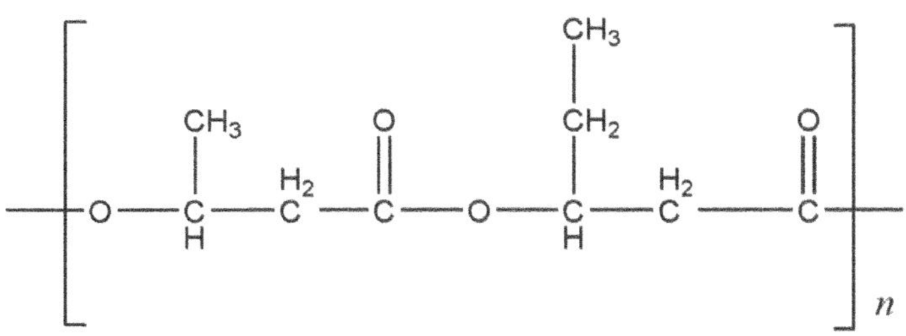


**Fig. 17.17** Structure of PHB

**Fig. 17.18** Structure of PLA

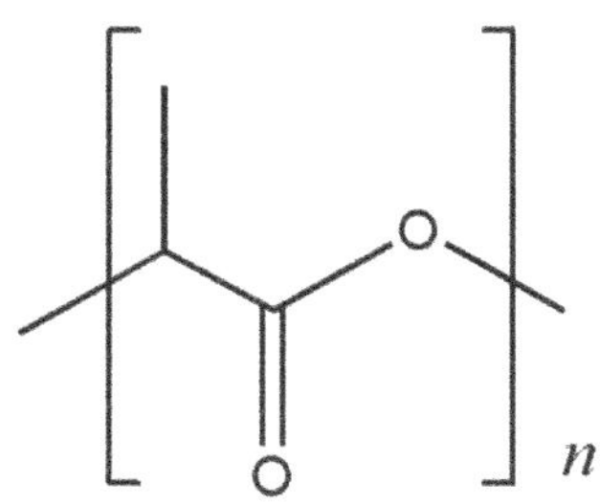

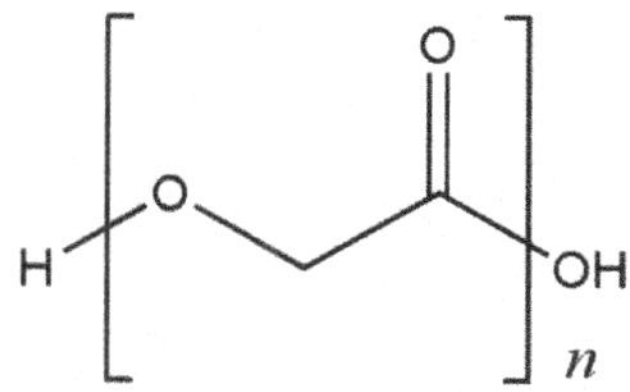


**Fig. 17.19** Structure of PGA

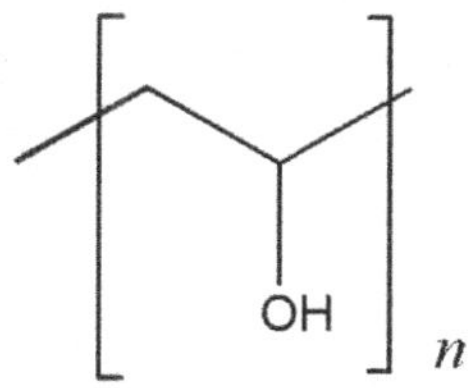


**Fig. 17.20** Structure of PVA

produced by microorganisms such as *Bacillus subtilis*, *B. anthracis*, *B. halodurans*, *B. licheniformis*, *Natrinococcus occultus*, *Natrialba aegyptiaca* [80, 81], etc. (Fig. 17.19).

PLA-based nanoparticles can be used as drug nano-carriers. There are PLA based micelles with huge potential as a vaccine carrier [82, 83]. For instance, Jimenez-Sanchez and coworkers have designed these structures from a PLA-b-P(*N*-acryloxysuccinimide-co–*N*-vinylpyrrolidone) block copolymer [83]. There are PLA and PGA based scaffolds which are useful in the application of biomedicine [84].

*Polyvinyl alcohol (PVA)*: With the chemical structure of $[CH_2CH(OH)]_n$, polyvinyl alcohol is another synthetic biopolymer with wide range of applications (Fig. 17.20).

Nguyen and coworkers silver loaded PVA nano-fibrous mats used as promoter in skin healing [85]. Investigation showed a significant acceleration in diabetic wound healing on the rats when they were treated with PVA/chitosan nano-fiber [86]. PVA and PCL nano-fibres patch produced through electrospinning can be used for ocular drug delivery including glaucoma [87] (Fig. 17.21).

## Applications of Biopolymer-Based Nanostructured Materials

Many scientists and researchers are interested in the study of biopolymer-based nano-structured materials because they have a wide range of applications and huge potential for the future of medicine, science, and technology (Fig. 17.22).

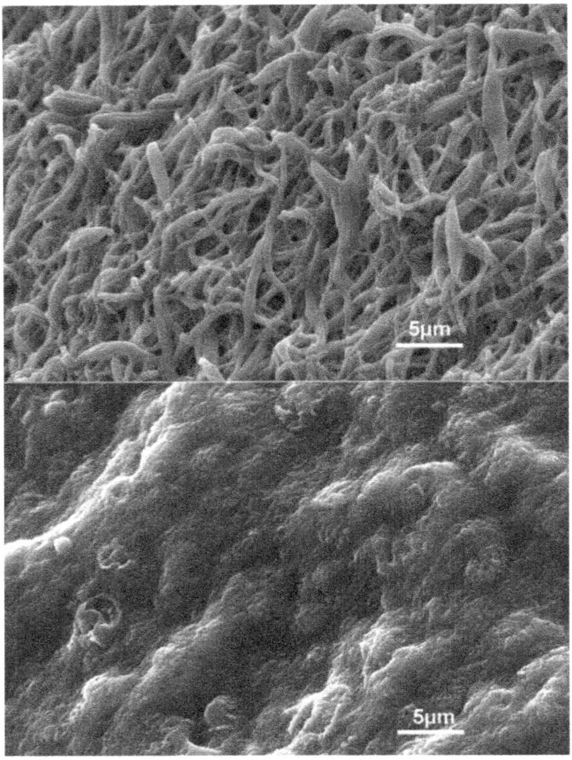


**Fig. 17.21** Scanning electron of nanofiber with poly(3-hydroxybutyrate-*co*-3-hydroxyhexanoate/PVA (in ratio of 30/70) (top), and (in ratio of 10/90) (bottom) without heat treatment after immersing in water for a day. They have a magnification of ×5000 (with permission from Ref. [88])

## *Drug Delivery and Tissue Engineering*

The important factors that account for the therapeutic applications of biopolymer-based nanomaterials in drug delivery and tissue engineering systems include their non-toxic nature, biodegradability, and biocompatibility [89]. We can treat tissue engineering as a case of drug delivery where specific cells are delivered at specific sites in a controlled manner. Nanoencapsulation is one of the most promising approaches in the field of nanomedicine. Many of the biopolymers used in these fields are produced by bacteria (Fig. 17.23).

For instance, let us consider poly-$\gamma$-glutamic acid (PGA) which is an ideal candidate for a drug nano-carrier due to its non-toxicity, biocompatibility, biodegradability, and its solubility in water [80, 81, 89]. Exploiting the molecular weight of PGA as a selectivity parameter, encapsulated drugs can be delivered to tumor tissues and not the ordinary ones [89, 90]. PLA-based micelles are under study for their potential application as vaccine carriers [82, 83]. In cardiac tissue engineering, researchers have synthesized PLA and PGA based scaffolds [84].

Polyhydroxyalkanoates (PHAs) have numerous applications such as plastic fabrication, materials for printing, biofuel, and many more. Its ability to turn into films

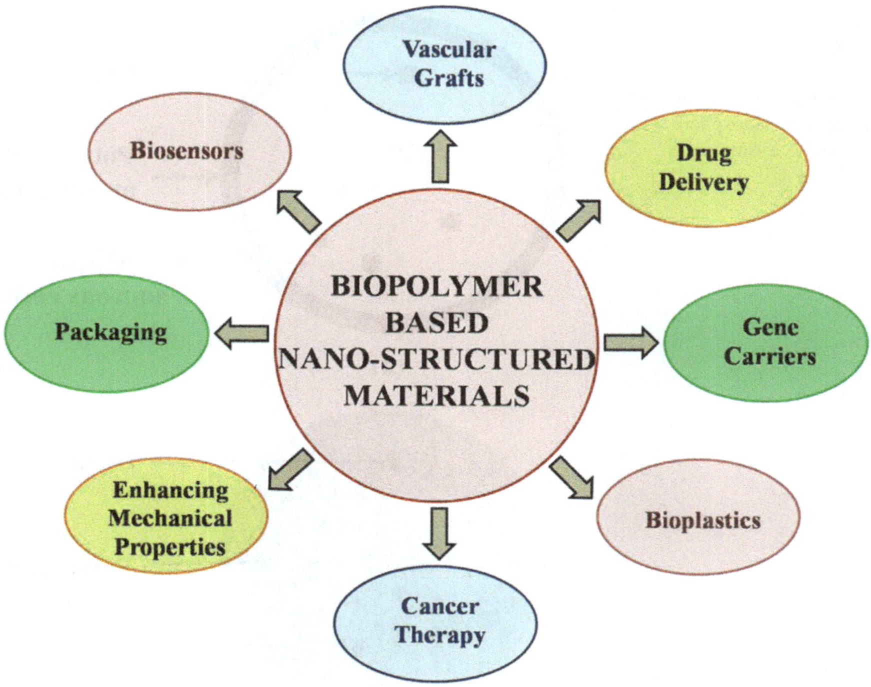


**Fig. 17.22** Some of the applications of biopolymer-based nano-structured materials

and porous matrix is useful in drug delivery systems. There are PHA copolymers with low melting point and lower degree of crystallinity. Its biodegradability can be altered through different fabrication processes. There are ways to increase its porosity like emulsion method [91]. All these factors make PHAs suitable for drug delivery systems. This biopolymer also has its uses in tissue engineering because its biocompatibility can be controlled. Moreover, it has other features like cytological compatibility, and it is not susceptible to hydrolytic degradation. Thus, it finds various applications in cardiovascular tissue engineering in which it focuses on the study of vascular patches and even the implantation of heart valve, bone tissue engineering which demands high-quality mechanical properties of the material to be engineered, cartilage tissue engineering in which making porous scaffolds comes to play, and nerve conduits tissue engineering that requires the matching of nerve degradation rate and nerve regeneration rate [91]. Xu et al. studied the behaviour of neural stem cells on PHA nanofiber scaffolds [72]. These matrices (nanofiber scaffolds) supported the growth of neural stem cells and so they can be developed for treating central nervous system injury (Figs. 17.24, 17.25, and 17.26).

Protein-based nanocarriers can also be used for tissue engineering. In addition to their biodegradability and high nutritional value, these proteinb-ased nanomaterials like collagen have low toxicity and are quite stable [94]. Albumin, gelatin, elastin,

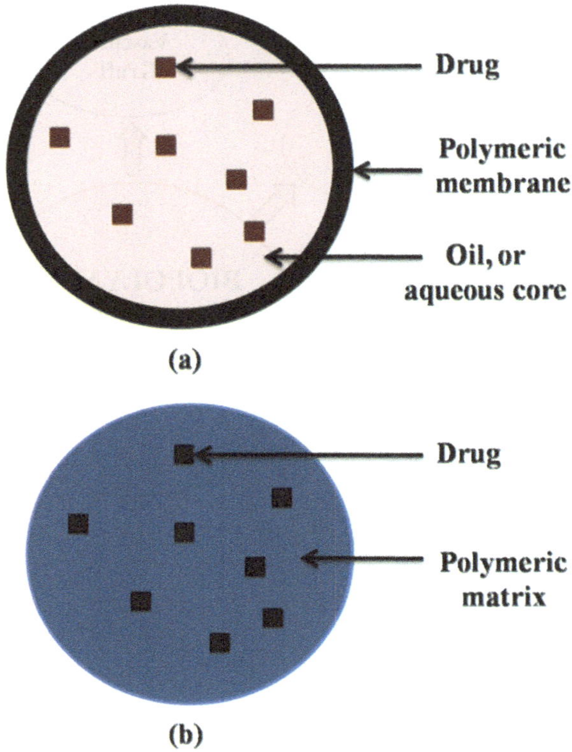


**Fig. 17.23** **a** Nanocapsule, and **b** Nanosphere. They are drug delivery systems where nanocapsule acts as reservoir system and nanosphere serve as a matrix system

zein, gliadin, and legumin are some of the proteins from where the nanoparticles are synthesized and used for cancer therapy [95]. Another fibrous protein called keratin is a biological material that can be extracted from human hairs and wool. It can be synthesized into fibres, films, gels, and so on for repairing and delivery applications [96].

Chitosan-based magnetic nanoparticles can be used in magnetic drug delivery methods. The chitosan nanomaterials carry the agents to be delivered and when they are administered, they move towards the target sites with the help of external high-gradient magnetic field and discharge the therapeutic agents [97]. Because silk nanocomposites are quite tunable, they are also used in biomedical applications like this. Enzymes like lipase are stabilized with the help of these silk nanostructures [98].

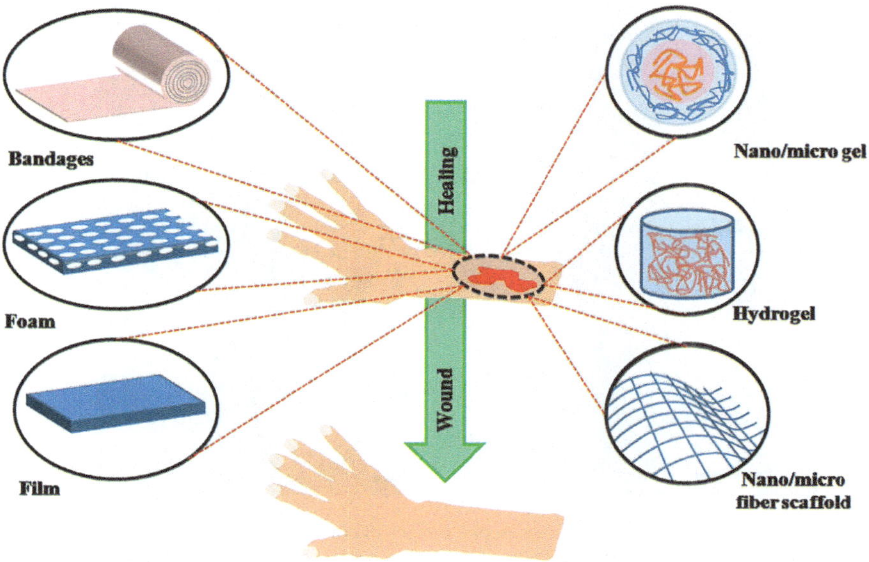


**Fig. 17.24** Different types of wound dressing. The biopolymer-based nanomaterials can be used in the treatment of wounds (with permission from Ref. [92])

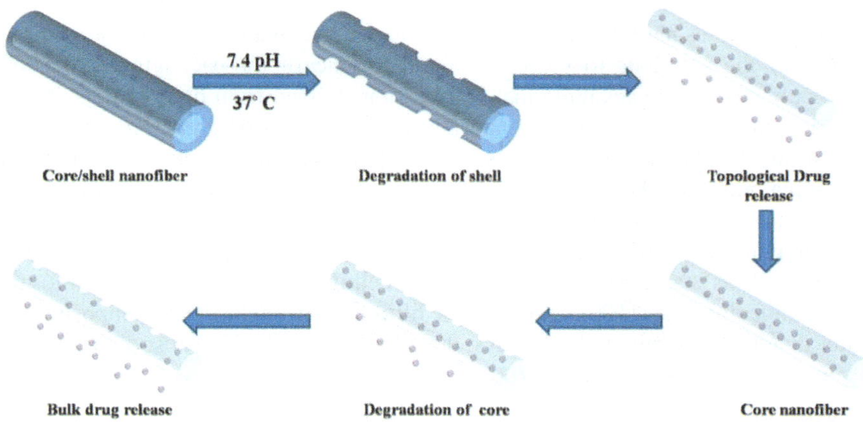


**Fig. 17.25** A method of drug delivery. Fang et al. also studied biodegradable core-shell electrospun nanofibres based on PLA and $\gamma$-PGA to heal wounds [93] (with permission from Ref. [92])

## *Applications in Packaging*

It was estimated by 2015 that nearly 6300 metric tons of plastics had been produced, out of which 79% was in the natural environment or dumping-grounds [99]. And

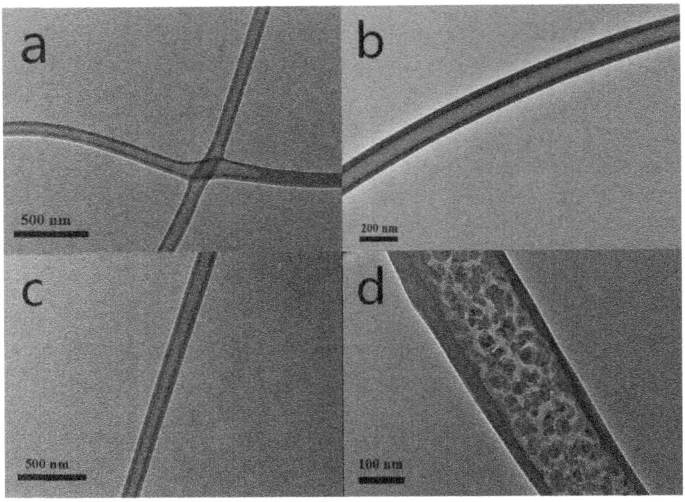


**Fig. 17.26** Transmission electron microscope images of core-shell nanofibers with $\gamma$-PGA as the core while PLA is the shell material. They are created by coaxial electrospinning method. Nanofibres with **a** different core–shell structures, **b** compact core structure, **c** porous core structure, and **d** porous core structure at high magnification (with permission from Ref. [93])

the largest single market for plastics is packaging [100]. Needless to say, our non-degradable wastes harm the very environments we live in and we should act now.

One of the solutions to the environmental concern for packaging is the use of biopolymer-based packaging materials—films derived from biopolymers. These materials can not only be degraded naturally, but they can also be renewed. Moreover, their ability to act as a barrier to gases, water vapor, and solutes makes them ideal for packaging, although there are some which have poor resistance to moisture. These biopolymer–based films can be prepared from polysaccharides, proteins and lipids exhibiting different properties. Likewise there are natural polymer-based nanocomposites such as starch-based, protein-based, cellulose-based [101], and chitosan-based nanocomposites [100]. Although the films made from bio-nanocomposites are better than the synthetic in many ways, there are drawbacks. The cost of production is very high and it also requires costly equipment.

## *Biosensors*

Biosensors are devices capable of giving quantitative analytical information with the use of biological recognition elements. The investigations carried out on biosensors are raising rapidly as they can be used in many areas. In biomedical applications, these sensors can act as glucose monitor for diabetic patients by detecting blood sugar levels. The qualities of air and water can be checked. They can detect toxicity

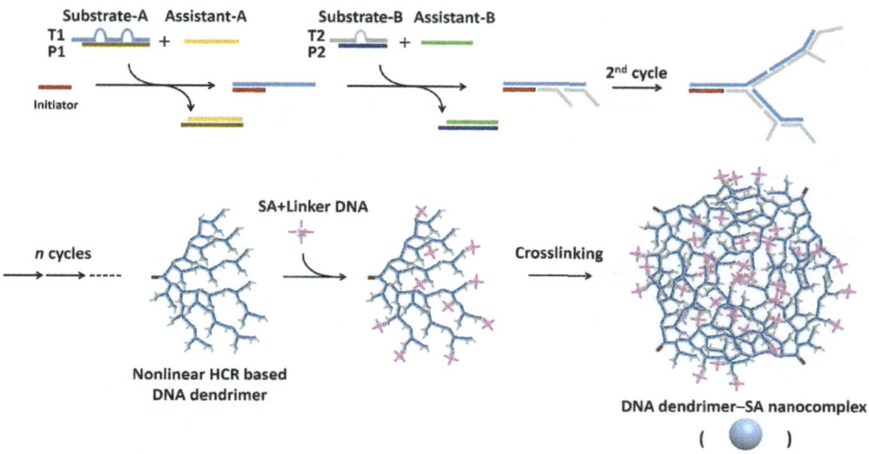


**Fig. 17.27** A representation of the assembly of a novel signal amplifier, DNA dendrimer-streptavidin (SA) nano-complex that can be used to make biosensing platforms for disease-related species (with permission from Ref. [104])

levels in them. In food industry, biosensors have applications in food ripening and fermentation process.

Biosensors can be fabricated from nanoparticle-biopolymer system. For instance, Yang and coworkers engineered biosensors from carbon nanotube/cobalt hexa-cyanoferrate nanoparticle-biopolymer system [102]. Hu et al. developed biosensors for detecting a lectin known as concanavalin A [103]. It is a multi-wall carbon nanotube-polyaniline biosensor based on lectin-carbohydrate biospecific interations (Figs. 17.27 and 17.28).

## *Therapy*

The self-assembly of proteins as well as peptides have also been widely studied, and it has been proved that the constructed supra-molecular nanostructures are useful in enhancing the therapeutic efficiency of phototherapies. Because of the controllability along with the miscellany of the building blocks of peptide and protein, and the non-covalent interactions, there are sufficient opportunities for controlling the protein-based nano-structured materials to meet the requirements of efficient photodynamic and photothermal therapy [37].

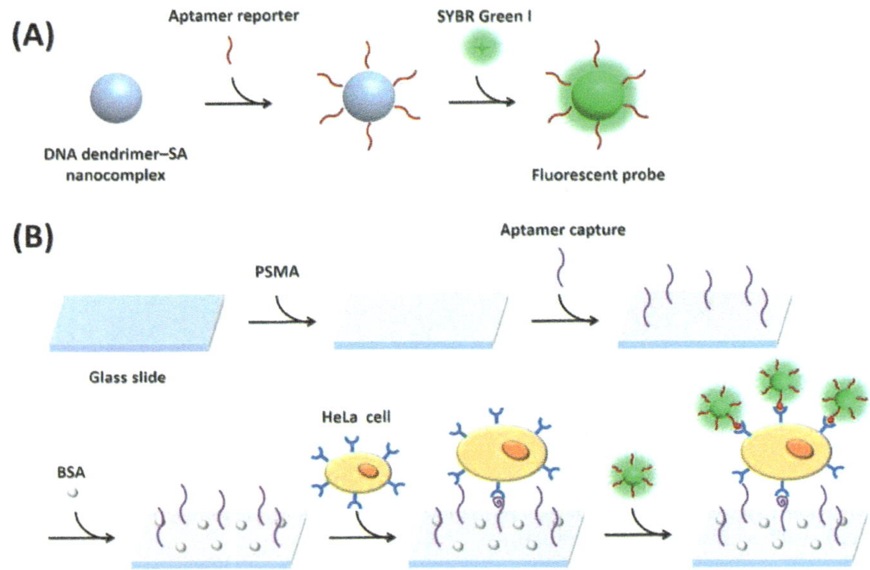


**Fig. 17.28** The making of an amplified fluorescence sensing platform for HeLa cells. **A** This is how the fluorescent probe is made. **B** This shows the sensing process. Here, PSMA means poly(styrene-co-maleic acid) and BSA means bovine serum albumin (with permission from Ref. [104])

## *Removal of Impurities*

Metals such as mercury and lead are toxic and their removal is of great interest. Copper may also be regarded as an impurity in food, and when accumulated in excess in the kidney and liver, may lead to liver and kidney-related diseases. Gandhi and coworkers studied polymeric composites of nano-hydroxyapatite along with chitin and chitosan that can be used for removal of copper(II) ions from aqueous solution. The sorption capacity of chitin/chitosan nano-hydroxyapatite is relatively higher than nano-hydroxyapatite, i.e., the nanocomposites without the biopolymers. The biopolymer-based nanocomposite is also advantageous as they can be shaped into many attractive forms owing to its biocompatibility [105].

The extensive use of chemical in the modern-day world has led to severe pollution of water around us. The removal of heavy metal pollutants, impurities, and toxics from water, therefore, attracts a lot of researchers and scientists alike [106]. Biopolymer based nanomaterials can also fulfill those intentions [107]. Sankar et al. verified that biopolymer-reinforced synthetic granular nanocomposites can be used for purification of water [108]. Chitosan as well as starch-based nanocomposites are also used for removal of different dyes [109, 110].

## Conclusion

In summary, we have discussed the classification of biopolymers depending on their origins and the biopolymers themselves that are both natural and synthetic. Nano-structured materials can be derived with high qualities in them. Furthermore, the applications provided by these nano-structured biomaterials are also considered. The need for these materials has several aspects.

Our only planet is polluted like never before. All of us bear responsibility for the protection of the environment. This is one vital reason we should focus on green technology. The biopolymer-based nano-structured materials offer wide range of applications that can bring about a greener future. Moreover, the nano-biomaterials provide us golden opportunities to have enhanced nanostructures that are capable of making our lives easier.

There are, neverthless, several challenges faced towards the practical applications of biopolymer based nano-structured materials. However good the product maybe, if it is extremely costly, it can't be commercialized for the common people. The high cost of production for many of these biomaterials remains a challenge. In the face of climate change and ever-increasing human population, the current agricultural methods are becoming unsustainable [111]. Biomaterials engineered through nanotechnology are a promising solution. For instance, nano-enabled additives could promote soil health, but there is always a concern for safety while using these materials. Chitosan-based nanomaterials have proven that they could purify water. The role of nanotechnology in tackling the global water crisis will be a huge task [112].

Every challenge is an opportunity. The fact that there is a spike in the research on biopolymer-based nano-structured materials means that the future will be more focused on green nanotechnology and hopefully new novel biomaterials are constantly synthesized with a better scope of applications.

**Acknowledgements** The authors are grateful to Maibam Biseshor Singh, Department of Chemistry, Manipur University, for his help in drawing the chemical structures.

## References

1. Gregory, M.R.: Environmental implications of plastic debris in marine settings—entanglement, ingestion, smothering, hangers-on, hitch-hiking and alien invasions. Philos. Trans. R. Soc. B Biol. Sci **364**(1526), 2013–2025 (2009)
2. Eriksen, M., Lebreton, L.C.M., Carson, H.S., Thiel, M., Moore, C.J., Borerro, J.C., Galgani, F., Ryan, P.G., Reisser, J.: Plastic pollution in the world's oceans: more than 5 trillion plastic pieces weighing over 250,000 tons afloat at sea. PLoS ONE **9**(12):e11913
3. Derraik, J.G.B.: The pollution of the marine environment by plastic debris: a review. Mar. Pollut. Bull. **44**, 842–852 (2002)
4. Padalkar, S., Capadona, J.R., Rowan, S.J., Weder, C., Won, Y.H., Stanciu, L.A., Moon, R.J.: Natural biopolymers: novel templates for the synthesis of nanostructures. Langmuir **26**(11), 8497–8502 (2010)

5. Singh, R.S., Saini, G.K., Kennedy, J.F.: Pullulan: microbial sources, production and applications. Carbohyd. Polym. **73**, 515–531 (2008)
6. Harada, T., Fujimori, F., Hirose, S., Masada, M.: Growth and $\beta$-glucan 10C3K production by a mutant of *Alcaligenes faecalis* var. *myxogenes* in defined medium. Agric. Biol. Chem. **30**, 764–769 (1966)
7. Singh, P., Parmar, N.: Isolation and characterization of two novel polyhydroxybutyrate (PHB)-producing bacteria. Afr. J. Biotech. **10**(24), 4907–4919 (2011)
8. Becker, A., Katzen, F., Puhler, A., Ielpi, L.: Xanthan gum biosynthesis and application: a biochemical/genetic perspective. Appl. Microbiol. Biotechnol. **50**, 145–152 (1998)
9. Sanda, F., Fujiyama, T., Endo, T.: Chemical synthesis of poly-$\gamma$-glutamic acid by polycondensation of $\gamma$-glutamic acid dimer: Synthesis and reaction of poly-$\gamma$-glutamic acid methyl ester. J. Polym. Sci., Part A Polym. Chem. **39**(5), 732–741 (2001)
10. Garlotta, D.: A literature review of poly(lactic acid). J. Polym. Environ. **9**(2), 63–84 (2002)
11. Cheng, Y., Deng, S., Chen, P., Ruan, R.: Polylactic acid (PLA) synthesis and modifications: a review. Front. Chem. China. **4**(3), 259–264 (2009)
12. Lasprilla, A.J.R., Martinez, G.A.R., Lunelli, B.H., Jardini, A.L., Filho, R.M.: Poly-lactic acid synthesis for application in biomedical devices—a review. Biotechnol. Adv. **30**, 321–328 (2012)
13. Peixoto, L.S., Silva, F.M., Niemeyer, M.A.L., Espinosa, G., Melo, P.A., Nele, M., Pinto, J,C.: Sunthesis of poly(vinyl alcohol) and/or poly(vinyl acetate) particles with spherical morphology and core-shell structure and its use in vascular embolization. Macromol. Symp. **243**, 190–199 (2006)
14. Ambre, S.G., Barchi, J.J.: Carbohydrate nanotechnology and its applications for the treatment of cancer. In: Carbohydrate Nanotechnology, pp 335–368 (2015)
15. Ranjbari, J., Mokhtarzadeh, A., Alibakhshi, A., Tabarzad, M., Hejazi, M., Ramezani, M.: Anticancer drug delivery using carbohydrate-based polymers. Curr. Pharm. Des. **23**, 6019–6032 (2017)
16. Jelinek, R., Kolusheva, S.: Carbohydrate biosensors. Chem. Rev. **104**, 5987–6015 (2004)
17. Gim, S., Zhu, Y., Seeberger, P.H., Delbianco, M.: Carbohydrate-based nanomaterials for biomedical applications. WIREs Nanomed. Nanobiotechnol., 1558 (2019)
18. Jebali, A., Nayeri, E.K., Roohana, S., Aghaei, S., Ghaffari, M., Daliri, K., Fuente, G.: Nano-carbohydrates: Synthesis and application in genetics, biotechnology, and medicine. Adv. Coll. Interface. Sci. **240**, 1–14 (2017)
19. Moon, R.J., Martini, A., Nairn, J., Simonsen. J., Youngblood, J.: Cellulose nanomaterials review: structure, properties and nanocomposites. Chem. Soc. Rev. **40**, 3941–3994 (2011)
20. Sajilata, M.G., Singhal, R.S., Kulkarni, P.R.: Resistant starch—a review. Compr. Rev. Food Sci. Food Saf. **5**, 1–17 (2006)
21. Abitbol, T., Rivkin, A., Cao, Y., Nevo, Y., Abraham, E., Ben-Shalom, T., Lapidot, S., Shoseyov, O.: Nanocellulose, a tiny fiber with huge applications. Curr. Opin. Biotechnol. **39**, 76–88 (2016)
22. Korhonen, J.T., Kettunen, M., Ras, R.H.A., Ikkala, O.: Hydrophobic nanocellulose aerogels as floating, sustainable, reusable, and recyclable oil absorbents. ACS Appl. Mater. Interfaces. **3**, 1813–1816 (2011)
23. Kargarzadeh, H., Sheltami, R.M., Ahmad, I., Abdullah, I., Dufresne, A.: Cellulose nanocrystal: a promising toughning agent for unsaturated polyester nanocomposites. Polymer., 1–12 (2014)
24. Ma, X., Jian, R., Chang, P.R., Yu, J.: Fabrication and characterization of citric acid-modified starch nanoparticles/plasticized-starch composites. Biomacromolecules **9**, 3314–3320 (2008)
25. Tan, Y., Xu, K., Li, L., Lui, C., Song, C., Wang, P.: Fabrication of size-controlled starch-based nanospheres by nanoprecipitation. ACS Appl. Mater. Interfaces. **1**(4), 956–959 (2009)
26. Kim, J.Y., Lim, S.T.: Preparation of nano-sized starch particles by complex formation with *n*-butanol. Carbohyd. Polym. **76**, 110–116 (2009)
27. Lui, D., Wu, Q., Chen, H., Chang, P.R.: Transitional properties of starch colloid with particle size reduction from micro to nanometer. J. Colloid Interface Sci. **339**, 117–124 (2009)

28. Xei, F., Pollet, E., Halley, P.J., Averous, L.: Starch-based nano-biocomposites. Progr. Polym. Sci. (2013)
29. Okoli, C.P., Naidoo, E.B., Ofomaja, A.E.: Role of synthesis process variables on magnetic functionality, thermal stability, and tetracycline adsorption by magnetic starch nanocomposites. Environ. Nanotechnol. Monit. Manag. **9**, 141–153 (2018)
30. Kumar, M.N.V.R.: A review of chitin and chitosan applications. React. Funct. Polym. **46**, 1–27 (2000)
31. Sung, H.W., Sonaje, K., Liao, Z.X., Hsu, L.W., Chuang, E.Y.: pH-responsive nanoparticles shelled with chitosan for oral delivery of insulin: from mechanism to therapeutic applications. Acc. Chem. Res. **45**(4), 619–629 (2012)
32. Yudin, V.E., Dobrovolskaya, I.P., Neelov, I.M., Dresvyanina, E.N, Popryadukhin, P.V., Ivan'kova, E.M., Elokhovskii, V.Y., Kasatkin, I.A., Okrugin, B.M., Morganti, P.: Wet spinning of fibres made of chitosan and chitin nanofibrils. Carbohydr. Polym. **108**, 176–182 (2014)
33. Floudas, C.A., Fung, H.K., McAllister, S.R., Monningmann, M., Rajgaria, R.: Advances in protein structure prediction and de novo protein design: a review. Chem. Eng. Sci. **61**, 966–988 (2006)
34. Goldstein, R.A., Luthey-Schulten, Z.A., Wolynes, P.G.: Protein tertiary structure recognition using optimized Hamiltonians with local interations. Proc. Natl. Acad. Sci. **89**(19), 9029–9033 (1992)
35. Janin, J., Bahadur, R.P., Chakrabarti, P.: Protein-protein interaction and quaternary structure. Q. Rev. Biophys. **41**(2), 133–180 (2008)
36. MaHam, A., Tang, Z., Wu, H., Wang, J., Lin, Y.: Protein-based nanomedicince platforms for drug delivery. Drug Deliv. Syst. **5**, 1706–1721 (2009)
37. Abbas, M., Zou, Q., Li, S., Yan, X.: Self-assembled peptide- and protein-based nanomaterials for antitumor photodynamic and photothermal therapy. Adv. Mater., **29**(12), 1605021 (2017)
38. Bourbon, A.I., Pereira, R.N., Pastrana, L.M., Vicente, A.A., Cerqueira, M.A.: Protein-based nanostructures for food applications. Gels **5**, 9–26 (2019)
39. Shoulders, M.D., Raines, R.T.: Collagen structure and stability. Annu. Rev. Biochem. **78**, 929–958 (2009)
40. Liao, S.S., Cui, F.Z., Zhang, W., Feng, Q.L.: Hierarchically biomimetic bone scaffold materials: Nano-HA/Collagen/PLA composite. J. Biomed. Mater. Res. B Appl. Biomater. **69B**, 158–165 (2004)
41. Du, C., Cui, F.Z., Feng, Q.L., Zhu, X.D., de Groot, K.: Tissue reponse to nano-hydroxyapatite/collagen composite implants in marrow cavity
42. Tomoaia, G., Soritau, O., Tomoaia-Cotisel, M., Pop, L.-B., Pop, A., Mocanu, A., Horovitz, O., Bobos, L.,-D.: Scaffolds made of nanostructured phosphates, collagen and chitosan for cell culture. Powder Technol. **238**, 99–107 (2013)
43. Hsia, C.C.W.: Respiratory function of hemoglobin. Mech. Dis. **338**(4), 239–247 (1998)
44. Bian, Y., Chang, T.M.S.: A novel nanobiotherapeutic poly-[hemoglobin-superoxide dismutase-catalase-carbonic anhydrase] with no cardiac toxicity for the resuscitation of a rat model with 90 minutes of sustained severe hemorrhagic shock with loss of 2/3 blood volume, artificial cells. Nanomed. Biotechnol. **43**(1), 1–9 (2014)
45. Chang, T.M.S.: Blood replacement with nanobiotechnologically engineered hemoglobin and hemoglobin nanocapsules. WIREs Nanomed. Nanobiotech. **2**(4), 418–430 (2010)
46. Chang, T.M.S.: Hemoglobin based red blood cell substitutes. Artif. Organs. **28**(9), 789–794 (2004)
47. Wei., X., Xiong, H., He., S., Wang, Y., Zhou, D., Jing, X., Huang, Y.: A facile way to prepare functionalized dextran nanogels for conjugation of hemoglobin. Coll. Surf. B Bioint. **115**, 440–448 (2017)
48. Watson, J.D., Crick, F.H.C.: Molecular structure of nucleic acids: a structure of deoxyribose nucleic acid. Nature **171**(4356), 737–738 (1953)
49. Portin, P.: The birth and development of the DNA theory of inheritance: sixty years since the discovery of the structure of DNA. J Genet. **93**(1), 293–302 (2014)

50. Chargaff, E.: Chemical specificity of nucleic acids and mechanism of their anzymatic degradation. Experientia **6**, 201–240 (1950)
51. Chargaff, E.: Essays on Nucleic Acids. Elsevier Publishing Company, Amsterdam, London, New York (1963)
52. Forsdyke, D.R., Mortimer, J.R.: Chargaff's legacy. Gene **261**, 127–137 (2000)
53. Jiang, Q., Zhao, S., Lui, J., Song, L., Wang, Z.-G., Ding, B.: Rationally designed DNA-based nanocarriers. Adv. Drug Deliv. Rev. **147**, 2–21(2019)
54. Turner, D.H., Sugimoto, N.: RNA structure prediction. Ann. Rev. Biophys. Chem. **17**, 167–192 (1988)
55. Deigan, K.E., Li, T.W., Mathews, D.H., Weeks, K.M.: Accurate SHAPE-drirected RNA structure determination. PNAS **106**(1), 97–102 (2008)
56. Dietz, H., Douglas, S.M.: Folding DNA into twisted and curved nanoscale shapes. Science **325**, 725–730 (2009)
57. Pei, H., Zuo, X., Zhu, D., Huang, Q., Fan, C.: Functional DNA nanostructures for theranostic applications. Acc. Chem. Res. **47**(2), 550–559 (2014)
58. Wen, Y., Lui, G., Pei, H., Li, L., Xu, Q., Liang, W., Li, Y., Xu, L., Ren, S., Fan, C.: DNA nanostructure-based ultrasensitive electrochemical microRNA biosensor. Methods **64**, 276–282 (2013)
59. Geary, C., Rothemund, P.W.K., Andersen, E.S.: A single-stranded architecture for contranscriptional folding of RNA nanostructures. Science **345**(6198), 799–804 (2014)
60. Pinheiro, A.V., Han, D., Shih, W.M., Yan, H.: Challenges and opportunities for structural DNA nanotechnology. Nat. Nanotechnol. **6**, 763–772 (2011)
61. Jaeger, L., Leontis, N.B.: Tecto-RNA: One-dimensional self-assembly through tertiary interactions. Angew. Chem. Int. Ed. **39**(14), 2521–2524 (2000)
62. Grabow, W., Jaeger, L.: RNA modularity for synthetic biology. F1000Prime Rep. **5**(46) (2013)
63. Lee, J., Kladwang, W., Lee, M., Cantu, D., Azizyan, M., Kim, H., Limpaecher, A., Yoon, S., Treuille, A., Das, R.: EteRNA participants: RNA design rules from a massive open laboratory. PNAS **111**(6), 2122–2127 (2014)
64. Wilner, O.I., Willner, I.: Functionalized DNA nanostructures. Chem. Rev. **112**, 2528–2556 (2012)
65. Rekha, M.R., Sharma, C.P.: Pullulan as a promising biomaterial applications: a perspective. Trends Biomater. Artif. Organs. **20**(2) (2007)
66. Marita, D., Clara, C., Carmen, R.L., Begona, S., Ana, M.R.C., Ana, G.: Pullulan-based nanoparticles as carriers of transmucosal protein delivery. Eur. J. Pharm. Sci. **50**, 102–113 (2013)
67. Lu, D., Wen, X., Liang, J., Gu, Z., Zhang, X., Fan, Y.: A pH-sensitive nano drug delivery system derived from pullulan/doxorubicin conjugate. J. Biomed. Mater. Res. Part B Appl. Biomater., 177–183 (2008)
68. Lu, J., Tappel, R.C., Noruma, C.T.: Mini-Review: Biosynthesis of poly(hydroxyalkanoates). Polym. Rev. **49**(3), 226–248 (2009)
69. Li, Z., Yang, J., Loh, X.J.: Polyhydroxyalkanoates: opening doors for a sustainable future. NPG Asia Mater. **8**, e256 (2016)
70. Raza, Z.A., Abid, S., Banat, I.M.: Polyhydroxyalkanoates: characteristics, production, recent developments and applications. Int. Biodeterior. Biodegradation **126**, 45–56 (2018)
71. Bugnicourt, E., Cinelli, P., Lazzeri, A., Alvarez, V.: Polyhydroxyalkanoate (PHA): Review of synthesis, characteristics, processing and potential applications in packaging. eXPRESS Polym. Lett. **8**(11), 791–808 (2014)
72. Xu, X.Y., Li, X.T., Peng, S.W., Xiao, J.F., Lui, C., Fang, G., Chen, K.C., Chen, G.Q.: The behaviour of neural stem cells on polyhydroxyalkanoate nanofiber scaffolds. Biomaterials **31**, 3967–3975 (2010)
73. Varishetty, M.M., Qui, W., Gao, Y., Chen, W.: Structure, electrical and optical properties of ($PVA/LiAsF_6$) polymer composite electrolyte films. Polym. Eng. Sci., 878–884 (2010)
74. Errico, C., Bartoli, C., Chiellini, F.: Chiellini, E: Poly(hydroxyalkanoates)-based polymeric nanoparticles for drug delivery. J. Biomed. Niotechnol. **10**, 1–10 (2009)

75. des Rieux, A., Fievez, V., Garinot, M., Schneider, Y.-J., Preat, V.: Nanoparticles as potential oral delivery systems of proteins and vaccines: a mechanistic approach. J. Controlled Release **116**, 1–27 (2006)
76. Ahmed, J., Varshney, S.K.: Polylactides—chemistry, properties and green packaging technology: a review. Int. J. Food Prop. **14**, 37–58 (2011)
77. Gupta, B., Revagade, N., Hilborn, J.: Poly(lactic acid) fiber: an overview. Prog. Polym. Sci. **32**, 455–482 (2007)
78. Singh, S., Ray, S.S.: Polylactide based nanostructured biomaterials and their applications. J. Nanosci. Nanotechnol. **7**, 2596–2615 (2007)
79. Bhardwaj, R., Mohanty, A.K.: Modification of Brittle polylactide by novel hyperbranched polymer-based nanostructures. Biomacromolecules **8**, 2476–2484 (2007)
80. Pereira, C.L., Antunes, J.C.: Biosynthesis of highly pure poly-$\gamma$-glutamate acid for biomedical applications. J. Mater. Sci.: Mater. Med. **23**, 1583–1591 (2012)
81. Ashiuchi, M.: Microbial production and chemical transformation of poly-$\gamma$-glutamate. Microb. Biotechnol. **6**, 664–674 (2013)
82. Trimaille, T., Verrier, B.: Micelle-based adjuvants for subunit vaccine delivery. Vaccines **3**(4), 803–813 (2015)
83. Jimenez-Sanchez, G., Pavot, V., Chane-Haong, C., Handke, N., Terrat, C., Gigmes, D., Trimaille, T., Verrier, B.: Preparation and in vitro evaluation of imiquimod loaded polylactide-based micelles as potential vaccine adjuvants. Pharm. Res. **32**(1), 311–320 (2014)
84. Zong, X., Bien, H., Chung, C.Y., Yin, L., Fang, D., Hsiao, B.S., Chu, B., Entcheva, E.: Electrospun fine-textured scaffolds for heart tissue constructs. Biomaterials **26**, 5330–5338 (2005)
85. Nguyen, T.H., Kim, Y.H., Song, H.Y., Lee, B.T.: Nano Ag loaded PVA nano-fibrous mats for skin applications. J. Biomed. Mater. Res. B Appl. Biomater. **96B**(2), 225–233 (2010)
86. Majd, S.A., Khorasgani, M.R., Moshtaghian, S.J., Talebi, A., Khezri, M.: Application of chitosan/PVA nano fiber as a potential wound dressing for Strptozotocin-induced diabetic rats. Int. J. Biol. Macromol. **92**, 1162–1168 (2016)
87. Gagandeep, Garg, T., Malik, B., Rath, G., Goyal, A.K.: Development and characterization of nano-fiber patch for the treatment of glaucoma. Eur. J. Pharm. Sci. **53**, 10–16 (2014)
88. Rebia, R.A., Rozet, S., Tamada, Y., Tanaka, T.: Biodegradable PHBH/PVA blend nanofibers: fabrication, characterization, in vitro degradation, and in vitro biocompatibility. Polym. Degrad. Stab. **154**, 124–136 (2018)
89. Mokhtarzadeh, A., Alibakhshi, A., Hejazi, M., Omidi, Y., Dolatabadi, J.E.N.: Advanced natural nanomaterials for drug delivery and tissue engineering. Trends Anal. Chem. **82**, 367–384 (2016)
90. Khalil, I.R., Burns, A.T.H., Radecka, I., Kowalczuk, M., Khalaf, T., Adamus, G., Johnston, B., Khechara, M.: Bacterial-derived polymer poly-$\gamma$-glutamic acid ($\gamma$-PGA)-based micro/nanoparticles as a delivery system for antimicrobials and other biomedical applications. Int. J. Mol. Sci. **18**, 313–331 (2017)
91. Wu, Q., Wang, Y., Chen, G.Q.: Medical application of microbial biopolyesters polyhydroxyalkanoates. Artif Cells Blood Substit. Biotechnol. **37**, 1–12 (2009)
92. Ambekar, R.S., Kandasubramanian, B.: Advancements in nanofibers for wound dressing: a review. Eur. Polymer J. **117**, 204–336 (2019)
93. Fang, Y., Zhu, X., Wang, N., Zhang, X., Yang, D., Nei, J., Ma, G.: Biodegradable core-shell electrospun nanofibers based on PLA and $\gamma$-PGA for wound healing. Eur. Polymer J. **116**, 30–37 (2019)
94. Nitta, S.K., Numata, K.: Biopolymer-based nanoparticles for drug/gene delivery and tissue engineering. Int. J. Mol. Sci. **14**, 1629–1654 (2013)
95. Lohcharoenkal, W., Wang, L., Chen, Y.C., Rojanasakul, Y.: Protein nanoparticles as drug delivery carriers for cancer therapy. BioMed. Res. Int. **14**:1–12 (2014)
96. Jao, D., Xue, Y., Medina, J., Hu, X.: Protein-based drug-delivery materials. Materials **10**, 517–540 (2017)

97. Park., J.H., Saravanakumar, G., Kim, K., Kwon, I.C.: Targeted delivery of low molecular drugs using chitosan and its derivatives. Adv. Drug Deliv. Rev. **62**, 28–41 (2010)
98. Lu, S., Wang, X., Lu, Q., Hu, Q., Uppal, N., Omenetto, F.G., Kaplan, D.: Stabilization of enzymes in silk films. Biomacromolecules **10**, 1032–1042 (2009)
99. Geyer, R., Jambeck, J.R., Law, K.L.: Production, use, and fate of all plastics ever made. Sci. Adv. **3** (2017)
100. Rhim, J.W., Ng, P.K.W.: Natural biopolymer-based nanocomposite films for packaging applications. Crit. Rev. Food Sci. Nutr. **47**, 411–433 (2007)
101. Li, F., Biagioni, P., Bollani, M., Maccagnan, A., Piergiovanni, L.: Multi-functional coating of cellulose nanocrystals for flexible packaging applications. Cellulose **20**, 2419–2504 (2013)
102. Yang, M., Jiang, J., Yang, Y., Chen, X., Shen, G., Yu, R.: Carbon nanotube/cobalt hexacyanoferrate nanoparticle-biopolymer system for the fabrication of biosensors. Biosens. Bioelectron. **21**, 1791–1797 (2006)
103. Hu, F., Chen, S., Wang, C., Yuan, R., Xiang, Y., Wang, C.: Multi-wall carbon nanotube-polyaniline biosensor based on lectin-carbohydrate affinity for ultrasensitive detection of Con A. Biosens. Bioelectron. **34**, 202–207 (2012)
104. Zhao, Y., Hu, S., Wang, H., Yu, K., Guan, Y., Lui, X., Li, N., Lui, F.: DNA dendrimer-streptavidin nanocomplex: an efficient signal amplifier for construction of biosensing platforms. Anal. Chem. **89**(12), 6907–6914 (2017)
105. Gandhi, M.R., Kousalya, G.N., Meenakshi, S.: Removal of copper(II) using chitin/chitosan nano-hydroxyapatite composite. Int. J. Biol. Macromol. **48**, 119–124 (2011)
106. Lin, Y.F., Chen, H.W., Chien, P.S., Chiou, C.S., Lui, C.C.: Application of bifunctional adsorbent to adsorb metal cations and anionic dyes in aqueous solution. J. Hazard. Mater. **185**, 1124–1130 (2001)
107. Chopra, H., Ruhi, G.: Eco friendly chitosan: An efficient material for water purification. Pharm. Innov. J. **5**(1), 92–95 (2016)
108. Pandey, N., Shukla, S.K., Singh, N.B.: Water purification by polymer nanocomposites: an overview. Nanocomposites **3**(2), 47–66 (2017)
109. Sankar, M.U., Aigal, S., Maliyekkal, S.M., Chaudhary, A., Anshup, Kumar, A.A, Chaudhari, K., Pradeep, T.: Biopolymer-reinforced synthetic granular nanocomposites for affordable point-of-use water purification. PNAS **110**(21), 8459–8464 (2013)
110. Olad, A., Azhar, F.F.: Eco-friendly biopolymer/clay/conducting polymer nanocomposites: characterization and its application in reactive dye removal. Fibers Polym. **15**(6), 1321–1329 (2014)
111. Lowry, G.V., Avellan, A., Gilbertson, L.M.: Opportunities and challenges for nanotechnology in the agri-tech revolution. Nat. Nanotechnol. **14**, 517–522 (2019)
112. Mauter, M.S., Zucker, I., Perreault, F., Werber, J.R., Kim, J.H., Elimelech, M.: The role of nanotechnology ini tackling global water challenges. Nat. Sustain. **1**, 166–175 (2018)

# Chapter 18
# Electrospinning of Graphene Oxide-Based Nanofibers for Supercapacitor Applications

**Wahengbam Ishwarchand Singh and Bibhu Prasad Swain**

## Introduction

Recently, the depletion of natural energy resources like fossil fuels has become a serious threat. To cope with this threat, researchers are required for subsequent developments in the energy storage device with the superior storage and retention of high capacity of electrical energy in the view of global economies growth [1]. An effective technique for storing energy would significantly reduce the scale of materials and would offer reasonable cost-effective energy for everybody. It seems that nearly most of the research people's attention and interest are directed toward energy generation rather than in the direction of the energy storage issue.

In order to achieve an ideal energy storage device having high power density and energy, many researchers have studied and proposed several designs and mechanisms. Later on with the sense of energy storage mechanisms and design, the term energy storage devices were introduced. However, most of these designs are based on nuclear energy and fossil fuel which was the only solution for delivering the needed energy from place to place where the energy is non-portable [2]. Generally, the main purpose of the energy storage device is to deliver the energy to a place where it is needed or stored at a place for later usage.

In 1800, various kinds of designs and mechanisms were proposed for storing electrical energy. In 1957, H.I. Becker claimed his patent of double-layer storage by employing porous carbon electrodes in the aqueous electrolyte. In late 1969, the first electric double-layer capacitor was manufactured for commercial purposes by SOHIO Corporation [3] that consists of non-aqueous electrolytes with a porous

W. I. Singh (✉) · B. P. Swain
Department of Physics, National Institute of Technology Manipur, Langol, Imphal 795004, India
e-mail: Ishorchand1993@gmail.com

B. P. Swain
e-mail: bibhuprasad.swain@gmail.com; bpswain@nitmanipur.ac.in



B. P. Swain (ed.), *Nanostructured Materials and their Applications*, Materials Horizons: From Nature to Nanomaterials, https://doi.org/10.1007/978-981-15-8307-0_18

carbon electrode. In 1970, another research fellow, Donald L. Boos, from the same corporation, designed the foundation for several upcoming patents and also reported for covering all essential features of electrochemical technology. In 1978, the term "supercapacitor" was tagged along with the electric double-layer capacitor (EDLC) by Nippon Electric Company (NEC) which is used in computers as a memory back up. At that time, commencements of electrochemical supercapacitors were measured to be lower specific energy values. In the same year, Panasonic manufactured "Gold-cap" their first stepped in manufacturing double-layer capacitor. The main differences between them are that the Goldcap (Panasonic product) used a non-pasted electrode in the non-aqueous electrolyte and supercapacitor (NEC products) used a pasted electrode immersed in the aqueous electrolyte with the construction of the bipolar cell. Overall, the electrochemical supercapacitor (ECs) is exclusively attributed to the complement of batteries and fuel cells' weaknesses. Therefore, many researchers have put efforts into the designing of the ECs to develop the higher efficiency electrochemical capacitor. Up to 1990, no reports have been observed for significant enhancement in the development and designing of a cost-effective electrochemical supercapacitor. Around 1991, a new type of capacitance was propounded by the Conway research group which is known as "pseudocapacitance" and reported of research work extensively on $RuO_2$ compound.

Generally, pseudocapacitance was observed as result of reversible redox and electrosorption process over the surface of certain metals oxide materials, e.g., $RuO_2$, $CO_3O_4$, $IrO_2$, and $MnO_2$,. Having reported that the performance of energy storage is 100 times higher than that of EDLCS. In pseudocapacitors, the occurrence of both types of non-faradic and faradic charge storage mechanisms is observed, which enhances the energy storage capacity. Then after 1990, there was a tremendous improvement in this field and development and very active research was taken to enrich more in designing direction and fundamental mechanisms. In the present scenario, various companies like ESMA Maxwell Technologies, CAP-XX, NessCap, and NCC have invested in the development of supercapacitors. Apart from the above materials, new types of electrodes like transition metal oxides and activated carbon are also utilized in the development of capacitor nowadays. The more and more demand for high-powered energy storage devices leads to the development of hybrid energy storage and other system obtained by integrating with battery and fuel cell mechanisms.

Generally, supercapacitors are classified as three types: (1) electrochemical double-layer capacitors, (2) pseudo-capacitors, and (3) hybrid capacitors, based on electrical charge storage process (Fig. 18.1).

In 1879, a new type of capacitance was developed with the mechanism of double layer by Helmholtz, consisting of electrolytes with ions and charged electrodes [4]. This mechanism flows non-faradic mechanisms of electrical energy, having no limit of electrochemical kinetic. EDLCs have certain advantageous characteristics like higher power density and high energy density over the pseudocapacitors and electrostatic capacitors. In this system, the accumulation charge occurs on the interface of the electrolytes and electrode. The formation of a double layer involved the separation of each layer by a few angstroms.

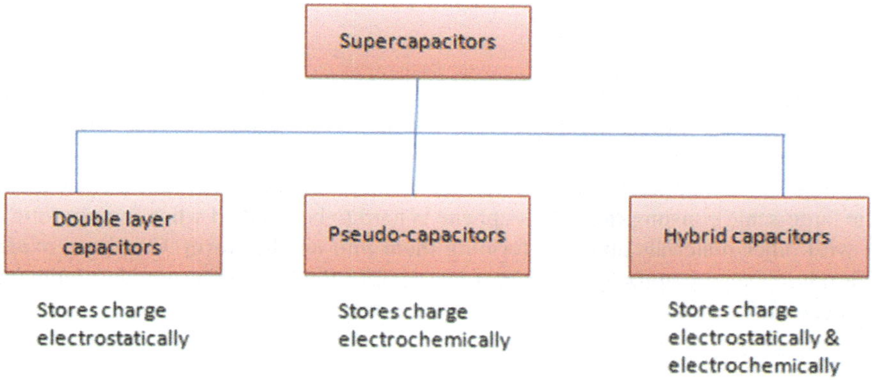


**Fig. 18.1** Types of supercapacitor

The charging process involved liberations of the electron by the negative electrodes, which travels towards the positive electrode by using an external load. At the same moment, the anions and cations are moving in the opposite direction inside the electrodes. Usually, the EDLC electrode materials comprise of the high surface area of porous carbon materials.

Pseudocapacitors are developed on the mechanisms of a redox-based electrochemical capacitor [5]. In this system, the mechanism is quite different from that of EDLC, as the pseudocapacitors follow redox-based capacitance mechanism. In the EDLC, a deficiency of electrostatic charge is observed at the electrode interface, while the redox reaction is observed in the electrode of the pseudocapacitor, which is reversible and fast, involving the passage of electrostatic charge. In 1991, B.E. Conway reported the effect on some faradic mechanisms that resulted in pseudocapacitance features, such as redox pseudocapacitance, underpotential deposition, and intercalation pseudocapacitance. The ion interpolation within the layers of the electrode material which is porous leads to the intercalation pseudocapacitance. The process of charged transfer occurs without crystalline phase change [6].

The hybrid supercapacitors as the name suggest it is the combination of the pseudocapacitors and double-layer supercapacitor. The hybrid supercapacitor has certain advantages such as high power density and energy than other existing supercapacitors (<3.5 V). Therefore, this supercapacitor having excellent characteristic is used for a high energy power system and electric system for a vehicle. This supercapacitor consists of two electrodes, which the positive one acted as a battery and a negative one as a double-layer capacitor [7].

## Graphene Oxide

Graphene has shown many promising characteristics due to its novel nanostructure of graphene in a variety of applications and not only on fundamental science. Its remarkable physical and chemical property leads to many attempts of the production of graphene on a large scale. However, it does not give adequate productivity to meet the large-scale consumption [8] Graphene is hard to be handled when incorporating it with other materials and transforming them into another form. Graphene oxide (GO) has been recognizing as the best materials in the field of current development by designing a structure that is more solvent friendly [9].

### *Structure*

Basically, GO is a graphene sheet consisting of oxygen functional groups, either carboxyl, epoxy, or hydroxyl. Through the process of sonication and oxidation of graphite, it can be easily exfoliated. Since graphene is a one-atomic layer of graphite, the structure of GO also consists of exfoliated layers of graphite oxide. The structural process can be simplistically shown in Fig. 18.2. The structure of carbon compounds

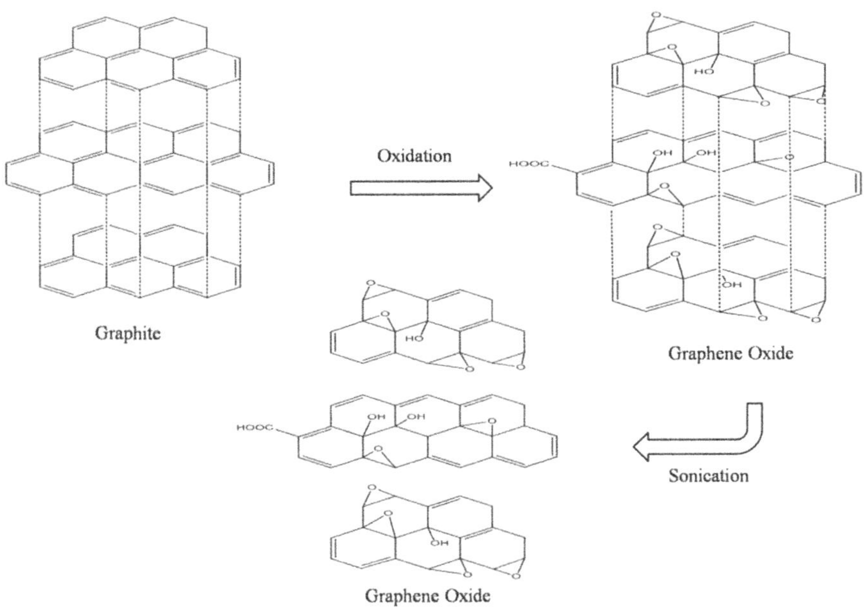


**Fig. 18.2** Structural illustration for the processing of GO

is usually organized in the form of the honeycomb structure, which is in the hexagonal structure with six atoms of carbon bonded by strong $\sigma$ and $\pi$ bonds.

Generally, the synthesis process for GO comprises of oxidation process which converts graphite into graphite oxide and follows by obtaining graphene oxide from GO by exfoliating the GO. Out of the several routes for the utilization of GO, the easiest routes so developed are the Brodie method, Brodie–Staudenmaier method, and Hummers method but they are not so friendly as they release gases like $NO_x$ and $ClO_x$ which are toxic in nature [10]. Many literary works have been reported on the introduction of several modified Hummers method [11–13]. Generally, the conversion process of graphite–graphene oxide can be precisely summarized into three steps [14]. The first step involves the processing of the intercalation compound of graphite ($H_2SO_4$–GIC) from graphite. The second process involved the conversion of $H_2SO_4$–GIC into pristine graphite oxide (PGO). Finally, the third step comprises converting PGO into GO by reacting PGO with water.

GO having complexity in its non-stoichiometric composition, many materials scientists have taken interest in studying the structure of GO [15]. Klinowski and Lerf constructed a model of GO which became the most recent and acceptable model. Their structure mainly focuses on the non-stoichiometric-amorphous alternative ignoring the model based on the lattice. It consists of six-membered aliphatic rings with hydroxyl, epoxy, and carboxyl or carbonyl groups on the edges or on the graphene sheet. It also contains benzene rings which are unoxidized. The formation of the epoxy ring is also observed as two carbon atoms are covalently bonded with each oxygen atom. Both $sp^2$ and $sp^3$ hybridized carbons are found in reduced graphene oxide (rGO) structure. The $sp_2$ carbons are bind to the other four atoms, while $sp^2$ carbons bind with three other atoms, double bonds being shared with one of them. Moreover, the carbon–oxygen ratio (C:O) is quite high in rGO.

In 2009, Mkhoyan et al. conducted a study on the electronic and atomic structure of GO using electron energy loss spectroscopy (EELS) and annular dark field (ADF) imaging [16]. They reported that the graphene sites are randomly attached by oxygen atoms and two carbon atoms bonded with an oxygen atom is pulled up the graphene plane by ab initio calculated structure. The absorption energy of this configuration is $-3.12$ eV, which makes is fairly stable. There is an expansion between the two carbon atoms making the bond length 1.514 Å larger than graphene whose bond length is 1.54 Å [16]. This becomes comparable with that of diamond $sp^3$ bond length which is 1.54. The unique chemical and physical properties of graphene oxide are due to its $\pi$ and $\sigma$ bonds and $sp^3$ hybridization. Mainly, the time, temperature, and quantity of oxidizing agents used in the oxidation reaction are responsible for spreading the polar groups on the edges and the surface of the graphene oxide.

## *Properties*

Around 2004, many researchers have started the close study of graphene and many interesting properties were subsequently reported one by one. Graphene is very light and thin as it is the single layer of graphite. It is reported to have an intrinsic strength of 42 N $m^{-1}$ making the strongest material that has been discovered [17]. Its Young's modulus is about 1 TPa, and it is approximately 200 times stronger than steel and hardness comparable with that of the diamond. It has pliable and elastic properties but also impermeable to liquids (except water) and gases. Graphene is also reported to show high thermal conductivity ranging from ~(4.84 ± 0.44) × 103 to (5.30 ± 0.48) × 103 W/mK, which exceed the thermal conductivity of carbon nanotubes [17–20]. Moreover, another interesting property of graphene is its electrical properties, it has high intrinsic mobility of 2 × 105 $cm^2$ $v^{-1}$ $S^{-1}$ which is comparable to the highest known value of carbon nanotubes (inorganic semiconductor), having approximately $1 \times 10^5$ $cm^2$ $v^{-1}$ $S^{-1}$ [21, 22] The graphene which is the building block of graphite layer is not synthesized by straightforward exfoliation process. In 2004, the graphene sheet was discovered by using cellophane tape but it was not considered as a readily scalable process. Production of graphene sheets sufficiently to meet its needs is still a challenge, despite being available at low cost and its existence [10]. The deposition of graphene on arbitrary substrates by using chemical deposition (CVD) is an incompatible process for mass production, and it is impossible to perform at low temperatures. What more concerning is that when it came to incorporating other polymer matrices graphene is not a suitable material. Firstly, we need to obtain a stable dispersion for graphene into other polymer matrices for the preparation of hybrid materials of graphene-based and composites of graphene polymer. The above disadvantages of graphene lead to the introduction of GO for its certain usefulness over graphene. The GO has a larger quantity of free bonds which is due to $sp_3$ hybridization of GO attributed to increasing its reactivity [23]. GO is hydrophilic in nature due to its functional groups as it is obtained from strong oxidation of graphite, improving the exfoliation in organic solutions and aqueous media. Another critical factor is the solvents and surface functionalization to determine the interaction between the polymer functional groups. Paredes et al. [24] studied the dispersibility of GO in methanol, acetone, ethanol, ethylene glycol, 1-propanol, dimethyl sulfoxide (DMSO), methyl-2-pyrrolidone (NMP), *N*,*N*-dimethylformamide (DMF), *N*-pyridine, tetrahydrofuran (THF), dichloromethane (DCM), o-xylene, and n-hexane [9].

GO contains a large amount of oxygen atom which makes it easier to modify. The electronic properties can be controlled by maintaining the oxidation, even graphene with zero bandgaps is made possible. Electrical conductivity is attributed to the $\pi$-network. GO becomes non-conductive when the *C*/*O* ratio is 2/1, where the conductivity is less than 1 S/m. Generally, GO undergoes reduction to removed oxygen-containing groups to make it more conductive and through reducing the *C*/*O* atomic ratio of GO to 6/1 to obtain conjugate structure and $\pi$—a network of the GO are restored. Basically, conversion $sp^3$ to $sp^2$ carbon occurred when GO is reduced which enhanced its electrical properties. Uses of conducting polymers like polyaniline to binds with oxygenated functionalities of GO with its functional groups are reported by some researchers to enhance its electrical conductivity [25].

## *Synthesis of Graphene Oxide (GO)*

The synthesis of GO was done by employing a modified Hummer's method [26, 27]. In this method, oxidation of graphite flake was carried out, taking 3 g of graphite flakes, along with the acid mixture $H_2SO_4$:$H_3PO_4$ mixed in the ratio 9:1 and also adding 18 g $KMnO_4$. This oxidation process is carried out for 5 min with continuous stirring. Later, in order to stop the oxidation process $H_2O_2$ solution was added to the mixture, then a change in color of the mixture will be observed from dark brown to milky yellow. Finally, using a solution of 1 M HCL the obtained mixture was washed using a centrifuge. The final product is repeatedly washed with deionized water to achieve its pH to 4–5 [28].

### Functionalization of GO

UV-assisted liquid-phase polymerization technique was used for surface functionalization of the GO. Basically, the coupling reaction of chloromethyl naphthalene and sodium diethyldithiocarbamate was carried out for the synthesis of photo-initiator (iniferter) and was synthesized via a coupling reaction of [29]. Taking 2 mg of GO and photo-initiator of the fixed amount which was then added to tetrahydrofuran (THF) solution. Sonication of this mixture was carried out for 2 h, allowing the formation of pi-pi stacking between the GO planar rings and naphthalene part from the photo-initiator. The mixture was sonicated and then centrifuged at 3000 rpm for 15 min. Later, discarded the obtained supernatant liquid followed by the addition of 2 mL of THF solution. Again sonication of the mixture was carried out for 5 min and then centrifuged for about 15 min. The above step was repeated at least three times in order to discard the unreacted initiator. Then, the mixture was sonicated for 2 h after re-dispersing the sediment in THF. Finally, 2 mL of MMA monomers were added to the GO/photo-initiator mixture in a quartz flask maintained at an inert atmosphere. The mixture is then subjected to UV irradiation inside a reaction chamber kept at a fixed temperature of 40 °C. Removal of unreacted MMA monomers was done by precipitation in methanol [28].

## Electrospinning

Nanoscale materials can be classified as zero, one, two, and three dimensions materials. Several techniques are employed for the synthesis of one dimensional, such as phase separation, self-assembly, template synthesis, and electrospinning. Out of these techniques, electrospinning has been considered to be a facile process for producing polymeric fibers in nano-size [30]. Due to its potential applications in various fields, it has gained the attention of many researchers for various purposes like tissue engineering, biological and chemical sensing, textiles and energy generation [25].

## *Basics*

Formhals was the first person who patented the experimental setup based on the electrostatic force in the production of polymer filaments [31]. Formhals published a patent in 1934, in which described the fabrication of textile yarn using cellulose acetate [32]. Later, many researchers have carried out their research in assembling the fiber processing by employing electrostatic force. In 1969, Taylor started studying the electrically driven jet which became a basic foundation for understanding the process of electrospinning. The term electrospinning was started using only in 1994. Since then, its rapid implementation in various fields leads to upgrading and modification of electrospinning to increases its productivity and effectiveness in the production of versatile nanofibers. According to meet the certainly required characterizations of the nanofibers, several advanced setup and techniques of electrospinning are employed. Multilayer, coaxial and mixed electrospinning, air-gap electrospinning forced and air-assisted electrospinning are a number of innovative processing techniques to enhance the function of electrospun fibers [33] Typically, the basic components for the setup of electrospinning are the spinneret (normally syringe with a blunt-tip needle), high-voltage power supply, and the collector (normally grounded). Figure 18.3 shows the typical basic components of the experimental setup for electrospinning. A positive electrode with a high-voltage supply is connected to the tip of the needle of the syringe which is filled with the polymer solution or melts. The syringe pump is held by a retort stand for horizontal and vertical setups. From the syringe, the polymer

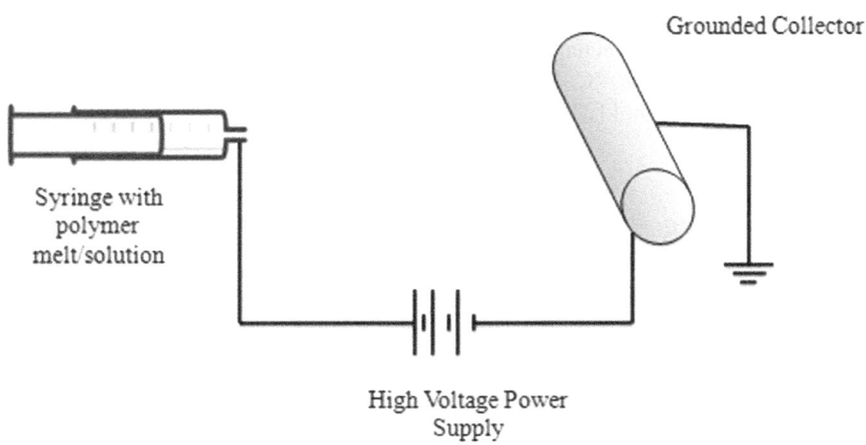


**Fig. 18.3** Experimental setup of the electrospinning technique

liquid is pumped out through the needle and electrostatically attracted towards the collector which is placed horizontally or beneath the syringes.

The basic working principle for the production of this unique continuous fiber is based on the surface tension that exerts on the polymer drops. The fluid drop becomes positively charged by gaining electrostatic charges from the supplied electricity. The formation of cone shape from the polymer drop is observed at the tip of the flat needle due to the interaction between two forces, which later term as Taylor cone. After increasing the high-voltage supply up to certain voltage till a critical point is reached at which the electrostatic repulsion began to overcome the surface tension acting on the polymer droplet, as a result, the liquid is started to eject as a spray toward the collector due to the electrostatic force (Fig. 18.4).

Figure 18.5 shows two segments of the polymer jet ejected from the spinneret targeting toward the grounded collector, and there are whipping jet and streaming jet. The whipping jet is not possible to see by the naked eye; however, with proper lighting, the streaming jet could be seen by the naked eye. Another interesting fact is that the morphology of the fabricated nanofibers could be predicted up to some extend

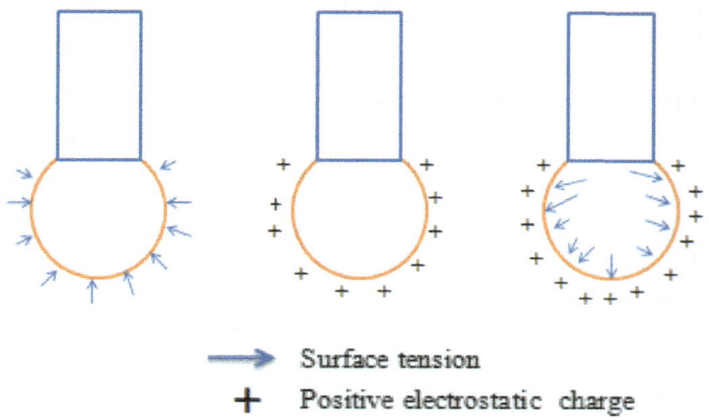


**Fig. 18.4** Description of electrospinning principle

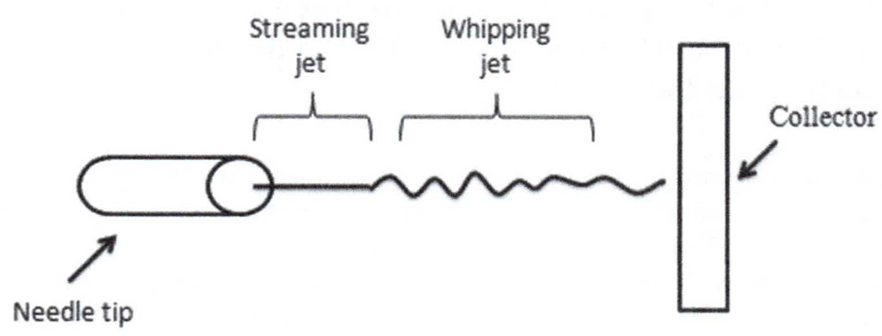


**Fig. 18.5** Streaming and whipping jets

by observing the streaming jet flow. In 2011, Leach et al. reported the possibility of poor alignment, wavy pattern, splattering, and beading in the collected fiber might be due to non-uniformity, inconsistency, and oscillating in the flow of the stream [34]. As the nature of fiber produced through electrospinning depends on certain experimental factor, researchers are further engaged in the study of various parameters to achieve optimized nanofibers. Therefore, the dependence of various factors on the properties of the nanofibers will be discussed in the next section [25].

### *Working Parameter*

In order to optimize the fabricated nanofibers to achieve the required properties and morphology, the processing of electrospinning has to deal with various working. Apart from the experimental setup for the electrospinning or employing different types of spinnerets and collectors, the main factors affecting the electrospun nanofibers are its working parameter. Mostly, the adjustment is done on the factor like ambient parameters, operational properties, and the properties of the polymer melt or solutions [30, 34]. The common parameter responsible for affecting the nanofibres traits is mainly operational properties and polymer solutions. Other factors that are widely investigated by the researchers for optimization of the nanofiber formation are surface tension, conductivity solution viscosity, electric field strength, flow rate, and also the distance between spinneret tip with the collector. Mostly, many researchers are preferred to keep some of the ambient parameters as constant such as humidity, solution temperature, and air velocity of the surrounding for their study. Another concern is that the chances of getting beads are higher when smaller diameter or finer fibers are spun. Thus, more precautions should be taken by the researcher while determining and recognizing the appropriate optimized condition for fabricating the electrospun fibers. Some parameters like tip-to-collector distance and applied voltage are different for every electrospinning material [30, 35].

## Functionalized Graphene Oxide-Reinforced Electrospun Carbon Nanofibers Supercapacitor Electrode and Its Electrochemical Performance

Figure 18.6a shows non-woven polymeric PAN electrospun fibers [36]. Figure 18.6b shows the increase in width of nanofibers with adding the fGO, having a width of approximately 957.35 nm. Figure 18.6c shows the uniform arrangement in CNF/fGO membrane due to carbonization, Fig. 18.6d with a higher quantity of strayed fGO. Figure 18.6e depicts the homogeneous morphology in the CNF/fGO membrane with noticeable strayed fGO. Figure 18.6f shows the astray fGO incorporating inside the CNF/fGO nanofibers which contribute to the enhancement of electrochemical

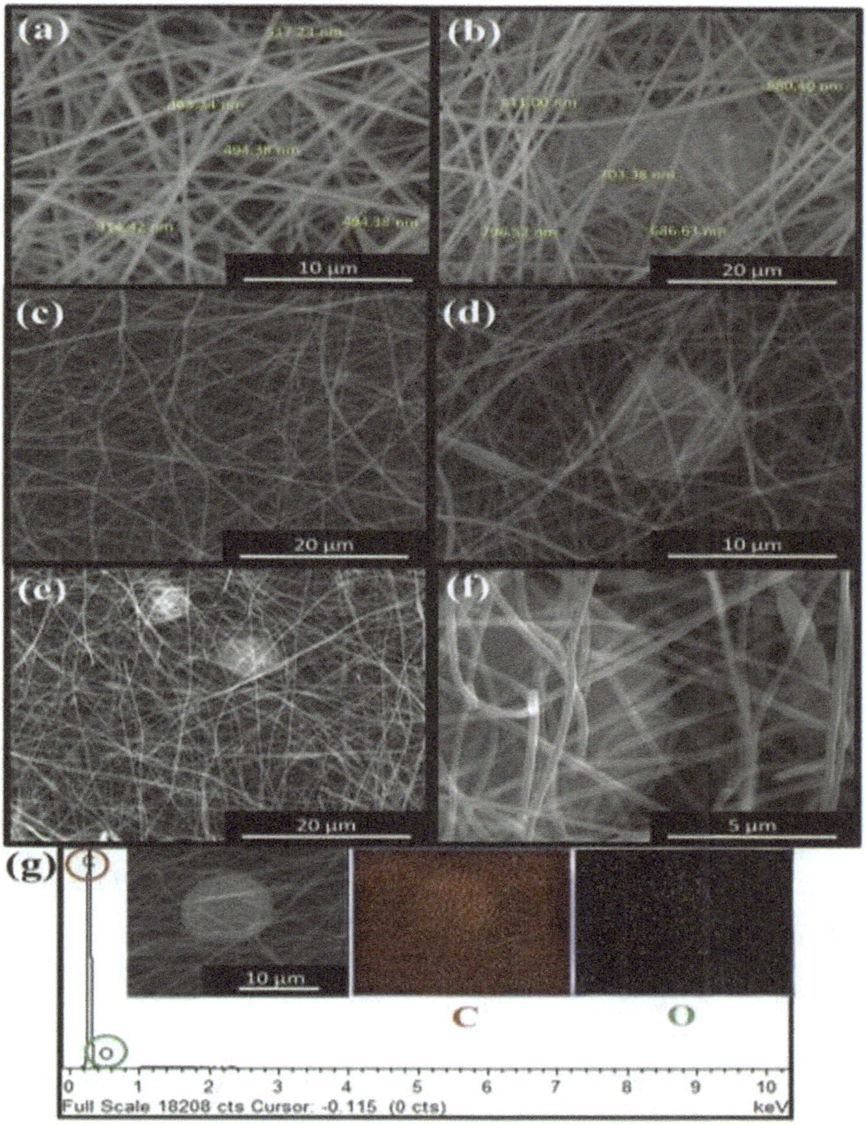


**Fig. 18.6** Surface morphology of **a** PAN, **b** PAN/fGO; and **c** CNF/fGO sheets intercalated within the nanofibers, **e** presence of the higher concentration fGO in the nanofibers, **f** high magnification of a single fGO sheet, and **g** EDX spectra of CNF/fGO. Printed with permission from Ref. [28]. Copyright (2017) Elsevier

properties. Figure 18.6g portrays the EDX analysis which results in the high intensity of carbon in the CNF/fGO membrane with a much lesser amount of oxygen.

Figure 18.7a shows the cyclic voltammetry (CV) curve which depicts as a skewed rectangular profile for pure CNF showing the resistivity characteristic of the electrode, which attributed to the oxygenated functional groups in the nanofibers. However in Fig. 18.7c, with the introduction of fGO, pseudo-rectangular behavior was observed in the CV profiles. Figure 18.7e shows more rectangular like profile when the concentration of the fGO is increased to double (CNF/fGO-2), proving that

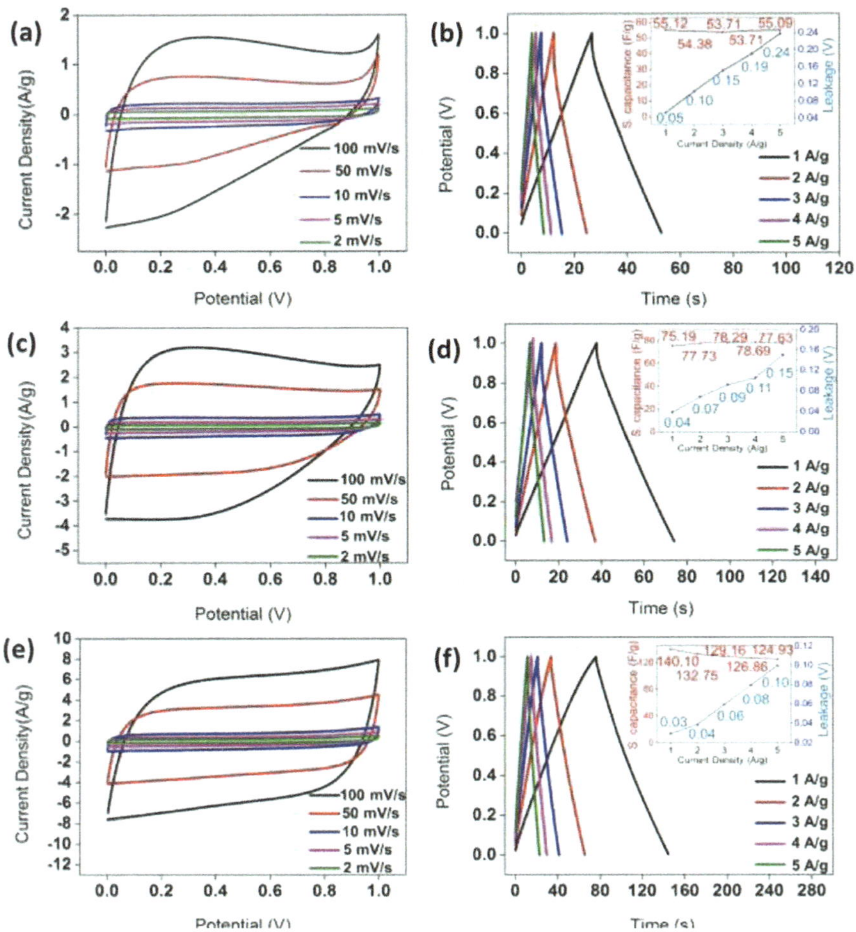


**Fig. 18.7** Cyclic voltammetry profiles for various scan rates of **a** pure CNF, **c** CNF/fGO-1, and **e** CNf/fGO-2; galvanostatic charge/discharge analysis of **b** pure CNF, **d** CNF/fGO-1, and **f** CNF/fGO-2. Insets illustrate the capacitance and leakage drop as the function of current density was applied. Printed with permission from Ref. [28]. Copyright (2017) Elsevier

the introduction of graphene structure into the nanofibers leads to the improvement in the performance [37]. In actual supercapacitor, the current involved in the system is static; however, for studies of its performance analysis of charge/discharge was performed to simulate the actual performance of a supercapacitor [38]. The specific capacitance of a pure CNF is 55.11 F/g when the current density is at 1 A/g. The current rate fluctuates in order to retain its capacitance. From Fig. 18.7b, it was found that the peak voltage drop was observed at 0.24 V and corresponding current density was found to be 5 A/g. The main reason behind this was due to the presence of an oxygenated group within the nanofiber having an insulating property. This developed a resistance between the interface of the electrolyte and electrode. Further, increases in the specific capacitance up to 79.19 F/g were observed with the addition of fGO, improving from that of pure. Figure 18.7d shows a high-voltage leakage along with the increases in current rate up to the highest point, where the leakage was reported to be 0.15 V. The specific capacitance was reported to achieve at 140.10 F/g, and the corresponding current rate was 1 A/g; this was observed by increasing the concentration of fGO to double. Figure 18.7f depicts the retaining of high specific capacitance of about 124.93 F/g which is 89%. Another interesting fact is that this was obtained even at the current rate of 5 A/g, with only 0.099 V of leakage. From the above performance, it can be concluded that the introduction of fGO into the nanostructure leads to the enhancement in the capacitive behavior. With the introduction of fGO, the adsorption of electrolyte ions increases as fGO serves as a template providing larger surface area. This finding is quite relevant to the CV curves of CNF/fGO-2 device [28].

Figure 18.8a shows the Nyquist plot illustrating the electrochemical impedance spectra of pure CNF, CNF/fGO-1, and CNF/fGO-2. Typically, this plot depicts two interesting features, a process of charge-limiting at the interface of the electrolyte and electrode affecting the resistivity ($R_{ct}$); this is illustrated as arc region of high frequency in the plot, and the resistivity is given by the diameter of the arc [39–41]. From the Nyquist plot, by obtaining the intercept in the first $X$-axis from the

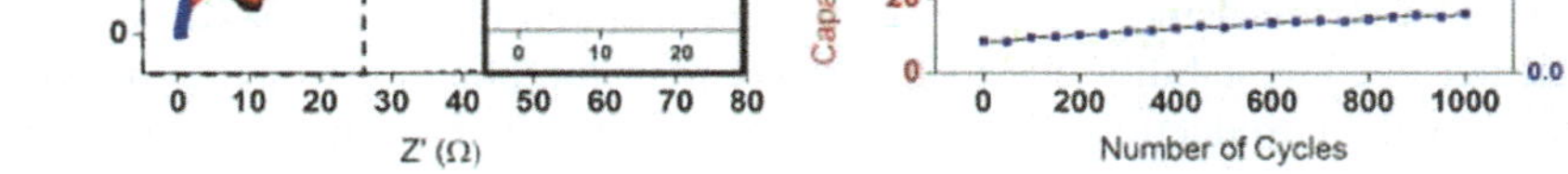


**Fig. 18.8** **a** Nyquist plots of pure (i) CNF, (ii) CNF/fGO-1, and (iii) CNF/fGO-2. The magnified the high-frequency region of the spectra; **b** apacitance retention and volt drop of CNF/fGO-2 as the function of continuous charge/discharge cycles. Printed with permission from Ref. [28] Copyright (2017) Elsevier

high-frequency region, the internal resistance of the electrode and equivalent series resistance (ESR) corresponding to the resistance of the solution was obtained [42, 43]. Figure 18.8a illustrates the significant amount of resistance to the charge transfer in CNF/fGO-2 as the highest semicircle arc was observed with the largest diameter. However, the semicircle recorded for CNF and CNF/fGO-1 recorded was also almost similar in diameter, which shows the system possess lower electrode/electrolyte resistance. It was reported that the addition of fGO leads to an increase in $R_{ct}$ (12.78 Ω), which is comparable to that of pure CNF (10.39 Ω). This may be possible because of the functionalization, as PMMA branching may be a presence on the surface, inducing the electron transfer mechanisms, contributing to the resistivity of the electrodes. The values of the $R_{ct}$ can be increased up to 22.45 Ω by increasing the fGO concentration in the carbon nanofibers. However, minimal increases in the ESR of carbon nanofibers were reported with a higher concentration of fGO, whose value is 0.288 Ω for CNF/fGO-2 which is comparable with that of pure CNF (0.22 Ω). The lowest value recorded was for CNF/fGO-1 (0.17 Ω). Generally, every supercapacitor based on nanofiber is reported to possess the same vertical behaviors for the region behind the Warburg region. This was mainly due to the total internal structure of the carbon electrode that was fully wetted by the electrolytes as it is of porous materials [44]. Figure 18.8b shows the plot for the cycle performance of supercapacitors based on CNF/fGO-2, respectively. Another concern is that for CNF/fGO-2, despite after charge/discharge continuously for 1000 cycles at 1 A/g, the capacitance retention is very high (96.2%), although the $R_{ct}$ recorded by was quite high. The voltage drop was recorded to be 0.04, which is very minimal drop and at initial the value was 0.02 V. The main reason for such unique properties is due to the effective ion transport across the interface of the electrolyte and electrode nanofibers facilitate by the introduction of fGO into the nanofibers. Moreover, it is reported that the introduction of fGO into the system leads to the prevention of degradation of the electrode materials as the influx and outflow of the electrolyte ions occurred at a vigorous rate [28].

## Conclusion

In this chapter, we focused on graphene oxide-based electrospun nanofibers as an electrode for supercapacitors applications. Graphene oxide-reinforced carbon nanofibers electrode was fabricated via two processes: firstly, the surface functionalization of graphene oxide and secondly, the electrospinning technique. The introduction of functionalized graphene oxide demonstrated high specific capacitance and better cycling stability due to its unique fiber morphology which is directly affected by the introduction of GO-based material into the nanofibers electrode. In the first part of the paper, we have discussed emergence of a supercapacitor as a means for better energy storage purposes and followed by a brief discussion on the structure and properties of GO and also give a detailed explanation of the experimental procedure of electrospinning technique for fabrication of nanofibers. Finally, the electrochemical performance of functionalized graphene oxide-reinforced electrospun carbon nanofibers supercapacitor electrode has been discussed adequately. Indeed, the fabricated fGO-reinforced CNF reveals a remarkable performance by achieving a specific capacitance of 140.10 F/g, which is almost three times higher than that of a pure CNF. Not only that, but the stability capacitance retention even after 1000 cycles of charge/discharges was also recorded to be 96.2% at a current density of 1 A/g.

**Acknowledgements** One of the authors Mr. Wahengbam Ishwarchand Singh acknowledges Salam Sujata Devi, Department of Chemistry, Manipur University, for her help in drawing the chemical structures.

## References

1. Pushparai, V.L., Sreekala, S., Nalamasu, O., Ajayan, P.M.: Flexible energy storage devices using nanomaterials. In: Semiconductor Nanomaterials for Flexible Technologies. William Andrew Publishing. pp. 227–245 (2010)
2. Becker, H.I., General Electric Co.: Low voltage electrolytic capacitor. U.S. Patent 2,800, 616 (1957)
3. Samantara, A.K., Ratha, S.: Historical background and present status of the supercapacitors. In: Materials Development for Active/Passive Components of a Supercapacitor, pp. 9–10. Spinger, Singapore
4. Sharma, P., Bhatti, T.S.: A review on electrochemical double-layer capacitors. Energy Conver.s Manag. **51**(12), 2901–2912 (2010)
5. Lu, Q., Chen, J.G., Xiao, J.Q.: Nanostructured electrodes for high-performance pseudocapacitors. Angew Ch Int Ed **52**(7), 1882–1889 (2013)
6. Conway, B.E.: Transition from “supercapacitor” to “battery” behavior in electrochemical energy storage. J. Electrochem. Soc. **138**(6), 1539–1548 (1991)
7. Muzaffar, A., Ahamed, M.B., Deshmukh, K., Thirumalai, J.: A review on recent advances in hybrid supercapacitors: design, fabrication and applications. Renew. Sustain. Energy Rev. **101**, 123–145 (2019)
8. Li, D., Mueller, M.B., Gilje, S., Kaner, R.B., Wallace, G.G.: Processable aqueous dispersions of graphene nanosheets. Nat. Nanotechnol. **3**(2), 101–105 (2008)
9. Konios, D., Stylianakis, M.M., Stratakis, E., Kymakis, E.: Dispersion behaviour of graphene oxide and reduced graphene oxide. J. Colloid Interface Sci. **430**, 108–112 (2014)
10. Li, J., Zeng, X., Ren, T., van der Heide, E.: The preparation of graphene oxide and its derivatives and their application in bio-tribological systems. Lubricants **2**, 137–161 (2014)
11. Li, G.L., Liu, G., Li, M., Wan, D., Neoh, K.G., Kang, E.T.: Organo- and waterdispersible graphene oxide-polymer nanosheets for organic electronic memory and gold nanocomposites. J. Phys. Chem. C **114**(29), 12742–12748 (2010)
12. Kuilla, T., Bhadra, S., Yao, D.H., Kim, N.H., Bose, S., Lee, J.H.: Recent advances in graphene based polymercomposites. Prog. Polym. Sci. **35**(11):1350–1375 (2010)
13. Fazal, A., Fancey, K.S.: Performance enhancement of nylon kevlar fiber composites through viscoelastically generated pre-stress. Polym. Polym. Compos. **16**(2), 101–113 (2008)
14. Dimiev, A.M., Tour, J.M.: Mechanism of graphene oxide formation. ACS Nano **8**(3), 3060–3068 (2014)
15. Dreyer, D.R., Park, S., Bielawski, C.W., Ruoff, R.S.: The chemistry of graphene oxide. Chem. Soc. Rev. **39**(1), 228–240 (2010)

16. Mkhoyan, K.A., Contryman, A.W., Silcox, J., Stewart, D.A., Eda, G., Mattevi, C., Miller, S.: Atomic and electronic structure of graphene-oxide. NANO Lett. **9**(3), 1058–1063 (2009)
17. Lee, C., Wei, X., Kysar, J.W., Hone, J.: Measurement of the elastic properties and intrinsic strength of monolayer graphene. Science **321**, 385–388 (2008)
18. Nair, R.R., Grigorenko, A.N., Blake, P., Novoselov, K.S., Booth, T.J., Peres, N.M.R., Stauber, T., Geim, A.K.: Fine structure constant defines visual transparency of graphene. Science **320**(5881), 1308 (2008)
19. Bunch, J.S., Verbridge, S.S., Alden, J.S., Van Der Zande, A.M., Parpia, J.M., Craighead, H.G., Mceuen, P.L.: Impermeable atomic membranes from graphene sheets. NANO Lett. **8**(8), 2458–2462 (2008)
20. Balandin, A.A., Ghosh, S., Bao, W., Calizo, I., Teweldebrhan, D., Miao, F., Lau, C.N.: Superior thermal conductivity of single-layer graphene. Nano Lett. **8**, 902–907 (2008)
21. Chen, J.H., Jang, C., Xiao, S., Ishigami, M., Fuhrer, M.S.: Intrinsic and extrinsic performance limits of graphene devices on $SiO_2$. Nat. Nanotechnol. **3**, 206–209 (2008)
22. Dürkop, T., Getty, S.A., Cobas, E., Fuhrer, M.S.: Extraordinary mobility in semiconducting carbon nanotubes. Nano Lett. **4**, 35–39 (2004)
23. Peschel, G.: Carbon-carbon bonds: hybridization. Obtained online from: https://www.physik.fu-berlin.de/einrichtungen/ag/agreich/lehre/Archiv/ss2011/docs/Gina_Peschel-Handout.pdf. Published on, 5(5) (2011)
24. Paredes, J.I., Villar-Rodil, S., Martínez-Alonso, A., Tascón, J.M.D.: Graphene oxide dispersions in organic solvents. Langmuir **24**, 10560–10564 (2008)
25. Wahab, I.F., Razak, S.I.A., Azmi, N.S., Dahli, F.N., Yusof, A.H.M., Nayan, N.H.M.: Electrospun graphene oxide-based nanofibres. In: Advances in Carbon Nanostructures, pp. 101–120 (2016)
26. Huang, N., Lim, H., Chia, C., Yarmo, M., Muhamad, M.: Simple room-temperature preparation of high-yield large-area graphene oxide. Int. J. Nanomed. **6**, 3443 (2011)
27. Chee, W.K., Lim, H.N., Huang, N.M.: Electrochemical properties of free-standing polypyrrole/graphene oxide/zinc oxide flexible supercapacitor. Int. J. Energy Res. **39**, 111–119
28. Chee, W.K., Lim, H.N., Andou, Y., Zainal, Z., Hamra, A.A.B, Harrison, I., Altarawneh, M., Jiang, Z.T., Huang, N.M.: Functionalized graphene oxide-reinforced electrospun carbon nanofibers as ultrathin supercapacitor electrode. J Energy Chem **26**, 790–798 (2017)
29. Andou, Y., Yasutake, M., Nishida, H., Endo, T.; Designed surface modification by photo-induced vapor phase assisted surface polymerization of vinyl monomers. J. Photopolym. Sci. Technol. **20**, 523–528 (2007)
30. Zhang, C.L., Yu, S.H.: Nanoparticles meet electrospinning: recent advances and future prospects. Chem. Soc. Rev. **43**, 4423–4448 (2014)
31. Bhardwaj, N., Kundu, S.C.: Electrospinning: a fascinating fiber fabrication technique. Biotechnol. Adv. **28**, 325–347 (2010)
32. Formhals, A.: Process and apparatus for preparing artificial threads. US Patent, 1975504 (1934)
33. Garg, K., Bowlin, G.L.: Electrospinning jets and nanofibrous structures. Biomicro-Fluidics **5**, 1–19 (2011)
34. Leach, M.K., Feng, Z.-Q., Tuck, S.J., Corey, J.M.: Electrospinning fundamentals: optimizing solution and apparatus parameters. J. Vis. Exp. **47**, e2 494 (2011)
35. Huang, Z.M., Zhang, Y.Z., Kotaki, M., Ramakrishna, S.: A review on polymer nanofibers by electrospinning and their applications in nanocomposites. Compos. Sci. Technol. **63**, 2223–2253 (2003)
36. Gu, S.Y., Wu, Q.L., Ren, J.: Preparation and surface structures of carbon nanofibers produced from electrospun PAN Precursors. New Carbon Mater. **23**, 171–176 (2008)
37. Zhi, M., Manivannan, A., Meng, F., Wu, N.: Highly conductive electrospun carbon nanofiber/$MnO_2$ coaxial nano-cables for high energy and power density supercapacitors. J Power Sour **208**, 345–353 (2012)
38. Chee, W.K., Lim, H.N., Harrison, I., Chong, K.F., Zainal, Z., Ng, C.H., Huang, N.M.: Performance of flexible and binderless polypyrrole/graphene oxide/zinc oxide supercapacitor electrode in a symmetrical two-electrode configuration. Electrochim. Acta **157**, 88–94 (2015)

39. Chen, Y.-L., Hu, Z.-A., Chang, Y.-Q., Wang, H.-W., Zhang, Z.-Y., Yang, Y.-Y., Wu, H.-Y.: Zinc oxide/reduced graphene oxide composites and electrochemical capacitance enhanced by homogeneous incorporation of reduced graphene oxide sheets in zinc oxide matrix. J. Phys. Chem. C **115**, 2563–2571 (2011)
40. Dubal, D.P., Lee, S.H., Kim, J.G., Kim, W.B., Lokhande, C.D.: Porous polypyrrole clusters prepared by electropolymerization for a high-performance supercapacitor. J. Mater Chem. **22**, 3044–3052 (2012)
41. Hassan, M., Reddy, K.R., Haque, E., Faisal,S.N., Ghasemi, S., Minett, A.I., Gomes, V.G.: Hierarchical assembly of graphene/polyaniline nanostructures to synthesize free-standing supercapacitor electrode. Composit. Sci. Tech. **98**, 1–8 (2014)
42. Lim, Y.S., Tan, Y.P., Lim, H.N., Huang, N.M., Tan, W.T., Yarmo, M.A., Yin, C.-Y.: Fabrication of carbon-based nanostructured flexible supercapacitor. Cer. Inter. **40**, 3855–3864 (2014)
43. Li, Y., Ye, K., Cheng, K., Yin, J., Cao, D., Wang, G.: Electrodeposition of nickel sulfide on graphene-covered make-up cotton as a flexible electrode material for high-performance supercapacitors. J. Power Sour. **274**, 943–950 (2015)
44. Ra, E.J., Raymundo-Piñero, E., Lee, Y.H., Béguin, F.: High power supercapacitors using polyacrylonitrile-based carbon nanofiber paper. Carbon **47**, 2984–2992 (2009)

# Chapter 19
# Nanowire Nanosensors for Biological and Medical Application

S. Nonganbi Chanu and Bibhu Prasad Swain

## Introduction

Nanotechnology devoted to the studies of the synthesis, properties, characterization, production, and application of structures, devices, and systems by controlling at least on critical dimension less than scale to 100 nm approximately. The size range of few nanometers to 100 nm is one where many interesting things all sorts of physical properties changes also many biological systems function occurred [1–3]. Nowadays, nanotechnology becomes an emerging concept and its applications are not limited to the physical sciences and also increasing in the field of life sciences and medicine. It plays an important role in advanced life science and medical research particularly in the development of drug delivery systems with greater efficiencies. In the field of drug delivery, nanotechnology has been seen to be suitable for treating chronic intracellular infection [4]. This system of drug delivery has been seen to have distinct advantages over traditional drug carriers. The development of the process of cancer nanotechnology gives rise to exciting opportunities in the diagnosis and treatment of cancer.

The straightforward connection between nanotechnology and life science as well as in medical field is to consider the size and organization of common structures shown in Fig. 19.1. Here quantum dots, nanowires, and proteins built from atoms have similar average diameters. These basic structures can be further organized into large structure such as viruses and nano-electronic circuit and ultimately further elaborated into functional system such as cells or computer chips [5]. The similarity in size of synthetic and natural nanostructures makes nanotechnology an observable choice for creating analyses and other tools that can enable detection and treatment of disease in powerful new ways. Detection and quantification of biological and

S. N. Chanu (✉) · B. P. Swain
Department of Physics, National Institute of Technology Manipur, Langol, Manipur 795004, India
e-mail: nonganbisagolsem@gmail.com



B. P. Swain (ed.), *Nanostructured Materials and their Applications*, Materials Horizons: From Nature to Nanomaterials, https://doi.org/10.1007/978-981-15-8307-0_19

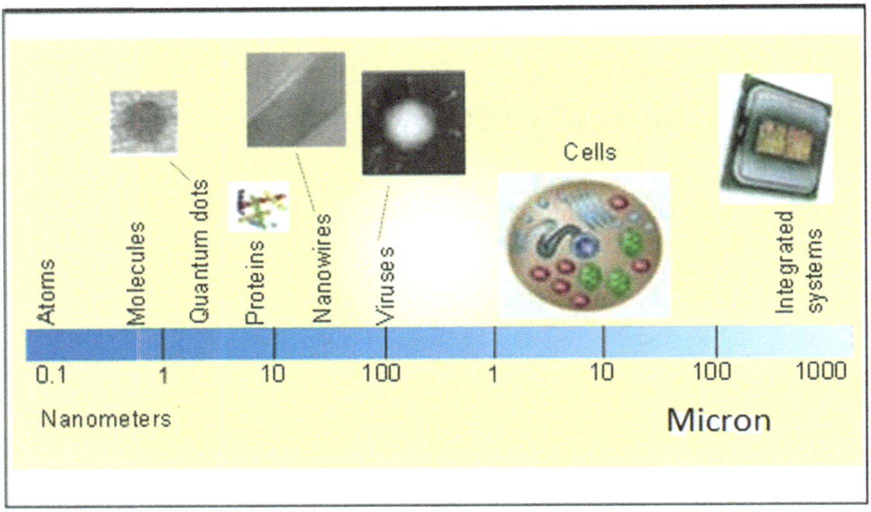


**Fig. 19.1** Comparison of the sizes of biological, chemical and nanoscale structures, assembles and systems

chemical species are central to many area of medicine, ranging from diagnosing disease to the discovery and screening of new drug nucleus. Nanostructures like nanowires, carbon nanotubes as well as nanoparticles offer new and sometimes unique opportunities for such analysis.

The development of nanotechnology in the field of diagnostic and therapeutic applications aims to reduce the toxicity or any adverse action of the agents by making the more target specific and hence reduce its doses [6, 7]. This technology has a potential application in the stem research which includes tracking of stem cell surface molecules, noninvasive tracking of stem cells and progenitor cells transplanted, tracking stem cell delivery system and intracellular delivery of DNA, ribonucleic and interference (RNAi), proteins, peptides, and small drugs for stem cell differentiation [8, 9].

Inorganic nanowires, nanocrystals, and carbon nanotubes exhibit unique electrical, optical, and magnetic properties that can be exploited for sensing and imaging [10]. The reproducible and turnable conducting properties of semiconducting nanowires combined with surface binding provide very different and powerful approach to Nanomedicine, specifically nanowires enable detection and sensing modality. Electronic nanowire devices are readily integrated into miniaturized systems, and moreover, direct electrical detection disperses with time-consuming labeling chemistry. These characteristics together with ultrahigh sensitivity suggest that nanowire devices could revolutionize may aspects of sensing and detection medicine, ranging from the diagnosis and monitoring of disease treatment to the discovery and screening of new drug molecules [5]. Silicon nanowires are widely used in biosensing application. They represent the most important and versatile nanometer scale wire structures synthesized as single crystals with controlled synthesis functionalization and electronic properties [11].

## Biomedical Applications of Nanowire Nanosensors

### *Drug Discovery*

Drug development and discovery process involve rigorous testing and optimization of selected compounds to identify the drug that is most effective this testing is done in cells (in vitro) and in animals (in vivo) to study [12]. In addition, the identification of organic molecules binds specifically to proteins and fundamental to the discovery and development of pharmaceuticals and to chemical genetic approaches for elucidating complex pathway in biological system [13]. A representative example of this area is the identification of molecular inhibitors to tyrosine kinases. Successful kinase inhibitor drug discovery relies heavily on the structural knowledge of the interaction of inhibitors with the target structural biology of kinases, and in particular of Tyrosine Kinases has given detailed insights into the intrinsic flexibility of the catalytic domain and has provided a rational basis for obtaining are proteins that mediate signal transduction in mammalian cell through phosphorylation of a tyrosine residue using adenosine triphosphate (ATP) [5]. Moreover, protein tyrosine kinases (PTKs) are essential enzymes in cellular signaling that regulate cell growth, proliferation, metabolism, differentiation, and migration [14–16]. Deregulation of the tyrosine kinase activity by various mechanisms leading to gain or loss of function have been observed in a large number of PTKs and shown to be associated with different human diseases [14, 15] including cancer. To organized nanowire nanosensors devices for screening small-molecule inhibitors to tyrosine kinases. The kinase Ab1 to the surface of nanowire FETs and investigated the binding of ATP and competitive inhibition of ATP binding with organic molecules, such as the drug Gleevec. In this configuration, binding or inhibition of binding of the negative charged ATP to Ab1 linked at the silicon nanowire surface can be detected simply as an increase or decrease in the conductance of the p-type nanowire nanosensor device. The direct yet simple nature of this approach enabled by nanotechnology contrasts conventional indirect assays used to study kinases [17, 18]. Time-dependent data recorded from Ab1-modified p-type Si nanowire devices have been found to exhibit reversible, concentration-dependent increases in conductance upon introducing solutions containing ATP. The increase in conductance was consistent with the binding of negatively charged ATP to Ab1 [19]. Plots of the normalized conductance recorded from Ab1-modified p-type Si nanowire nanosensors devices exhibit reversible decreases in conductance because of competitive inhibition of ATP. Notably, the conductance decreases at constant small-molecule concentration, which demonstrates that the degree of inhibition depends strongly on molecular structure (Fig. 19.2).

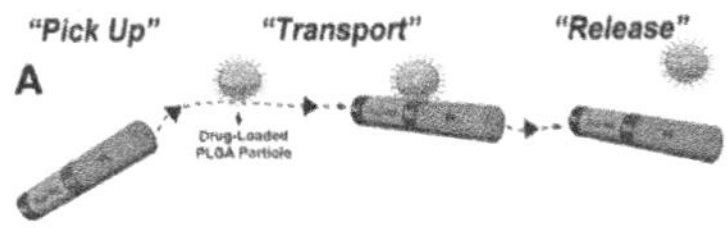


**Fig. 19.2** Translocation of model drug carriers by catalytic nanowire motors. **a** Scheme depicting the dynamic pickup, transport and release of drug-loaded PLGA particles using a nanoshuttle.

## *Detection of Diabetes*

Nanotechnology has impact these efforts by increasing the surface area of sensors, improving the catalytic properties of electrodes and providing nanoscale sensors. The developments of nanotechnology in the past several years on both nanosensors that directly measure level of glucose and nanomaterials improve glucose sensor function. Diabetes is a rapidly growing problem that is managed at the individual level by monitoring and controlling blood glucose level to minimize the negative effects of the disease. Traditional monitoring of blood glucose uses discrete blood sampling time points during the cause of a day. Such procedures are extremely uncomfortable and inconvenient. Nowadays, constant glucose monitoring was performed using medical nanorobotic. In addition, the innovative nanorobot architecture for diabetes controls currently developments in nanoelectronics, biochemistry, and information technologies are providing feasible development pathways to allow the creation of nanorobots. To envision how actual and upcoming stages of nanotechnology can be utilize in medicine, numerical analysis and computational nanotechnology are used to illustrate the proposed nanorobot performance with the bloodstream, using a 3D vessel as test bed for diabetes control [20] (Fig. 19.3).

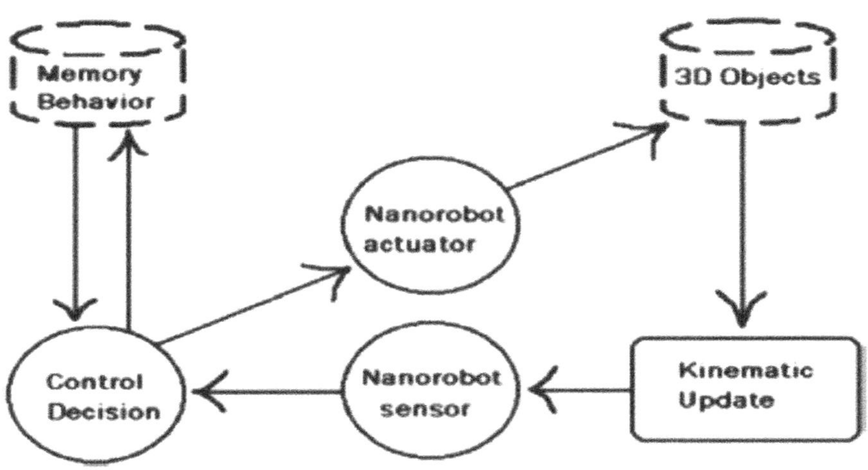


**Fig. 19.3** Multithread implementation allows real-time high performance, helping in multirobot sensing capability [20]

The nanorobot design includes integrated nanoelectronics as a valuable tool for diabetes control; nanorobots are used for automatically monitoring glucose levels. Nanorobots inside human body observe and diagnose and even treat diseases much earlier and less painful. The exterior of a nanorobot will likely be constructed of carbon atom in a diamonded structure because it possesses atomic smoothness, chemical inertness, and hardness properties close to those of diamond [21]. It has an artificial glycocalyx surface [22, 23] which minimize adsorption and bioactivity in relation of fibrinogen as well as other blood proteins, ensuring sufficient biocompatibility to avoid immune system attack [24, 25]. Typically, carbon's favorable characteristics such as large motilities and high transconductance promote it as an ideal component for integrated nanoelectronics, providing the suitable properties for successful radiofrequency applications [26, 27]. The nanorobot uses a radiofrequency identification device (RFID) CMOS transponder system for in vivo positioning [28, 29].

## *Virus Detection*

Viruses are the most serious cause of human disease [30], and there is increasing concern about as agents for biological weapons [31, 32]. Nanowire biosensor will play an important role in diagnostic of epidemic diseases. In addition, Si nanowire sensors have successfully detected many dangerous viruses, including Influenza AH3N2 [33], H1N1 [34], HIV [35], Dengue [36]. Silicon nanowire sensors also possible to detect two distinct viruses at the single virus level [37]. Its surface is functionalized with antibodies that specifically bind to the target viruses, affecting the conductivity of the nanowire. Ibarlucea et al. [38] introduced Si nanowires to detect Ebola VP40 matrix protein. The limit of detection was seen to be around 6.25 nM after 30 min incubation, outperforming the ELISA technique by six orders, Shen et al. [39], developed a Si nanowire-based biosensor for the rapid diagnosis of flu that could detect as few as 29 viruses/μL from exhaled breath condensate sample (Fig. 19.4).

## *Cancer Application*

The detection of telomerase activity in cancer tissues as well as cancer cells is quit valuable for practical diagnosis, prognosis, and monitoring treatment of cancer patients. In addition, telomerase is an enzyme that synthesizes telomere, which are a repetitive nucleotide sequence consisting of short tandem repeats (TTAGGG) and an important organismal aging-related DNA [40, 41]. In normal cells, a critical telomere length is eventually reached, thereby including cellular senescence and finally leading to apoptosis [42]. Wang [43] reported miniaturized devices that enable rapid and direct analysis of the specific binding of small molecules to proteins

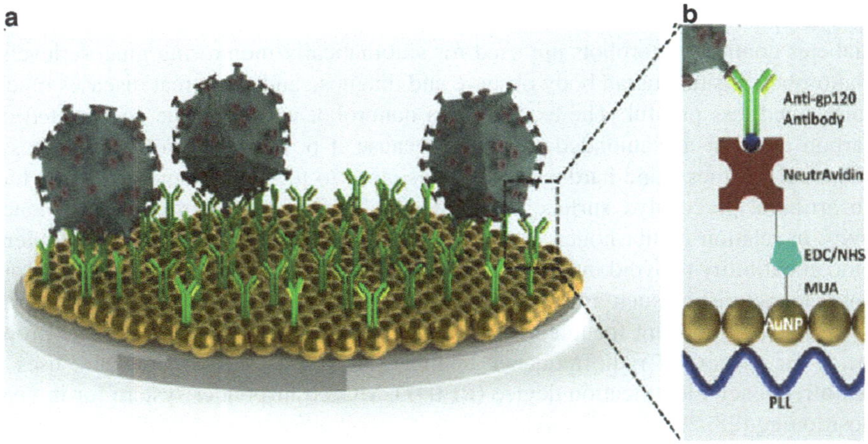


**Fig. 19.4** Nanoplasmonic viral load detection platform. **a** HIV was captured on the antibody immobilized biosensing surface. **b** To capture HIV on the biosensing surface, polystyrene surfaces were first modified by poly-l-lysine (PLL) generating amine groups [35]

using silicon nanowire (Si NW) field-effect transistor devices. Detection of small-molecule inhibitors of long from adenosine triphosphate (ATP) binding to tyrosine kinase (Ab1) with constitutive expression that was responsible for chronic myelogenous leukemia. Direct label-free detection of tumor markers is a hallmark of silicon nanowire sensor functionalized mostly with specific antibodies to a detected soluble antigen (Table 19.1).

A simple and low-cost device to fabricate poly Si NW-FET (polycrystalline silicon nanowire field-effect transistor) for biosensing application was published [44]. Recently, it was also reported a detection method for vascular endothelial growth factor (VEGF) for cancer diagnoses using an anti-VEGF aptamer [45] (Fig. 19.5).

**Table 19.1** Nanowire sensors for the detection of cancer and tumor markers

| Sensor type[a] | Wire synthesis | Tumor or cancer marker detected | Direct/Indirect detection | Multiplex detection | Tissue | Refs. |
|---|---|---|---|---|---|---|
| FET Si NW | Bottom-up | VEGF-A, aptamer | Direct | No | Buffer | [46] |
| FET Si NW | Top-down | VEGF-A | Direct | No | Buffer | [47] |
| Memristor Si NW | Top-down | VEGF-A | Direct | No | Buffer | [48] |
| Memristor Si NW | Top-down | VEGF-A | Direct | No | Buffer | [49] |

[a]Unless indicated otherwise, the wires were functionalized with antibodies; NW, nanowire

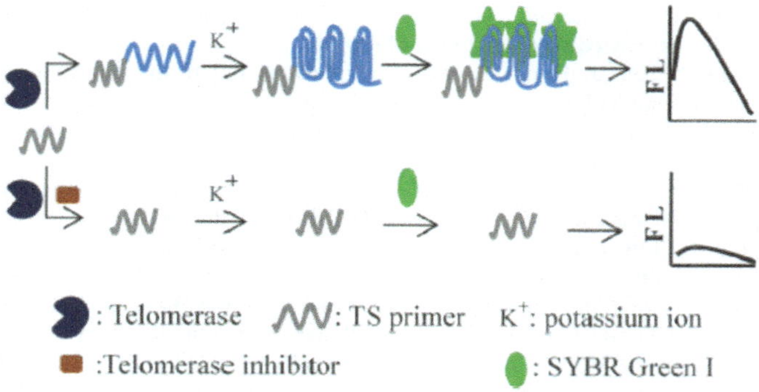


**Fig. 19.5** Illustration of the telomerase activity sensing mechanism [40]

## Conclusion

Nanomaterials such as nanowires and nanoparticles have already led to advances in the field of biological and medical sensors. The quantum size effects of the nanomaterials and their similarity in size to natural and synthetic nanomaterials are expected to improve sensor sensitivity dramatically. As chemical-friendly surface for various sensing applications and their electrical controllability for accurate measurement, nanowires are considered as key nanomaterials in the field of treatment of several illnesses. Nanowire sensors are also surveyed to highlight recent progress in the applications for the detection of viruses, biomarkers, and DNA, as well as for drug discovery. Advances in the performance and functionality of nanowire sensors are also include the improvements in reusability, sensitivity in high ionic strength solvent, long-term stability, and self-powering. Overall, with the advantages of ultra-sensitivity and the ease of fabrication, it is expected that nanowires will contribute significantly to the development of biological and medical sensors in the immediate future.

## References

1. Navalakhe, R.M., Nandedkar, T.D.: Application of nanotechnology in biomedicine. Indian J. Exp. Biol. **45**, 160–165 (2007)
2. Sahoo, S.K., Parveen, S., Panda, J.J.: The present and future of nanotechnology in human health care. Nanomedicine **3**, 20–31 (2007)
3. Drabu, S., Khatri, S., Babu, S., Verma, D.: Nanotechnology: an introduction to future drug delivery system. J. Chem. Pharm. Res. **2**, 171–179 (2010)
4. Putheti, R.R., Okigbo, R.N., Saiadvanapu, M., Chavanpati, S.: Nanotechnology importance in the pharmaceutical industry. Afr. J. Pure Appl. Chem. **2**, 27–31 (2008)
5. Patolsky, F., Zheng, G., Lieber, C.M.: Nanowire sensors for medicine and the life sciences. Nanomedicine **1**, 51–65 (2006)

6. Ma, Y.L., Henry, J.A.: The antidotal effect of a acid glycoprotein on amitriptyline toxicity in cardiac myocytes. Toxicology **169**, 133–144 (2001)
7. Gabel, A., Hinkelbein, J.: Hypotensive cardio-circulatory failure and metabolic acidosis after suicidal intoxication with trmipramine and quetiapine. Anaesthesist **53**, 53–58 (2004)
8. Moghimi, S.M., Hunter, A.C., Murray, J.C.: Nanomedicine: current status and future prospects. FASEB J **19**, 311–330 (2005)
9. Ferreira, L., Karp, J.M., Nobre, L., Langer, R.: New opportunities: the use of nanotechnologies to manipulate and track stem cells. Cell Stem Cell **3**, 136–146 (2008)
10. Wu, X., Liu, H., Liu, J., et al.: Immuofluorescent labelling of cancer marker Her2 and other cellular targets with semiconductor quantum dots. Nat. Biotechnol. **21**, 41–46 (2003)
11. Doucey, M.A., Carrara, S.: Nanowire sensors in cancer. Trends Biotechnol. **37**, 86–99 (2019)
12. Elhassa, G.O., Alfarouk, K.O.: (2015) Stem cell therapy in drug discovery and development. J. Pharmacovigilance **3**, e140 (2015). https://doi.org/10.4172/2329-6887.1000e141
13. Agrawal, S., Prajapati, R.: Nanosensors their pharmacentical Applications. Int. J. Pharm. Sci. Nanotechnol. **4**, 1528–1535 (2012)
14. Blume-Jensen, P., Henter, T.: Oncogenic kinase signalling. Nature **41**, 355–365 (2001)
15. Cohen, P.: Protein Kinases—the major drug targets of the twenty-first century? Nat. Rev. Drug Discov. **1**, 309–315 (2002)
16. Gschwind, A., Fischer, O.M., Ullrich, A.: The discovery of receptor tyrosine: targets for cancer therapy. Nat. Rev. Cancer **4**, 361–370 (2004)
17. Peck, S.C.: Analysis of protein phosphorylation: methods and strategies for studying kinases and substracts. Plant J. **45**, 512–522 (2006)
18. Tagliatic, F., Bottoni, A., Bosetti, A., Zatelli, M.C., Uberti, E.C.D.: Utilization of luminescent technology to develop a kinase assay: Cdk4 as a model system. J. Pbarm. Biomed Anal. **39**, 811–814 (2005)
19. Wang, W.U., Chen, C., Lin, K.H., Fang, Y., Lieber, C.M.: Label-free detection of small-molecule-protein interaction by using nanowire nanosensors. Proc. Natl. Acad. Sci USA **2**, 3208–3212 (2005)
20. Cavalcanti, A., Shirinzadeh, B., Kretly, L.C.: Medical nanorobotics for diabetes control. Nanomedicine **4**, 127–138 (2008)
21. Narayan, R.J.: Pulsed laser deposition of functionally gradient diamond like carbon-metal nanocomposites. Diam. Relat. Mater. **14**, 1319–1330 (2005)
22. Freitas, R.A., Jr.: Nanotechnology, nanomedicine and nanosurgery. Int. J. Surg. **3**, 1–4 (2005a)
23. Marchant, R.E., Zhang, T., Qiu, Y., Ruegsegger, M.A.: Surfactants that mimic the glycocalyx. United States patent US 6759388 (1999)
24. Freitas, R.A., Jr.: What is Nanomedicine? Nanomedicine **1**, 2–9 (2005b)
25. Freitas, R.A., Jr.: Nanotechnology, Nanomedicine and Nanosurgery. Int. J. Surg. **3**, 243–247 (2005c)
26. Musa, G., Mustata, I., Ciupina, V., Vladoiu, R., Prodan, G., Vasile, E., Ehrich, H.: Diamond—like nanostructured carbon film using thermionic vacuum arc. Diamond Rel. Mater. **13**, 1398–1401 (2004)
27. Rutherglen, C., Burke, P.: Carbon nanotube radio. Nano Lett. **7**, 3296–3299 (2007)
28. Ricciardi, L., Pitz, I., Sarawi, S.F.A., Varadan, V., Abbott, D.: Investigation into the future of RFID in biomedical applications. Proc. SPIE-Int. Soc. Opt. Eng. **5119**, 199–209 (2003)
29. Ricci, A., Grisanti, M., De Munari, I., Ciampolini, P.: Improved pervasive sensing with RFID: an ultra-low power basedband processor for UHF tags. IEEE Trans. Very Large Scale Integr (VLSI) Syst **17**, 1719–1729 (2009)
30. Stadler, K., Masignani, V., Eickmann, M., Becker, S., Abrignani, S., Klenk, H.D., Rappuoli, R.: SARS—beginning to understand a new virus. Nat. Rev. Microbiol. **1**, 209–218 (2003)
31. Kim, K., Park, C., Kwon, D., Kim, D., Meyyappan, M., Jeon, S., Lee, J.S.: Silicon nanowire biosensors for detection of cardiac troponin I (cTnI) with high sensitivity. Biosens. Bioelectron. **77**, 695–701 (2016)
32. Atlas, R.M.: Bioterrorism and biodefence research: Changing the focus of microbiology. Nat. Rev. Microbiol. **1**, 70–74 (2003)

33. Shen, F., Wang, J., Xu, Z., Wu, Y., Chen, Q., Li, X., Jie, X., Li, L., Yao, M., Guo, X., et al.: Rapid flu diagnosis. Using silicon nanowire sensor. Nano Lett. **12**, 3722–3730 (2012)
34. Kao, L.T.H., Shankar, L., Kang, T.G., Zhang, G., Tay, G.K.I., Rafei, S.R.M., Lee, C.W.H.: Multiplexed detection and differentiation of the DNA strains for influenza A (H1N1 2009) using a silicon-based microfluidic system. Biosens. Bioelectron **26**, 2006–2011 (2011)
35. Inci, F., Tokel, O., Wang, S., Gurkan, U.A., Tasoglu, S., Kuritzkes, D.R., Demirci, U.: Nanoplasmonic quantitative detection of intact viruses from unprocessed whole blood. ACS Nano **7**, 4733–4745 (2013)
36. Zhang, G.J., Zhang, L., Huang, M.J., Luo, Z.H.H., Tay, G.K.I., Lim, E.J.A., Kang, T.G., Chen, Y.: Silicon nanowire biosensor for highly sensitive and rapid detection of Dengue virus. Sens. Actuators B Chem. **146**, 138–144 (2010)
37. Patolsky, F., Zheng, G., Hayden, O., Lakadamyali, M., Zhuang, X., Lieber, C.M.: Electrical detection of single viruses. Proc. Natl. Acad. Sci. USA **101**, 14017–14022 (2004)
38. Ibarlucea, B., Akbar, T.F., Kim, K., Rim, T., Baek, C.K., Ascoli, A., Tetzlaff, R., Baraban, L., Cuniberti, G.: Ultrasensitive detection of Ebola matrix protein in a memristor mode. Nano Res. **11**, 1057–1068 (2018)
39. Shen, F., Wang, J., Xu, Z., Wu, Y., Chen, Q., Li, X., Jie, X., Li, L., Yao, M., Guo, X., et al.: Rapid flu diagnosis using silicon nanowire sensor. Nano Lett. **12**, 3722–3730 (2012)
40. Quach, Q.H., Jung, J., Kim, H., Chung, B.H.: A simple, fast and highly sensitive assay for the detection of telomerase activity. Che. Commun. **49**, 6596–6598 (2013)
41. Blackburn, E.H.: Structure and function of telomeres. Nature **350**, 569–573 (1991)
42. Cong, Y., Shay, J.W.: Actions of human telomerase beyond telomeres. Cell Res. **18**, 725–732 (2008)
43. Wang, W.U., Chen, C., Lin, K.H., Fang, Y., Lieber, C.M.: Label—free detection of small-molecule-protein interactions by using nanowire nanosensors. Proc Natl Acad Sci USA **102**, 3208–3218 (2005)
44. Hsiao, C.Y., Lin, C.H., Hung, C.H., Su, C.J., Lo, Y.R., Lee, C.C., Lin, H.C., Ko, F.H., Huang, T.Y., Yang, Y.S.: Novel poly-silicon nanowire field effect transistor for biosensing application. Biosens. Bioelectron. 24, 1223–1229 (2009)
45. Lee, H.S., Kim, K.S., Kim, C.J., Hahn, S.K., Jo, M.H.: Electrical detection of Vegts for cancer diagnoses using anti-vascular endothermial growth factor aptamer-modified Si nanowire fets. Biosens. Bioelectron. **24**, 1801–1805 (2009)
46. Lee, H.S., et al.: Electrical detection of VEGFs for cancer diagnoses using anti-vascular endotherial growth factor aptamer-modified Si nanowire FETs. Biosens. Bioelectron. **24**, 1801–1805 (2009)
47. Puppo, F., et al.: Femto-molar sensitive field effect tran-sistor biosensors based on silicon nanowires and antibodies. In: SENSORS, pp. 1–4. IEEE (2013)
48. Puppo, F., et al.: Memristor-based devices for sensing. In: 2014 IEEE international symposium on circuits and systems (ISCAS), pp. 2257–2260. IEEE (2014)
49. Puppo, F., et al.: Memristive biosensors under varying humidity conditions. IEEE Trans. Nanobiosci. **13**, 19–30 (2014)

# Chapter 20
# Nanowire for Diagnostic Tool for Doctors

**Pukhrambam Sushma Devi and Bibhu Prasad Swain**

## Introduction

Nanotechnology is becoming a perfect field for the advancement of science and technology. It is a field which has its root and application in various research field involving chemistry, engineering, biology, and medicine [1]. The field has been rapidly evolving and expanding, and worldwide it has earned public and media interest [2]. Nanoparticles offer remarkable interactions with biomolecules on both the surface of and inside the cells, which can revolutionize diagnosis and treatment.

Starting from disclosing and diagnosing disease to the discovery and screening of new drug molecules, the detection of biological and chemical species is the primary focus for several areas of healthcare and the life sciences. Hence, it is the need of the generation to undergo the development of new devices that could enable direct, sensitive, and rapid analysis of these biological and chemical species. It is evidently reliable that devices based on nanowires are emerging as a powerful and perfect class of ultrasensitive, electrical sensors for the direct detection of these biological and chemical species. Nanowire devices can impact humankind in significant ways [3].

P. S. Devi (✉) · B. P. Swain
Department of Physics, National Institute of Technology Manipur, Imphal, Manipur 795004, India
e-mail: pukhrambamsushma@gmail.com

B. P. Swain
e-mail: bibhuprasad.swain@gmail.com



B. P. Swain (ed.), *Nanostructured Materials and their Applications*, Materials Horizons: From Nature to Nanomaterials, https://doi.org/10.1007/978-981-15-8307-0_20

## Nanoparticles

The creation and/or manipulation of materials in the nanometer range, either by refining or reducing bulk materials into nanoparticles (NPs) or by scaling up from single groups of atoms, are the expected strategies of nanotechnology [1, 4]. Typically, NPs are several hundreds of nanometers in size and have the ability to offer unprecedented interactions with biomolecules on cell surfaces or inside the cell [1, 5]. Extensive varieties of nanoparticles composed of various materials, shapes, and sizes, and with different chemical and surface properties, have already been engineered (Table 20.1).

**Table 20.1** Comparison of various materials of nanodevices showing the shapes, the sizes, the properties, and the use of various materials for diagnosis and/or therapeutics [1]

| Material | Size range | Shape | Properties | Applicability | References |
|---|---|---|---|---|---|
| Fullerene carbon nanotubes | nm | Nanotubes | Carbon nanomaterials in molecular electronics | Nanowire and biosensor for diagnosis | [6, 7] |
| Polylactic acid<br>Poly(cyano)acrylates<br>Polyethylenimine<br>block copolymers<br>Polycaprolactone | 1–1000 nm | Spherical | Biodegradable<br>Biocompatible<br>Smart material (external stimuli degradation pH, temperature, etc.) | Drug/gene delivery | [8–14] |
| Gold nanoparticles | 3–100 nm | Spherical | Electronic, optical, and thermal properties | Diagnostics and detection of biological molecules at low concentration | [15–17] |
| Magnetic nanoparticles | 3–100 nm | Spherical | Magnetic properties | Magnetic immunoassays, drug delivery, cell separation, purification, and tissue repair | [18–20] |
| Quantum dots | 1–10 nm | Spherical | Cd/Zn-selenides | In vitro diagnostic imaging | [21–23] |
| Dendrimers | 10–200 nm | Complex branched polymer | | Drug delivery systems | [24–26] |

## Nanowire Sensor

Several sensors have been created by researchers by bonding antibodies to nanowires [27]. The tunable colors and size of semiconductor nanocrystals, along with their highly robust emission properties, are disclosing the opportunities for optical-based detection and labeling of biological species that provide advantages compared with conventional organic molecular dyes widely used today [3, 28–32]. Semiconducting nanowires are electronically switchable and thus can provide a sensing modality—direct and label-free electrical readout. Due to this, semiconducting nanowires are exceptionally attractive for several applications [3, 33–41]. These electrical-based devices will generate signals and can be directly routed to the outer world. Compared to time-consuming labeling chemistry, direct electrical detection with electronic nanodevices are more effective and are readily integrated into miniaturized systems. These characteristics, along with ultrahigh sensitivity, suggest that many aspects of sensing and detection in biology and medicine could be revolutionized by nanowire devices.

### *Nanowire Field-Effect Sensor*

The primary underlying mechanism for nanowire sensors is the field-effect which is transduced using field-effect transistors (FETs) [3, 33], the extensively used switches of the microelectronics industry. A standard FET is illustrated in Fig. 20.1a. In it, *p*-type semiconductor such as silicon (p-Si) is connected to the metal source and drain electrodes through which a current will be injected and collected, respectively. The gate electrode is capacitively coupled through a very thin layer of dielectric. It will be switched on and off the conductance of the semiconductor between source and drain [3, 42]. In any *p*-type semiconductor, when a positive voltage is applied on the gate, the carriers will deplete and the conductance will reduce. And when a negative voltage is applied on the gate, there will be an accumulation of carriers and eventually conductance will increase. Since the conductance of the FET devices is dependent on the gate voltage, FETs are natural candidates for electrically based sensing. The electric field resulting due to the binding of the charged species to the gate dielectric is similar to applying a voltage using the gate electrode. This possibility for sensing with FETs was introduced several years ago [3, 43–45], but these planar devices have limited sensitivity, and due to this, they have been precluded from having an outsized impact. Semiconductor nanowires comprising of Si and other materials can also be designed to function as FET devices [3, 46–57]. Si nanowires (Fig. 20.1b), one among the well-studied examples, can be prepared as single-crystal structures with diameters as small as 2–3 nm [3, 58–63], and it has also been shown that, for both *n*-type and *p*-type materials, it shows performance characteristics comparable to or better than the best achieved within the industry of microelectronics [3, 50, 51, 54–57]. These robust performance characteristics are also carried through with high

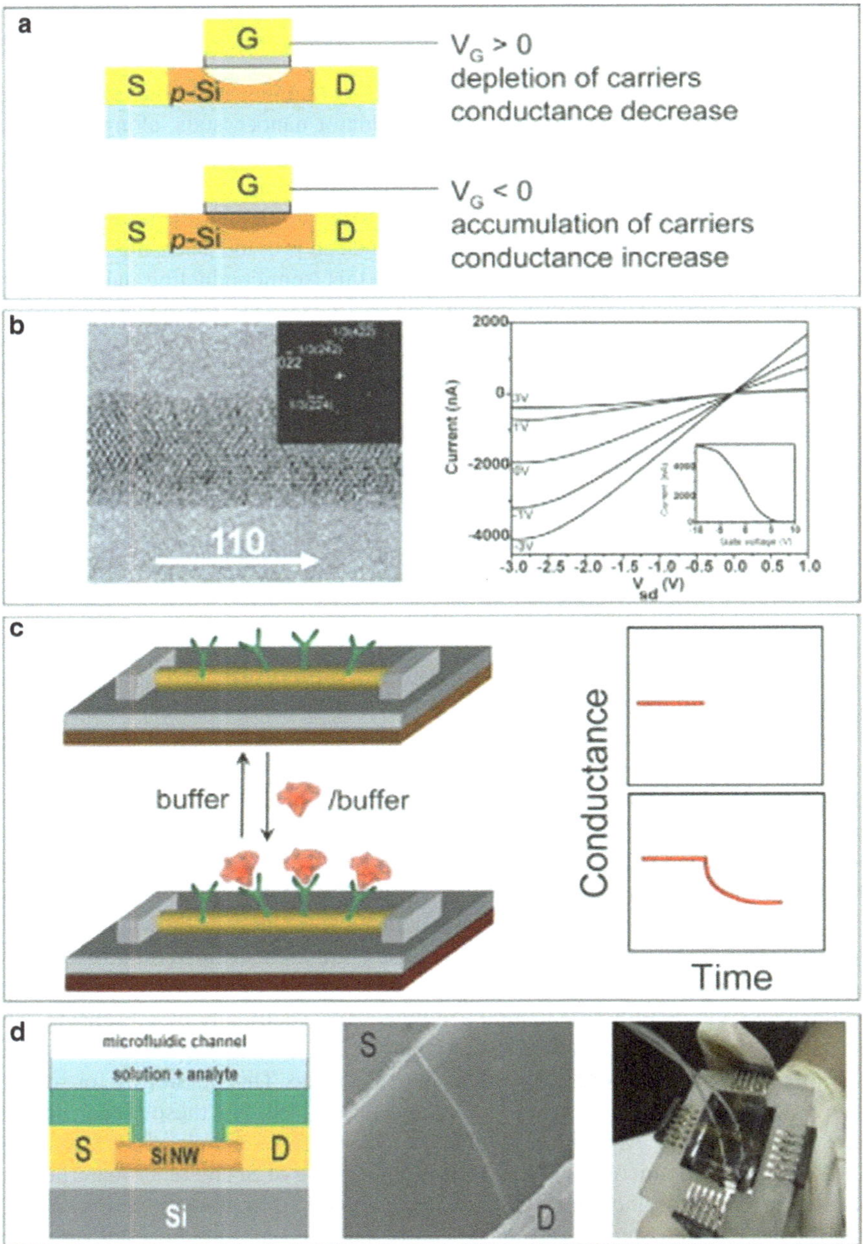
a
G
S
p-Si
D
$V_G > 0$
depletion of carriers
conductance decrease
G
S
p-Si
D
$V_G < 0$
accumulation of carriers
conductance increase
b
110
Current (nA)
2000
0
-2000
-4000
-3.0 -2.5 -2.0 -1.5 -1.0 -0.5 0.0 0.5 1.0
$V_{sd}$ (V)
c
buffer
/buffer
Conductance
Time
d
microfluidic channel
solution + analyte
S
Si NW
D
Si
S
D

◂**Fig. 20.1** Nanowire FET sensor. **a** Schematic of a regular planar FET device, where *S*, *D*, and *G* correspond to the source, drain, and gate, respectively. **b** High-resolution transmission electron microscopy image and electron diffraction pattern for a 4.5 nm diameter single-crystal Si nanowire with <110> growth axis, and electrical transport data for a typical *p*-type nanowire that is characteristic of a FET. **c** Schematic of a-Si nanowire-based FET device configured as a sensor with antibody receptors (green), where binding of a protein with a net positive charge (red) yields a decrease in the conductance. **d** Cross-sectional diagram and scanning electron microscopy image of a single Si nanowire sensor device and a photograph of a prototype nanowire sensor biochip with integrated microfluidic sample delivery [3]

reproducibility [3, 54]; that is, in contrast to carbon nanotubes, during the growth the electronic characteristics of nanowires are well controlled. The one-dimensional morphology of these nanoscale structures is the more important factor in overcoming the sensitivity limitation of previous planar FET sensors since, as compared to the only surface region of planar devices, the binding to the surface of a nanowire will lead to depletion or accumulation of carriers in the 'bulk' of the nanometer-diameter structure [3, 33] and sensitivity is increased to the point that single-molecule detection becomes possible.

As illustrated in Fig. 20.1c, from the high-performance field-effect nanowire transistors, a general sensing device with specific sensing can be configured by linking a recognition group to the surface of the nanowire. There already exist extensive data for the chemical modification of silicon oxide or glass surfaces from planar chemical and biological sensors. Si nanowires have natural oxide coating, and this makes the receptor linkage straightforward [3, 64, 65]. When the sensor device mechanized with the surface receptor is allowed to freely exposed to a solution containing a macromolecule such as the protein that carries a net positive charge in aqueous solution, there will occur specific binding to the receptor and this will lead to an increase of the positive charge in the surface and eventually conductance will decrease for a *p*-type nanowire device. As shown in Fig. 20.1d, a very flexible and reliable integrated nanowire sensor device has been developed, incorporating a Si nanowire with well-defined *n*- or *p*-type doping, insulated source-drain electrodes, from the environment, and a microfluidic device for delivery of solutions to be examined. Source and drain electrodes are well insulated from the environment. In order, electrical signals are contributed only by processes occurring at the Si nanowire surface, and source and drain electrodes are well insulated from the environment.

## Nanowire in Diagnostic

For the direct electrical ultrasensitive detection of biological and chemical species, devices based on nanowires are providing powerful general platforms [1, 4]. A microfluidic channel maybe uses to lay down nanowires (Fig. 20.2), and when the particles flow through the microfluidic channel, the nanowire sensors will pick up the molecular signatures of these particles and transmit the data to a signal analyzer.

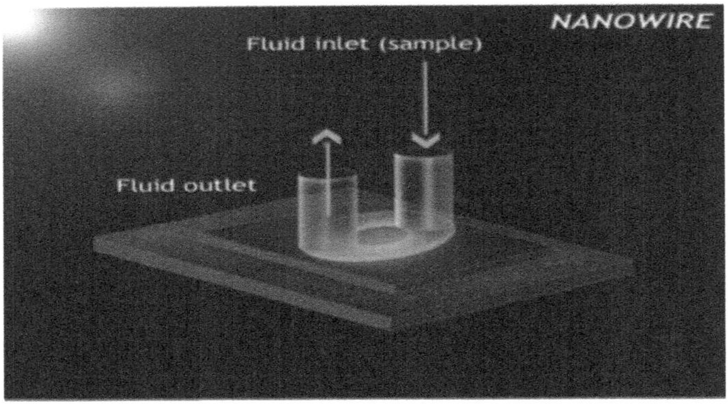


**Fig. 20.2** Schematic of a regular planar nanowire sensor biochip with integrated microfluidic sample delivery. The schematic represents the a-Si nanowire-based device configured as a sensor potentially covered with receptors and where binding of ligands yields a decrease in the conductance [1]

Such sensitive systems can detect the presence of altered genes related to the disease and can help researchers to identify the position of these genetic changes [1, 66]. Zheng et al. had reported that a silicon nanowire (SiNW) biosensor array has been prepared for the simultaneous detection of multiple cancer biomarkers in a very single versatile detection platform [1, 67].

Using silicon nanowire biosensors functionalized with three cognate antibodies, the real-time detection of three cancer markers namely prostate-specific antigen, carcinoembryonic antigen, and mucin-1 was demonstrated [1, 68]. The early detection of cancer could also be facilitated by the simultaneous high-sensitivity analysis of multiple biomarkers [1, 69, 70].

The synthesis of aligned ZnO nanowire arrays employing a vapor–solid process has been described in their study. The expansion of aligned arrays of nanowires plays a very important role in nanobiotechnology. It can be used for cellular manipulation, biosensing, and also the conversion of energy into electricity for powering nanodevices. By incorporating distinct nanowires and surface receptors into the array, a silicon nanowire field-effect devices were developed [1, 67]. Silicon nanowire probes have the capacities for multiplexed real-time monitoring of protein markers in relevant clinical samples with high selectivity and sensitivity. These capabilities offer the potential for diagnosis and treatment of cancers [1, 70]. Distinctive spectroscopic and electrical properties including strong Raman scattering and near-infrared photoluminescence, which will be useful in imaging application and biological detection, are exhibited by single-walled carbon nanotubes (SWCNTs) [1, 71].

The silicon sensor, one of growing nanotechnology-based promising diagnostic tools, represents the first example to detect DNA directly electrically using nanotechnology. Studies demonstrating the capability of the new sensor, the researchers showed that more efficiently than conventional tests, the silicon sensor has the potential to detect the gene for cystic fibrosis [72].

Charles M. Lieber, Ph.D., a professor of chemistry at Harvard and one of the leading researchers in nanotechnology, says that for medical diagnostics, this tiny sensor could represent a new future. He further states that the benefits and comfort of the new technology will be that by just visiting the doctor's office and giving a drop of blood by pinpricking on the finger, and within minutes, one can find out drug interactions, the presence of a genetic disease, particular virus, or risk for different diseases. Having very high sensitivity, the sensor may be able to detect diseases that were never possible before with conventional tests. If all goes well in future studies, he predicts that ultimately, an array of sensors can be configured to a small computer or handheld PDA-type device, which will allow almost instant test results during a doctor's visit or possibly even by the patient at home. Potentially, it could be used to screen for disease markers in any bodily fluid including urine, saliva, and tears [72]. The researcher also claimed that the sensor also shows promise for early detection of bioterrorism threats such as viruses. An experimental version of the technology has also been developed which consists of a thin plate about the size of a small business card containing the tiny nanowire sensor. A working prototype device suitable for testing of body fluids and human blood could be five years away, Lieber estimates [72].

In an approach to demonstrate the effectiveness of the sensor device, the researches grafted nucleic acids, the building blocks of DNA, to a silicon nanowire. The nucleic acid is designed specifically to recognize a particular mutation site in the cystic fibrosis (CF) gene that is responsible for most fatal cases of the genetic disease. Then the researchers exposed the nanowire to fragments of the CF gene with a lethal mutation and to fragment of CF gene some without it. The researcher found that successfully, they could be distinguished between the two types of gene fragments, even down to an extremely low level that would have been missed by conventional DNA sensors, according to Lieber [72]. The conventional DNA detection method requires a complex procedure called PCR amplification to view the results. Unlike conventional DNA detection methods, the nanowire sensor does not need such sophisticated and expensive techniques, which could speed up genetic testing ultimately while reducing costs [72].

## Conclusion

Several researches in the field of designing diagnostic and therapeutic tools and devices have been carried out, and nanostructures and nanotechnology-based devices are under active development. The size of nanoparticles is in order of 1-100 nm and can be functionalized to show specific properties at atomic, molecular, and cellular levels. The field has been through continual changes due to rapid innovations and improvements. The field of nanomedicine, which has the potential to have a major impact on human health, has been defined by the use of nanotechnology in biomedical research and clinical practice.

Diagnostic techniques, based on nanotechnology, currently under development may provide two major advantages:

- With the possibility of rapid testing, potentially in a doctor's office, will allow complete diagnosis and start of treatment within one visit to the doctor.
- Possibility of the detection of diseases at an earlier stage than possible with current techniques offers the potential of stopping a disease as earlier as possible, possibly with less damage to the patient.

In the near future, diagnostics and medical approach will be dramatically changed. Nanoscale devices and sensors may provide a personal and cost-effective continuous medical monitoring of the patient's health. Nanoscale devices and sensors may also support an advanced prediction in the early stage of diseases such as inflammations or cancer. In the future, sensor systems may be able to use multiple physical phenomena to detect many biomarkers simultaneously. A tiny sensitive nanowire sensor—thinner than the width of a human hair, 1000 times more sensitive than current DNA tests, with capabilities of producing results in minutes rather than days or weeks—could pave the way for more accurate and faster medical diagnostic tests for countless conditions and may ultimately save lives by allowing earlier disease detection and intervention.

## References


1. Laroui, H., Rakhya Bo Xiao, P., Viennois, E., Merlin, D.: Nanotechnology in diagnostics and therapeutics for gastrointestinal disorders digestive and liver disease. Off. J. Ital. Soc. Gastroenterol. Ital. Assoc. Study Liver **45** (2013). 10.1016/j.dld.2013.03.019
2. Cai, W., Gao, T., Hong, H., Sun, J.: Applications of Gold Nanoparticles in Cancer Nanotechnology Nanotechnology, Science, and Applications (2008) 10.2147/NSA.S3788.NNP
3. Patolsky, F., Lieber, C.: Nanowire nanosensors. Mater. Today Mater Today **8**, 20–28 (2005). https://doi.org/10.1016/S1369-7021(05)00791-1
4. Jabir, N.R., Tabrez, S., Ashraf, G.M., Shakil, S., et al.: Nanotechnology-based approaches in anticancer research. Int J Nanomed. **7**, 4391–4408 (2012)
5. Bloomington, (ed.).: An E World of Nanobioengineering: Potential Big Ideas for the Future. AuthorHouse (2010)
6. Balasubramanian, K., Burghard, M.: Biosensors based on carbon nanotubes. Anal Bioanal Chem. **385**, 452–468 (2006) ([PubMed: 16568294])
7. Gruner, G.: Carbon nanotube transistors for biosensing applications. Anal. Bioanal. Chem. **384**, 322–335 (2006) ([PubMed: 16132132])
8. Laroui, H., Dalmasso, G., Nguyen, H.T., Yan, Y., et al.: Drug-loaded nanoparticles targeted to the colon with polysaccharide hydrogel reduce colitis in a mouse model. Gastroenterology. **138**, 843, 53 e1–2 (2010). [PubMed: 19909746]
9. Laroui, H., Gross, L., Leonard, M., Stoltz, J.F., et al.: Hyaluronate-covered nanoparticles for the therapeutic targeting of cartilage. Biomacromolecule **8**, 3879–3885 (2007) ([PubMed: 18039001])
10. Laroui, H., Theiss, A.L., Yan, Y., Dalmasso, G., et al.: Functional TNFalpha gene silencing mediated by polyethyleneimine/TNFalpha siRNA nanocomplexes in the inflamed colon. Biomaterials **32**, 1218–1228 (2011) ([PubMed: 20970849])

11. Theiss, A.L., Laroui, H., Obertone, T.S., Chowdhury, I., et al.: Nanoparticle-based therapeutic delivery of prohibitin to the colonic epithelial cells ameliorates acute murine colitis. Inflamm. Bowel Dis. **17**, 1163–1176 (2011) ([PubMed: 20872832])
12. Zhao, Z.X., Gao, S.Y., Wang, J.C., Chen, C.J., et al.: Self-assembly nanomicelles based on cationic mPEGPLA-b-Polyarginine(R15) triblock copolymer for siRNA delivery. Biomaterials **33**, 6793–6807 (2012) ([PubMed: 22721724])
13. Shi, S., Zhu, X., Guo, Q., Wang, Y., et al.: Self-assembled mPEG-PCL-g-PEI micelles for simultaneous codelivery of chemotherapeutic drugs and DNA: synthesis and characterization in vitro. Int. J. Nanomed. **7**, 1749–1759 (2012) ([PubMed: 22619525])
14. Acharya, G., Lee, C.H., Lee, Y.: Optimization of cardiovascular stent against restenosis: factorial design-based statistical analysis of polymer coating conditions. PLoS ONE **7**, e43100 (2012) ([PubMed: 22937015])
15. Ma, X., Sim, S.J.: Femtomolar detection of single mismatches by discriminant analysis of DNA hybridization events using gold nanoparticles. Analyst (2013)
16. Mancuso, M., Jiang, L., Cesarman, E., Erickson, D.: Multiplexed colorimetric detection of Kaposi's sarcoma associated herpesvirus and Bartonella DNA using gold and silver nanoparticles. Nanoscale **5**, 1678–1686 (2013) ([PubMed: 23340972])
17. Zhu, Y., Chandra, P., Shim, Y.B.: Ultrasensitive and selective electrochemical diagnosis of breast cancer based on a hydrazine-au nanoparticle-aptamer bioconjugate. Anal. Chem. **85**, 1058–1064 (2013) ([PubMed: 23215018])
18. Sensenig, R., Sapir, Y., MacDonald, C., Cohen, S., et al.: Magnetic nanoparticle-based approaches to locally target therapy and enhance tissue regeneration in vivo. Nanomedicine (Lond.) **7**, 1425–1442 (2012) ([PubMed: 22994959])
19. Hsieh, W.J., Liang, C.J., Chieh, J.J., Wang, S.H., et al.: In vivo tumor targeting and imaging with antivascular endothelial growth factor antibody-conjugated dextran-coated iron oxide nanoparticles. Int. J. Nanomed. **7**, 2833–2842 (2012) ([PubMed: 22745546])
20. Fan, Z., Senapati, D., Singh, A.K., Ray, P.C.: Theranostic magnetic core-plasmonic shell star shape nanoparticle for the isolation of targeted rare tumor cells from whole blood, fluorescence imaging, and photothermal destruction of cancer. Molec. Pharm. (2012)
21. Ma, Q., Nakane, Y., Mori, Y., Hasegawa, M., et al.: Multilayered, core/shell nanoprobes based on magnetic ferric oxide particles and quantum dots for multimodality imaging of breast cancer tumors. Biomaterials **33**, 8486–8494 (2012)
22. Poselt, E., Schmidtke, C., Fischer, S., Peldschus, K., et al.: Tailor-made quantum dot and iron oxide-based contrast agents for in vitro and in vivo tumor imaging. ACS Nano **6**, 3346–3355 (2012)
23. Erogbogbo, F., Yong, K.T., Roy, I., Hu, R., et al.: In vivo targeted cancer imaging, sentinel lymph node mapping, and multi-channel imaging with biocompatible silicon nanocrystals. ACS Nano **5**, 413–423 (2011)
24. Kambhampati, S.P., Kannan, R.M.: Dendrimer nanoparticles for ocular drug delivery. J. Ocul. Pharmacol. Ther. (2013)
25. Gras, R., Relloso, M., Garcia, M.I., de la Mata, F.J., et al.: The inhibition of Th17 immune response in vitro and in vivo by the carbosilane dendrimer 2G-NN16. Biomaterials **33**, 4002–4009 (2012)
26. Fleming, C.J., Yin, N.N., Riechers, S.L., Chu, G., et al.: High-resolution imaging of the intramolecular structure of indomethacin-carrying dendrimers by scanning tunneling microscopy. ACS Nano **5**, 1685–1692 (2011)
27. https://www.dummies.com/education/science/nanotechnology/diagnostic-nano-sensors-using-nanowire/
28. Niemeyer, C.M.: Angew. Chem. Intl. Ed. **40**(22), 4128 (2001)
29. Chan, W.C.W., et al.: Curr. Opin. Biotechnol. **13**(1), 40 (2002)
30. West, J.L., Halas, N.J.: Annu. Rev. Biomed. Eng. **5**, 285 (2003)
31. Alivisatos, P.: Nat. Biotechnol. **22**, 47 (2004)
32. Gould, P.: Mater. Today **7**(2), 36 (2004)
33. Cui, Y., et al.: Science **293**, 1289 (2001a)

34. Comini, E., et al.: Appl. Phys. Lett. **81**(10), 1869 (2002)
35. Zhou, H.T., et al.: Chem. Phys. Lett. **369**(1–2), 220 (2003)
36. Li, C., et al.: Appl. Phys. Lett. **82**(10), 1613 (2003)
37. Hahm, J., Lieber, C.M.: Nano Lett. **4**(1), 51 (2004)
38. Wang, W., et al.: Proc. Natl. Acad. Sci. USA **102**, 3208 (2005)
39. Patolsky, F., et al.: Proc. Natl. Acad. Sci. USA **101**, 14017 (2004)
40. Kolmakov, A., Moskovits, M.: Ann. Rev. Mater. Res. **34**, 151 (2004)
41. Wan, Q., et al.: Appl. Phys. Lett. **84**(18), 3654 (2004)
42. Sze, S.M.: Physics of Semicondutor Devices, p. 431. Wiley, New York (1981)
43. Bergveld, P.: IEEE Trans. Biomed. Eng. **19**, 342 (1972)
44. Blackburn, G.F.: In: Turner, A.P.F., et al. (eds.) Biosensors: Fundamentals and Applications, vol. 481. Oxford University Press, Oxford (1987)
45. Hafeman, D.G., et al.: Science **240**, 1182 (1988)
46. Cui, Y., et al.: J. Phys. Chem. B **104**(22), 5213 (2000)
47. Duan, Y., et al.: Nature **409**, 66 (2001)
48. Cui, Y., Lieber, C.M.: Science **291**, 851 (2001)
49. Lauhon, L.J., et al.: Nature **420**, 57 (2002)
50. Cui, Y., et al.: Nano Lett. **3**(2), 149 (2003)
51. McAlpine, M.C., et al.: Nano Lett. **3**(11), 1531 (2003)
52. Zhong, Z., et al.: Science **302**, 1377 (2003)
53. Panev, N., et al.: Appl. Phys. Lett. **83**(11), 2238 (2003)
54. Jin, S., et al.: Nano Lett. **4**(5), 915 (2004)
55. Greytak, A.B., et al.: Appl. Phys. Lett. **84**(21), 4176 (2004)
56. Wu, Y., et al.: Nature **430**, 61 (2004a)
57. Zheng, G., et al.: Adv. Mater. **16**(21), 1890 (2004)
58. Morales, A.M., Lieber, C.M.: Science **279**, 208 (1998)
59. Hu, J., et al.: Acc. Chem. Res. **32**(5), 435 (1999)
60. Lieber, C.M.: MRS Bull. **28**(7), 486 (2003)
61. Cui, Y., et al.: In: Wang, Z.L., (ed.) Nanowires and Nanobelts—Materials, Properties and Devices, vol. 3. Kluwer Academic Publishers (2003)
62. Cui, Y., et al.: Appl. Phys. Lett. **78**(15), 2214 (2001b)
63. Wu, Y., et al.: Nano Lett. **4**(3), 433 (2004b)
64. Iler, R.K.: The Chemistry of Silica. Wiley, New York (1979)
65. Bartlett, P.N., In: Taylor, R.F., Schultz, J.S. (eds.) Handbook of Chemical and Biological Sensors, vol. 139. IOP Publishing, Philadelphia (1996)
66. Madani, S.Y., Naderi, N., Dissanayake, O., Tan, A., et al.: A new era of cancer treatment: carbon nanotubes as drug delivery tools. Int. J. Nanomed. **6**, 2963–2979 (2011)
67. Zheng, G., Patolsky, F., Cui, Y., Wang, W.U., et al.: Multiplexed electrical detection of cancer markers with nanowire sensor arrays. Nat. Biotechnol. **23**, 1294–1301 (2005)
68. Zhang, G.J., Ning, Y.: Silicon nanowire biosensor and its applications in disease diagnostics: a review. Anal Chim. Acta. **749**, 1–15 (2012)
69. Etzioni, R., Urban, N., Ramsey, S., McIntosh, M., et al.: The case for early detection. Nat. Rev Cancer **3**, 243–252 (2003)
70. Stern, E., Vacic, A., Rajan, N.K., Criscione, J.M., et al.: Label-free biomarker detection from whole blood. Nat. Nanotechnol. **5**, 138–142 (2010)
71. Sun, Y.P., Fu, K.F., Lin, Y., Huang, W.J.: Functionalized carbon nanotubes: properties and applications. Acc. Chem Res. **35**, 1096–1104 (2002)
72. American Chemical Society.: Tiny nanowire could be next big diagnostic tool for doctors. Science Daily (2003). Retrieved April 2, 2020 from www.sciencedaily.com/releases/2003/12/031217072033.htm

# Chapter 21
# Ternary Metal Oxides/Graphene Hybrids for Lithium-Ion Batteries

**Sumitra Nongthombam and Bibhu Prasad Swain**

## Introduction

Energy has become the greatest concern globally owing to the continuous exploitation of fossil fuels in order to meet the increasing demands of the energy of our ever-growing population, which on the other hand, has been imposing a great threat to the environment causing pollution, whereas there is rapidly increased harnessing of renewable energy sources from wind energy and solar energy, the intermittent nature of these energy sources urges for the development of energy storage devices as an equally important one. Lithium-ion batteries are rechargeable clean energy storage devices that can store energy from these renewable energy sources and reversibly generate energy to ground maintaining sustainability and environmental friendliness. These rechargeable batteries have become indispensable in the day-to-day mobile technology advancing in portable electronics—mobile phones, laptops, pacemakers, cameras, etc. because of its significant characteristics such as high energy density, long cycle life, no memory effect, and stable cyclicity [1]. Nevertheless, the extended demand for use in electric vehicles (EV) and hybrid electric vehicles (HEV) still necessitates improved capacity, rate capability, high energy density and power density, durability with increased safety and effective cost [2]. The rechargeable lithium-ion battery was first commercialized by Sony and Asahi Kasei team in 1991 after Akira Yoshino practically developed a prototype lithium-ion battery in 1985 based on earlier John B. Goodenough and Stanley Whittingham theories. After this, LIBs have become undisputedly popular with millions of research articles being published on the development of high-performance LIBs. LIBs have currently

S. Nongthombam (✉) · B. P. Swain
Department of Physics, National Institute of Technology Manipur, Imphal, Manipur 795004, India
e-mail: sumitranongs@gmail.com

B. P. Swain
e-mail: bibhuprasad.swain@gmail.com



B. P. Swain (ed.), *Nanostructured Materials and their Applications*, Materials Horizons: From Nature to Nanomaterials, https://doi.org/10.1007/978-981-15-8307-0_21

impacted the world tremendously toward a new rechargeable world, that in 2019, Akira Yoshino, John B. Goodenough. and Stanley Whittingham received the Nobel prize together for the development of LIBs. The commercial lithium-ion battery basically consists of two electrode materials—lithium cobaltite ($LiCoO_2$) as the cathode and graphitic anode dipped in an organic electrolyte containing lithium ions as a medium for the exchange of lithium ions, and a separator to insulate the electrical contact of the two electrodes. Electrode materials utilized in LIBs play a crucial role in the development of high-performance LIBs. A series of research studies have been devoted to the exploration of new cathodic and anodic materials to stimulate the efficacy of LIBs. In the quest of advance anode materials for LIBs, metal oxides are widely investigated because of their high theoretical specific capacity (>700 mAh $g^{-1}$) and easy maintenance with greater lithium incorporation capability as alternatives to graphite which has the comparatively less theoretical capacity (372 mAh $g^{-1}$) [3]. However, their advantages are still plagued by large volume change and agglomeration leading to capacity fading and also suffer from poor electrical conductivity. In this regard, metal oxides are incorporated with graphene to mitigate the problems addressed as graphene is an excellent candidate fascinated with its high electrical conductivity, large surface area, flexibility and chemical stability for energy storage applications. Graphene-based metal oxides nanocomposites received substantial attention owing to their synergistic effect enhancing the performance which is reviewed in Sun et al. and Kucinskis et al. [1, 4]. Although the problems of electrical conduction are improved, electrode pulverization still cannot be avoided. With the introduction of two kinds of metal oxides, that is, ternary metal oxides/graphene nanocomposites offer huge advancements in LIBs as anode materials lately. Recent advances in the LIBs energy storage applications encompass binary/ternary metal oxides/graphene nanocomposites as anodic material. While electrolytes, conducting agents, separators have specific roles in battery performance, the present book chapter is focused on the exploration of potential anode materials of LIBs.

## Lithium-Ion Batteries

Lithium-ion batteries are rechargeable batteries based on the conduction of electricity through the transport of lithium ions back and forth between the positive electrode and negative electrode. These batteries not only serve as the basis for the future fossil-fuel-free society but also provide large energy storage tools for renewable energy sources, maintaining eco-friendliness. So, for the generation of electricity, anode materials having a high capacity to release electrons is a requisite. Lithium, being the most electropositive metal, has the greatest tendency to lose electrons and since it is the lightest metal, it offers the advantage for realizing light, wearable rechargeable batteries. It consists of three main components—electrodes, electrolyte, and separator. The basic working of LIBs is illustrated based on the first commercialized battery by Sony as depicted in Fig. 21.1.

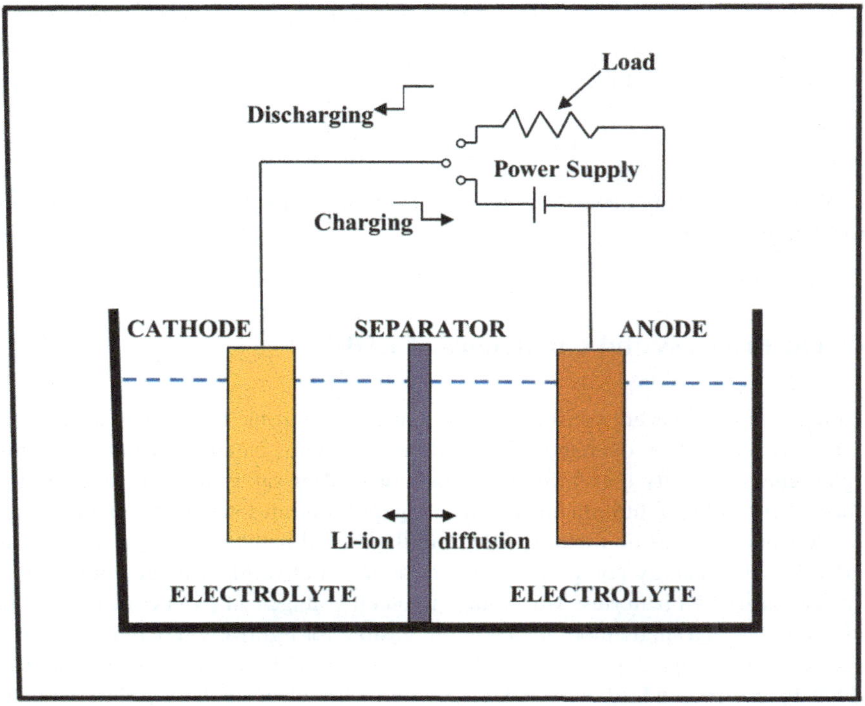


**Fig. 21.1** Electrochemical cell of a Li-ion battery

Lithium cobaltite ($LiCoO_2$) and graphite serve as the cathode and anode, respectively, dipped in an organic electrolyte composing Li-ions which are separated by a permeable membrane called separator which allows only the flow of Li-ions and not electrons and can insulate the electrical contact of the two electrodes.

In charging mode, when a power supply is connected externally to the electrode, the positive side of the external source will attract the electrons detached from the lithium metal in the compound and pass to the anode while lithium ions on the other hand also flow simultaneously through the electrolyte and gather the graphite sheet. Here, electrolyte acts as a medium for the flow of lithium ions and block the flow of electrons. Both the lithium ions and electrons combine together at the cathode and are intercalated between the graphite layers. Again, during discharging, the power supply is removed and connected through the load. Lithium being unstable on its own traced back to its form of a compound internally while the electrons flow back in the opposite direction (from anode to cathode) externally. In this manner, electrical energy is stored electrochemically in the electrodes in the form of chemical energy. The electrochemical reactions occurring at the electrodes are summarized as follows:

The half-reaction at the anode is

$$LiC_6 \leftrightarrow C_6 + Li^+ + e^-$$

And, half-reaction at the cathode is as follows:

$$CoO_2 + Li^+ + e^- \leftrightarrow LiCoO_2$$

When the charging occurs, the reverse reaction occurs where the electrons flow from cathode to anode.

## Recent Progress and Challenges of LIBs

Lithium-ion batteries are constantly developing in electronic technology as it offers numerous advantages over the other batteries. Lithium batteries although having higher energy density than lithium-ion batteries suffer safety hazards utilizing the inherent instability of lithium metal. Advantages of lithium-ion batteries include easy maintenance as it has no memory effect unlike Ni–Cd batteries, higher gravimetric and volumetric energy compared to other batteries delivering voltage almost three times in case of Ni batteries. This high single-cell voltage can reduce the number of cells required in a single module making it feasible for lightweight battery products. Also, energy storage depends on the lithiation process in contrast to other batteries which in case degrade the electrodes promotes the clean rechargeable reversible battery. Despite the advantages of LIBs, safety issues and high cost per unit energy stored are two critical factors that need to be taken into account most particularly, in electric vehicles (EV) and hybrid electric vehicles (HEV) because a large number of cells are required to be packed in a single module for these applications. The commercial LIBs employing lithium compound and graphite as cathode and anode suffer certain drawbacks. LIBs are usually packed with electronic protective circuits to avoid thermal runaway so as to maintain safe operation. However, this protective circuit limits the peak voltage of each cell during charge/discharge lowering the energy density of the battery. In addition, the cell temperature is monitored to prevent extreme temperatures as at high internal temperatures; otherwise, the liquid electrolyte may dry up and short-circuited leading to fire/explosion. For utilization in EV and HEV, battery capacity, cycle life, energy density, and power density still need improvement. As the parameters of battery performance such as cyclability, safety, charge/discharge rate, and specific energy are largely dependent on electrode materials, it is a prerequisite to explore and develop advanced cathodic and anodic materials. For anode materials, tin, silicon, germanium, sulfides, nitrides and metal oxide-based materials are developed as an alternative to commercial carbon anodes owing to their higher theoretical capacities in comparison to graphite and greater Li incorporation capability [5]. However, they face capacity fading and electrode pulverization caused by large volume alteration during Li insertion/extraction process, aggregation, and poor electrical conductivity. Binary hybrids have been investigated as an advancement to mitigate these issues and enhance battery performance as reviewed in

Deng's work [6]. Similarly, graphene is studied as an excellent additive for cathode materials due to its various desirable properties. Although $LiMPO_4$ (M=Fe, Mn, Co, Vn) are extensively investigated as alternatives to the commercial cathode, their poor electrical conductivity limits the rate capability of batteries. This encourages the introduction of carbonaceous materials, particularly graphene to obtain graphene-based composites cathodes. Other advancements in graphene-based composites cathodes include lithium metal oxide/graphene composites, sulfur/graphene composites, metal oxide/graphene composites [7]. Another notable challenge is in the replacement of organic electrolytes used in commercial LIBs. Organic electrolytes are highly flammable and volatile by which its damage can cause fire/explosion. In the last few decades, numerous studies on the structural characteristics and modifications of electrolytes along with alternatives to organic electrolyte—aqueous electrolytes, ceramic electrolytes, solid electrolytes, polymer electrolytes, and heavily fluorinated systems have been proposed for better safety and enhanced performance [8, 9].

## Overview of Graphene-Based Anode Materials

Graphene is a promising material possessing excellent physical and chemical properties that are inherently multifunctional. In recent years where graphene has been widely explored for numerous applications including energy storage applications like supercapacitors, likewise, in lithium-ion batteries, graphene-based composites anodes are extensively studied as anode materials. Among the graphene-based composites, graphene/metal oxides composites have drawn remarkable consideration as the hybrid material yields enhanced battery performance than their individual counterparts. In LIBs, graphene acts as an excellent substrate to host active materials used in Li insertion/extraction processes nanomaterials for energy applications because of its large surface area, flexibility, and exceptionally high electrical conductivity.

### *Fe-Based Oxides/Graphene Hybrids*

$Fe_3O_4$ is an appealing candidate owing to its prominent features such as high specific capacity (580 mAh $g^{-1}$), low cost, natural abundance, and eco-friendliness [10]. Notwithstanding, high-capacity LIBs based on anodic $Fe_3O_4$ are hindered with the issues of large volume change and agglomeration causing electrode pulverization leading to the rapid fading of capacity after repeated charging/discharging. Strategies that have been proposed to mitigate the electrode pulverization and specific capacity of electrode materials include designing nanostructured materials, porosity control, designing unique configuration, and a combination of iron oxides with carbonaceous materials [11]. Addressing this strategy can be expected to increase the good cyclic performance as well as to maintain high structural stability. Zhou et. al synthesized

$Fe_3O_4$/graphene nanocomposite by an in situ reduction method which delivers the specific capacity to nearly 1026 mAh $g^{-1}$ and 580 mAh $g^{-1}$ at the current density of 35 and 700 mA $g^{-1}$ after 30 cycles and 100 cycles, respectively, with enhanced cyclability and high rate capability compared to bare $Fe_3O_4$ particles when used as anode materials for LIBs [12]. Fu et al. demonstrated a solvothermal prepared $Fe_3O_4$/graphene composite for LIB applications which exhibited a reversible specific capacity above 963 mAh $g^{-1}$ after 100 cycles at 100 mA $g^{-1}$ current density and reported capacity retentivity after 50 cycles at increasing current densities [13].

The importance of graphene incorporation with metal oxides in LIBs can be stated pointwise.

(1) Graphene nanosheets (GNSs) possess porous structure which aids to act as a flexible confinement matrix providing remarkable buffering capacity. This can reduce electrode pulverization.
(2) GNSs act as a conductive media as it possesses high electrical conductivity and can facilitate ion transport.

On the other hand, $Fe_3O_4$ prevents the restacking of GNSs, and the porosity formed by lateral GNSs and $Fe_3O_4$ improves ion transportation. With this, the composite offers great advantages to enhance the lithium storage capacity, cyclic stability, and rate capability.

## *Mn-Based Oxides/graphene Hybrids*

Among the transition metal oxides that can be pursued as an alternative to the commercial graphite as anode materials for LIBs, $Mn_3O_4$ is a potential material because of its natural abundance, low cost, high theoretical specific capacity (936 mAh $g^{-1}$) and non-toxicity. Nevertheless, it has very low electrical conductivity (~$10^{-7}$ to $10^{-10}$ S/cm) and also practically, limited specific capacity can be realized much below the theoretical capacity. Wang et al. reported a two-step synthesis of $Mn_3O_4$/graphene hybrids affording a high specific capacity of 900 mAh $g^{-1}$ approaching the theoretical capacity as well as cycling stability and good rate capability. The unique performance of the insulating $Mn_3O_4$ nanoparticles (NPs) grown on the graphene sheets can be ascribed to the well dispersion of $Mn_3O_4$ NPs over graphene sheets preventing agglomeration [14]. Another form of appealing manganese oxide is manganese dioxide ($MnO_2$) which has a high theoretical specific capacity (1230 mAh $g^{-1}$), low cost, natural abundance and eco-benignity. However, low electrical conductivity ($10^{-5}$ to $10^{-6}$ S/cm), capacity loss due to huge volume alteration and aggregation hinders the wide utilization of $MnO_2$-based LIBs. Xu et al. discussed various approaches to enhance the $MnO_2$-based LIBs that have been made by several researchers (1) to design nanostructured materials, (2) to incorporate with carbonaceous materials and (3) to improve the interfacing interactions between the nanostructured materials and the conductive substrate [15]. A facile synthesis of nanostructured $MnO_2$ grown over graphene sheets forming interfacial interactions is able to achieve enhanced electrochemical performance with the boosted capacity of 236 mAh $g^{-1}$ which is nearly double times to that of the capacity of bare MnO2 [16]. Another unique hierarchical structure of $MnO_2$ wrapped between graphene and graphene nanoribbons is demonstrated in Li et al. work reporting well-improved cycling stability, rate capability and reversible capacity compared to individual $MnO_2$, $MnO_2$/graphene-based anodes [17].

## *Sn-Based Oxides/Graphene Hybrids*

Sn- and $SnO_2$-based hybrids are one of the most commonly studied areas as anode materials in LIBs and its developments as anode materials in LIBs are being reviewed in Deng et al. works [18]. In spite of the low cost, eco-friendly, exceptionally high theoretical capacity (782 mAh $g^{-1}$), and lowest operating voltages compared to other transition metal oxides makes it a promising material for energy storage applications. However, high-capacity $SnO_2$-based LIBs cannot be realized practically owing to three major problems: (1) large volume variation (~300%) during lithiation/delithiation processes causing electrode pulverization which in turn leads to capacity fading [19]; (2) aggregation of $SnO_2$ phases; (3) formation of a very thick solid interface layer (SEI). To address this issue, $SnO_2$/graphene hybrids are investigated as anode materials for LIBs to enhance electrochemical performance [20]. Improved results of reversible capacity, enhance cyclicity and rate capability are received when $SnO_2$ is hybridized with graphene in comparison to pure $SnO_2$ [20, 21].

## *Ti-Based Oxides/graphene Hybrids*

$TiO_2$, being a low cost, naturally abundant, and non-toxic material is particularly interested to be used as anode materials in LIBs. Although the theoretical capacity of $TiO_2$ is comparatively low (336 mAh $g^{-1}$) to other metal oxides, the most interesting feature of $TiO_2$ is its ability to maintain structural stability during lithium intercalation which makes it feasible for use as anodes in LIBs [22]. However, the extremely low electrical conductivity ($10^{-16}$ S/cm) limits its rate capability which encourages to combine $TiO_2$ with a highly conductive material such as graphene to improve the $TiO_2$-based anodes for LIBs applications [23]. The $TiO_2$/graphene composite developed by Xin et al. is able to obtain a capacity of 230 mAh $g^{-1}$ at 17 mA $g^{-1}$ current density and superior rate capability. The hybridization of $TiO_2$ with graphene can simultaneously avoid aggregation as well as boosted the conductivity [24]. In another work, a hydrothermally prepared robust $TiO_2$/graphene by Qiu et al. deliver a discharge capacity of 141.7 mAh $g^{-1}$ at 5000 mA $g^{-1}$ current density after 100 cycles without compromising the structural integrity during lithium intercalation. Another strategy to improve the mobility of Li-ion is proposed by Zhou et al. by introducing surface amorphization. A resilient hierarchical surface amorphized $TiO_2$/graphene composites are developed which exhibited a substantial capacity of 103 mAh $g^{-1}$ at 50 C, 9.3 times improvement to the capacity of $TiO_2$-based battery performance. The improved performance is attributed to the combining advantages of surface amorphization as well as graphene incorporation providing (1) large contact area, (2) rapid electronic/ionic diffusion channels and (3) formation of interfacial conductive layer between $TiO_2$ and graphene [25].

## Ternary Graphene-Based Metal/oxides Nanocomposites

Although graphene/metal oxides hybrids have received remarkable enhancement in electrochemical performance, they are still plagued by electrode pulverization. Recently, to address the problem, ternary graphene-based metal/metal oxides nanocomposites are investigated as new advancements to anode materials for next-generation LIBs. The contributions from each individual component aids in the enhancement of the battery performance of the ternary nanocomposites.

### *Ternary $SnO_2$/$TiO_2$/graphene Nanocomposites*

In the trend to explore ternary metal oxides nanocomposites, $SnO_2$/$TiO_2$ composite received substantial attention. This is because $SnO_2$/$TiO_2$ has the same lattice structure and the only slight difference in the lattice which can be easily made possible for the formation of lattice mismatch nanostructure. The induction of $TiO_2$ in the $SnO_2$/graphene composite can impact the electrochemical performance of LIBs. $TiO_2$ plays a vital role in maintaining the structural stability of the electrode during lithium ion insertion/extraction and can prevent the agglomeration of $SnO_2$ NPs. In addition, $SnO_2$/$TiO_2$/graphene nanocomposites can be expected to pursue a synergistic effect because of the following reasons:

(1) $SnO_2$, being high theoretical capacity and low discharge potential, is limited to high capacity, suffering from poor cycling performance and large irreversible capacity due to the dramatic volume change caused during the repeated lithium insertion/extraction process, which in turn leads to electrical conductivity loss and severe electrode pulverization.
(2) Graphene, excellent conductive material with a large surface area when introduced hybridization with $SnO_2$, can buffer large volume change and provide fast electronic transport. However, cracking and pulverization of the electrode cannot be avoided.
(3) To address the above problems, combining with other metal oxides are encouraged and evidenced better performance of Li-ion as it can prevent agglomeration

of $SnO_2$ phases with the existence of extra oxides. $TiO_2$ has very low volume expansion (~4%) which can act as stable barriers preventing agglomeration of $SnO_2$ NPs during the charge/discharge cycle.

Jiao et al. demonstrated a two-step preparation of $SnO_2/TiO_2$/graphene nanocomposites and attain a high capacity of 1276 mAh $g^{-1}$ after 200 cycles at the current density of 200 mA $g^{-1}$. Also, the composite acquires a specific capacity of 611 mAh $g^{-1}$ at an ultrahigh current density of 2000 mA $g^{-1}$ superior to that of $SnO_2$ and $SnO_2$/graphene hybrids [26]. In another report by Jiang et al., $SnO_2/TiO_2$/graphene nanocomposites can obtain good cycling performance of 573 mAh $g^{-1}$ at 50 mA $g^{-1}$, good reversible capacities of 250 mAh $g^{-1}$ at 1000 mA $g^{-1}$ current density in the voltage range 0.01–3 V and coulombic efficiency of 97% after 50 cycles [27]. The overall performance of the hybrid nanocomposite portraits good cycle stability and excellent lithium storage capacity owing to the synergistic effect, large surface area, intergrown $SnO_2$–$TiO_2$ structure, and conductive graphene matrix.

## *Ternary $SnO_2/Fe_2O_3$/ Graphene Nanocomposites*

In preparation of ternary $SnO_2/Fe_2O_3$/graphene hybrids as anode materials, $Fe_2O_3$, $SnO_2$, and graphene have contemporary roles in the enhancement of overall electrochemical performance.

Graphene serves as a matrix for Li storage due to its high electrical conductivity, flexibility, and large surface area due to which Li storage capacity has increased, enhanced electronic transport, and an effective elastic buffer that can prevent agglomeration of metal oxide particles. However, one disadvantage of graphene is the restacking of its layers that lead to irreversible capacity loss.

$SnO_2$ has a high theoretical capacity, low working potential (~0.6 vs. $Li/Li^+$) which helps to improve the lithium storage capability and also prevents the restacking of graphene. One significant problem of $SnO_2$ is the large volume change during Li insertion/extraction leading to capacity loss. $SnO_2$ can also help in preventing the aggregation of $Fe_2O_3$ particles.

$Fe_2O_3$ has a high theoretical capacity of 1005 mAh $g^{-1}$, and its introduction in the composite can prevent aggregation of $SnO_2$ NPs which otherwise would cause structural degradation leading to capacity fading.

Xia et al. prepared $SnO_2/Fe_2O_3$/graphene nanocomposites where the novel composite attains initial charge and discharge capacities of 746 mAh $g^{-1}$ and 1179 mAh $g^{-1}$, respectively, at a current density of 400 mA $g^{-1}$ and nearly 100% charge/discharge efficiency are maintained during 100 cycles with capacities above 700 mAh $g^{-1}$ [28].

### *Ag/TiO$_2$/Graphene Nanocomposites*

$TiO_2$ is studied as anodic material for LIBs owing to its distinct characteristics such as low volume expansion (~4%) during Li-ion intercalation, low cost, high rate capability, and eco-friendliness. However, high performance in $TiO_2$-based LIBs is hindered by low theoretical specific capacity and poor ionic and electrical conductivity. To improve the capacity, some strategies that can be proposed to alleviate the above issues are as follows:

(1) To synthesize a nanostructured material so as to enhance the kinetics of the electrode.
(2) To combine $TiO_2$ with other metals/metal oxides to achieve high capacity.
(3) To incorporate with carbonaceous material as a conductive media to increase the conductivity.

Yu et al. demonstrated a solvothermal preparation of Ag/$TiO_2$/graphene nanocomposite which shows a good reversible capacity of 191.6 mAh $g^{-1}$ after 100 cycles at 1C, and cyclability of 144.2 mAh $g^{-1}$ after 3000 cycles at 5C with almost 100% coulombic efficiency [29].

The contribution from each individual in the achievements of the overall performance is

(1) rGO serves as a conductive substrate and can prevent the agglomeration of $TiO_2$ NPs.
(2) Ag NPs can increase intra- or intergrain connectivity.
(3) $TiO_2$ maintains the structural stability of the electrode during the charging/discharging process.

## Different Preparation Methods of Ternary Material Hybrids

### *SnO$_2$/TiO$_2$/Graphene Nanocomposites*

Jiao et al. proposed a hydrothermal method to synthesize ternary $SnO_2$/$TiO_2$/graphene nanocomposites. In a typical procedure, reduced graphene oxide (rGO) and ethanol are dispersed in water using an ultrasonicator. In the mixture, CTAB and HCl solution was added and stirred after which $SnCl_2.2H_2O$ and $TiOSO_4$ were introduced to the mixture and kept stirring under vigorous conditions. The whole mixture is subsequently transferred to a stainless steel reactor, and the reactor is maintained at 180 °C for a day. The resultant product is centrifuged and washed thoroughly with ethanol and water and finally dried at 60 °C. To prepare SnO2/$TiO_2$/graphene nanocomposite Jiang et al. used $SnCl_4.5H_2O$ as the source for Sn which is added to dispersed $TiO_2$ solution. The mixture is then transferred to an autoclave reactor and maintained the temperature at 180 °C for 12 h. The final precipitate is collected and rinsed properly with water and ethanol followed by overnight drying at 50 °C to obtain the ternary composite [26].

### *Ag/$TiO_2$/Graphene Nanocomposite*

In brief, a few mg of GO is taken and dispersed in DI water to form a uniform GO suspension. Two different solutions are made-in one beaker, $CH_3COOAg$ is dissolved in DI water, and in another beaker, SDS is mixed with KCl in the ratio (2:1). The former solution is initially added slowly to the GO solution and kept stirring for 12 h using a magnetic stirrer. Then, the later solution is added to the above mixture under continuous stirring for 0.5 h. Again, a mixture of TTIP in ethanol is poured into the mixture and kept stirring for 0.5 h. The obtained mixture is shifted to a hydrothermal chamber and heated at 180 °C for around 10 h. A black composite is obtained then followed by the subsequent centrifugation and rinsing with DI water and dried overnight at 45 °C. Thus, the final Ag/$TiO_2$/graphene nanocomposite is obtained [29].

### *$Fe_2O_3$/$SnO_2$/Graphene Nanocomposite*

To prepare $Fe_2O_3$/$SnO_2$/graphene nanocomposite, initially, 0.8 g of $FeCl_3.H_2O$ and 1.5 g of urea are poured into dispersed GO solution and sonicated for 30 min. The whole mixture is then transferred to a hydrothermal chamber and reacted at 120 °C for 4 h. Subsequently, the obtained product is taken out in a beaker and added with $SnCl_2.H_2O$ to reduce GO by stirring using a magnetic stirrer. After the mixture is kept stirring for an hour, the reaction is stopped and left to cooled and filtrated. The obtained precipitate is rinsed several times with water and subsequently by ethanol and placed in an oven at 80 °C for some hours for drying. Finally, the product is annealed at 400 °C for an hour to enhance the crystallization of $Fe_2O_3$ and $SnO_2$ in the ternary composite [28].

## Conclusion

In summary, lithium-ion batteries serve as indispensable power sources in portable electronics and its applications are rapidly extending into transportation technology. In spite of the excellent characteristics of LIBs, applications in large-scale electric vehicles demand the improvement of battery in terms of high energy density, power density and capacity, coupling with safety maintenance and low cost. As electrode materials utilized in LIBs play a critical role in enhancing the performance of batteries, this chapter deals with the recent studies in the advancement of anodic materials. A brief understanding of LIBs-working, advantages, recent progress, and its challenges—is also summarized in this chapter. Although the other components of LIBs-electrolytes, separators, cathode materials, binders, protective circuits etc—are accountable for the better performance of LIBs, the present chapter focuses on ternary graphene/metal oxides hybrids as anode materials and discussed their electrochemical performance. Graphene-based ternary hybrids emerge as a potential candidate for next-generation LIBs exhibiting excellent rate capability, reversible capacity, and stable cycling owing to their synergistic effect between the metal oxides and graphene and also between the two kinds of metal oxides.

## References

1. Sun, W., Wang, Y.: Graphene-based nanocomposite anodes for lithium-ion batteries. Nanoscale. **6**(20), 11528–21152 (2014)
2. Lu, L., Han, X., Jianqiu, L., Li, J., Hua, J., Ouyang, M.: A review of the key issues for lithium-ion battery management in electric vehicles. J. Power Sources. **226**, 272–288 (2013)
3. Zheng, Z., Yi, Z., Zhang, Q., Cheng, Y., Zao, Y., Chen, H., Zhang, K., Wang, M.S., Peng, D.L.: Robust erythrocyte-like $Fe_2O_3$ @carbon with Yolk-shell structures as high-performance anode for lithium-ion batteries. Chem. Eng. J. **347**, 1385–8947 (2018)
4. Kucinskis, G., Bajars, G., Kleperis, J.: Graphene in lithium-ion battery cathode materials: a review. J. Power Sources **240**, 66–79 (2013)
5. Mauger, A., Xie, H., Julien, C.M.: Composite anodes for lithium-ion batteries: status and trends. AIMS Mater. Sci. **3**, 1054–1106 (2016)
6. Deng, D.: Li-ion batteries: Basics, progress, and challenges. Energy Sci Eng. **3**, 385–418 (2015)
7. Chen, L..., Zhang, M., Wei, W.: Graphene-based composites as cathode materials for lithium-ion batteries. J. Nanomater. 1–8 (2013)
8. Zhang, H., Zhao, H., Khan, M.A., Zou, W., Xu, J., Zhang, L., Zhang, J.: Recent progress in advanced electrode materials, separators and electrolytes for lithium batteries. J. Mater. Chem. A **6**, 20564–20620 (2018)
9. Amereller, M., Schedlbauer, T., Moosbauer, D., Schreiner, C., Stock, C., Wudy, F., Zugmann, S., Hammer, H., Maurer, A., Gschwind, R.M., Wiemhöfer, H.D., Winter, M., Gores, H.J.: Electrolytes for lithium and lithium ion batteries: from synthesis of novel lithium borates and ionic liquids to development of novel measurement methods. Prog. Solid. State Ch **42**, 39–56 (2014)
10. He, Y., Huang, L., Cai, J.S., Zheng, X.M., Sun, S.G.: Structure and electrochemical performance of nanostructured $Fe_3O_4$/carbon nanotube composites as anodes for lithium-ion batteries. Electrochim. Acta **55**, 1140–1144 (2010)
11. Yang, L., Hu, J., Dong, A., Yang, D.: Novel $Fe_3O_4$-CNTs nanocomposite for Li-ion batteries with enhanced electrochemical performance. Electrochim. Acta **144**, 235–242 (2014)
12. Zhou, G., Wang, D.W., Li, F., Zhang, L., Li, N., Wu, Z.S., Wen, L., Lu, G.Q., Cheng, H.M.: Graphene-wrapped $Fe_3O_4$ anode material with improved reversible capacity and cyclic stability for lithium-ion batteries. Chem. Mater. **22**, 5306–5313 (2010)
13. Fu, C., Zhao, G., Zhang, H., Li, S.: A Facile route to controllable synthesis of $Fe_3O_4$/graphene composites and their application in lithium-ion batteries. Int. J. Electrochem. Sci. **9**, 46–60 (2014)
14. Wang, H., Cui, L.F., Yang, Y., Sanchez, C.H., Robinson, J.T., Liang, Y., Cui, Y., Dai, H.: $Mn_3O_4$-graphene hybrid as a high-capacity anode material for lithium-ion batteries. J. Am. Chem. Soc. **132**, 13978–13980 (2010)
15. Xu, T., Meng, Q., Fan, Q., Yang, M., Zhi, W., Cao, B.: Electrophoretic deposition of binder-free $MnO_2$/graphene films for lithium-ion batteries. Chin. J. **35**, 1575–1585 (2017)

16. Zhao, Y., Chen, G., Song, C., Xie, L., Chang, L., Wang, R., Zhong, N.: Facile synthesis of $MnO_2$ nanoparticles well-dispersed on graphene for the enhanced electrochemical performance. Int. J. Electrochem. Sci. **11**, 2525–2533 (2016)
17. Li, L., Raji, A.R.O., Tour, J.M.: Graphene-wrapped $MnO_2$-graphene nanoribbons as anode materials for high-performance lithium-ion batteries. Adv. Mater. **25**, 6298–6302 (2013)
18. Deng, Y., Fang, C., Chen, G.: The developments of $SnO_2$/graphene nanocomposites as anode materials for high-performance lithium-ion batteries: a review. J. Power Sources **304**, 81–101 (2016)
19. Huang, X., Zhou, X., Zhou, L., Qian, K., Wang, Y., Liu, Z., Yu, C.: A facile one-step solvothermal synthesis of $SnO_2$/graphene nanocomposite and its application as an anode material for lithium-ion batteries. Chem. Phys. Chem **12**, 278–281 (2011)
20. Yang, S., Yue, W., Zhu, J., Ren, Y., Yang, X.: Graphene-based mesoporous $SnO_2$ with the enhanced electrochemical performance for lithium-ion batteries. Adv. Funct. Mater. **23**, 3570–3576 (2013)
21. Prabakar, S.J.R., Hwang, Y.H., Bae, E.G., Shim, S., Kim, D., Lah, M.S., Sohn, K.S., Pyo, M.: $SnO_2$/Graphene Composites with self-assembled alternating oxide and amine layers for high li-storage and excellent stability. Adv. Mater. **25**, 3307–3312 (2013)
22. Ren, H.M., Ding, Y.H., Chang, F.H., He, X., Feng, J.Q., Wang, C.F., Jiang, Y., Zhang, P.: Flexible free-standing $TiO_2$/graphene/PVdF films as anode materials for lithium-ion batteries. Appl. Surf. Sci. **263**, 54–57 (2012)
23. Xin, X., Zhou, X., Wu, J., Yao, X., Liu, Z.: Scalable synthesis of $TiO_2$/graphene nanostructured composite with the high-rate performance for lithium-ion batteries. ACS Nano **6**, 11035–11043 (2012)
24. Qiu, J., Lai, C., Wang, Y., Li, S., Zhang, S.: Resilient mesoporous $TiO_2$/graphene nanocomposite for high rate performance lithium-ion batteries. Chem. Eng. J. **256**, 247–254 (2014)
25. Zhou, T., Zheng, Y., Gao, H., Min, S., Li, S., Liu, H.K., Guo, Z.: Surface engineering and design strategy for surface-amorphized $TiO_2$@graphene hybrids for high power Li-ion battery electrodes. Adv Sci. **2**, 1500027 (2015)
26. Jiao, Z., Gao, R., Tao, H., Yuan, S., Xu, L., Xia, S., Zhang, H.: Intergrown $SnO_2$–$TiO_2$@graphene ternary composite as high-performance lithium-ion battery anodes. J Nanopart Res. **18**, 307 (2016)
27. Jiang, X., Yang, X., Zhu, Y., Fan, K., Zhao, P., Li, C.: Designed synthesis of graphene-$TiO_2$-$SnO_2$ ternary nanocomposites as lithium-ion anode materials. New J. Chem. **37**, 3671–3678 (2013)
28. Xia, G., Li, N., Li, D., Liu, R., Wang, C., Li, Q., Lü, X., Spendelow, J.S., Zhang, J., Wu, G.: Graphene/$Fe_2O_3$/$SnO_2$ ternary nanocomposites as a high-performance anode for lithium-ion batteries. ACS Appl. Mater. Interfaces. **5**, 8607–8614 (2013)
29. Yu, J., Huang, D., Liu, Y., Luo, H.: A ternary Ag–$TiO_2$/reduced graphene oxide nanocomposite as the anode material for lithium-ion batteries. Inorg. Chem. Front. **6**, 2126–2134 (2019)

# Chapter 22
# Cotton Coated with Graphene-Based Nanomaterials for Designing the Next-Generation Flexible Supercapacitor Electrodes

**Naorem Aruna Devi and Bibhu Prasad Swain**

## Introduction

At present, there is a great challenge in developing lightweight, thin, and flexible portable electronic devices such as roll-up displays, mobile phones, notebooks, cameras, and wearable systems for personal multimedia [1–3] but there are still some drawbacks where the today's technologies have some limitations such as having short cycle life, and longer charging and discharging time, low energy density. To overcome these challenges, some researchers have turned the focus of their research and development to flexible and safe energy devices, based on supercapacitors and thin-film batteries, to meet the various requirements of modern electronic industry and environmental degradation. All these electronic applications require cheap, flexibility, lightweight, wearable power conversion, and storage devices. Among the energy storage devices, supercapacitors are considered as a potential candidate due to owing high power density, good cyclic stability, and eco-friendly nature that charge and discharge quickly, safe and can be packaged easily by sandwiching an active layer between two electrodes [1]. A supercapacitor is typically composed of current collectors, two active electrodes, electrolyte, and separator. Textiles are flexible and porous materials made by weaving or pressing natural or synthetic fibers, which gives them important properties such as flexibility, stretchability, and lightweight. Thus, the ideal wearable power device would incorporate a textile as a component. There are two types of supercapacitors based on their charge storage mechanism, one is the electrical double-layer capacitors (EDLCs) and another is the pseudocapacitors. In EDLC, the energy is stored between the electrolyte and electrode through the charge

N. A. Devi (✉) · B. P. Swain
Department of Physics, National Institute of Technology Manipur, Imphal, Manipur 795004, India
e-mail: aruna.nrm94@gmail.com

B. P. Swain
e-mail: bibhuprasad.swain@gmail.com



B. P. Swain (ed.), *Nanostructured Materials and their Applications*, Materials Horizons: From Nature to Nanomaterials, https://doi.org/10.1007/978-981-15-8307-0_22

accumulation occurs towards the interface and it employed activated carbon materials like graphene, carbon nanotubes as an electrode while in pseudocapacitor stores charge by reversible Faradaic redox reaction and it uses a conductive polymer or metal oxide as an electrode. EDLCs are able to achieve high power density while pseudocapacitor exhibits good electrochemical stability and high energy. Hybrid supercapacitors use both the faradaic and non-faradaic charge storage mechanism which enhances conductivity, power density, the energy density at once. Recently, many researchers have reported on the preparation of textiles-based composite in an efficient, economic, and effective manner for the utilization of supercapacitor application which is emerging as an innovative technology with great interest. Among textiles, cotton is a cheap natural fiber, which is highly hydrophilic and light. To further enhance the characteristics of a supercapacitor, extensive efforts have been dedicated to the facile fabrication of supercapacitors with advanced materials. Cotton-based flexible supercapacitors are a lightweight, flexible, economical, and viable alternative to their conventional rigid and bulky counterparts. Due to having congenital flexibility and small volume, textile-based can be transformed into various shapes and structures, which in turn supports its integration with miniaturized futuristic wearable electronic devices and gadgets through well-established textile manufacturing technologies. The construction of a cotton-based supercapacitor was very easy and cost-effective, and it is blended with other conducting materials such as graphene or its derivative onto it which will enhance its conductivity and electrochemical stability. However, the direct conversion of cotton textiles into electrically active textiles for constructing supercapacitors remains a great challenge. Graphene has its great attention due to its unique properties which include high electron mobility, high thermal conductivity, ability to sustain high electric currents densities, and easy chemical functionalization. Graphene is 2-D carbon nanostructure having a hexagonal honeycomb packed structure with $sp^2$ hybridized single layer carbon atom [4] which is considered as a potential candidate for various technical fields such as lithium-ion batteries [5], supercapacitors [6, 7], solar cells [8], energy storage devices [9]. Various methods have been used for the synthesis of graphene for both energy storage and other applications, where some of the commonly used methods were chemical vapor deposition (CVD), chemically reduction method of GO, and mechanical exfoliation method. Among these methods, the chemical reduction method was the most commonly used method for various applications. In addition, the hybrid composite electrodes have great attention as it enhances the electrochemical performance of the electrode material. Recently, cotton@graphene-based metal oxide nanomaterials and conducting polymers composite attracts many researchers as it achieved a high specific capacitance improving the electrochemical performance of the electrode due to the synergistic effect of graphene with the metal oxide nanomaterials, which enhances the electrical conductivity along with owing pseudocapacitance behavior occurred due to the presence of the metal oxide and the conducting polymers [10]. Hence, in this developing world we need an energy storage device which stores a greater amount of energy and easily gets charged and discharged, but the conventional charge storage devices like batteries, usually cannot provide the required energy and power storage without becoming too heavy or bulky, so it is required to develop a flexible, stretchable, and lightweight materials for the utilization in various technologies fields. Thus, developing a facile synthesis method of cheap, flexible, lightweight supercapacitor based on cotton@graphene-based nanomaterial possessing foldable and reliable flexibility is still a great challenge. In this chapter, attempts are made in reviewing the previously reported work on the cotton coated with graphene-based nanomaterials for the flexible supercapacitor applications.

## *Overview of the Cotton@graphene-Based Nanomaterials for Flexible Supercapacitor Electrodes*

Nowadays, the development of multifunctional and portable electronic devices with high stability, low cost, lightweight, and high mechanical flexibility is in great demand. The soft and porous nature of the cotton fabrics with the incorporation of conductive elements such as carbon nanotubes, carbon, and graphene into it has been considered as a good candidate for energy storage devices which will enhance the conductivity of the material thus representing a type of lightweight material possessing foldable and reliable flexibility. Cotton is not expensive; it is the make-up of cellulose fibers. Cotton coated with graphene has a great application potential due to the combined merits of tiny volume, high flexibility, and weave-ability, which promise some applications in the next-generation supercapacitors for wearable and portable devices and electric vehicles [11]. The addition of pseudocapacitance materials such as metal oxides, hydroxides, and conducting with the cotton@graphene composite will further enhance the conductivity and chemical stability along with exhibiting fast faradaic reaction. Therefore, supercapacitors have been used in a variety of applications ranging from portable consumer electronics, computer memory backup systems, to hybrid electric vehicles and next-generation all-electric vehicles, and to large industrial scale power and energy management. Many researchers were also have reported the enhancement of cotton supercapacitor electrode by coating the cotton fabrics with the conductive materials such as carbon, graphene [12], and the materials with high electrochemical properties which includes metal oxide nanoparticles such as $Fe_3O_4$, NiO, $MnO_2$, $TiO_2$, $Cu_2O$ [13] and conducting polymers like PANI, PPy, and polythiophene. [14–16]. Xu et al. investigated the electrochemical performance of cotton@graphene composite fabrics where the specific capacitance was found to be 40 $Fg^{-1}$ within the potential range along with the cyclic retentivity of 90% over 1000 cycles showing good capacitance, excellent mechanical flexibility, and good stability [17]. Etana et al. reported the ternary composite based cotton fabric electrode material for the first time in the supercapacitor application which consists of reduced graphene oxide (rGO), $MnO_2$ and PANI which exhibits high specific capacitance values of 888 $Fg^{-1}$ and 252 $Fg^{-1}$ at a discharge current density of 1 $Ag^{-1}$ and 25 $Ag^{-1}$ in 1 $MH_2SO_4$ electrolyte solution along with cyclic retention as 70% of specific capacitance after 3000 cycles at the charge–discharge current density of 15 $Ag^{-1}$ showing good cyclic stability [11].

Yu et al. reported the hybrid graphene@$MnO_2$-based textile which exhibits a high specific capacitance up to 315 $Fg^{-1}$ with cyclic retentivity ~95% even after 5000 cycles showing excellent cycling stability [3]. Li et al. synthesized the graphene nanosheets with nickel sulfide-based make-up cotton electrode using the dip and the dry process followed by electrodeposition method and its specific capacitance was achieved as 775 $Fg^{-1}$ at current density 0.5 $Ag^{-1}$ with 88.1% cyclic retention over 1000 cycles at 2 $Ag^{-1}$ showing good cyclic stability [18]. Liu et al. fabricated the graphene sheets/cotton cloth composite for the utilization of supercapacitor electrodes which exhibited the specific capacitance of 81.7 $Fg^{-1}$ [7]. Pasta et al. fabricated the wearable supercapacitor using the cotton fibers coated with carbon nanotubes and using the aqueous lithium sulfate as the electrolyte in the devices which further exhibits ~ 70–80 $Fg^{-1}$ specific capacitance at 0.1 $Ag^{-1}$ with good cyclic stability which shows negligible decay over 35,000 cycles [19]. Zhang et al. synthesized cotton@graphene composite for the lithium battery application and was found that cotton@graphene composite achieved a higher reversible discharge capacity of 160 mAh $g^{-1}$ than that of the cotton alone which exhibits 115 mAh $g^{-1}$ after 100 cycles respectively [5]. Zhou et al. reported 3-D flexible rGO with carbonized cotton supercapacitor electrodes achieving a high specific capacitance of 87.53 mF $cm^{-2}$ at 2 mV $s^{-1}$ along with excellent cyclic retention of 89.82% after 1000 cycles and 90.5% over 100 cycles [12]. Jost et al. fabricated a flexible and lightweight supercapacitor electrode using a printing technique that achieved 85 F $g^{-1}$ as a specific capacitance [2]. However, most reported methods for preparing flexible electrode materials consisted of the cotton fabrics and carbon materials always involve the use of an organic surfactant or a high-temperature treatment process, which inevitably influences the electrochemical performance or the practical production of flexible electrode materials. Thus, the development of an easy synthesis method of flexible cotton@graphene composite is still a great challenge.

## Experimental Details

### *Synthesis of Graphene Oxide*

Graphene oxide (GO) is synthesized using various chemical methods which include Hummer's method, modified hummer's method, Brodie's method, Staudenmaier's method, Tour's method, and Hoffman's method. Among the above methods, Hummer's method is the most commonly used method for the preparation of GO. In this synthesis process, we take the graphite powder, sodium nitrate, and potassium permanganate ($KMnO_4$) in the ratio of 1:0.5:3, respectively, and stirred it together along with the sulfuric acid ($H_2SO_4$) concentration of 98%. The temperature should be maintained below 20 °C by keeping the mixture containing beaker into the cold water bath to stay safe from the explosion and to avoid overheating. After that, the temperature is to be raised up to 35 °C and stirring is done continuously for about

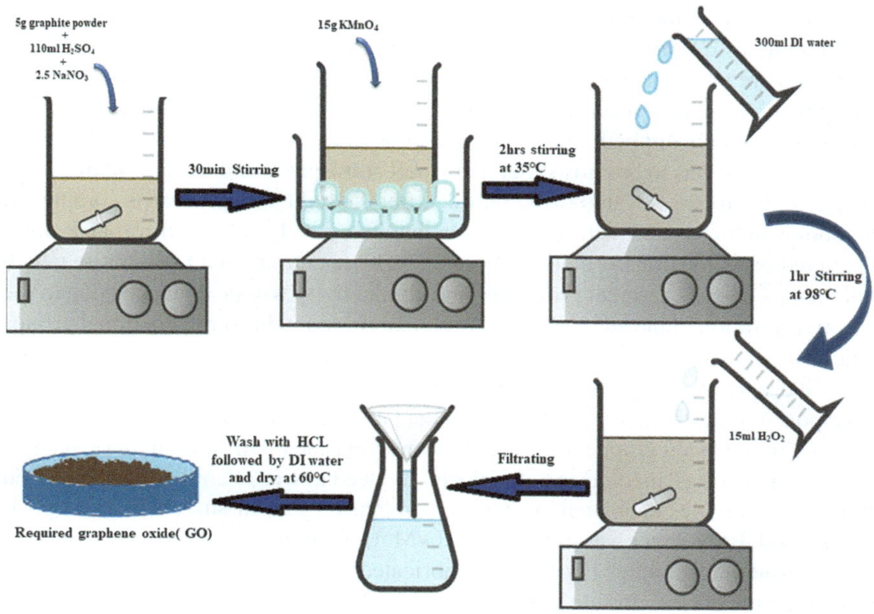


**Scheme 22.1.** Diagrammatic representation for the synthesis of GO. Reprinted with permission from Ref. [6]. Copyright (2020) Elsevier

1 h and then slowly adding the deionized water into the mixture the temperature is then increased at 98 °C and kept it to stir for another 1 h. When 1 h is completed, hydrogen peroxide ($H_2O_2$) is added in order to end the reaction and to remove the metal ions from the obtained results it is to be washed several times using HCl solution followed by deionized water. Then it should be kept to dry to obtained GO powder [6] (Scheme 22.1).

## *Different Techniques for the Preparation of Cotton@graphene-Based Nanomaterials*

There are various preparation techniques for the production of cotton@graphene-based composites, four of which are (i) printmaking technique, (ii) brush-coating and drying technique, (iii) dipping and drying method, and (iv) dipping and the freeze-drying process followed by the reduction process of as-prepared cotton coated with GO composite. There is a various chemical reducing agent which are employed to transform cotton@GO composite to cotton@rGO composite which includes sodium borohydride ($NaBH_4$), hydrazine hydrate, Titanium(III) chloride ($TiCl_3$), hydroiodic acid, ascorbic acid, sodium hydroxide (NaOH), hydroxylamine hydrochloride, or ammonium hydroxide. Further, other reduction techniques were also been reported which includes UV, electrochemical or thermal reduction process where no harmful chemicals are used for the reduction of cotton@rGO composite. The electrochemical reduction requires a conductive substrate to carry the current flow, and for this reason, it is mainly applied on carbon cloth substrates. In the case of thermal reduction, the process has to be carried out in an inert atmosphere and damage to the fibers can occur due to the elevated temperature. After the reduction process is completed, the cotton fabric coated with GO will change its color from brown to black color which indicates the transformation of cotton@GO to cotton@rGO composite which will then its electrochemical properties and conductivity will be enhanced due to the partial restoration of the $sp^2$ structure. Cotton fabrics contain some functional groups which support their strong interaction. The hydrophobic interactions, $\pi$–$\pi$ interactions, hydrogen bonding, electrostatic interaction, van der Waals forces, etc., might be responsible for the interaction between the GO and the cotton fabrics. Etana et al. prepared the reduced graphene oxide/$MnO_2$/polyaniline on the cotton fabric using a simple "dip and dry" method and chemically reduced into rGO/cotton fabric and then $MnO_2$ nanoparticles deposited on rGO/cotton fabric by an in-situ chemical deposition method. PANi layer was coated on rGO/$MnO_2$/cotton fabric by in-situ oxidative polymerization technique [10]. Gu et al. fabricated graphene-modified cotton textiles by soaking the cotton cloth in graphene oxide colloid followed by hydrazine reduction method [20]. Xu et al. synthesized the graphene/cotton composite fabrics using the dipping and drying process followed by a $NaBH_4$ reduction method [17]. Li et al. fabricated the hybrid graphene nanosheets/nickel sulfide cotton-based material for high-performance supercapacitor using a simple dip and dry method along with electrodeposition method [18]. Liu et al. prepared graphene@cotton cloth using a facile brush-coating and drying method followed by annealing process [7]. Pasta et al. reported the synthesis of conductive cotton fabric using an SWNT ink method [19]. Zhang et al. fabricated the graphene@cotton composite using a simple dipping and freeze-drying method for utilization of the lithium-ion battery suggesting that it as simple, cheap, and suitable large-scale production method [5]. Some of the methods, which are commonly used, are explained in detail below.

### Printmaking Technique (Screen Printing)

In this printmaking process, firstly the GO suspension was prepared by dispersing the as-prepared GO powder in DI water under ultrasonication. Then, the GO suspension is centrifuged in order to remove any unexfoliated GO impurity. Then, the obtained GO ink is then filled into the cleaned Epson ink cartridges through a syringe and kept it for some time to allow it to equalize before conducting the printing. After the equilibration is completed, printing is performed which will be printed onto the cleaned cotton cloth fabrics [2, 21]. Ink-jet printing has the advantage of high precision ink droplet spacing. Screen printing is material-efficient and can coat large surface areas quickly, though it may have a small variation in the carbon mass that is impregnated into the fabrics.

### Brush-Coating and Drying Technique

Firstly in this technique, the GO ink was prepared by dispersing the as-synthesized GO powder in water with the concentration of 2 mg $mL^{-1}$ and probe-ultra-sonication is done for about 20 min to produce a uniform GO suspension. Afterward, the GO suspension was painted onto the clean cotton cloth using a brush-coating technique. Consequently, the wet cotton cloth absorbed with GO ink was subjected to a drying process for the removal of water by putting it in an oven at 150 °C for 10 min. During the drying process, the water evaporates and the GO sheets form an entangled random network on the textile fibers. Repeating this simple brush-coating and drying process as often as necessary, a large-area GO-coated cotton cloth textile was readily fabricated. Thereafter, the cotton@GO composite fabric was transformed into cotton@rGO composite fabric through annealing at 300 °C for 2 h in an argon atmosphere.

### Dipping and Drying Method

In this dip and dry method, firstly the cotton fabric is to be cleaned to remove the unwanted impurities which are attached to the cotton fabric. The cotton fabrics are marinated by dipping in a boiled NaOH aqueous solution for 1 h. On the other side, the GO suspension ink was prepared by dissolving the as-prepared GO powder in de-ionized water under ultra-sonication for 30 min. After that, the preprocess cotton fabric is then dipped into a GO suspension ink and soaked for 30 min to coat GO on to the cotton fabric. The dipping and coating process should be repeated several times to achieve more GO adsorption on cotton fabric and then vacuum dried at 60 °C for 1 h. The more dipping will result in loading a higher amount of GO ink to the cotton. Afterward, the reduction of obtained cotton@GO fabric is carried out by dipping in an aqueous $NaBH_4$ solution of hydrazine solution which acts as a reducing agent and stirred it for about 5 h. After that, the obtained rGO/cotton fabric was collected and washed with DI water and then kept it to dry. This method is the most commonly used method due to its simple, cheap, and eco-friendly nature but the amount of carbon impregnated into the fabric depends greatly on the hydrophilicity of the material, which can also vary from section to section of the same fabric.

### Dipping and Freeze-Drying Process

In the dipping and freeze-drying process, the GO suspension was first prepared under ultrasonication for about 1 h and after that into it the cleaned cotton fabrics are immersed and is kept for around 20–30 min so that the GO ink absorbs in the cotton fabric; afterward, the wet cotton fabric is kept in the cryogenic refrigerator for cooling down to a temperature − 50 °C for 1 day. Subsequently, it is kept for about 2 days to freeze-dry then the cotton@GO composite is obtained and furthermore to

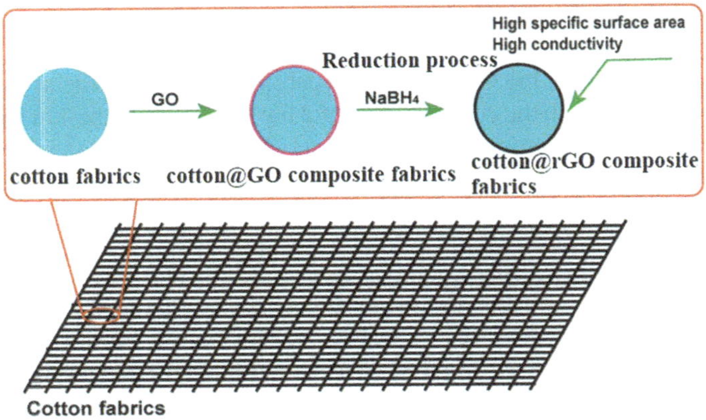


**Scheme 22.2.** Schematic diagram for the preparation of cotton@graphene nanomaterials. Reprinted with permission from Ref. [17]. Copyright (2015) Royal Society of Chemistry

achieve cotton@rGO composite, the as-prepared cotton@GO composite is further annealed at 1000 °C for about 3 h (Scheme 22.2).

## *Electrochemical Measurement*

Electrochemical measurements were performed in a three-electrode configuration—reference electrode (RE), working electrode (WE), and counter electrode (CE) using Ag/AgCl, glassy carbon electrode and platinum wire, respectively. The potential of WE is measured against RE which maintains a constant potential while current flows between WE and CE. In the CV measurement, we set the potential window in the range where CV curves are widely open. And while scanning cathodically, the analytic ions from the electrolyte get reduced in the working electrode producing a reducing current. During this reverse scan, the working electrode acts as a negative electrode. Whereas during the forward scan, oxidation occurs in the working electrode resulting in the generation of oxidation current (anodic current) which flows in the opposite direction. The CV and galvanostatic charge–discharge (GCD) measurements were carried out to investigate the electrochemical performance of cotton@rGO nanomaterials using electrochemical analyzer CHI608E CH Instruments. The specific capacitance was calculated from the CV and curves based on the following equations given below

$$C = A\left(\int_{V_a}^{V_b} I\mathrm{d}V\right)/(m \times v(V_b - V_a)) \tag{22.1}$$

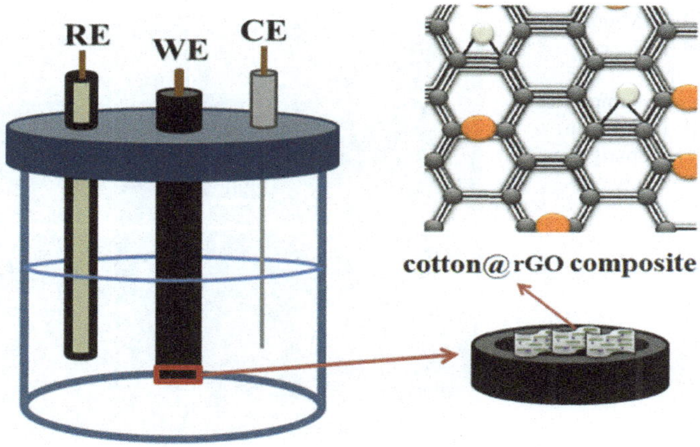


**Fig. 22.1** Diagrammatic representation of the electrochemical measurement of cotton coated with graphene

where $C$ is the specific capacitance, $I$ is the current (in A) of the CV curve, $m$ is the mass of active electrode materials which is 0.1 mg, $v$ is the scan rate, $V_b$ and $V_a$ indicate the high and low potential limit of the CV measurements.

$$C_{sc} = \frac{I\Delta t}{m\Delta v} \tag{22.2}$$

where $I$ is the discharge current, $\Delta t$ is the discharge time after IR drop, and $m$ is the mass of active sample on the electrode in the cell (Fig. 22.1).

## Results and Discussion Obtained from Different Characterization Techniques

### *Scanning Electron Microscopse (SEM)*

Figure 22.2 shows the SEM image of the pure cotton fabrics (a–b) and cotton@graphene composite (c–d) at different magnification. A porous hierarchy nature like structure was observed from the image of cotton fabrics. The SEM images obtained after the reduction of cotton coated with GO shows a wavy wrinkled structured nanosheet that confirming the interaction of rGO with cotton fabrics.

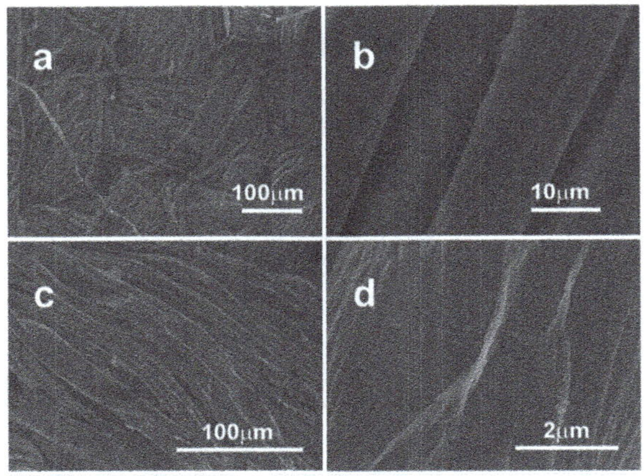


**Fig. 22.2** **a–b** SEM images of pure cotton fabric. **c–d** SEM images of cotton@graphene composite. Reprinted with permission from Ref. [17]. Copyright (2015) Royal Society of Chemistry

## *Fourier Transform Infrared Spectroscopy (FTIR)*

Figure 22.3 shows the FTIR spectrum of cotton, cotton@GO composite, and cotton@graphene composite fabrics. The chemical bonding due to OH stretching, CH

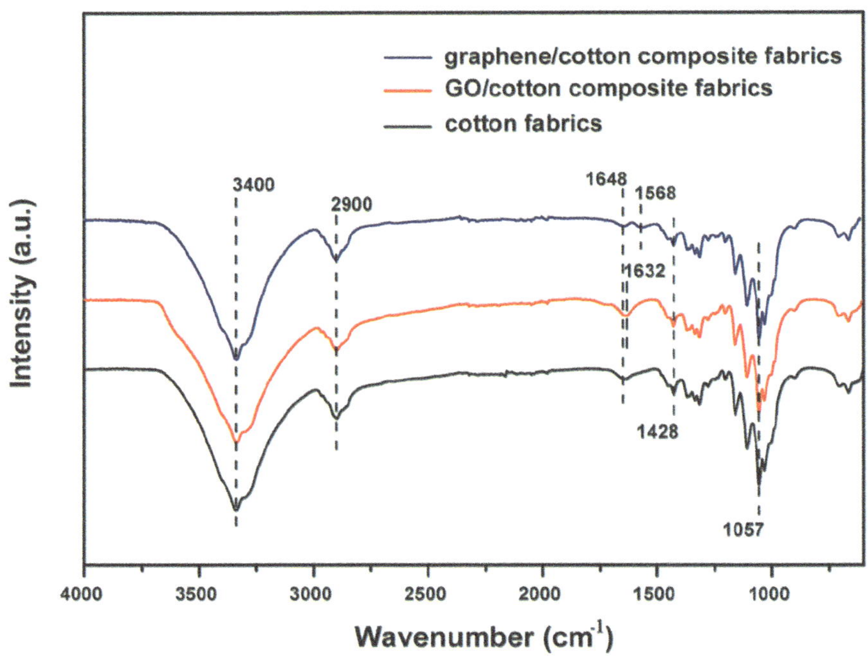


**Fig. 22.3** FTIR spectra of cotton fabrics, cotton@GO, and cotton@rGO nanomaterials. Reprinted with permission from Ref. [17]. Copyright (2015) Royal Society of Chemistry

stretching, OH of water absorbed from cellulose, $CH_2$ symmetric bending, and C–O stretching was observed at 3340, 2900, 1648, 1428, and 1057 $cm^{-1}$, respectively, for the pure cotton fabrics [17]. An additional peak at 1632 $cm^{-1}$ appeared in the FTIR spectra of cotton coated with GO composite which appears due to the attachment of carbon vibration of graphene with the cotton fabrics. On the other side, in the FTIR spectra of cotton@graphene composite, there was the absence of one peak that was located at 1648 $cm^{-1}$. The appearance of a peak at around 1568 $cm^{-1}$ is attributed to C = C skeletal vibration of graphene which confirms the formation of cotton with rGO sheets [17].

## *Raman Spectroscopy*

Raman spectrum of pure cotton, cotton@GO composite, and cotton@graphene composite were shown in Fig. 22.4. Figure 22.4a–b shows the Raman spectra of pure cotton (a) before and (b) after annealing, respectively, which shows strong fluorescence with a large featureless background. The direct conversion of cotton into carbonized cotton by a simple annealing process increases its conductivity and also maintained structural and mechanical properties. The D band and G band were obtained in all the Raman spectrum at 1345 $cm^{-1}$ and 1570 $cm^{-1}$, respectively

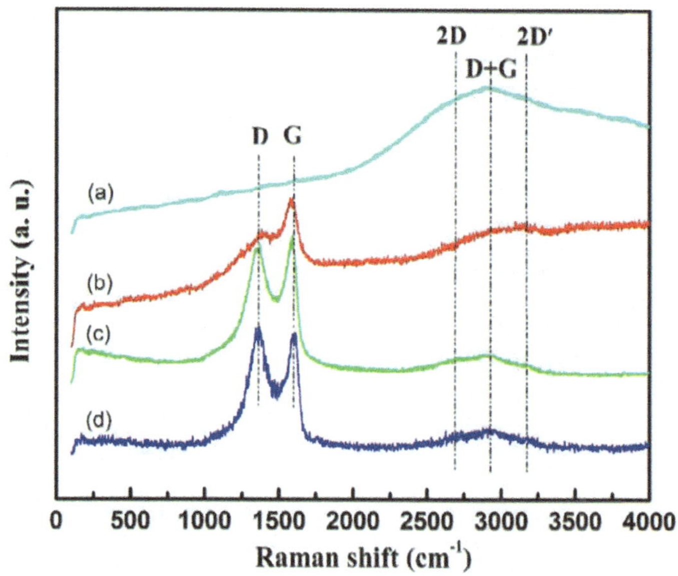


**Fig. 22.4** Raman spectrum of pure cotton **a** before annealing and **b** after annealing, **c** Raman spectra of GO@cotton composite **c** before and **d** after annealing. Reprinted with permission from Ref. [7]. Copyright (2012) Royal Society of Chemistry

[7] where D band is appeared due to the defects and disorders in the layers of the graphene sheets, whereas G band attributed an $E_{2g}$ mode of graphite which is related to the vibration of $sp^2$ bonded carbon atoms. The appearance of D and G peaks in the spectrum of the annealed cotton fabric is due to carbonation of the pure cotton. Further, the intensity of the D and G band was observed to increase in the FTIR spectrum of the annealed cotton coated with GO composite (Fig. 22.4d) as compared to that of before annealing composite (Fig. 22.4c) which indicates that cotton@GO composite was reduced properly into cotton@rGO composite after the annealing process. Moreover, the Raman $I_D/I_G$ ratio gives the idea about the defects density in graphene-based materials and the graphitization degree of carbonaceous materials.

## *X-ray Photoelectron Spectroscopy (XPS)*

The XPS analysis is carried out to identify the compositional and elemental properties of the pure cotton, cotton@GO, and cotton@rGO composites which are shown in Fig. 22.5a. The binding energies of C 1 s and O 1 s orbitals were appeared at around

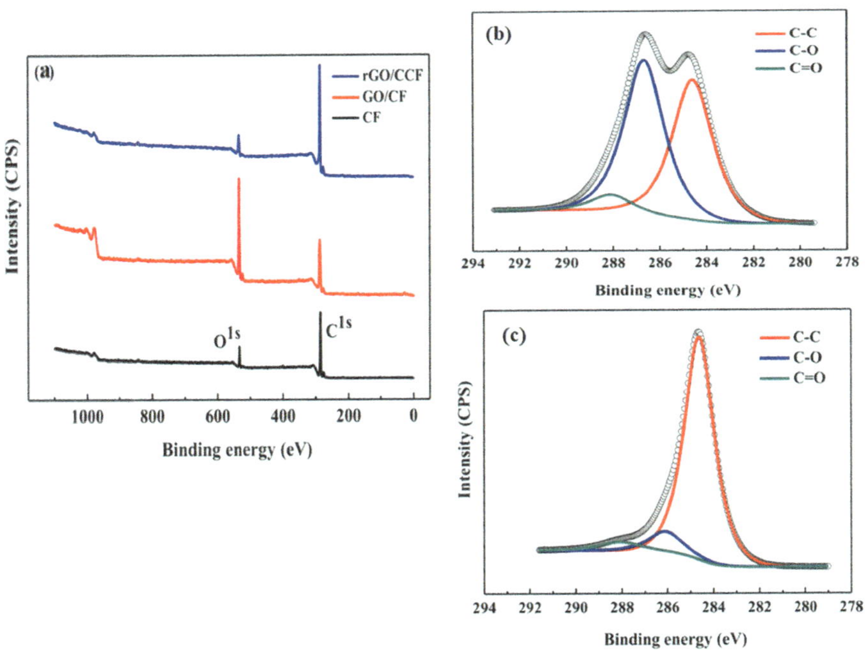


**Fig. 22.5** **a** XPS survey of pure cotton fabric (CF), cotton@GO composite fabric (GO/CF), and cotton@rGO composite fabric, **b** deconvolution of C1s core orbital for cotton@GO composite fabric, **c** deconvolution of C1s core orbital for cotton@rGO composite fabric. Reprinted with permission from Ref. [12]. Copyright (2015) Elsevier

286 and 531.0 eV, respectively [12]. It was observed that the peak intensity of O 1 s in the XPS spectra of cotton@GO composite increased as compared to that of pure cotton fabrics that indicate the attachment of GO with cotton fabrics. However, in the XPS spectra of cotton@rGO composite, the O1s peak was observed to decrease quickly, while the intensity of the C1s peak significantly increased, which indicates that most of the oxygen-containing groups in the cotton@GO/composite have been successfully removed. Further, to find out the various functional groups of carbon that are present in the cotton@GO and cotton@rGO composite, the deconvolution of the core orbital of C 1 s was done which are shown in Fig. 22.5b–c, respectively. The peaks were observed at the binding energies, 286.69 eV, 288.09 eV, and 284.58 eV which corresponds to C–O, C=O, and C–C, respectively [12]. Moreover, it was also observed that the XPS spectra of O 1 s core orbital decrease dramatically after the reduction of cotton@GO composite which signifies the formation of rGO sheets with the cotton fabrics.

## *Cyclic Voltammetry*

The cyclic voltammetry (CV) and galvanostatic charge–discharge (GCD) measurement were also carried out to investigate the electrochemical performance of the pure cotton, cotton@GO composite, and cotton@rGO composite fabrics. Figure 22.6a shows the CV curves of cotton coated with graphene composite at various scan rates which varies from 2 to 20 $mVs^{-1}$ at the potential window range 0–1 V and the comparison of the CV curves obtained for pure cotton with that of cotton@rGO composite at 5 $mVs^{-1}$ were also shown in Fig. 22.6b. It was found that the cotton@rGO composite exhibits a high specific capacitance of 81.4 $F\,g^{-1}$ at 2 $mVs^{-1}$ scan rate [12]. The CV curves are in a rectangular-like shape which signifies that the cotton@rGO exhibits EDLC behavior while there also observed some weak redox peaks which attribute that it also exhibits a small amount of pseudocapacitance behavior. Further, it was also observed that with the increase of the scan rate, there is a slight change in the CV curves which can be explained as when the scan rate increases, ions will migrate faster from the bulk electrolyte solution to the electrode/electrolyte interface, while the diffusion rate of the ions migrating from the electrode/electrolyte interface into the internal electrode material is relatively slow, thus resulting in the accumulation of a large amount of ions near the electrode/electrolyte interface and the polarization of the electrodes. From Fig. 22.6b, we can observe that the area of the pure cotton fabric CV curve is very much smaller than that of the CV curve's area of cotton@rGO composite fabrics. Figure 22.6c shows the GCD curves of the cotton@rGO composite at different current densities in the potential window range from 0–1 V which is in a triangular-shape like structure showing good capacitive behavior. Figure 22.6d shows the cyclic stability curve obtained from the GCD analysis of cotton@rGO composite which achieves a cyclic retention of 89.82% over 1000 cycles showing good electrochemical stability.

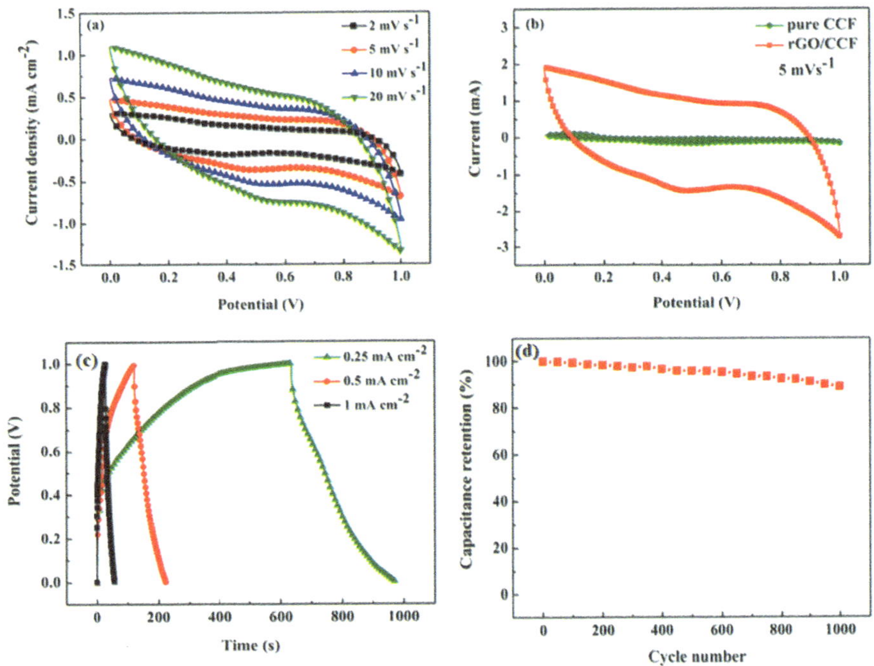


**Fig. 22.6** **a** Cyclic-voltammetry curves of cotton@rGO supercapacitor electrode at different scan rates varying from 2 to 20 $mVs^{-1}$. **b** Cyclic-voltammetry curves of pure cotton fabrics and cotton@rGO composite supercapacitor electrode at a scan rate of 5 $mVs^{-1}$. **c** Galvanostatic charge–discharge (GCD) curves of cotton coated graphene composite at different densities. **d** Cyclic stability plot of cotton coated with graphene electrode. Reprinted with permission from Ref. [12]. Copyright (2015) Elsevier

## Conclusions

In this chapter, the previously reported work on cotton@graphene-based nanomaterials was overviewed. Different techniques for the preparation of cotton@graphene were highlighted. The morphological, chemical, compositional, electrochemical properties of the pure cotton, cotton@GO, and cotton@rGO composite were also presented. Cotton coated with graphene composite shows a good electrochemical performance with excellent cyclic stability and capacitive behavior which can be further used for the next-generation flexible supercapacitor application. Recently, cotton-based graphene has gained a great interest in order to develop lightweight and flexible supercapacitor with large energy storage. The coating of cotton with conducting materials such as graphene improves its electrical conductivity and can be used for designing various electronic devices such as lightweight solar cells and batteries. Rapidly developing of the modern and smart technologies for various energy storage applications has extended requirements against the process of designing and developing the more effective and strength energy storage devices. Thus, it currently needs to develop a cost-effective and better supercapacitor electrode by taking into account the recent available technology positions that can provide high power and high electrochemical performance with lightweight and cheap cotton fabrics which can store a large amount of charge and also in low cost.

**Acknowledgements** The author would like to thank the Department of Physics, NIT Manipur.

## References

1. Xue, J., Zhao, Y., Cheng, H., Hu, C., Hu, Y., Meng, Y., Shao, H., Zhang, Z., Qu, L.: An all-cotton-derived, arbitrarily foldable, high-rate, electrochemical supercapacitor. Phys. Chem. Chem. Phys. **15**, 8042 (2013). https://doi.org/10.1039/C3CP51571K
2. Jost, K., Perez, C.R., McDonough, J.K., Presser, V., Heon, M., Dion, G., Gogotsi, Y.: Carbon coated textiles for flexible energy storage. Energy Environ. Sci. **4**, 5060 (2011). https://doi.org/10.1039/c1ee02421c
3. Yu, G., Hu, L., Vosgueritchian, M., Wang, H., Xie, X., McDonough, J.R., Cui, X., Cui, Y., Bao, Z.: Solution-processed graphene/$MnO_2$ nanostructured textiles for high-performance electrochemical capacitors, Nano Lett. **11**, 2905–2911 (2011). doi.org/https://doi.org/10.1021/nl2013828
4. Novoselov, K.S., Geim, A.K., Morozov, S.V., Jiang, D., Zhang, Y., Dubonos, S.V., Grigorieva, I.V., Firsov, A.A.: Electric Field Effect in Atomically Thin Carbon Films. Science **306**, 666–669 (2004). https://doi.org/10.1126/science.1102896
5. Zhang, X., Huang, X., Zhang, X., Zhong, B., Xia, L., Liu, J., Pan, H., Wen, G.: A facile method to prepare graphene-coat cotton and its application for lithium battery. J. Solid State Electrochem. **20**, 1251–1261 (2016). https://doi.org/10.1007/s10008-016-3118-6
6. Devi, N.A., Nongthombam, S., Sinha, S., Bhujel, R., Rai, S., Singh, W.I., Dasgupta, P., Swain, B.P.: Investigation of chemical bonding and supercapacitivity properties of $Fe_3O_4$-rGO nanocomposites for supercapacitor applications. Diam. Relat. Mater. **104**, 107756 (2020). https://doi.org/10.1016/j.diamond.2020.107756
7. Liu, W., Yan, X., Lang, J., Peng, C., Xue, Q.: Flexible and conductive nanocomposite electrode based on graphene sheets and cotton cloth for supercapacitor. J. Mater. Chem. **22**, 17245–17253 (2012). https://doi.org/10.1039/C2JM32659K
8. Ghasemi, S., Hosseini, S.R., Kazemi, Z.: Electrophoretic preparation of graphene-iron oxide nanocomposite as an efficient Pt-free counter electrode for dye-sensitized solar cell. J. Solid State Chem. **22**, 245–253 (2018). https://doi.org/10.1007/s10008-017-3741-x
9. Queraltó, A., Pérez del Pino, A., Logofatu, C., Datcu, A., Amade, R., Bertran-Serra, E., György, E.: Reduced graphene oxide/iron oxide nanohybrid flexible electrodes grown by laser-based technique for energy storage applications. Ceram. Int. **44**, 20409–20416 (2018). https://doi.org/10.1016/j.ceramint.2018.08.034
10. Etana, B.B., Ramakrishnan, S., Dhakshnamoorthy, M., Saravanan, S., Ramamurthy, P.C., Demissie, T.A.: Functionalization of Textile cotton fabric with reduced graphene oxide/MnO2/polyaniline based electrode for supercapacitor. Mater. Res. Express **6**, 125708 (2019). https://doi.org/10.1088/2053-1591/ab669d
11. Meng, Y., Zhao, Y., Hu, C., Cheng, H., Hu, Y., Zhang, Z., Shi, G., Qu, L.: Graphene core-sheath microfibres for all-solid-state, stretchable fibriform supercapacitors and wearable electronic textiles. Adv. Mater. **25**, 2326–2331 (2013). https://doi.org/10.1002/adma.201300132
12. Zhou, Q., Ye, X., Wan, Z., Jia, C.: A three-dimensional flexible supercapacitor with enhanced performance based on lightweight, conductive graphene-cotton fabric electrode. J. Power Sources. **296**, 186–196 (2015). https://doi.org/10.1016/j.jpowsour.2015.07.012

13. Liu, X.Y., Wang, K.X., Chen, J.S.: Template-directed metal oxides for electrochemical energy storage. Energy Stor. Mater. **3**, 1–17 (2016). https://doi.org/10.1016/j.ensm.2015.12.002
14. Ling, B.K., Jing, Z., Jing, J., Yong, C.L., Long, K.: MWNTs/PANI composite materials prepared by in-situ chemical oxidative polymerization for supercapacitor electrode. J. Mater. Sci. **43**, 3664–3669 (2008). https://doi.org/10.1007/s10853-008-2586-1
15. Jie, X., Daxiang, W., Ye, Y., Wei, W., Shaojin, G., Ruina, L., Xiaojun, W., Li, L., Weilin, X.: Polypyrrole-coated cotton fabrics for flexible supercapacitor electrodes prepared using CuO nanoparticles as template. Cellulose **22**, 1355–1363 (2015). https://doi.org/10.1007/s10570-015-0546-x
16. Ji, E.L., Seon, J.P., Oh, S.K., Hyeon, W.S., Jyongsik, J., Hyeonseok, Y.: Systematic investigation on charge storage behaviour of multidimensional poly (3,4ethylenedioxythiophene) nanostructures. RSC Adv. **4**, 37529–37535 (2014). https://doi.org/10.1039/C4RA06161F
17. Xu, L.L., Guo, M.X., Liu, S., Bian, S.W.: Graphene/cotton composite fabrics as flexible electrode materials for electrochemical capacitors. RSC Adv. **5**, 25244 (2015). https://doi.org/10.1039/c4ra16063k
18. Li, Y., Ye, K., Cheng, K., Yin, J., Cao, D., Wang, G.: Electrodeposition of nickel sulfide on graphene-covered make-up cotton as a flexible electrode material for high-performance supercapacitors. J. Power Sources **274**, 943–950 (2015). https://doi.org/10.1016/j.jpowsour.2014.10.156
19. Pasta, M., Mantia, F.L., Hu, L., Deshazer, H.D., Yi, C.: Aqueous supercapacitors on conductive cotton. Nano Res. **3**, 452–458 (2010). https://doi.org/10.1007/s12274-010-0006-8
20. Gu, W., Zhao, Y.: Graphene modified cotton textiles. Adv. Mat. Res. **331**, 93–96 (2011). https://doi.org/10.4028/www.scientific.net/AMR.331.93
21. Chen, P., Chen, H., Qiu, J., Zhou, C.: Inkjet printing of single-walled carbon nanotube/$RuO_2$ nanowire supercapacitors on cloth fabrics and flexible substrates. Nano Res. **3**, 594–603 (2010). https://doi.org/10.1007/s12274-010-0020-x

The manufacturer's authorised representative in the EU is Springer Nature Customer Service Centre GmbH, Europaplatz 3, 69115 Heidelberg, Germany. If you have any concerns regarding our products, please contact ProductSafety@springernature.com

Printed and bound by CPI Group (UK) Ltd, Croydon, CR0 4YY
30/09/2026
02225288-0013